Springer Monographs in Mathematics

This series publishes advanced monographs giving well-written presentations of the "state-of-the-art" in fields of mathematical research that have acquired the maturity needed for such a treatment. They are sufficiently self-contained to be accessible to more than just the intimate specialists of the subject, and sufficiently comprehensive to remain valuable references for many years. Besides the current state of knowledge in its field, an SMM volume should also describe its relevance to and interaction with neighbouring fields of mathematics, and give pointers to future directions of research. The volumes in the series are in hardcover.

More information about this series at http://www.springer.com/series/3733

John F. Jardine

Local Homotopy Theory

John F. Jardine
University of Western Ontario Mathematics Department
London, Ontario
Canada

ISSN 1439-7382   ISSN 2196-9922 (electronic)
Springer Monographs in Mathematics
ISBN 978-1-4939-2299-4   ISBN 978-1-4939-2300-7 (eBook)
DOI 10.1007/978-1-4939-2300-7

Library of Congress Control Number: 2015933625

Springer New York Heidelberg Dordrecht London
© Springer-Verlag New York 2015
This work is subject to copyright. All rights are reserved by the Publisher, whether the whole or part of the material is concerned, specifically the rights of translation, reprinting, reuse of illustrations, recitation, broadcasting, reproduction on microfilms or in any other physical way, and transmission or information storage and retrieval, electronic adaptation, computer software, or by similar or dissimilar methodology now known or hereafter developed.
The use of general descriptive names, registered names, trademarks, service marks, etc. in this publication does not imply, even in the absence of a specific statement, that such names are exempt from the relevant protective laws and regulations and therefore free for general use.
The publisher, the authors and the editors are safe to assume that the advice and information in this book are believed to be true and accurate at the date of publication. Neither the publisher nor the authors or the editors give a warranty, express or implied, with respect to the material contained herein or for any errors or omissions that may have been made.

Printed on acid-free paper

Springer is part of Springer Science+Business Media (www.springer.com)

Preface

The subject of this monograph is the homotopy theory of diagrams of spaces, chain complexes, spectra, and generalized spectra, where the homotopy types are determined locally by a Grothendieck topology.

The main components of the theory are the local homotopy theories of simplicial presheaves and simplicial sheaves, local stable homotopy theories, derived categories, and non-abelian cohomology theory. This book presents formal descriptions of the structures comprising these theories, and the links between them. Examples and sample calculations are provided, along with some commentary.

The subject has broad applicability. It can be used to study presheaf and sheaf objects which are defined on the open subsets of a topological space, or on the open subschemes of a scheme, or on more exotic covers. Local homotopy theory is a foundational tool for motivic homotopy theory, and for the theory of topological modular forms in classical stable homotopy theory. As such, there are continuing applications of the theory in topology, geometry, and number theory. The applications and extensions of the subject comprise a large and expanding literature, in multiple subject areas.

Some of the ideas of local homotopy theory go back to the work of the Grothendieck school in the 1960s. The present form of the theory started to emerge in the late 1980s, as part of a study of cohomological problems in algebraic K-theory. Within the framework of this theory, these K-theory questions have now been almost completely resolved, with a fusion of ideas from homotopy theory and algebraic geometry that represents the modern face of both subjects. The theory has broadened the scope of the applications of homotopy theory, while those applications have led to a rethinking of what homotopy theory and particularly stable homotopy theory should be. We also now have a good homotopy theoretic understanding of non-abelian cohomology theory and its applications, and this theory has evolved into the modern theories of higher stacks and higher categories.

The foundational ideas and results of local homotopy theory have been established in a series of well-known papers which have appeared over the last 30 years, but have not before now been given a coherent description in a single source.

This book is designed to rectify this difficulty, at least for basic theory. It is intended for the members of the Mathematics research community, at the senior graduate student level and beyond, with interests in areas related to homotopy theory and algebraic geometry. The assumption is that the reader either has a basic knowledge of these areas, or has a willingness to acquire it.

This project was initiated as a result of a series of conversations with Ann Kostant at Springer-Verlag during the spring of 2009, and I am grateful for her encouragement and support.

I would like to thank Cihan Okay for a careful reading of the manuscript during the summer of 2014, and for finding many typographical errors. I would also like to thank Karol Szumiłło and Kris Kapulkin for a series of helpful questions and remarks.

Basically, I have worked on this book everywhere that I have gone over the past 5 years. Most of the work was done at the University of Western Ontario, but I would like to give special thanks to the Fields Institute and to the Pacific Institute for the Mathematical Sciences for the hospitality and stimulating environments that they provided during multiple visits.

This work was made possible by the generous support of the Natural Sciences and Engineering Research Council of Canada and the Canada Research Chairs program.

I would, finally, like to thank my wife Catharine Leggett, for her encouragement and patience through another book writing adventure.

London, Canada   J. F. Jardine

October, 2014

Contents

1   Introduction 1

Part I Preliminaries

2   **Homotopy Theory of Simplicial Sets** 15
     2.1   Simplicial Sets 15
     2.2   Model Structure for Simplicial Sets 21
     2.3   Projective Model Structure for Diagrams 25

3   **Some Topos Theory** 29
     3.1   Grothendieck Topologies 31
     3.2   Exactness Properties 38
     3.3   Geometric Morphisms 42
     3.4   Points 46
     3.5   Boolean Localization 49

Part II Simplicial Presheaves and Simplicial Sheaves

4   **Local Weak Equivalences** 59
     4.1   Local Weak Equivalences 60
     4.2   Local Fibrations 69
     4.3   First Applications of Boolean Localization 77

5   **Local Model Structures** 91
     5.1   The Injective Model Structure 93
     5.2   Injective Fibrations 100
     5.3   Geometric and Site Morphisms 107
     5.4   Descent Theorems 116
     5.5   Intermediate Model Structures 126
     5.6   Postnikov Sections and n-Types 131

6	**Cocycles**	139
	6.1 Cocycle Categories	142
	6.2 The Verdier Hypercovering Theorem	150

7	**Localization Theories**	159
	7.1 General Theory	161
	7.2 Localization Theorems for Simplicial Presheaves	174
	7.3 Properness	185

Part III Sheaf Cohomology Theory

8	**Homology Sheaves and Cohomology Groups**	191
	8.1 Chain Complexes	194
	8.2 The Derived Category	202
	8.3 Abelian Sheaf Cohomology	207
	8.4 Products and Pairings	223
	8.5 Localized Chain Complexes	227
	8.6 Linear Simplicial Presheaves	235

9	**Non-abelian Cohomology**	247
	9.1 Torsors	251
	9.2 Stacks and Homotopy Theory	267
	9.3 Groupoids Enriched in Simplicial Sets	280
	9.4 Presheaves of Groupoids Enriched in Simplicial Sets	304
	9.5 Extensions and Gerbes	318

Part IV Stable Homotopy Theory

10	**Spectra and T-spectra**	337
	10.1 Presheaves of Spectra	344
	10.2 T-spectra and Localization	360
	10.3 Stable Model Structures for T-spectra	368
	10.4 Shifts and Suspensions	383
	10.5 Fibre and Cofibre Sequences	391
	10.6 Postnikov Sections and Slice Filtrations	405
	10.7 T-Complexes	412

11	**Symmetric T-spectra**	431
	11.1 Symmetric Spaces	436
	11.2 First Model Structures	441
	11.3 Localized Model Structures	447
	11.4 Stable Homotopy Theory of Symmetric Spectra	451
	11.5 Equivalence of Stable Categories	461
	11.6 The Smash Product	472
	11.7 Symmetric T-complexes	483

References 499

Index 505

Chapter 1
Introduction

In the broadest terms, classical homotopy theory is the study of spaces and related objects such as chain complexes or spectra, and equivalence relations between them, which are defined by maps called weak equivalences. These maps are typically defined by inducing isomorphisms of homotopy groups and homology groups.

A homotopy theory also comes equipped with classes of cofibrations or fibrations (usually both), which are families of maps that react with weak equivalences in ways that specify solutions to particular, universally defined obstruction problems. Such a theory is usually encoded in a Quillen closed model structure, but variants of that concept occur.

The list of definitions, axioms and formal results which make up a closed model structure is the barest of beginnings, on the way to the calculational results that are typical of the homotopy theory of spaces and spectra. The point of the axiomatic approach is to make basic obstruction theory easy and formal—more interesting calculations and theoretical statements require further input, from either geometry or algebra. Calculations also usually involve structures that are constructed from the basic building blocks of homotopy theory.

Diagrams of spaces are everywhere. A space can be identified with a homotopy colimit of a diagram of its universal covers, indexed on its fundamental groupoid, while the universal covers are often studied as homotopy inverse limits of their Postnikov towers. An action of a group G on a space (or spectrum) X is a diagram of spaces (or spectra) which consists of the actions $g : X \to X$ by the various elements g of G. It is also a standard practice to study the corresponding diagram of fixed point spaces.

Homotopy colimits are derived colimits and homotopy inverse limits are derived inverse limits, according to a construction that has been familiar to homotopy theorists at least since the 1970s. Explicitly, the homotopy inverse limit of an I-diagram X is an inverse limit of a fibrant model of X, in an "injective" model structure for I-diagrams of spaces for which the cofibrations and weak equivalences are defined sectionwise. The homotopy colimit of X is a colimit of a cofibrant model of X, in a "projective" model structure for I-diagrams, for which the fibrations and weak equivalences are defined sectionwise. Here, X is a space-valued functor which is defined on a small index category I, and the collection of all such functors and the natural transformations between them is the category of I-diagrams. One makes analogous definitions for diagrams of spectra and for diagrams of chain complexes.

Local homotopy theory is the study of diagrams of spaces or spectrum-like objects and weak equivalences, where the weak equivalences are determined by a topology on the underlying index category.

The choice of underlying topology determines the local nature of the resulting homotopy theory. For example, a simplicial presheaf X (or presheaf of spaces) on a topological space T is a contravariant diagram

$$X : (op|_T)^{op} \to s\mathbf{Set}$$

that is defined on the category $op|_T$ of open subsets of T, and which takes values in simplicial sets. We already have injective and projective model structures, and hence homotopy theories, for the category of such diagrams in which a weak equivalence of simplicial presheaves $X \to Y$ is a *sectionwise weak equivalence*, meaning a map which induces weak equivalences $X(U) \to Y(U)$ in sections for all open subsets U of T. On the other hand, a map $X \to Y$ is a *local weak equivalence* (for the topology on the space T) if it induces weak equivalences $X_x \to Y_x$ in stalks for all $x \in T$. Every sectionwise weak equivalence is a local weak equivalence, but local weak equivalences certainly fail to be sectionwise equivalences in general.

There is a model structure on the category of simplicial presheaves on the space T, for which the weak equivalences are the local weak equivalences and the cofibrations are defined sectionwise. Contrast this with the diagram-theoretic injective model structure described above, for which both the weak equivalences and the cofibrations are defined sectionwise. These two homotopy theories, while different, are still avatars of the same phenomenon; the locally defined homotopy theory is determined by the usual topology on the category $op|_T$ of open subsets of the space T, while the sectionwise theory is defined by the "chaotic" topology on $op|_T$, which imposes no local conditions.

These examples are special cases of an extremely general and widely applicable theorem: the category of simplicial presheaves on an *arbitrary* small Grothendieck site has a model structure for which the cofibrations are defined sectionwise and the weak equivalences are defined locally by the topology on the site. This result is Theorem 5.8 of this book. Applications of this result have accumulated since its discovery in the 1980s, in algebraic K-theory, algebraic geometry, number theory and algebraic topology.

The groundwork for Theorem 5.8 is a bit technical, and is dealt with in Chaps. 2–4. The technicality primarily arises from specifying what is meant by locally defined weak equivalences, here called *local weak equivalences*, for general Grothendieck sites. In cases where there is a theory of stalks, this is no problem; a local weak equivalence is a stalkwise weak equivalence, as in the example above for a topological space T.

In the language of topos theory, having a theory of stalks means that the Grothendieck topos of sheaves on the site $\mathcal{C}$ has "enough points". Not all Grothendieck toposes have enough points—the flat sites from algebraic geometry are examples—and we

Introduction

have to be much more careful in such cases. There are multiple approaches, and all of them are useful: one can, for example, specify local weak equivalences to be maps that induce isomorphisms on locally defined sheaves of homotopy groups for all local choices of base points (this was the original definition), or one can use the theorem from topos theory which says that every Grothendieck topos has a Boolean localization. This theorem amounts to the existence of a "fat point" for sheaves and presheaves, taking values in a sheaf category which behaves very much like the category of sets. In general, a local weak equivalence is a map which induces a sectionwise equivalence of associated fat points, although there is still some fussing to say exactly what that means.

The Boolean localization theorem (or Barr's theorem, Theorem 3.31) is a major result, and it is used everywhere. There are workarounds which avoid Boolean localization—they have the flavour of the definition of local weak equivalence that is based on sheaves of homotopy groups [7] and can be similarly awkward. A quick proof of the Boolean localization theorem appears in Chap. 3, after a self-contained but rapid discussion of the basic definitions and results of topos theory that one needs for this type of homotopy theory. There is also a proof of the theorem of Giraud (Theorem 3.17) which identifies Grothendieck toposes by exactness properties. Some basic definitions and results for simplicial sets are reviewed in the first chapter.

From a historical point of view, these last paragraphs about simplicial presheaves are quite misleading, because the original focus of local homotopy theory was on simplicial sheaves. The homotopy theory of simplicial sheaves had its origins in algebraic geometry and topos theory, and the applications of the theory, in algebraic geometry, number theory and algebraic K-theory, are fundamental.

The local weak equivalence concept evolved, over the years, from Grothendieck's notion of quasi-isomorphism of sheaves of chain complexes [36] of 1957. Illusie [47] extended this idea to morphisms of simplicial sheaves [47] in 1971, in his description of a quasi-isomorphism of simplicial sheaves. We now say that such a map is a local weak equivalence of simplicial sheaves.

Illusie conjectured that a local weak equivalence of simplicial sheaves induces a quasi-isomorphism of associated free integral chain complexes—this is easy to prove in the presence of stalks, but it resisted proof in general at the time. Illusie's conjecture was proved by van Osdol [101], in 1977, in the first homotopy theoretic application of Boolean localization.

Joyal showed, in a letter to Grothendieck of 1984, that the category of simplicial sheaves on an arbitrary small site has a closed model structure with sectionwise cofibrations and local weak equivalences. Joyal's proof again used Boolean localization, and was completely topos theoretic. His result generalized a theorem of Brown and Gersten [17] from algebraic K-theory, which treated the case of simplicial sheaves for the Zariski topology on a Noetherian scheme. Joyal's result also specializes to the injective model structure for diagrams which is described above. The first published appearance of the injective model structure for diagrams is found in work of Heller [37].

A half model structure (meaning, in this case, a category of fibrant objects structure) of local weak equivalences and local fibrations for simplicial sheaves was introduced in [49], along with a homotopy theoretic definition of the cohomology of a simplicial sheaf. This was done to initiate a method of tackling various questions of algebraic K-theory which relate discrete and topological behaviour.

The main example of the time was this: suppose that k is an algebraically closed field and let ℓ be a prime which is distinct from the characteristic of k. Let $(Sm|_k)_{et}$ be the site of smooth k-schemes (of high bounded cardinality), endowed with the étale topology. The general linear groups Gl_n represent sheaves of groups on this site, and one can form the corresponding simplicial sheaves BGl_n by applying the nerve functor sectionwise. Form the simplicial sheaf BGl by taking the filtered colimit of the objects BGl_n along the standard inclusions in the sheaf category. The étale cohomology groups $H_{et}^*(BGl, \mathbb{Z}/\ell)$ are defined, and can be computed from classical topological results by using base change theorems—this graded ring is a polynomial ring over $\mathbb{Z}/\ell$ in Chern classes. There is a canonical map of simplicial sheaves $\epsilon : \Gamma^* BGl(k) \to BGl$, where Γ^* is the constant simplicial sheaf functor and the simplicial set $BGl(k)$ is global sections of the simplicial sheaf BGl. The map ϵ induces a ring homomorphism

$$\epsilon^* : H_{et}^*(BGl, \mathbb{Z}/\ell) \to H^*(BGl(k), \mathbb{Z}/\ell)$$

relating the étale cohomology of BGl to the cohomology of the simplicial set $BGl(k)$. Here is a theorem: this map ϵ^* is an isomorphism. This statement is equivalent to the rigidity theorems of Suslin which compute the mod ℓ algebraic K-theory of the field k [95, 98], but it has an alternate proof that uses Gabber rigidity [29] to show that ϵ induces an isomorphism in mod ℓ homology sheaves.

This alternate proof of the Suslin theorems for algebraically closed fields was one of the first calculational successes of the homotopy theory of simplicial sheaves (and presheaves) in algebraic K-theory. The statement is a special case of the isomorphism conjecture of Friedlander and Milnor, which asserts that the map ϵ^* is an isomorphism when the sheaf of groups Gl is replaced by an arbitrary reductive algebraic group G. The Friedlander–Milnor conjecture is still open, but it is the subject of vigorous study [81].

The passage from the local homotopy theory of simplicial sheaves to the local homotopy theory of simplicial presheaves was predicated on a simple observation: where one has stalks for sheaves, one also has stalks for presheaves, and the canonical map $F \to \tilde{F}$ from a presheaf F to its associated sheaf $\tilde{F}$ induces an isomorphism on all stalks. Thus, if you can talk about stalkwise weak equivalences for simplicial sheaves, you can do the same for simplicial presheaves, in which case the canonical map $X \to \tilde{X}$ from a simplicial presheaf X to its associated simplicial sheaf $\tilde{X}$ should be a local weak equivalence. This is certainly true in general.

There are many circumstances where we care about presheaves more than their associated sheaves. For example, the mod ℓ K-theory presheaf of spectra K/ℓ can be defined on the étale site $et|_L$ of a sufficiently nice field L, and we might want to use étale (or Galois) cohomological techniques to compute the stable homotopy groups

$$K_s(L, \mathbb{Z}/\ell) = \pi_s K/\ell(L)$$

of the spectrum $K/\ell(L)$ in global sections of the presheaf of spectra K/ℓ. Specifically, there is a map

$$K/\ell(L) \to K/\ell^{et}(L)$$

taking values in the étale K-theory presheaf of spectra, and it is a special case of the Lichtenbaum–Quillen conjecture that this map induces an isomorphism in stable homotopy groups in a range of degrees. The étale K-theory presheaf of spectra K/ℓ^{et} can be taken to be a sheaf of spectra, but we do not assume that the presheaf of spectra K/ℓ has anything to do with its associated sheaf, except locally for the étale topology. This is an example of a descent problem. Such problems are about presheaves rather than sheaves, and the associated sheaf just gets in the way.

Joyal's theorem for simplicial sheaves follows from the simplicial presheaves result, and it appears as Theorem 5.9 here. The observation that the associated sheaf map is a local weak equivalence further implies that the forgetful and associated sheaf functors together determine a Quillen equivalence between the model structures for simplicial presheaves and simplicial sheaves. This means that there is no difference between the *local* homotopy theories for simplicial presheaves and simplicial sheaves, for a given topology on a site. What matters, in practice, is the behaviour of the local homotopy theories as one varies the topology of the underlying site, or more generally as one base changes along a geometric morphism. The behaviour of the corresponding direct image functors creates descent problems.

The descent concept and descent theory are discussed in some detail in Chap. 5. Descent theorems almost always have interesting proofs, which typically require substantial geometric input. The Brown–Gersten descent theorem for the Zariski topology and the Morel–Voevodsky descent theorem for the Nisnevich topology are two of the most striking and useful examples. Proofs of these results are presented in Sect. 5.4; they appear as Theorems 5.33 and 5.39, respectively. The Morel–Voevodsky theorem is often called "Nisnevich descent"—it is a key foundational result for much of motivic homotopy theory.

The locally defined model structure for simplicial presheaves on a site of Theorem 5.8, here called the injective model structure for simplicial presheaves, is a basis for all that follows. There are many variants of this theme, including the projective local model structure of Blander [9] which has been used frequently in motivic homotopy theory, and a plethora of model structures between the local projective structure and the injective structure, in which "intermediate structures" are characterized by their classes of cofibrations. These model structures all have the same class of weak equivalences, namely the local weak equivalences, so that they all define models for local homotopy theory—see Theorem 5.41. Intermediate model structures have been used in complex analytic geometry, in the study of Gromov's Oka principle [25].

Classically [4], a hypercover of a scheme S is a resolution $\pi : U \to S$ of S by a simplicial scheme U which is rather strongly acyclic. In modern terms, this means that, if we identify the map π with a map $U \to *$ of simplicial sheaves on the étale site $et|_S$ for S, then this map is a local trivial fibration. There are various ways to say what this means: the map π is a hypercover if and only if all induced simplicial set maps $U_x \to *$ in stalks are trivial Kan fibrations, or alternatively π is a hypercover if and only if it has a local right lifting property with respect to all simplicial set inclusions $\partial \Delta^n \subset \Delta^n$, $n \geq 0$. This local lifting concept generalizes immediately to that of a local trivial fibration $X \to Y$ of simplicial presheaves on an arbitrary site, and one often says that such a map is a hypercover over the simplicial presheaf Y.

There is a corresponding notion of local fibration, which is defined by a local right lifting property with respect to all inclusions $\Lambda^n_k \subset \Delta^n$, $n \geq 1$, of horns in simplices. Theorem 4.32 says that a map is a local trivial fibration if and only if it is both a local fibration and a local weak equivalence—the proof of this result uses a Boolean localization argument.

Every fibration for the injective model structure on a simplicial presheaf category is a local fibration, but the converse assertion is false. The injective fibrations exist formally, and are usually mysterious. In particular, injective fibrant simplicial presheaves behave somewhat like injective resolutions, hence the name.

The Verdier hypercovering theorem, in its classical form [4], says that if A is a sheaf of abelian groups on a site $\mathcal{C}$ having sufficiently many points, then the sheaf cohomology group $H^n(\mathcal{C}, A)$ can be calculated by hypercovers in the sense that there is an isomorphism

$$H^n(\mathcal{C}, A) = \varinjlim_U H^n(A(U)),$$

where U varies over simplicial homotopy classes of maps between hypercovers $U \to *$, and $A(U)$ is the cosimplicial abelian group which is defined by evaluating A on the simplicial object U. This theorem is extended to a result about locally fibrant simplicial sheaves in [49]: there is an isomorphism

$$[Y, K(A, n)] \cong \varinjlim_{[\pi]: X \to Y} \pi(Y, K(A, n)). \tag{1.1}$$

Here, $[Y, K(A, n)]$ denotes morphisms in the homotopy category arising from a category of fibrant objects structure, while $\pi(Y, K(A, n))$ is defined to be simplicial homotopy classes of maps between simplicial sheaves. The colimit is indexed over the category of simplicial homotopy classes of hypercovers $\pi : Y \to X$ and the simplicial homotopy classes of maps between them. The colimit is filtered, by a calculus of fractions argument.

The cohomology group $H^n(A(U))$ can be identified with the group $\pi(U, K(A, n))$ of simplicial homotopy classes of maps. What we end up with is a general definition

$$H^n(Y, A) = [Y, K(A, n)]$$

of cohomology for locally fibrant simplicial sheaves Y that specializes to a homotopy theoretic description

$$H^n(\mathcal{C}, A) = [*, K(A, n)]$$

of sheaf cohomology. One defines the cohomology groups $H^n(X, A)$ for general simplicial presheaves X in the same way: we define $H^n(X, A)$ by setting

$$H^n(X, A) = [X, K(A, n)],$$

where $[X, K(A, n)]$ denotes morphisms in the homotopy category which is associated to the injective model structure.

The isomorphism (1.1) leads to a different interpretation of morphisms in the homotopy category: the displayed colimit is the group of path components of a category whose objects are the pictures

$$X \xleftarrow{[\pi]} Y \xrightarrow{[g]} K(A, n)$$

where the square brackets indicate simplicial homotopy classes of maps and π is a hypercover. The morphisms of this category are the commutative diagrams

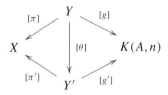

in simplicial homotopy classes of maps. The fact that the colimit in (1.1) is filtered is incidental from this point of view, but the filtered colimit is useful in calculations.

One generalizes this observation as follows: for simplicial presheaves X and Y, say that a cocycle from X to Y is pair of simplicial presheaf maps

$$X \xleftarrow[\simeq]{\sigma} U \xrightarrow{g} Y$$

such that the map σ is a local weak equivalence. A morphism of cocycles from X to Y is a commutative diagram

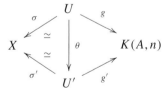

The corresponding category is denoted by $h(X, Y)$ and is called the category of cocycles from X to Y. It is a consequence of Theorem 6.5 that there is an isomorphism

$$\pi_0 h(X, Y) \cong [X, Y], \qquad (1.2)$$

which identifies the set of morphisms in the homotopy category for simplicial presheaves with the set of path components for the cocycle category $h(X, Y)$.

Theorem 6.5 is a general phenomenon, which applies to all model structures, which are right proper and whose classes of weak equivalences are closed under finite products. The theory of cocycle categories is a particularly useful technical device for the discussion of cohomology, both abelian and non-abelian, which follows in Chaps. 8 and 9. The power of the theory comes from the generality of the definition: there are no conditions on the objects X, U, and Y in the definition of cocycle above, and the map σ is a local weak equivalence which is not required to be a fibration in any sense.

The isomorphism (1.2) specializes to the isomorphism (1.1) of the Verdier hypercovering theorem, with a little work. Variations on this theme form the subject of Sect. 6.2.

The motivic homotopy theory of Morel and Voevodsky [82] is a localized model structure, in which one takes the injective model structure for the category of simplicial presheaves on the smooth Nisnevich site of a scheme S and collapses the affine line $\mathbb{A}^1$ over S to a point by formally inverting a rational point $* \to \mathbb{A}^1$. The choice of rational point does not matter, though one normally uses the 0-section. This is an important special case of a construction which formally inverts a cofibration $f : A \to B$ in the injective model structure for simplicial presheaves, to produce an f-local model structure.

Chapter 7 gives a self-contained account of a general machine, which applies in all homotopy theoretic contexts described in this book, for formally inverting a set of cofibrations of simplicial presheaves or of presheaves of spectrum-like objects. Localizations of simplicial presheaf categories are formally constructed in Sect. 7.2. Other applications include the construction of stable categories and localized stable categories for categories of spectrum-like objects, and localizations for categories of presheaves of chain complexes.

The approach that is taken here is axiomatic; it starts with a closed model category **M** and a functor $L : \mathbf{M} \to \mathbf{M}$ which together have a certain list of properties. One shows (Theorem 7.5) that there is a model structure on the category **M** which has the same cofibrations, and has L-equivalences for weak equivalences. An L-equivalence is the obvious thing: it is a map $X \to Y$ such that the induced map $L(X) \to L(Y)$ is a weak equivalence in the original model structure on **M**.

The functor L is constructed, in all cases, as the solution to a lifting problem for a set of cofibrations, following a prototype which is first displayed in Sect. 7.2. The most interesting condition on the functor L that one needs to verify, in practice, is that the L-equivalences should satisfy a *bounded monomorphism condition*. This is always done with some kind of argument on cardinality, hence the name. The bounded monomorphism condition is one of the most important technical devices of the theory.

The homotopy theoretic approach to the homological algebra of abelian sheaves and presheaves is the subject of Chap. 8. The starting point is an injective model structure for presheaves of simplicial R-modules (or, via the Dold–Kan correspondence, presheaves of chain complexes), which is defined by transport of structure from the injective model structure on simplicial presheaves, in Theorem 8.6. The weak equivalences in this case are quasi-isomorphisms in the traditional sense, and

Introduction

the fibrations are those maps $A \to B$ whose underlying simplicial presheaf maps are injective fibrations. Here, R is a presheaf of commutative rings with identity. For the proof of Theorem 8.6, we need to know that the free simplicial R-module functor takes local weak equivalences to quasi-isomorphisms—this statement is the "Illusie conjecture" which is discussed above, and it is a consequence of Lemma 8.2.

The preservation of local weak equivalences by the free R-module functor sets up a Quillen adjunction, through which we identify sheaf cohomology with morphisms in the simplicial presheaf homotopy category in Theorem 8.26. The proof of this result involves chain homotopy calculations which are effected by using variants of cocycle categories. The universal coefficients spectral sequence of Lemma 8.30 is a byproduct of this approach, as is the fact that the good truncations of chain complexes of injectives satisfy descent. Cocycle categories are also used to give an elementary description of cup product structures for the cohomology of simplicial presheaves in Sect. 8.4.

Chapter 8 concludes with a discussion of localizations in the injective model structure for presheaves of chain complexes. In all cases, the technique is to formally invert a cofibration $f : A \to B$ to construct an f-local model structure, by using the methods of Chap. 7.

This technique is applied, in Sect. 8.6, to categories of presheaves of chain complexes which are specializations of linear functors that are defined on an additive category. The main example is the category of simplicial presheaves with transfers which is defined on the category Cor_k of additive correspondences on a perfect field k—the result is a model structure of chain complex objects whose derived category is Voevodsky's category of effective motives over the field k.

Chapter 9 is an extended discussion of non-abelian cohomology theory. The injective model structure for simplicial presheaves specializes to an injective model structure for presheaves of groupoids, for which a map $G \to H$ is a local weak equivalence, respectively fibration, if and only if, the induced map $BG \to BH$ of simplicial presheaves given by the nerve construction is a local weak equivalence, respectively, injective fibration—this is Proposition 9.19, which is due to Hollander [43]. From this point of view, stacks are sheaves (or presheaves) of groupoids which satisfy descent for this injective model structure—this follows from Proposition 9.28. Stacks are thus identified with homotopy types of sheaves or presheaves of groupoids. Further, an injective fibrant model of a sheaf of groupoids can be identified with its stack completion.

The examples are classical: they include the presheaf of G-torsors for a sheaf of groups G and the quotient stack for an action $G \times X \to X$ of G on a sheaf X. The quotient stack X/G. for such an action is identified with the homotopy type of the Borel construction $EG \times_G X$ in Lemma 9.24.

These identifications are consequences of cocycle category calculations. A similar calculation identifies the stack completion of a sheaf of groupoids H with a presheaf of H-torsors, suitably defined. In all cases, torsors are defined by the local acyclicity of certain homotopy colimits. This homotopy colimit description coincides with the classical description of G-torsors as principal G-bundles, for sheaves of groups G.

A description of higher stack theory, as the local homotopy theory of presheaves of groupoids enriched in simplicial sets, appears in Sects. 9.3 and 9.4.

Preliminary results, for "ordinary" groupoids enriched in simplicial sets are discussed in Sect. 9.3. These include the Dwyer–Kan model structure for these objects in Theorem 9.30, for which the weak equivalences are those maps $G \to H$ whose associated bisimplicial set map $BG \to BH$ is a diagonal weak equivalence, and the natural identification of the diagonal $d(BG)$ with the Eilenberg–Mac Lane construction $\overline{W}G$. The description of the object $\overline{W}G$ which is presented here, as the Artin–Mazur total simplicial set $T(BG)$ of the bisimplicial set BG, has only been properly exposed recently, in [94].

This is the enabling technology for a model structure (Theorem 9.43) on groupoids enriched in simplicial sets, for which the fibrations (respectively, weak equivalences) are those maps $G \to H$ which induce Kan fibrations (respectively, weak equivalences) $\overline{W}G \to \overline{W}H$ of simplicial sets. This model structure is Quillen equivalent to the standard model structure on simplicial sets, via the left adjoint $X \mapsto G(X)$ (the loop groupoid functor) for the Eilenberg–Mac Lane functor $\overline{W}$.

The model structure of Theorem 9.43 is promoted to a local model structure for presheaves of groupoids enriched in simplicial sets in Theorem 9.50, which is Quillen equivalent to the injective model structure on simplicial presheaves. This model structure specializes to a homotopy theory of n-types in Theorem 9.56, and to a homotopy theory for presheaves of 2-groupoids in Theorem 9.57.

The presheaves of 2-groupoids which satisfy descent for the structure of Theorem 9.57 are the 2-stacks. The model structure for 2-groupoids is also susceptible to a cocycle calculus, which is used to give a homotopy classification of gerbes locally equivalent to objects in a family of sheaves of groups in Corollary 9.68, and a homotopy classification of families of extensions of a sheaf of groups in Corollary 9.72. These are modern expressions of classical results from higher non-abelian cohomology theory.

The final two chapters of this book give an account of the basic results of local stable homotopy theory. The discussion includes the stable homotopy theory of presheaves of spectra on arbitrary sites, the various forms of motivic stable homotopy theory, and abstractions of these theories.

The stable homotopy theory of presheaves of spectra has its own rich set of applications, first in algebraic K-theory, and then most recently in "traditional" stable homotopy theory via the theory of topological modular forms. This flavour of stable homotopy theory is relatively simple to derive in isolation, and this is done in Sect. 10.1.

Voevodsky's motivic stable category [102] is a more complex object, for a collection of reasons.

The first of these is a departure from the use of the topological circle S^1 as a parameter space. An ordinary spectrum E consists of pointed spaces E^n, $n \geq 0$, pointed maps $\sigma : S^1 \wedge E^n \to E^{n+1}$. A presheaf of spectra E is a simple extension of this concept: it consists of pointed simplicial presheaves E^n and pointed simplicial presheaf maps σ. We continue to use the circle S^1 as a parameter object in the definition, and for this reason, most of the usual rules of stable homotopy theory apply to presheaves of spectra.

The objects which define the motivic stable category are T-spectra, which are defined on the smooth Nisnevich site of a scheme: a T-spectrum X, in this context, consists of pointed simplicial presheaves X^n, with bonding maps $\sigma : T \wedge X^n \to X^{n+1}$, where the parameter space T can be viewed as either the projective line $\mathbb{P}^1$, or equivalently a topological suspension $S^1 \wedge \mathbb{G}_m$ of the multiplicative group $\mathbb{G}_m$. The group structure on $\mathbb{G}_m$ is irrelevant for this theory—we care more about its underlying scheme $\mathbb{A}^1 - \{0\}$—but the $\mathbb{G}_m$ notation is convenient. The parameter object $S^1 \wedge \mathbb{G}_m$ is not a circle: it is a topological suspension of a geometric circle, and should perhaps be thought of as a twisted 2-sphere.

The other major complication in the setup for motivic stable homotopy theory is the requirement of working within the motivic model structure, which is a localization of the base injective model structure that is constructed by formally inverting a rational point of the affine line.

The original construction of the motivic stable model structure [57] used the methods of Bousfield and Friedlander. That construction requires properness of the underlying motivic model structure for simplicial presheaves, and then Nisnevich descent is required at multiple points in the development to get traditional stabilization constructions to work out correctly. When one looks at the overall argument, two major features emerge, namely a compactness property for the parameter object and a strong descent property for the underlying motivic model structure. To go beyond the base construction of the motivic stable model structure, for example to show that smashing with T is invertible on the motivic stable category, one also needs to know that the cycle permutations act trivially on the threefold smash $T^{\wedge 3}$, and one uses a geometric argument to show this. Finally, to show that cofibre sequences coincide with fibre sequences in the motivic stable category (so that, for example, finite wedges coincide with finite products, as in the ordinary stable category), one uses the fact that T is the topological suspension $S^1 \wedge \mathbb{G}_m$ in the motivic homotopy category.

Motivic stable homotopy theory and motivic cohomology theory [78] together form the setting for many of the recent calculational advances in algebraic K-theory, including the proofs of the Bloch–Kato conjecture [107] and the Lichtenbaum–Quillen conjecture [96].

The major features of motivic stable homotopy theory can be abstracted, to form the f-local stable homotopy theory of presheaves of T-spectra on an arbitrary site, where f is a cofibration of simplicial presheaves and T is an arbitrary parameter object. The overall model structure is constructed in a different way, following Theorem 10.20, by inverting the map f and the stablization maps all at once. One then has to assume a few things to show that this f-local stable model structure behaves like ordinary stable homotopy theory: these assumptions amount to compactness of T, a strong descent property for the f-local model structure, and the cycle triviality of the 3-fold smash $T^{\wedge 3}$. If the parameter object T is a topological suspension, then we have coincidence of cofibre and fibre sequences in the f-local model structure, as in ordinary stable homotopy theory—see Lemma 10.62.

This technique specializes to a different construction of the motivic stable model structure. Other applications arise from the fact that the class of pointed simplicial presheaves T which are compact and cycle trivial is closed under finite smash products.

Unlike the Bousfield–Friedlander stabilization method, one does not need properness of the f-local model structure on simplicial presheaves to initiate the construction. Further, with the assumptions on the parameter object and the underlying f-local structure described above in place, the f-local stable model structure on $(S^1 \wedge T)$-spectra is proper (Theorem 10.64), without an assumption of properness for the underlying f-local model structure for simplicial presheaves. The f-local stable model structure for $(S^1 \wedge T)$-spectra also has slice filtrations in the parameter T, which are constructed with localization techniques in Sect. 10.6.

The same general themes and constraints continue in the study of symmetric T-spectra in f-local settings in Chap. 11. The concept of symmetric T-spectrum is a generalization of that of symmetric spectrum: a symmetric T-spectrum is a T-spectrum X with symmetric group actions $\Sigma_n \times X^n \to X^n$, which actions are compatible with twists of the smash powers $T^{\wedge k}$ under iterated bonding maps. One constructs the f-local stable model structure for symmetric T-spectra by localizing at f and the stabilization maps simultaneously, by using Theorem 11.13. In the presence of the three main assumptions just discussed, the resulting model structure is a model for f-local stable homotopy theory in the sense that it is Quillen equivalent to the f-local stable model structure for T-spectra (Theorem 11.36). As in the case of ordinary symmetric spectra, there is a symmetric monoidal smash product for symmetric T-spectra which is the basis of a theory of products for these objects—this smash product is the subject of Sect. 11.6. The theory is a generalization of the theory of motivic symmetric spectra [27].

These ideas carry over, in Sects. 10.7 and 11.7, to stable homotopy constructions in various abelian settings. There are categories of of T-complexes and symmetric T-complexes in presheaves of simplicial modules for each parameter object T, which categories have stable f-local model structures which are Quillen equivalent under the usual assumptions (Theorem 11.61). These ideas and results specialize to f-local stable model structures for S^1-complexes (or unbounded chain complexes) and symmetric S^1-complexes, and to Quillen equivalences between them. An f-local stable model structure can also be defined for T-complex objects in enriched linear settings (Theorem 10.96), such as the category of simplicial presheaves with transfers. The f-local stable model structure, for the rational point $f : * \to \mathbb{A}^1$ and $(S^1 \wedge \mathbb{G}_m)$-complexes, has an associated stable category which is equivalent to Voevodsky's big category of motives—see Example 10.97.

Part I
Preliminaries

Chapter 2
Homotopy Theory of Simplicial Sets

We begin with a brief description of the homotopy theory of simplicial sets, and a first take on the homotopy theory of diagrams of simplicial sets. This is done to establish notation, and to recall some basic constructions and well-known lines of argument. A much more detailed presentation of this theory can be found in [32].

The description of the model structure for simplicial sets which appears in the second section is a bit unusual, in that we first show in Theorem 2.13 that there is a model structure for which the cofibrations are the monomorphisms and the weak equivalences are defined by topological realization. The proof of the fact that the fibrations for this theory are the Kan fibrations (Theorem 2.19) then becomes a somewhat delicate result whose proof is only sketched. For the argument which is presented here, it is critical to know Quillen's theorem [88] that the realization of a Kan fibration is a Serre fibration.

The proof of Theorem 2.13 uses the "bounded monomorphism property" for simplicial sets of Lemma 2.16. This is a rather powerful principle which recurs in various guises throughout the book.

The model structure for diagrams of simplicial sets (here called the projective model structure) appears in the third section, in Proposition 2.22. This result was first observed by Bousfield and Kan [14] and has a simple proof with a standard method of attack. In the context of subsequent chapters, Proposition 2.22 gives a preliminary, essentially non-local model structure for all simplicial presheaf categories.

2.1 Simplicial Sets

The finite ordinal number **n** is the set of counting numbers

$$\mathbf{n} = \{0, 1, \ldots, n\}.$$

There is an obvious ordering on this set which gives it the structure of a poset, and hence a category. In general, if C is a category then the functors $\alpha : \mathbf{n} \to C$ can be identified with strings of arrows

$$\alpha(0) \to \alpha(1) \to \cdots \to \alpha(n)$$

of length n. The collection of all finite ordinal numbers and all order-preserving functions between them (or poset morphisms, or functors) form the *ordinal number category* $\mathbf{\Delta}$.

Example 2.1 The ordinal number monomorphisms $d^i : \mathbf{n-1} \to \mathbf{n}$ are defined by the strings of relations

$$0 \leq 1 \leq \cdots \leq i-1 \leq i+1 \leq \cdots \leq n$$

for $0 \leq i \leq n$. These morphisms are called *cofaces*.

Example 2.2 The ordinal number epimorphisms $s^j : \mathbf{n+1} \to \mathbf{n}$ are defined by the strings

$$0 \leq 1 \leq \cdots \leq j \leq j \leq \cdots \leq n$$

for $0 \leq j \leq n$. These are the *codegeneracies*.

The cofaces and codegeneracies together satisfy the following relations

$$\begin{aligned} d^j d^i &= d^i d^{j-1} \text{ if } i < j, \\ s^j s^i &= s^i s^{j+1} \text{ if } i \leq j \\ s^j d^i &= \begin{cases} d^i s^{j-1} & \text{if } i < j, \\ 1 & \text{if } i = j, j+1, \\ d^{i-1} s^j & \text{if } i > j+1. \end{cases} \end{aligned} \quad (2.1)$$

The ordinal number category $\mathbf{\Delta}$ is generated by the cofaces and codegeneracies, subject to the *cosimplicial identities* (2.1) [76]. In effect, every ordinal number morphism has a unique epi-monic factorization, and has a canonical form defined in terms of strings of codegeneracies and strings of cofaces.

A *simplicial set* is a functor $X : \mathbf{\Delta}^{op} \to \mathbf{Set}$, or a contravariant set-valued functor on the ordinal number category $\mathbf{\Delta}$. Such things are often written as $\mathbf{n} \mapsto X_n$, and X_n is called the set of *n-simplices* of X. A *map of simplicial sets* $f : X \to Y$ is a natural transformation of such functors. The simplicial sets and simplicial set maps form the category of simplicial sets, which will be denoted by $s\mathbf{Set}$.

A simplicial set is a simplicial object in the set category. Generally, $s\mathbf{A}$ denotes the category of simplicial objects $\mathbf{\Delta}^{op} \to \mathbf{A}$ in a category $\mathbf{A}$. Examples include the categories $s\mathbf{Gr}$ of simplicial groups, $s(R-\mathbf{Mod})$ of simplicial R-modules, $s(s\mathbf{Set}) = s^2\mathbf{Set}$ of bisimplicial sets, and so on.

Example 2.3 The topological standard n-simplex is the space

$$|\Delta^n| := \{(t_0, \ldots, t_n) \in \mathbb{R}^{n+1} \mid 0 \leq t_i \leq 1, \sum_{i=0}^{n} t_i = 1\}$$

The assignment $\mathbf{n} \mapsto |\Delta^n|$ is a cosimplicial space, or a cosimplicial object in spaces. A covariant functor $\mathbf{\Delta} \to \mathbf{A}$ is a *cosimplicial object* in the category $\mathbf{A}$.

2.1 Simplicial Sets

If X is a topological space, then the *singular set* or *singular complex* $S(X)$ is the simplicial set which is defined by

$$S(X)_n = \hom(|\Delta^n|, X).$$

The assignment $X \mapsto S(X)$ defines a functor

$$S : \mathbf{CGHaus} \to s\mathbf{Set},$$

and this functor is called the *singular functor*. Here, **CGHaus** is the category of compactly generated Hausdorff spaces, which is the usual category of spaces for homotopy theory [32, I.2.4].

Example 2.4 The ordinal number **n** represents a contravariant functor

$$\Delta^n = \hom_{\mathbf{\Delta}}(\ , \mathbf{n}),$$

which is called the *standard n-simplex*. Write

$$\iota_n = 1_{\mathbf{n}} \in \hom_{\mathbf{\Delta}}(\mathbf{n}, \mathbf{n}).$$

The n-simplex ι_n is often called the *classifying n-simplex*, because the Yoneda Lemma implies that there is a natural bijection

$$\hom(\Delta^n, Y) \cong Y_n$$

that is defined by sending the simplicial set map $\sigma : \Delta^n \to Y$ to the element $\sigma(\iota_n) \in Y_n$. One usually says that a simplicial set map $\Delta^n \to Y$ is an n-simplex of Y.

In general, if $\sigma : \Delta^n \to X$ is a simplex of X, then the ith *face* $d_i(\sigma)$ is the composite

$$\Delta^{n-1} \xrightarrow{d^i} \Delta^n \xrightarrow{\sigma} X,$$

while the jth *degeneracy* $s_j(\sigma)$ is the composite

$$\Delta^{n+1} \xrightarrow{s^j} \Delta^n \xrightarrow{\sigma} X.$$

Example 2.5 The simplicial set $\partial \Delta^n$ is the subobject of Δ^n which is generated by the $(n-1)$-simplices d^i, $0 \le i \le n$, and Λ^n_k is the subobject of $\partial \Delta^n$ which is generated by the simplices d^i, $i \ne k$. The object $\partial \Delta^n$ is called the *boundary* of Δ^n, and Λ^n_k is called the kth *horn*.

The faces $d^i : \Delta^{n-1} \to \Delta^n$ determine a covering

$$\bigsqcup_{i=0}^n \Delta^{n-1} \to \partial \Delta^n,$$

and for each $i < j$, there are pullback diagrams

$$\begin{array}{ccc} \Delta^{n-2} & \xrightarrow{d^{j-1}} & \Delta^{n-1} \\ d^i \downarrow & & \downarrow d^i \\ \Delta^{n-1} & \xrightarrow{d^j} & \Delta^n. \end{array}$$

It follows that there is a coequalizer

$$\bigsqcup_{i<j, 0 \leq i,j \leq n} \Delta^{n-2} \rightrightarrows \bigsqcup_{0 \leq i \leq n} \Delta^{n-1} \to \partial \Delta^n$$

in $s\mathbf{Set}$. Similarly, there is a coequalizer

$$\bigsqcup_{i<j, i,j \neq k} \Delta^{n-2} \rightrightarrows \bigsqcup_{0 \leq i \leq n, i \neq k} \Delta^{n-1} \to \Lambda^n_k.$$

Example 2.6 Suppose that a category C is *small* in the sense that the morphisms $\mathrm{Mor}(C)$ of C form a set. Examples of such things include all finite ordinal numbers **n**, all monoids (small categories having one object), and all groups.

If C is a small category then there is a simplicial set BC with

$$BC_n = \hom(\mathbf{n}, C),$$

meaning the functors $\mathbf{n} \to C$. The simplicial structure on BC is defined by pre-composition with ordinal number maps. The object BC is called, variously, the *classifying space* or *nerve* of the category C.

Note that the standard n-simplex Δ^n is the classifying space $B\mathbf{n}$ in this notation.

Example 2.7 Suppose that I is a small category, and that $X : I \to \mathbf{Set}$ is a set-valued functor. The *translation category*, or *category of elements*

$$*/X = E_I(X)$$

associated to X has for objects all pairs (i, x) with $x \in X(i)$, or equivalently all functions

$$* \xrightarrow{x} X(i).$$

A morphism $\alpha : (i, x) \to (j, y)$ is a morphism $\alpha : i \to j$ of I such that $\alpha_*(x) = y$, or equivalently a commutative diagram

2.1 Simplicial Sets

The simplicial set $B(E_I X)$ is often called the *homotopy colimit* for the functor X, and one writes

$$\underrightarrow{\operatorname{holim}}_I X = B(E_I X).$$

There is a canonical functor $E_I X \to I$ which is defined by the assignment $(i, x) \mapsto i$, and induces a canonical simplicial set map

$$\pi : B(E_I X) = \underrightarrow{\operatorname{holim}}_I X \to BI.$$

The functors $\mathbf{n} \to E_I X$ can be identified with strings

$$(i_0, x_0) \xrightarrow{\alpha_1} (i_1, x_1) \xrightarrow{\alpha_2} \ldots \xrightarrow{\alpha_n} (i_n, x_n).$$

Such a string is uniquely specified by the underlying string $i_0 \to \cdots \to i_n$ in the index category Y and $x_0 \in X(i_0)$. It follows that there is an identification

$$(\underrightarrow{\operatorname{holim}}_I X)_n = B(E_I X)_n = \bigsqcup_{i_0 \to \cdots \to i_n} X(i_0).$$

This construction is functorial with respect to natural transformations in X. Thus, a diagram $Y : I \to s\mathbf{Set}$ in simplicial sets determines a bisimplicial set with (n, m) simplices

$$B(E_I Y)_m = \bigsqcup_{i_0 \to \cdots \to i_n} Y(i_0)_m.$$

The *diagonal* $d(Z)$ of a bisimplicial set Z is the simplicial set with n-simplices $Z_{n,n}$. Equivalently, $d(Z)$ is the composite functor

$$\mathbf{\Delta}^{op} \xrightarrow{\Delta} \mathbf{\Delta}^{op} \times \mathbf{\Delta}^{op} \xrightarrow{Z} \mathbf{Set}$$

where Δ is the diagonal functor.

The diagonal $dB(E_I Y)$ of the bisimplicial set $B(E_I Y)$ is the *homotopy colimit* $\underrightarrow{\operatorname{holim}}_I Y$ of the diagram $Y : I \to s\mathbf{Set}$ in simplicial sets. There is a natural simplicial set map

$$\pi : \underrightarrow{\operatorname{holim}}_I Y \to BI.$$

Example 2.8 Suppose that X and Y are simplicial sets. There is a simplicial set $\mathbf{hom}(X, Y)$ with n-simplices

$$\mathbf{hom}(X, Y)_n = \mathrm{hom}(X \times \Delta^n, Y),$$

called the *function complex*.

There is a natural simplicial set map

$$ev : X \times \mathbf{hom}(X, Y) \to Y,$$

which sends the pair $(x, f : X \times \Delta^n \to Y)$ to the simplex $f(x, \iota_n)$. Suppose that K is another simplicial set. The function

$$ev_* : \hom(K, \mathbf{hom}(X, Y)) \to \hom(X \times K, Y),$$

which is defined by sending the map $g : K \to \mathbf{hom}(X, Y)$ to the composite

$$X \times K \xrightarrow{1 \times g} X \times \mathbf{hom}(X, Y) \xrightarrow{ev} Y,$$

is a natural bijection, giving the *exponential law*

$$\hom(K, \mathbf{hom}(X, Y)) \cong \hom(X \times K, Y).$$

This natural isomorphism gives $s\mathbf{Set}$ the structure of a cartesian closed category. The function complexes also give $s\mathbf{Set}$ the structure of a category enriched in simplicial sets.

The *simplex category* Δ/X for a simplicial set X has for objects all simplices $\Delta^n \to X$. Its morphisms are the incidence relations between the simplices (meaning all commutative diagrams).

The *realization* $|X|$ of a simplicial set X is defined by

$$|X| = \varinjlim_{\Delta^n \to X} |\Delta^n|,$$

where the colimit is defined for the functor $\Delta/X \to \mathbf{CGHaus}$, which takes a simplex $\Delta^n \to X$ to the space $|\Delta^n|$.

The space $|X|$ is constructed by glueing together copies of the topological standard simplices of Example 2.3 along the incidence relations of the simplices of X.

The assignment $X \mapsto |X|$ defines a functor

$$|\ | : s\mathbf{Set} \to \mathbf{CGHaus}.$$

The proof of the following lemma is an exercise:

Lemma 2.9 *The realization functor $|\ |$ is left adjoint to the singular functor S.*

Example 2.10 The realization $|\Delta^n|$ of the standard n-simplex is the space $|\Delta^n|$ described in Example 2.3, since the simplex category Δ/Δ^n has a terminal object, namely $1 : \Delta^n \to \Delta^n$.

Example 2.11 The realization $|\partial \Delta^n|$ of the simplicial set $\partial \Delta^n$ is the topological boundary $\partial|\Delta^n|$ of the space $|\Delta^n|$. The space $|\Lambda^n_k|$ is the part of the boundary $\partial|\Delta^n|$

with the face opposite the vertex k removed. To see this, observe that the realization functor is a left adjoint and therefore preserves coequalizers and coproducts.

The *n-skeleton* $\mathrm{sk}_n X$ of a simplicial set X is the subobject generated by the simplices X_i, $0 \leq i \leq n$. The ascending sequence of subcomplexes

$$\mathrm{sk}_0 X \subset \mathrm{sk}_1 X \subset \mathrm{sk}_2 X \subset \ldots$$

defines a filtration of X, and there are pushout diagrams.

$$\begin{array}{ccc} \bigsqcup_{x \in NX_n} \partial \Delta^n & \longrightarrow & \mathrm{sk}_{n-1} X \\ \downarrow & & \downarrow \\ \bigsqcup_{x \in NX_n} \Delta^n & \longrightarrow & \mathrm{sk}_n X \end{array}$$

Here, NX_n denotes the set of non-degenerate n-simplices of X. An n-simplex x of X is non-degenerate if it is not of the form $s_i y$ for some $(n-1)$-simplex y.

It follows that the realization of a simplicial set is a CW-complex. Every monomorphism $A \to B$ of simplicial sets induces a cofibration $|A| \to |B|$ of spaces, since $|B|$ is constructed from $|A|$ by attaching cells.

The realization functor preserves colimits (is right exact) because it has a right adjoint. The realization functor, when interpreted as taking values in compactly generated Hausdorff spaces, also has a fundamental left exactness property:

Lemma 2.12 *The realization functor*

$$| \ | : s\mathbf{Set} \to \mathbf{CGHaus}.$$

preserves finite limits. Equivalently, it preserves finite products and equalizers.

This result is proved in [30].

2.2 Model Structure for Simplicial Sets

This section summarizes material which is presented in some detail in [32].

Say that a map $f : X \to Y$ of simplicial sets is a *weak equivalence* if the induced map $f_* : |X| \to |Y|$ is a weak equivalence of **CGHaus**. A map $i : A \to B$ of simplicial sets is a *cofibration* if and only if it is a monomorphism, meaning that all functions $i : A_n \to B_n$ are injective. A simplicial set map $p : X \to Y$ is a *fibration* if and only if it has the right lifting property with respect to all trivial cofibrations.

As usual, a *trivial cofibration* (respectively *trivial fibration*) is a cofibration (respectively fibration) which is also a weak equivalence.

In all that follows, a *closed model category* will be a category **M** equipped with three classes of maps, called cofibrations, fibrations and weak equivalences such that the following axioms are satisfied:

CM1 The category **M** has all finite limits and colimits.

CM2 Suppose given a commutative diagram

in **M**. If any two of the maps f, g and h are weak equivalences, then so is the third.

CM3 If a map f us a retract of g and g is a weak equivalence, fibration or cofibration, then so is f.

CM4 Suppose given a commutative solid arrow diagram

where i is a cofibration and p is a fibration. Then the dotted arrow exists, making the diagram commute, if either i or p is a weak equivalence.

CM5 Every map $f : X \to Y$ has factorizations $f = p \cdot i$ and $f = q \cdot j$, in which i is a cofibration and a weak equivalence and p is a fibration, and j is a cofibration and q is a fibration and a weak equivalence.

The definition of closed model category which is displayed here is the traditional one, which is due to Quillen [86]. There are variants in the literature, which involve either removing the finiteness condition from the limits and colimits in **CM1**, or insisting that the factorizations of **CM5** are functorial. These conditions almost always hold in practice, and in particular they hold for all model structures that we use.

There are common adjectives which decorate closed model structures. For example, one says that the model structure on **M** is *simplicial* if the category can be enriched in simplicial sets in a way that behaves well with respect to cofibrations and fibrations, and the model structure is *proper* if weak equivalences are preserved by pullback along fibrations and pushout along cofibrations. A model structure is *cofibrantly generated* if its classes of cofibrations and trivial cofibrations are generated by sets of maps in a suitable sense. The factorizations of **CM5** can be constructed functorially in a cofibrantly generated model structure, with a *small object argument*. Much more detail can be found in [32] or [44].

Theorem 2.13 *With the definitions given above of weak equivalence, cofibration and fibration, the category s**Set** of simplicial sets satisfies the axioms for a closed model category.*

Here are the basic ingredients of the proof:

Lemma 2.14 *A map $p : X \to Y$ is a trivial fibration if and only if it has the right lifting property with respect to all inclusions $\partial \Delta^n \subset \Delta^n$, $n \geq 0$.*

2.2 Model Structure for Simplicial Sets

The proof of Lemma 2.14 is formal. If p has the right lifting property with respect to all inclusions $\partial \Delta^n \subset \Delta^n$ then it is a homotopy equivalence. Conversely, p has a factorization $p = q \cdot j$, where j is a cofibration and q has the right lifting property with respect to all maps $\partial \Delta^n \subset \Delta^n$, so that j is a trivial cofibration, and then p is a retract of q by a standard argument.

The following result can be proved with simplicial approximation techniques [62].

Lemma 2.15 *Suppose that a simplicial set X has at most countably many non-degenerate simplices. Then the set of path components $\pi_0|X|$ and all homotopy groups $\pi_n(|X|, x)$ are countable.*

The following *bounded monomorphism property* for simplicial sets is a consequence.

Lemma 2.16 *Suppose that $i : X \to Y$ is a trivial cofibration and that $A \subset Y$ is a countable subcomplex. Then there is a countable subcomplex $B \subset Y$ with $A \subset B$ such that the map $B \cap X \to B$ is a trivial cofibration.*

Lemma 2.16 implies that the set of countable trivial cofibrations generates the class of all trivial cofibrations, while Lemma 2.14 implies that the set of all inclusions $\partial \Delta^n \subset \Delta^n$ generates the class of all cofibrations. Theorem 2.13 then follows from small object arguments.

A *Kan fibration* is a map $p : X \to Y$ of simplicial sets which has the right lifting property with respect to all inclusions $\Lambda^n_k \subset \Delta^n$. A *Kan complex* is a simplicial set X for which the canonical map $X \to *$ is a Kan fibration.

Every fibration is a Kan fibration, and every fibrant simplicial set is a Kan complex.

Kan complexes Y have combinatorially defined homotopy groups: if $x \in Y_0$ is a vertex of Y, then

$$\pi_n(Y, x) = \pi((\Delta^n, \partial \Delta^n), (Y, x))$$

where $\pi(\ ,\ )$ denotes simplicial homotopy classes of maps and pairs. The path components of any simplicial set X are defined by the coequalizer

$$X_1 \rightrightarrows X_0 \to \pi_0 X,$$

where the maps $X_1 \to X_0$ are the face maps d_0, d_1. Say that a map $f : Y \to Y'$ of Kan complexes is a *combinatorial weak equivalence* if it induces isomorphisms

$$\pi_n(Y, x) \xrightarrow{\cong} \pi_n(Y', f(x))$$

for all $x \in Y_0$, and

$$\pi_0(Y) \xrightarrow{\cong} \pi_0(Y').$$

Going further requires the following major theorem, due to Quillen [88],[32]:

Theorem 2.17 *The realization of a Kan fibration is a Serre fibration.*

The proof of this result requires much of the classical homotopy theory of Kan complexes (in particular the theory of minimal fibrations), and will not be discussed here—see [32].

Here are the consequences:

Theorem 2.18 *[Milnor theorem] Suppose that Y is a Kan complex and that $\eta : Y \to S(|Y|)$ is the adjunction homomorphism. Then η is a combinatorial weak equivalence.*

It follows that the combinatorial homotopy groups of $\pi_n(Y, x)$ coincide up to natural isomorphism with the ordinary homotopy groups $\pi_n(|Y|, x)$ of the realization, for all Kan complexes Y. The proof of Theorem 2.18 is a long exact sequence argument that is based on the path-loop fibre sequences in simplicial sets. These are Kan fibre sequences, and the key is to know, from Theorem 2.17 and Lemma 2.12, that their realizations are fibre sequences.

Theorem 2.19 *Every Kan fibration is a fibration.*

Proof [Sketch] The key step in the proof is to show, using Theorem 2.18, that every map $p : X \to Y$ which is a Kan fibration and a weak equivalence has the right lifting property with respect to all inclusions $\partial \Delta^n \subset \Delta^n$. This is true if Y is a Kan complex, since p is then a combinatorial weak equivalence by Theorem 2.18. Maps which are weak equivalences and Kan fibrations are stable under pullback by Theorem 2.17 and Lemma 2.12. It follows from Theorem 2.18 that all fibres of the Kan fibration p are contractible. It also follows, by taking suitable pullbacks, that it suffices to assume that p has the form $p : X \to \Delta^k$. If F is the fibre of p over the vertex 0, then the Kan fibration p is fibrewise homotopy equivalent to the projection $F \times \Delta^k \to \Delta^k$ [32, I.10.6]. This projection has the desired right lifting property, as does any other Kan fibration in its fibre homotopy equivalence class—see [32, I.7.10]. □

Remark 2.20 Theorem 2.19 implies that the model structure of Theorem 2.13 consists of cofibrations, Kan fibrations and weak equivalences. This is the standard, classical model structure for simplicial sets. The identification of the fibrations with Kan fibrations is the interesting part of this line of argument.

The realization functor preserves cofibrations and weak equivalences, and it follows that the adjoint pair

$$| \ | : s\mathbf{Set} \leftrightarrows \mathbf{CGHaus} : S,$$

is a *Quillen adjunction*. The following is a consequence of Theorem 2.18:

Theorem 2.21 *The adjunction maps $\eta : X \to S(|X|)$ and $\epsilon : |S(Y)| \to Y$ are weak equivalences, for all simplicial sets X and spaces Y, respectively.*

In particular, the standard model structures on the categories $s\mathbf{Set}$ of simplicial sets and $\mathbf{CGHaus}$ of compactly generated Hausdorff spaces are Quillen equivalent.

2.3 Projective Model Structure for Diagrams

Suppose that I is a small category, and let $s\mathbf{Set}^I$ denote the category of I-diagrams of simplicial sets. The objects of this category are the functors $X : I \to s\mathbf{Set}$, and the morphisms $f : X \to Y$ are the natural transformations of functors. One often says that the category $s\mathbf{Set}^I$ is the *I-diagram category*.

There is a model structure on the I-diagram category, which was originally introduced by Bousfield and Kan [14], and for which the fibrations and weak equivalences are defined sectionwise. This model structure is now called the *projective model structure*. Cofibrant replacements in this structure are like projective resolutions of chain complexes.

Explicitly, a weak equivalence for this category is a map $f : X \to Y$ such that the simplicial set maps $f : X(i) \to Y(i)$ (the components of the natural transformation) are weak equivalences of simplicial sets for all objects i of I. One commonly says that such a map is a *sectionwise weak equivalence*. A map $p : X \to Y$ is said to be a *sectionwise fibration* if all components $p : X(i) \to Y(i)$ are fibrations of simplicial sets. Finally, a *projective cofibration* is a map which has the left lifting property with respect to all maps which are sectionwise weak equivalences and sectionwise fibrations, or which are sectionwise trivial fibrations.

The function complex $\mathbf{hom}(X, Y)$ for I-diagrams X and Y is the simplicial set whose n-simplices are all maps $X \times \Delta^n \to Y$ of I-diagrams. Here the n-simplex Δ^n has been identified with the constant I-diagram which takes a morphism $i \to j$ to the identity map on Δ^n.

The i-sections functor $X \mapsto X(i)$ has a left adjoint

$$L_i : s\mathbf{Set} \to s\mathbf{Set}^I,$$

which is defined for simplicial sets K by

$$L_i(K) = \hom(i, \ ) \times K,$$

where $\hom(i, \ ) : I \to \mathbf{Set}$ is the functor which is represented by i.

We then have the following:

Proposition 2.22 *The I-diagram category $s\mathbf{Set}^I$, together with the classes of projective cofibrations, sectionwise weak equivalences and sectionwise fibrations defined as above, satisfies the axioms for a proper closed simplicial model category.*

Proof A map $p : X \to Y$ is a sectionwise fibration if and only if it has the right lifting property with respect to all maps $L_i(\Lambda^n_k) \to L_i(\Delta^n)$ which are induced by inclusions of horns in simplices. A map $q : Z \to W$ of I-diagrams is a sectionwise fibration and a sectionwise weak equivalence if and only if it has the right lifting property with respect to all maps $L_i(\partial \Delta^n) \to L_i(\Delta^n)$.

Every cofibration (monomorphism) $j : A \to B$ of simplicial sets induces a projective cofibration $j_* : L_i(A) \to L_i(B)$ of I-diagrams, and that this map j_* is a sectionwise cofibration. If j is a trivial cofibration then j_* is a sectionwise weak equivalence.

It follows, by a standard small-object argument, that every map $f : X \to Y$ of I-diagrams has factorizations

 (2.2)

where i is a projective cofibration and a sectionwise weak equivalence, p is a sectionwise fibration, j is a projective cofibration and q is a sectionwise trivial fibration. We have therefore proved the factorization axiom **CM5** for this structure.

The maps i and j in the diagram (2.2) are also sectionwise cofibrations, by construction, and the map i has the left lifting property with respect to all sectionwise fibrations.

In particular, if $\alpha : A \to B$ is a projective cofibration and a sectionwise weak equivalence, then α has a factorization

where i is a projective cofibration and a sectionwise weak equivalence, and has the left lifting property with respect to all sectionwise fibrations, and p is a sectionwise fibration. The map p is also a sectionwise weak equivalence so the lift exists in the diagram

$$\begin{array}{ccc} A & \xrightarrow{i} & C \\ \alpha \downarrow & \nearrow & \downarrow p \\ B & \xrightarrow[1]{} & B \end{array}$$

It follows that α is a retract of the map i, and therefore has the left lifting property with respect to all projective fibrations. This proves the axiom **CM4**.

All of the other closed model axioms are easily verified.

Suppose that $j : K \to L$ is a cofibration of simplicial sets. The collection of all sectionwise cofibrations $\alpha : A \to B$, such that the induced map

$$(\alpha, j) : (B \times K) \cup (A \times L) \to B \times L$$

is a projective cofibration, is closed under pushout, composition, filtered colimits, retraction and contains all maps $L_i M \to L_i N$ which are induced by cofibrations

2.3 Projective Model Structure for Diagrams

$M \to N$ of simplicial sets. This class of cofibrations α therefore contains all projective cofibrations. Observe further that the map (α, j) is a sectionwise weak equivalence if either α is a sectionwise equivalence or j is a weak equivalence of simplicial sets.

The I-diagram category therefore has a simplicial model structure in the sense that, if $\alpha : A \to B$ is a projective cofibration and $j : K \to L$ is a cofibration of simplicial sets, then the map (α, j) is a projective cofibration, which is a sectionwise weak equivalence if either α is a sectionwise weak equivalence or j is a weak equivalence of simplicial sets.

All projective cofibrations are sectionwise cofibrations. Properness for the I-diagram category therefore follows from properness for simplicial sets.

The model structure for the I-diagram category $s\mathbf{Set}^I$ is *cofibrantly generated*, in the sense that the class of projective cofibrations, respectively trivial projective cofibrations, is generated by the set of maps

$$L_i(\partial \Delta^n) \to L_i(\Delta^n), \tag{2.3}$$

respectively the set of maps

$$L_i(\Lambda_k^n) \to L_i(\Delta^n). \tag{2.4}$$

This means that the projective cofibrations form the smallest class of maps which contains the set (2.3) and is closed under pushout, composition, filtered colimit and retraction. The class of trivial projective cofibrations is similarly the smallest class containing the set of maps (2.4) that has the same closure properties. The claim about the cofibrant generation is an artifact of the proof of Proposition 2.22.

The category $s\mathbf{Set}$ of simplicial sets is also cofibrantly generated. The simplicial set category is a category of I-diagrams, where I is the category with one morphism.

The proof of Proposition 2.22 follows a standard pattern. One starts with a set I of cofibrations and a set J of trivial cofibrations and verifies that the set J determines the fibrations and that the set I determines the trivial fibrations, both via lifting properties. Small-object arguments are then used to construct the factorizations of **CM5**, provided one knows (and this is a key point) that the pushouts of members of the set J are weak equivalences. With that construction for the factorization axiom **CM5** in place (which argument produces functorial factorizations), the lifting axiom **CM4** is a formal consequence: one shows that every trivial cofibration is a retract of a member of the saturation of the set of cofibrations J.

This last step, along with the analogous argument for cofibrations, also implies that the model structure is cofibrantly generated.

Chapter 3
Some Topos Theory

This chapter contains a rapid introduction to the geometric flavour of topos theory that is used in this book. There has been an effort to make this material as self-contained as possible.

A Grothendieck topology is a formal calculus of coverings which generalizes the algebra of open covers of a topological space, and can exist on much more general categories. A category equipped with such a Grothendieck topology is a Grothendieck site. The examples of most interest for our purposes come from algebraic geometry, including the Zariski, flat, étale and Nisnevich topologies. All of these are discussed in the first section of this chapter, along with the basic notions of presheaves, sheaves and the associated sheaf functor.

A presheaf is just a functor, but the term "presheaf" is used to flag the existence of a Grothendieck topology. A sheaf is a presheaf which satisfies patching conditions arising from the Grothendieck topology, and applying the associated sheaf functor to a presheaf forces compliance with these conditions. This forcing can be rather brutal, in that a presheaf and its associated sheaf can be very different objects.

A simplicial presheaf is a functor, or diagram, which is defined on a site and takes values in simplicial sets (or "spaces", according to the standard abuse). The main idea behind the theory that is presented here is that a Grothendieck topology defines a *local* homotopy type for that diagram, through the associated simplicial sheaf construction or otherwise. The local homotopy type of a simplicial presheaf depends on the choice of topology, and there can be many topologies on a fixed index category. Whatever the topology, one retains the underlying diagram of spaces or simplicial presheaf as an explicit object of study.

A category of sheaves for a Grothendieck site is a Grothendieck topos. The remaining sections of this chapter cover the main constructions and results for Grothendieck topoi (or toposes) that we shall use, with rapid proofs. We shall not discuss elementary toposes, as they do not explicitly occur in geometric settings.

Giraud's theorem (Theorem 3.17) gives a recognition principle that says that a category is a Grothendieck topos if it satisfies a certain list of exactness properties. It implies, in particular, that the category of sheaves admitting an action by a sheaf of groups G is a Grothendieck topos; this is the classifying topos for G, which is a traditional geometric substitute for the classifying space of a group. This set of ideas applies also to profinite groups—see Examples 3.19 and 3.20.

Geometric morphisms are discussed in Sects. 3.4–3.6. These are adjoint pairs of functors between topoi (inverse and direct image functors) that satisfy certain exactness properties.

In particular, a point of a Grothendieck topos $\mathcal{E}$ is a geometric morphism $x :$ **Set** $\to \mathcal{E}$. Points are generalized stalks, and a set I of points x_i can be assembled to define a geometric morphism

$$(x_i) : \sqcup_I \mathbf{Set} \to \mathcal{E}.$$

One says that the topos $\mathcal{E}$ has enough points if one can find a set of points $\{x_i\}$ such that the inverse image part of the morphism (x_i) is fully faithful. This means that the behaviour of morphisms of $\mathcal{E}$, whether they be monomorphisms, epimorhisms or isomorphisms, is completely determined by their inverse images in the disjoint union of copies of the set category.

The categories of sheaves on a topological space and the categories of sheaves on the Zariski and étale sites of a scheme all have enough points (stalks) in this sense. Other commonly occurring geometric toposes, such as the flat topos on a scheme, do not.

There is, however, a good generalization of this concept: Barr's theorem (Theorem 3.31) says that there is a geometric morphism

$$\pi : \mathbf{Shv}(\mathcal{B}) \to \mathcal{E}$$

for every Grothendieck topos $\mathcal{E}$ with fully faithful inverse image, and such that $\mathbf{Shv}(\mathcal{B})$ is the category of sheaves on a complete Boolean algebra $\mathcal{B}$.

The existence of the morphism π generalizes the notion of "enough points", in that the topos $\sqcup_I$ **Set** can be identified with the category of sheaves $\mathbf{Shv}(\mathcal{P}(I))$ on the Boolean algebra given by the power set $\mathcal{P}(I)$ for the set I that indexes the points.

Barr's theorem is commonly called the principle of Boolean localization, and the morphism π is said to be a *Boolean localization*. The point is that the inverse image functor π^* is fully faithful, while the category of sheaves on a complete Boolean algebra satisfies the axiom of choice, with the result that epimorphisms and isomorphisms in $\mathbf{Shv}(\mathcal{B})$ are defined sectionwise. The Boolean localization morphism π is a "fat point", and Barr's theorem says that every Grothendieck topos has one.

A sketch proof of Barr's theorem appears in the final section of this chapter. This proof is abstracted from a much longer presentation that appears in the Mac Lane–Moerdijk book [77]. The reader will note that some of the ideas from the logical side of topos theory (e.g. frames and locales) make an appearance in the argument.

Barr's theorem is perhaps the most important result for homotopy theoretic applications that has so far been found in the topos theory, and it is used heavily. In this, we are following the methods of Joyal's letter to Grothendieck [70], as well as Van Osdol's proof of the Illusie conjecture [101]. This general approach, for simplicial presheaves, was also described in [55]. It is possible to get by without Barr's theorem, as in [50], but the quantifiers which are involved in that approach can be a bit complicated.

3.1 Grothendieck Topologies

A *Grothendieck site* is a small category $\mathcal{C}$ that is equipped with a topology $\mathcal{T}$. A *Grothendieck topology* $\mathcal{T}$ consists of a collection of subfunctors

$$R \subset \hom(\ , U), \quad U \in \mathcal{C},$$

called *covering sieves*, such that the following axioms hold:

1) (*base change*) If $R \subset \hom(\ , U)$ is covering and $\phi : V \to U$ is a morphism of $\mathcal{C}$, then the subfunctor

$$\phi^{-1}(R) = \{\gamma : W \to V \mid \phi \cdot \gamma \in R\}$$

   is covering for V.
2) (*local character*) Suppose that $R, R' \subset \hom(\ , U)$ are subfunctors and R is covering. If $\phi^{-1}(R')$ is covering for all $\phi : V \to U$ in R, then R' is covering.
3) $\hom(\ , U)$ is covering for all $U \in \mathcal{C}$

Typically, Grothendieck topologies arise from covering families in sites $\mathcal{C}$ having pullbacks. Covering families are sets of functors which generate covering sieves.

More explicitly, suppose that the category $\mathcal{C}$ has pullbacks. Since $\mathcal{C}$ is small, a *pretopology* (equivalently, a topology) $\mathcal{T}$ on $\mathcal{C}$ consists of families of sets of morphisms

$$\{\phi_\alpha : U_\alpha \to U\}, \quad U \in \mathcal{C},$$

called *covering families* , such that the following axioms hold:

1) Suppose that $\phi_\alpha : U_\alpha \to U$ is a covering family and that $\psi : V \to U$ is a morphism of $\mathcal{C}$. Then the collection $V \times_U U_\alpha \to V$ is a covering family for V.
2) If $\{\phi_\alpha : U_\alpha \to V\}$ is covering, and $\{\gamma_{\alpha,\beta} : W_{\alpha,\beta} \to U_\alpha\}$ is covering for all α, then the family of composites

$$W_{\alpha,\beta} \xrightarrow{\gamma_{\alpha,\beta}} U_\alpha \xrightarrow{\phi_\alpha} U$$

   is covering.
3) The family $\{1 : U \to U\}$ is covering for all $U \in \mathcal{C}$.

Example 3.1 Let X be a topological space. The site op$|_X$ is the poset of open subsets $U \subset X$. A covering family for an open subset U is an open cover $V_\alpha \subset U$.

Example 3.2 Suppose that S is a scheme (which is a topological space with sheaf of rings locally isomorphic to affine schemes $\mathrm{Sp}(R)$). The underlying topology on X is the Zariski topology. The *Zariski site* $Zar|_S$ is the poset whose objects are the open subschemes $U \subset S$. A family $V_\alpha \subset U$ is covering if $\cup V_\alpha = U$ as sets.

A scheme homomorphism $\phi : Y \to X$ is *étale at* $y \in Y$ if

a) $\mathcal{O}_y$ is a flat $\mathcal{O}_{f(y)}$-module (ϕ is flat at y).
b) ϕ is unramified at y: $\mathcal{O}_y/\mathcal{M}_{f(y)}\mathcal{O}_y$ is a finite separable field extension of $k(f(y))$.

Say that a map $\phi : Y \to X$ is *étale* if it is étale at every $y \in Y$ and locally of finite type (see [79], for example).

Example 3.3 Let S be a scheme. The *étale site* $et|_S$ has as objects all étale maps $\phi : V \to S$ and all diagrams

for morphisms (with ϕ, ϕ' étale). A covering family for the étale topology is a collection of étale morphisms $\phi_\alpha : V_\alpha \to V$ such that $V = \cup \phi_\alpha(V_\alpha)$ as a set. Equivalently every morphism $\mathrm{Sp}(\Omega) \to V$ lifts to some V_α if Ω is a separably closed field.

Example 3.4 The *Nisnevich site* $Nis|_S$ has the same underlying category as the étale site, namely all étale maps $V \to S$ and morphisms between them. A Nisnevich cover is a family of étale maps $V_\alpha \to V$ such that every morphism $\mathrm{Sp}(K) \to V$ lifts to some V_α, where K is any field. Nisnevich originally called this topology the "completely decomposed topology" or "*cd*-topology" [83], because of the way it behaves over fields—see [56].

Example 3.5 A flat covering family of a scheme S is a set of flat morphisms $\phi_\alpha : S_\alpha \to S$ (i.e. mophisms that are flat at each point) such that $S = \cup \phi_\alpha(S_\alpha)$ as a set (equivalently the scheme homomorphism $\sqcup S_\alpha \to S$ is faithfully flat). The category $(Sch|_S)_{fl}$ is the "big" flat site. Pick a large cardinal κ; then $(Sch|_S)$ is the category of S-schemes $X \to S$ such that the cardinality of both the underlying point set of X and all sections $\mathcal{O}_X(U)$ of its sheaf of rings are bounded above by κ. Then the flat site $(Sch|_S)_{fl}$ is the (small, but big) category $Sch|_S$, equipped with the flat topology.

Example 3.6 There are corresponding big sites $(Sch|_S)_{Zar}$, $(Sch|_S)_{et}$, $(Sch|_S)_{Nis}$, and one can play similar games with big sites of topological spaces.

Example 3.7 Suppose that $G = \{G_i\}$ is profinite group such that all $G_j \to G_i$ are surjective group homomorphisms. Write also $G = \varprojlim G_i$. A *discrete G-set* is a set X with G-action which factors through an action of G_i for some i. Write $G - \mathbf{Set}_{df}$ for the category of G-sets which are both discrete and finite. A family $U_\alpha \to X$ in this category is covering if and only if the induced map $\bigsqcup U_\alpha \to X$ is surjective.

Example 3.8 Suppose that $\mathcal{C}$ is any small category. Say that $R \subset \mathrm{hom}(\ ,x)$ is covering if and only if $1_x \in R$. This is the *chaotic topology* on $\mathcal{C}$.

3.1 Grothendieck Topologies 33

Example 3.9 Suppose that $\mathcal{C}$ is a site and that $U \in \mathcal{C}$. Then the *slice category* $\mathcal{C}/U$ has for objects all morphisms $V \to U$ of $\mathcal{C}$, and its morphisms are the commutative triangles

The category $\mathcal{C}/U$ inherits a topology from $\mathcal{C}$: a collection of maps $V_\alpha \to V \to U$ is covering if and only if the family $V_\alpha \to V$ covers V.

There is a canonical functor

$$q : \mathcal{C}/U \to \mathcal{C}$$

that takes the object $V \to U$ to the object V. This functor q and the topology on the site $\mathcal{C}$ together create the topology on $\mathcal{C}/U$, in the same way that the topology on a topological space creates a topology on each of its open subsets.

A *presheaf* (of sets) on a Grothendieck site $\mathcal{C}$ is a set-valued contravariant functor defined on $\mathcal{C}$, or equivalently a functor $\mathcal{C}^{op} \to \mathbf{Set}$ defined on the opposite category $\mathcal{C}^{op}$, where $\mathcal{C}^{op}$ is the category $\mathcal{C}$ with its arrows reversed. The presheaves on $\mathcal{C}$ form a category whose morphisms are natural transformation, which is often denoted by $\mathbf{Pre}(\mathcal{C})$ and is called the *presheaf category* for the site $\mathcal{C}$.

One can similarly define presheaves taking values in any category $\mathbf{E}$, and following [72] (for example) we can write $\mathbf{Pre}(\mathcal{C}, \mathbf{E})$ for the corresponding category of $\mathbf{E}$-presheaves on $\mathcal{C}$, or functors $\mathcal{C}^{op} \to \mathbf{E}$ and their natural transformations. The shorter notation

$$s\mathbf{Pre}(\mathcal{C}) := \mathbf{Pre}(\mathcal{C}, s\mathbf{Set})$$

denotes the category of presheaves $\mathcal{C}^{op} \to s\mathbf{Set}$ on $\mathcal{C}$ taking values in simplicial sets—this is the category of *simplicial presheaves* on $\mathcal{C}$. One views simplicial presheaves as either simplicial objects in presheaves, or as presheaves in simplicial sets.

The *constant presheaf* that is associated to a set K is the functor which takes an object U of the site $\mathcal{C}$ to the set K, and takes all morphisms $V \to U$ of $\mathcal{C}$ to the identity function on K. We shall use the notation K for both the set K and its associated constant presheaf. It is easily seen that there is a natural bijection

$$\hom(K, F) \cong \hom(K, \varprojlim_{U \in \mathcal{C}} F(U))$$

that identifies presheaf morphisms $K \to F$ with functions $K \to \varprojlim_U F(U)$.

In particular, let $*$ denote the one-point set. Then, there is an identification

$$\hom(*, F) \cong \varprojlim_{U \in \mathcal{C}} F(U).$$

A *global section* of a presheaf F is either a presheaf map $* \to F$ or an element of the inverse limit

$$\varprojlim_{U \in \mathcal{C}} F(U).$$

One often writes $\Gamma_*(F)$ for the set of global sections of the presheaf F.

It is an exercise to show that, if the site $\mathcal{C}$ has a terminal object t, then the global sections of a presheaf F coincides naturally with the set $F(t)$ of sections at the object t. This is the classical description of global sections—sites having terminal objects are ubiquitous in geometry and topology.

Constant presheaves in other categories are similarly just constant functors.

For example, if L is a simplicial set, then the constant simplicial presheaf L is the functor on $\mathcal{C}$ which takes all objects U to L, and takes all morphisms to the identity on L. We use the notation $* = \Delta^0$ to denote the one-point constant simplicial presheaf—this is the terminal object in the simplicial presheaf category on the site $\mathcal{C}$. It is an exercise to show that there is a natural identification

$$\hom(\Delta^n, X) = \Gamma_* X_n$$

for all simplicial presheaves X.

Suppose that $R \subset \hom(\ , U)$ is a sieve, and F is a presheaf on $\mathcal{C}$. There is an isomorphism

$$\hom(R, F) \cong \varprojlim_{V \xrightarrow{\phi} U \in R} F(V). \tag{3.1}$$

One often writes $\{x_\phi\}$ for the collection of component elements of a fixed element x of the inverse limit. We say that such a collection is an *R-compatible family* in F. The isomorphism (3.1) allows us to identify such R-compatible families with natural transformations $x : R \to F$. The natural transformation x takes the morphism $\phi : V \to U$ in $R(V)$ to the element $x_\phi \in F(V)$.

A *sheaf* (of sets) on $\mathcal{C}$ is a presheaf $F : \mathcal{C}^{op} \to \mathbf{Set}$ such that the canonical map

$$F(U) \to \varprojlim_{V \to U \in R} F(V) \tag{3.2}$$

is an isomorphism for each covering sieve $R \subset \hom(\ , U)$. Equivalently, all induced functions

$$\hom(\hom(\ , U), F) \to \hom(R, F)$$

should be bijections.

Exercise 3.10 If the topology on $\mathcal{C}$ is defined by a pretopology (so that $\mathcal{C}$ has all pullbacks), show that F is a sheaf if and only if all diagrams

$$F(U) \to \prod_\alpha F(U_\alpha) \rightrightarrows \prod_{\alpha, \beta} F(U_\alpha \times_U U_\beta)$$

3.1 Grothendieck Topologies

arising from covering families $U_\alpha \to U$ are equalizers.

Remark 3.11 In many cases of interest, the objects of the site $\mathcal{C}$ have underlying sets, and the empty set $\emptyset$ is the initial object of $\mathcal{C}$, while coverings are described by existence of lifting of points. This is true of sites originating in topology and algebraic geometry, as in Examples 3.1–3.7.

In all such cases, a sieve $R \subset \hom(\ , U)$ is covering if and only if the corresponding map

$$\bigsqcup_{V \to U \in R} V \to U$$

is an epimorphism on the underlying sets. The inclusion

$$\emptyset \subset \hom(\ , \emptyset)$$

defines a covering sieve in such cases, since an empty colimit must be the initial object. It follows that the induced map

$$\hom(\hom(\ , \emptyset), F) \to \hom(\emptyset, F)$$

is a bijection for all sheaves of sets F, and so there is a bijection

$$F(\emptyset) \cong *$$

and the set of sections $F(\phi)$ is a one-point set for all sheaves F on such sites $\mathcal{C}$.

Morphisms of sheaves are natural transformations—write $\mathbf{Shv}(\mathcal{C})$ for the corresponding category. The *sheaf category* $\mathbf{Shv}(\mathcal{C})$ is a full subcategory of $\mathbf{Pre}(\mathcal{C})$.

There is an analogous definition for sheaves in any complete category $\mathbf{E}$, and one would write $\mathbf{Shv}(\mathcal{C}, \mathbf{E})$ for the corresponding category. The assertion that the category $\mathbf{E}$ is complete means that it has all small limits, so it makes sense to require that the morphism (3.2) should be an isomorphism for the functor $F : \mathcal{C}^{op} \to \mathbf{E}$ to be a sheaf.

We use the notation

$$s\mathbf{Shv}(\mathcal{C}) := \mathbf{Shv}_\mathcal{C}(\mathcal{C}, s\mathbf{Set})$$

for the category of *simplicial sheaves* on the site $\mathcal{C}$.

Lemma 3.12

1) If $R \subset R' \subset \hom(\ , U)$ and R is covering then R' is covering.
2) If $R, R' \subset \hom(\ , U)$ are covering then $R \cap R'$ is covering.
3) Suppose that $R \subset \hom(\ , U)$ covering and that $S_\phi \subset \hom(\ , V)$ is covering for all $\phi : V \to U$ of R. Let $R * S$ be the sieve which is generated by the composites

$$W \xrightarrow{\gamma} V \xrightarrow{\phi} U$$

with $\phi \in R$ and $\gamma \in S_\phi$. Then $R * S$ is covering.

Proof For 1), one shows that $\phi^{-1}(R) = \phi^{-1}(R')$ for all $\phi \in R$, so that R' is covering by the local character axiom. The relation $\phi^{-1}(R \cap R') = \phi^{-1}(R')$ for all $\phi \in R$ implies that $R \cap R'$ is covering, giving 2). Statement 3) is proved by observing that $S_\phi \subset \phi^{-1}(R * S)$ for all $\phi \in R$.

If $S \subset R$ are sieves in hom(, U) and F is a presheaf, then there is an obvious restriction map

$$\hom(R, F) \to \hom(S, F).$$

Write

$$LF(U) = \varinjlim_{R} \hom(R, F),$$

where the colimit is indexed over all covering sieves $R \subset \hom(\ , U)$. This colimit is filtered by Lemma 3.12. Elements of $LF(U)$ are classes $[x]$ of morphisms $x : R \to F$. Then the assignment $U \mapsto LF(U)$ defines a presheaf, and there is a natural presheaf map

$$\nu : F \to LF.$$

Say that a presheaf G is *separated* if the map $\nu : G \to LG$ is a monomorphism in each section, i.e. if all functions $G(U) \to LG(U)$ are injective.

Lemma 3.13

1) *The presheaf LF is separated, for all presheaves F.*
2) *If G is a separated presheaf then LG is a sheaf.*
3) *If $f : F \to G$ is a presheaf map and G is a sheaf, then f factors uniquely through a presheaf map $f_* : LF \to G$.*

It follows from Lemma 3.13 that the object $L^2 F$ is a sheaf for every presheaf F, and the functor $F \mapsto L^2 F$ is left adjoint to the forgetful functor

Shv($\mathcal{C}$) $\subset$ **Pre**($\mathcal{C}$).

The unit of the adjunction is the composite

$$F \xrightarrow{\nu} LF \xrightarrow{\nu} L^2 F. \tag{3.3}$$

One often writes $\tilde{F} := L^2 F$ for the *sheaf associated to the presheaf F*. It is standard to write $\eta : F \to \tilde{F}$ for the composite (3.3).

The resulting associated sheaf functor $F \mapsto \tilde{F}$ is left adjoint to the forgetful functor, and the natural map $\eta : F \to \tilde{F}$ is the unit of the adjunction.

Proof [of Lemma 3.13] Suppose that $\psi^*([x]) = \psi^*([y])$ for all $\psi : W \to U$ in some covering sieve $S \subset \hom(\ , U)$, where $[x], [y] \in LF(U)$. We can assume that x and y are maps $R \to F$ that are defined on the same covering sieve $R \subset \hom(\ , U)$. By restricting to the intersection $S \cap R$ (Lemma 3.12), we can also assume that $S = R$.

3.1 Grothendieck Topologies 37

It follows that for each $\phi : V \to U$ in R, there is a covering sieve $T_\phi \subset \hom(\ , V)$ such that
$$x_{\phi\gamma} = \gamma^*(x_\phi) = \gamma^*(y_\phi) = y_{\phi\gamma}$$
for each $\gamma : W \to V$ in $T_\phi(W)$. The maps $x, y : R \to \hom(\ , U)$ therefore restrict to the same map $R * T \to F$ on the covering sieve $R * T \subset R$, so that $[x] = [y]$ in $LF(U)$, and we have proved statement 1).

If $\phi : V \to U$ is a member of a covering sieve $R \subset \hom(\ , U)$, then $\phi^{-1}(R) = \hom(\ , V)$ is the unique covering sieve for V which contains the identity $1_V : V \to V$. It follows that $\phi^*[x] = \nu(x_\phi)$ for all $\phi \in R$.

Suppose that G is a separated presheaf, and that $[\nu] : R \to LG$ is a map, where $R \subset \hom(\ , U)$ is a covering sieve of U. Then each ν_ϕ lifts locally to G along some covering sieve T_ϕ by the previous paragraph, so there is a refinement $R * T \subset R$ of covering sieves such that $\nu_\psi = \nu(x_\psi)$ for each $\psi : W \to U$ of $R * T$. The presheaf G is separated, so that the elements x_ψ define an element of $x : R * T \to G$ and an element $[x]$ of $LG(U)$. Then $\phi^*[\{x_\psi\}] = \nu_\phi$ for each $\phi \in R$ since LG is separated, and it follows that the canonical function
$$LG(U) \to \hom(R, LG)$$
is surjective. This function is injective since LG is separated. Thus, LG is a sheaf, giving statement 2).

If G is a sheaf, then the presheaf map $\nu : G \to LG$ is an isomorphism essentially by definition, and statement 3) follows.

The sheaf $L^2 K = \tilde{K}$ which is associated to the constant presheaf K is the *constant sheaf* associated to the set K.

One often writes K for the constant sheaf associated to K, but this can be a bit dangerous because the sheaf $\tilde{K}$ might be quite different from the constant functor K.

For example, it is an exercise to show that if et $|_S$ is the étale site of a scheme S, then there is a bijection
$$L^2(K)(\phi) \cong \prod_{\pi_0(U)} K$$
for étale maps $\phi : U \to S$, where $\pi_0(U)$ is the set of connected components of the topological space underlying the scheme U.

In general, the constant sheaf $L^2(*)$ is the terminal object in the sheaf category (see Lemma 3.14 below, or prove the claim directly), but this sheaf is not necessarily terminal in the presheaf category.

The constant sheaf construction still picks out global sections of sheaves F, by adjointness. There is a natural bijection
$$\hom(L^2(*), F) \cong \Gamma_*(F)$$
relating sheaf morphisms $L^2(*) \to F$ with elements of the inverse limit
$$\Gamma_*(F) = \varprojlim_{U \in \mathcal{C}} F(U).$$

For example, if F is a sheaf on the étale site $et|_S$, then there is an identification

$$\Gamma_* F \cong F(S)$$

(note the standard abuse of notation), since the identity map $S \to S$ is terminal in $et|_S$.

3.2 Exactness Properties

Lemma 3.14

1) The associated sheaf functor preserves all finite limits.
2) The sheaf category $\mathbf{Shv}(\mathcal{C})$ is complete and co-complete. Limits are formed sectionwise.
3) Every monomorphism in $\mathbf{Shv}(\mathcal{C})$ is an equalizer.
4) If the sheaf morphism $\theta : F \to G$ is both a monomorphism and an epimorphism, then θ is an isomorphism.

Proof Statement 1) is proved by observing that LF is defined by filtered colimits, and finite limits commute with filtered colimits.

If $X : I \to \mathbf{Shv}(\mathcal{C})$ is a diagram of sheaves, then the colimit in the sheaf category is $L^2(\varinjlim X)$, where $\varinjlim X$ is the presheaf colimit, giving statement 2).

If $A \subset X$ is a subset, then there is an equalizer

$$A \longrightarrow X \underset{*}{\overset{p}{\rightrightarrows}} X/A.$$

The same holds for subobjects $A \subset X$ of presheaves, and hence for subobjects of sheaves, since the associated sheaf functor L^2 preserves finite limits. Statement 3) follows.

For statement 4), observe that the map θ appears in an equalizer

$$F \overset{\theta}{\longrightarrow} G \underset{g}{\overset{f}{\rightrightarrows}} K$$

since θ is a monomorphism. But θ is an epimorphism, so $f = g$. But then $1_G : G \to G$ factors through θ, giving a section $\sigma : G \to F$. Finally, $\theta \sigma \theta = \theta$ and θ is a monomorphism, so $\sigma \theta = 1$.

Here are some fundamental definitions:

1) A presheaf map $f : F \to G$ is a *local epimorphism* if for each $x \in G(U)$ there is a covering $R \subset \hom(\ , U)$ such that $\phi^*(x) = f(y_\phi)$ for some y_ϕ, for all $\phi \in R$.
2) $f : F \to G$ is a *local monomorphism* if given $x, y \in F(U)$ such that $f(x) = f(y)$, there is a covering $R \subset \hom(\ , U)$ such that $\phi^*(x) = \phi^*(y)$ for all $\phi \in R$.

3.2 Exactness Properties

3) A presheaf map $f : F \to G$ which is both a local epimorphism and a local monomorphism is a *local isomorphism*.

Example 3.15 The canonical map $\nu : F \to LF$ is a local isomorphism for all presheaves F. The fact that ν is a local monomorphism is a consequence of the definitions, and the claim that ν is a local epimorphism appears as a detail in the proof of Lemma 3.13.

It follows that the associated sheaf map $\eta : F \to L^2 F$ is also a local isomorphism, for all presheaves F.

Lemma 3.16 *Suppose that $f : F \to G$ is a presheaf morphism. Then f induces an isomorphism (respectively monomorphism, epimorphism) $f_* : L^2 F \to L^2 G$ of associated sheaves if and only if f is a local isomorphism (respectively local monomorphism, local epimorphism) of presheaves.*

Proof It is an exercise to show that, given a commutative diagram

of presheaf morphisms, if any two of f, g and h are local isomorphisms, then so is the third. A sheaf map $g : E \to E'$ is a monomorphism (respectively epimorphism) if and only if it is a local monomorphism (respectively local epimorphism). Now use the comparison diagram

$$\begin{array}{ccc} F & \xrightarrow{\eta} & L^2 F \\ f \downarrow & & \downarrow f_* \\ G & \xrightarrow{\eta} & L^2 G \end{array}$$

to finish the proof of the Lemma.

A *Grothendieck topos* is a category $\mathcal{E}$ that is equivalent to a sheaf category $\mathbf{Shv}(\mathcal{C})$ on some Grothendieck site $\mathcal{C}$.

Grothendieck toposes are characterized by exactness properties

Theorem 3.17 (Giraud) *A category $\mathcal{E}$ having all finite limits is a Grothendieck topos if and only if it has the following properties:*

1) *The category $\mathcal{E}$ has all small coproducts; they are disjoint and stable under pullback.*
2) *Every epimorphism of $\mathcal{E}$ is a coequalizer.*
3) *Every equivalence relation $R \rightrightarrows E$ in $\mathcal{E}$ is a kernel pair and has a quotient.*

4) *Every coequalizer $R \rightrightarrows E \to Q$ is stably exact.*
5) *There is a set of objects that generates the category $\mathcal{E}$.*

A sketch proof of Giraud's theorem appears below, but the result is proved in many places—see, for example, [77, 91].

Here are the definitions of the terms appearing in the statement of Theorem 3.17:

1) The coproduct $\bigsqcup_i A_i$ is *disjoint* if all diagrams

$$\begin{array}{ccc} \emptyset & \longrightarrow & A_j \\ \downarrow & & \downarrow \\ A_i & \longrightarrow & \bigsqcup_i A_i \end{array}$$

are pullbacks for $i \neq j$. The coproduct $\bigsqcup_i A_i$ is *stable under pullback* if all diagrams

$$\begin{array}{ccc} \bigsqcup_i C \times_B A_i & \longrightarrow & \bigsqcup_i A_i \\ \downarrow & & \downarrow \\ C & \longrightarrow & B \end{array}$$

are pullbacks.

3) An *equivalence relation* is a monomorphism $m = (m_0, m_1) : R \to E \times E$ such that
   a) The diagonal $\Delta : E \to E \times E$ factors through m (ie. $a \sim a$),
   b) The composite $R \xrightarrow{m} E \times E \xrightarrow{\tau} E \times E$ factors through m (ie. $a \sim b \Rightarrow b \sim a$),
   c) The map

$$(m_0 m_{0*}, m_1 m_{1*}) : R \times_E R \to E \times E$$

   factors through m (this is transitivity) where the pullback is defined by

$$\begin{array}{ccc} R \times_E R & \xrightarrow{m_{1*}} & R \\ m_{0*} \downarrow & & \downarrow m_0 \\ R & \xrightarrow{m_1} & E \end{array}$$

The *kernel pair* of a morphism $u : E \to D$ is a pullback

$$\begin{array}{ccc} R & \xrightarrow{m_1} & E \\ m_0 \downarrow & & \downarrow u \\ E & \xrightarrow{u} & D \end{array}$$

3.2 Exactness Properties

It is an exercise to show that every kernel pair is an equivalence relation.

A *quotient* for an equivalence relation $(m_0, m_1) : R \to E \times E$ is a coequalizer

$$R \xrightarrow[m_1]{m_0} E \longrightarrow E/R.$$

4) A coequalizer $R \rightrightarrows E \to Q$ is *stably exact* if the induced diagram

$$R \times_Q Q' \rightrightarrows E \times_Q Q' \to Q'$$

is a coequalizer for all morphisms $Q' \to Q$.

5) Following [10], a *generating set* is a set of objects S that detects the difference between maps. This means precisely that the map

$$\bigsqcup_{x \to E} x \to E$$

which is defined by all maps $x \to E$ with $x \in S$, is an epimorphism, for all objects E of $\mathcal{E}$.

Exercise 3.18 Show that any category $\mathbf{Shv}(\mathcal{C})$ on a site $\mathcal{C}$ satisfies the conditions of Giraud's theorem. The family $L^2\hom(\ ,U)$, $U \in \mathcal{C}$ is a set of generators.

Proof [Sketch proof of Theorem 3.17] The key is to show that a category $\mathcal{E}$ that satisfies the conditions of the theorem is cocomplete. In view of the assumption that $\mathcal{E}$ has all small coproducts it is enough to show that $\mathcal{E}$ has all coequalizers. The coequalizer of the maps $f_1, f_2 : E' \to E$ is constructed by taking the canonical map $E \to E/R$, where R is the minimal equivalence relation which contains (f_1, f_2) in the sense that there is a commutative diagram

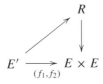

See also [77, p. 575].

Suppose that S is the set of generators for $\mathcal{E}$ that is prescribed by the statement of Giraud's theorem, and let $\mathcal{C}$ be the full subcategory of $\mathcal{E}$ on the set of objects A. A subfunctor $R \subset \hom(\ , x)$ on $\mathcal{C}$ is covering if the map

$$\bigsqcup_{y \to x \in R} y \to x$$

is an epimorphism of $\mathcal{E}$. In such cases, there is a coequalizer

$$\bigsqcup_{y_0 \to y_1 \to x} y_0 \rightrightarrows \bigsqcup_{y \to x \in R} y \to x$$

where the displayed strings $y_0 \to y_1 \to x$ are defined by morphisms between generators such that $y_1 \to x$ is in R.

It follows that every object $E \in \mathcal{E}$ represents a sheaf hom(, E) on $\mathcal{C}$, and a sheaf F on $\mathcal{C}$ determines an object

$$\varinjlim_{\text{hom}(\ ,y) \to F} y$$

of $\mathcal{E}$.

The adjunction

$$\text{hom}(\varinjlim_{\text{hom}(\ ,y) \to F} y, E) \cong \text{hom}(F, \text{hom}(\ , E))$$

determines an adjoint equivalence between $\mathcal{E}$ and **Shv**($\mathcal{C}$).

The strategy of the proof of Giraud's theorem is arguably as important as the statement of the theorem itself. Here are some examples;

Example 3.19 Suppose that G is a sheaf of groups, and let $G -$ **Shv**($\mathcal{C}$) denote the category of all sheaves X admitting G-action, with equivariant maps between them. $G -$ **Shv**($\mathcal{C}$) is a Grothendieck topos, called the *classifying topos* for G, by Giraud's theorem. The objects $G \times L^2\text{hom}(\ , x)$ form a generating set.

Example 3.20 If $G = \{G_i\}$ is a profinite group such that all transition maps $G_i \to G_j$ are surjective, then the category $G -$ **Set**$_d$ of discrete G-sets is a Grothendieck topos. A discrete G-set is a set X equipped with a pro-map $G \to \text{Aut}(X)$. The finite discrete G-sets form a generating set for this topos, and the full subcategory $G -$ **Set**$_{df}$ on the finite discrete G-sets (as in Example 3.7) is the site prescribed by Giraud's theorem.

If the profinite group G is the absolute Galois group of a field k, then the category $G -$ **Set**$_d$ of discrete G-sets is equivalent to the category **Shv**($et|_k$) of sheaves on the étale site for k.

More generally, if S is a locally Noetherian connected scheme with geometric point x, and the profinite group $\pi_1(S, x)$ is the Grothendieck fundamental group, then the category of discrete $\pi_1(S, x)$-sets is equivalent to the category of sheaves on the finite étale site $fet|_S$ for the scheme S. See [1], [79].

3.3 Geometric Morphisms

Suppose that $\mathcal{C}$ and $\mathcal{D}$ are Grothendieck sites. A *geometric morphism*

$$f : \mathbf{Shv}(\mathcal{C}) \to \mathbf{Shv}(\mathcal{D})$$

consists of functors $f_* : \mathbf{Shv}(\mathcal{C}) \to \mathbf{Shv}(\mathcal{D})$ and $f^* : \mathbf{Shv}(\mathcal{D}) \to \mathbf{Shv}(\mathcal{C})$ such that f^* is left adjoint to f_* and f^* preserves finite limits.

3.3 Geometric Morphisms

The left adjoint f^* is called the *inverse image* functor, while f_* is called the *direct image*. The inverse image functor f^* is left and right exact in the sense that it preserves all finite colimits and limits, respectively. The direct image functor f_* is usually not left exact (i.e. it may not preserve finite colimits), and therefore has derived functors.

Example 3.21 Suppose $f : X \to Y$ is a continuous map of topological spaces. Pullback along f induces a functor $\text{op}\,|_Y \to \text{op}\,|_X$ that takes an open subset $U \subset Y$ to $f^{-1}(U)$. Open covers pull back to open covers, so that if F is a sheaf on X, then composition with the pullback gives a sheaf f_*F on Y with $f_*F(U) = F(f^{-1}(U))$. The resulting functor

$$f_* : \mathbf{Shv}(\text{op}|_X) \to \mathbf{Shv}(\text{op}|_Y)$$

is the direct image. It extends to a direct image functor

$$f_* : \mathbf{Pre}(\text{op}|_X) \to \mathbf{Pre}(\text{op}|_Y)$$

between presheaf categories that is defined in the same way.

The left Kan extension

$$f^p : \mathbf{Pre}(\text{op}|_Y) \to \mathbf{Pre}(\text{op}|_X)$$

of the presheaf-level direct image is defined by

$$f^p G(V) = \varinjlim G(U),$$

where the colimit is indexed over all diagrams

in which the vertical maps are inclusions of open subsets. The category $\text{op}\,|_Y$ has all products (i.e. intersections), so the colimit is filtered. The functor $G \mapsto f^p G$ therefore commutes with finite limits. The sheaf theoretic inverse image functor

$$f^* : \mathbf{Shv}(\text{op}|_Y) \to \mathbf{Shv}(\text{op}|_X)$$

is defined by $f^*(G) = L^2 f^p(G)$. The resulting pair of functors forms a geometric morphism $f : \mathbf{Shv}(\text{op}|_X) \to \mathbf{Shv}(\text{op}|_Y)$.

Example 3.22 Suppose that $f : X \to Y$ is a morphism of schemes. Etale maps (respectively covers) are stable under pullback, and so there is a functor $\text{et}\,|_Y \to \text{et}\,|_X$ which is defined by pullback. If F is a sheaf on $\text{et}\,|_X$ then there is a sheaf f_*F on $\text{et}\,|_Y$ that is defined by $f_*F(V \to Y) = f(X \times_Y V \to X)$.

The restriction functor $f_* : \mathbf{Pre}(et|_X) \to \mathbf{Pre}(et|_Y)$ has a left adjoint f^p defined by
$$f^p G(U \to X) = \varinjlim G(V),$$
where the colimit is indexed over the category of all diagrams

$$\begin{array}{ccc} U & \longrightarrow & V \\ \downarrow & & \downarrow \\ X & \xrightarrow{f} & Y \end{array}$$

where both vertical maps are étale. The colimit is filtered, because étale maps are stable under pullback and composition. The inverse image functor
$$f^* : \mathbf{Shv}(et|_Y) \to \mathbf{Shv}(et|_X)$$
is defined by $f^* F = L^2 f^p F$, and so f induces a geometric morphism
$$f : \mathbf{Shv}(et|_X) \to \mathbf{Shv}(et|_Y).$$

A morphism of schemes $f : X \to Y$ induces geometric morphisms $f : \mathbf{Shv}(?|_X) \to \mathbf{Shv}(?|_Y)$ and $f : \mathbf{Shv}(Sch|_X)_? \to \mathbf{Shv}(Sch|_Y)_?$ for all of the geometric topologies (Zariski, flat, Nisnevich, qfh, ...), by similar arguments.

Example 3.23 A *point* of $\mathbf{Shv}(\mathcal{C})$ is a geometric morphism $\mathbf{Set} \to \mathbf{Shv}(\mathcal{C})$. Every point $x \in X$ of a topological space X determines a continuous map $\{x\} \subset X$ and hence a geometric morphism
$$\mathbf{Set} \cong \mathbf{Shv}(op|_{\{x\}}) \xrightarrow{x} \mathbf{Shv}(op|_X).$$

The set
$$x^* F = \varinjlim_{x \in U} F(U)$$
is the *stalk* of F at x.

Example 3.24 Suppose that k is a field. Any scheme homomorphism $x : Sp(k) \to X$ induces a geometric morphism
$$\mathbf{Shv}(et|_k) \to \mathbf{Shv}(et|_X).$$
If the field k happens to be separably closed, then there is an equivalence $\mathbf{Shv}(et|_k) \simeq \mathbf{Set}$ and the resulting geometric morphism $x : \mathbf{Set} \to \mathbf{Shv}(et|_X)$ is called a *geometric point* of X. The inverse image functor
$$F \mapsto x^* F = \varinjlim F(U)$$

3.3 Geometric Morphisms

is the stalk of the sheaf (or presheaf) F at x. The indicated colimit is indexed by the filtered category of diagrams

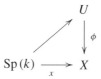

in which the vertical maps ϕ are étale. These diagrams are the *étale neighbourhoods* of the geometric point x.

The stalk x^*F coincides with global sections $\pi_x^* F(\mathcal{O}_{x,X}^{sh})$ of the inverse image π_x^* along the induced scheme homomorphism $\pi_x : \text{Sp}(\mathcal{O}_{x,X}^{sh}) \to X$, where $\mathcal{O}_{x,X}^{sh}$ is the strict henselization of the local ring $\mathcal{O}_{x,X}$ of the point x on the scheme X [79, II.3.2].

Example 3.25 Suppose that S and T are topologies on a site $\mathcal{C}$ and that $S \subset T$. In other words, T has more covers than S and hence refines S. Then every sheaf for T is a sheaf for S; write

$$\pi_* : \mathbf{Shv}_T(\mathcal{C}) \subset \mathbf{Shv}_S(\mathcal{C})$$

for the corresponding inclusion functor. The associated sheaf functor for the topology T gives a left adjoint π^* for the inclusion functor π_*, and the functor π^* preserves finite limits.

In particular, comparing an arbitrary topology with the chaotic topology on a site $\mathcal{C}$ gives a geometric morphism

$$\mathbf{Shv}(\mathcal{C}) \to \mathbf{Pre}(\mathcal{C})$$

for which the direct image is the inclusion of the sheaf category in the presheaf category, and the inverse image is the associated sheaf functor.

A *site morphism* is a functor $f : \mathcal{D} \to \mathcal{C}$ between Grothendieck sites such that

1) If F is a sheaf on $\mathcal{C}$, then the composite functor

$$\mathcal{D}^{op} \xrightarrow{f^{op}} \mathcal{C}^{op} \xrightarrow{F} \mathbf{Set}$$

is a sheaf on $\mathcal{D}$.

2) Suppose that f^p is the left adjoint of the functor

$$f_* : \mathbf{Pre}(\mathcal{C}) \to \mathbf{Pre}(\mathcal{D})$$

which is defined by precomposition with f^{op}. Then the functor f^p is left exact in the sense that it preserves all finite limits.

One often paraphrases the requirement 1) by saying that the functor f_* should be *continuous*: it restricts to a functor

$$f_* : \mathbf{Shv}(\mathcal{C}) \to \mathbf{Shv}(\mathcal{D}).$$

The left adjoint

$$f^* : \mathbf{Shv}(\mathcal{D}) \to \mathbf{Shv}(\mathcal{C})$$

is defined for a sheaf E by $f^*(E) = L^2 f^p(E)$. The functor f^* preserves finite limits since the presheaf-level functor f^p is required to have this property. It follows that every site morphism $f : \mathcal{D} \to \mathcal{C}$ induces a geometric morphism

$$f : \mathbf{Shv}(\mathcal{C}) \to \mathbf{Shv}(\mathcal{D}).$$

Suppose that $g : \mathcal{D} \to \mathcal{C}$ is a functor between Grothendieck sites such that

1') If $R \subset \hom(\ ,U)$ is a covering sieve for $\mathcal{D}$ then the image $g(R)$ of the set of morphisms of R in $\mathcal{C}$ generates a covering sieve for $\mathcal{C}$.
2') The sites $\mathcal{D}$ and $\mathcal{C}$ have all finite limits, and the functor g preserves them.

It is an exercise to show that such a functor g must satisfy the corresponding properties 1) and 2) above, and therefore defines a site morphism. The functor g is what Mac Lane and Moerdijk [77] would call a site morphism, while the definition in use here is consistent with that of SGA4 [2].

In many practical cases, such as Examples 3.21 and 3.22 above, geometric morphisms are induced by functors g which satisfy conditions 1') and 2').

3.4 Points

Say that a Grothendieck topos $\mathbf{Shv}(\mathcal{C})$ has *enough points* if there is a set of geometric morphisms $x_i : \mathbf{Set} \to \mathbf{Shv}(\mathcal{C})$ such that the induced functor

$$\mathbf{Shv}(\mathcal{C}) \xrightarrow{(x_i^*)} \prod_i \mathbf{Set}$$

is faithful.

Lemma 3.26 *Suppose that $f : \mathbf{Shv}(\mathcal{D}) \to \mathbf{Shv}(\mathcal{C})$ is a geometric morphism. Then the following are equivalent:*

a) The functor $f^* : \mathbf{Shv}(\mathcal{C}) \to \mathbf{Shv}(\mathcal{D})$ is faithful.
b) The functor f^* reflects isomorphisms
c) The functor f^* reflects epimorphisms
d) The functor f^* reflects monomorphisms

Proof Suppose that f^* is faithful, which means that if $f^*(g_1) = f^*(g_2)$ then $g_1 = g_2$. Suppose that $m : F \to G$ is a morphism of $\mathbf{Shv}(\mathcal{C})$ such that $f^*(m)$ is a monomoprhism. If $m \cdot f_1 = m \cdot f_2$ then $f^*(f_1) = f^*(f_2)$ so $f_1 = f_2$. The map m is therefore a monomorphism. Similarly, the functor f^* reflects epimorphisms and hence reflects isomorphisms.

3.4 Points

Suppose that the functor f^* reflects epimorphisms and suppose given morphisms $g_1, g_2 : F \to G$ such that $f^*(g_1) = f^*(g_2)$. We have equality $g_1 = g_2$ if and only if their equalizer $e : E \to F$ is an epimorphism. But f^* preserves equalizers and reflects epimorphisms, so e is an epimorphism and $g_1 = g_2$. The other arguments are similar.

Here are some basic definitions:

1) A *lattice* L is a partially ordered set that has all finite coproducts $x \vee y$ and all finite products $x \wedge y$.
2) A lattice L has 0 and 1 if it has an initial and terminal object, respectively.
3) A lattice L is said to be *distributive* if
$$x \wedge (y \vee z) = (x \wedge y) \vee (x \wedge z)$$
for all x, y, z.
4) A *complement* for x in a lattice L with 0 and 1 is an element a such that $x \vee a = 1$ and $x \wedge a = 0$. If L is also distributive then the complement, if it exists, is unique: if b is another complement for x, then
$$b = b \wedge 1 = b \wedge (x \vee a) = (b \wedge x) \vee (b \wedge a)$$
$$= (x \wedge a) \vee (b \wedge a) = (x \vee b) \wedge a = a.$$
One usually writes $\neg x$ for the complement of x.
5) A *Boolean algebra* $\mathcal{B}$ is a distributive lattice with 0 and 1 in which every element has a complement.
6) A lattice L is said to be *complete* if it has all small limits and colimits (or all small meets and joins).
7) A *frame* P is a lattice that has all small joins (and all finite meets) and which satisfies an infinite distributive law
$$U \wedge (\bigvee_i V_i) = \bigvee_i (U \wedge V_i).$$

Example 3.27

1) The poset $\mathcal{O}(T)$ of open subsets of a topological space T is a frame. Every continuous map $f : S \to T$ induces a morphism of frames $f^{-1} : \mathcal{O}(T) \to \mathcal{O}(S)$, defined by $U \mapsto F^{-1}(U)$.
2) The power set $\mathcal{P}(I)$ of a set I is a complete Boolean algebra.
3) Every complete Boolean algebra $\mathcal{B}$ is a frame. In effect, every join is a filtered colimit of finite joins.

Every frame A has a canonical Grothendieck topology: a family $y_i \leq x$ is covering if $\bigvee_i y_i = x$. Write **Shv**(A) for the corresponding sheaf category. Every complete Boolean algebra $\mathcal{B}$ is a frame, and therefore has an associated sheaf category **Shv**$(\mathcal{B})$.

Example 3.28 Suppose that I is a set. Then there is an equivalence

$$\mathbf{Shv}(\mathcal{P}(I)) \simeq \prod_{i \in I} \mathbf{Set}.$$

Any set I of points $x_j : \mathbf{Set} \to \mathbf{Shv}(\mathcal{C})$ assembles to give a geometric morphism

$$x : \mathbf{Shv}(\mathcal{P}(I)) \to \mathbf{Shv}(\mathcal{C}).$$

Observe that the sheaf category $\mathbf{Shv}(\mathcal{C})$ has enough points if there is such a set I of points such that the inverse image functor x^* for the geometric morphism x is faithful.

Lemma 3.29 *Suppose that F is a sheaf of sets on a complete Boolean algebra $\mathcal{B}$. Then the poset $Sub(F)$ of subobjects of F is a complete Boolean algebra.*

Proof The poset $Sub(F)$ is a frame, by an argument on the presheaf level. It remains to show that every object $G \in Sub(F)$ is complemented. The obvious candidate for $\neg G$ is

$$\neg G = \bigvee_{H \leq F,\ H \wedge G = \emptyset} H$$

and we need to show that $G \vee \neg G = F$.

Every $K \leq \hom(\ , A)$ is representable: in effect,

$$K = \varinjlim_{\hom(\ , B) \to K} \hom(\ , B) = \hom(\ , C),$$

where

$$C = \bigvee_{\hom(\ , B) \to K} B \in \mathcal{B}.$$

It follows that $Sub(\hom(\ , A)) \cong Sub(A)$ is a complete Boolean algebra.

Consider all diagrams

$$\begin{array}{ccc} \phi^{-1}(G) & \longrightarrow & G \\ \downarrow & & \downarrow \\ \hom(\ , A) & \xrightarrow{\phi} & F \end{array}$$

There is an induced pullback

$$\begin{array}{ccc} \phi^{-1}(G) \vee \neg\phi^{-1}(G) & \longrightarrow & G \vee \neg G \\ \cong \downarrow & & \downarrow \\ \hom(\ , A) & \xrightarrow{\phi} & F \end{array}$$

3.5 Boolean Localization

The sheaf F is a union of its representable subsheaves, since all ϕ are monomorphisms—in effect all hom(, A) are subobjects of the terminal sheaf. It follows that $G \vee \neg G = F$.

Lemma 3.30 *Suppose that $\mathcal{B}$ is a complete Boolean algebra. Then every epimorphism $\pi : F \to G$ in $\mathbf{Shv}(\mathcal{B})$ has a section.*

Lemma 3.30 asserts that the sheaf category on a complete Boolean algebra satisfies the axiom of choice.

Proof Consider the family of lifts

This family is non-empty, because every $x \in G(1)$ restricts along some covering $B \leq 1$ to a family of elements x_B that lift to $F(B)$.

All maps hom(, B) $\to G$ are monomorphisms, so that all such morphisms represent objects of $Sub(G)$, which is a complete Boolean algebra by Lemma 3.29.

Zorn's Lemma implies that the family of lifts has maximal elements. Suppose that N is maximal and that $\neg N \neq \emptyset$. Then there is an $x \in \neg N(C)$ for some C, and there is a covering $B' \leq C$ such that $x_{B'} \in N(B')$ lifts to $F(B')$ for all members of the cover. Then $N \wedge$ hom(, B') $= \emptyset$ so the lift extends to a lift on $N \vee$ hom(, B'), contradicting the maximality of N.

A *Boolean localization* for $\mathbf{Shv}(\mathcal{C})$ is a geometric morphism $p : \mathbf{Shv}(\mathcal{B}) \to \mathbf{Shv}(\mathcal{C})$ such that $\mathcal{B}$ is a complete Boolean algebra and p^* is faithful.

Theorem 3.31 (Barr) *Boolean localizations exist for every Grothendieck topos $\mathbf{Shv}(\mathcal{C})$.*

Theorem 3.31 is one of the big results of the topos theory and is proved in multiple places—see [5], [78], for example.

A Grothendieck topos $\mathbf{Shv}(\mathcal{C})$ may not have enough points, in general (example: sheaves on the flat site for a scheme), but Theorem 3.31 asserts that every Grothendieck topos has a "fat point" defined by a Boolean localization.

3.5 Boolean Localization

This section contains a relatively short proof of the Barr theorem (Theorem 3.31) which says that every Grothendieck topos has a Boolean cover.

The proof is in two steps, just as in the literature (eg. [77]):

1) Show that every Grothendieck topos has a localic cover.
2) Show that every localic topos has a Boolean cover.

We begin with the second step: the precise statement is Theorem 3.39 below. The first statement is Diaconescu's theorem, which appears here as Theorem 3.44.

Recall that a frame F is a lattice that has all small joins and satisfies an infinite distributive law. Recall also that every frame A has a canonical Grothendieck topology: say that a family $y_i \leq x$ is covering if $\bigvee_i y_i = x$. Write **Shv**(A) for the corresponding sheaf category.

Say that a Grothendieck topos B is *localic* if it is equivalent to **Shv**(A) for some frame A.

Theorem 3.32 *A Grothendieck topos $\mathcal{E}$ is localic if and only if it is equivalent to* **Shv**(P) *for some topology on a poset P.*

Proof [Outline] Starting with the poset P, the corresponding frame is the poset of subobjects of the terminal object $1 = *$. For the reverse implication, these subobjects generate $\mathcal{E}$, and then Giraud's theorem is used to finish the proof.

A more detailed proof of Theorem 3.32 can be found in [77, IX.5].

A *morphism of frames* is a poset morphism $f : A \to B$ that preserves structure, i.e. preserves all finite meets and all infinite joins, hence preserves both 0 and 1.

Lemma 3.33 *Every frame morphism $f : A \to B$ has a right adjoint $f_* : B \to A$.*

Proof Set $f_*(y) = \bigvee_{f(x) \leq y} x$.

Suppose that $i : P \to B$ is a morphism of frames. Then precomposition with i determines a functor $i_* : \mathbf{Shv}(B) \to \mathbf{Shv}(P)$, since i preserves covers. The left adjoint

$$i^* : \mathbf{Shv}(P) \to \mathbf{Shv}(B)$$

of i_* associates to a sheaf F the sheaf i^*F, which is the sheaf associated to the presheaf $i^p F$, where

$$i^p F(x) = \varinjlim_{x \to i(y)} F(y).$$

This colimit is filtered since i preserves meets.

Lemma 3.34 *Suppose that $i : P \to B$ is a morphism of frames and that F is a sheaf on P. Then the presheaf $i^p F$ is separated.*

Proof Suppose that $\alpha, \beta \in i^p F(x)$ map to the same element in $i^*F(x)$. Then there is a covering family $z_j \leq x$ such that α, β restrict to the same element of $i^p F(z_j)$ for all j.

Identify α and β with representatives $\alpha, \beta \in F(y)$ for some fixed $x \leq i(y)$. For each j there is a commutative diagram of relations

$$\begin{array}{ccc} z_j & \longrightarrow & i(v_j) \\ \downarrow & & \downarrow \\ x & \longrightarrow & i(y) \end{array}$$

3.5 Boolean Localization

such that α and β restrict to the same element of $F(v_j)$. But then α and β restrict to the same element of $F(\vee v_j)$ and $\vee z_j = x$ and there is a commutative diagram

Then F is a sheaf, so that α and β map to the same element of $F(\vee v_j)x$ and therefore represent the same element of $i^p F(x)$.

Lemma 3.35 *Suppose that the frame morphism $i : P \to B$ is a monomorphism. Then the functor $i^* : \mathbf{Shv}(P) \to \mathbf{Shv}(B)$ is faithful.*

Proof By Lemma 3.34, it is enough to show that the canonical map $\eta : F \to i_* i^p F$ is a monomorphism of presheaves for all sheaves F on P. For then $\eta : F \to i_* i^* F$ is a monomorphism, and so i^* is faithful (exercise).

The map
$$\eta : F(y) \to \varinjlim_{i(y) \leq i(z)} F(z)$$
is the canonical map into the colimit which is associated to the identity map $i(y) \leq i(y)$ of B.

The frame morphism i is a monomorphism, so that $x = i_* i(x)$ for all $x \in P$, where i_* is the right adjoint of $i : P \to B$. Thus, $i(y) \leq i(z)$ if and only if $y \leq z$, so that category of all morphisms $i(y) \leq i(z)$ has an initial object, namely the identity on $i(y)$. The map η is therefore an isomorphism for all y.

Suppose that P is a frame and $x \in P$. Write P_x for the subobject of P consisting of all y such that $x \leq y$. Then P_x is a frame with initial object x and terminal object 1. There is a frame morphism
$$\phi_x : P \to P_x$$
which is defined by $\phi_x(w) = x \vee w$.

Suppose that Q is a frame and that $x \in Q$. Write
$$\neg x = \bigvee_{x \wedge y = 0} y.$$

Note that $x \wedge \neg x = 0$ so that there is a relation (morphism)
$$\eta : x \leq \neg\neg x$$
for all $x \in Q$; this relation is natural in x. Further, the relation η induces the relation $\neg\eta : \neg\neg\neg x \leq \neg x$, while we have the relation $\eta : \neg x \leq \neg\neg\neg x$ for $\neg x$. It follows that the relation
$$\eta : \neg x \leq \neg\neg\neg x$$

is an equality (isomorphism) for all $x \in Q$.

Define a subposet $\neg\neg Q$ of Q by

$$\neg\neg Q = \{y \in Q |\ y = \neg\neg y\}.$$

There is a diagram of relations

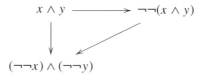

Thus, the element $x \wedge y$ is a member of $\neg\neg Q$ if both x and y are in $\neg\neg Q$, for in that case the vertical map in the diagram is an isomorphism. If the set of objects x_i are members of $\neg\neg Q$, then the element $\neg\neg(\vee_i x_i)$ is their join in $\neg\neg Q$. It follows that the poset $\neg\neg Q$ is a frame, and that the assignment $x \mapsto \neg\neg x$ defines a frame morphism

$$\gamma : Q \to \neg\neg Q.$$

Lemma 3.36 *The frame $\neg\neg Q$ is a complete Boolean algebra, for every frame Q.*

Proof Observe that $y \leq \neg z$ if and only if $y \wedge z = 0$. It follows that $\neg(\vee(\neg y_i))$ is the meet $\wedge y_i$ in $\neg\neg Q$, giving the completeness. Also, x is complemented by $\neg x$ in $\neg\neg Q$ since

$$x \vee (\neg x) = \neg\neg x \vee \neg\neg\neg x = \neg(\neg x \wedge x) = \neg 0 = 1.$$

Write ω for the composite frame morphism

$$P \xrightarrow{(\phi_x)} \prod_{x \in P} P_x \xrightarrow{(\gamma)} \prod_{x \in P} \neg\neg P_x,$$

and observe that the product $\prod_x \neg\neg P_x$ is a complete Boolean algebra.

Lemma 3.37 *The frame morphism γ is a monomorphism.*

Proof If $x \leq y$ then $\neg\neg \phi_x(y) = 0$ in P_x implies that there is a relation

$$x \vee y \leq \neg\neg(x \vee y) = x,$$

so that $x \vee y = x$ in P_x and hence $y = x$ in P. Thus, if $x \leq y$ and $y \neq x$ then x and y have distinct images $\omega(x) < \omega(y)$ in $\prod_x \neg\neg P_x$.

Suppose that y and z are distinct elements of P. Then $y \neq y \vee z$ or $z \leq y$ and $z \neq y$. Then $\omega(y) \neq \omega(y) \vee \omega(z)$ or $\omega(z) \leq \omega(y)$ and $\omega(z) \neq \omega(y)$. The assumption that $\omega(y) = \omega(z)$ contradicts both possibilities, so $\omega(y) \neq \omega(z)$.

3.5 Boolean Localization

Corollary 3.38 *Every frame P admits an imbedding $i : P \to B$ into a complete Boolean algebra.*

We have proved the following:

Theorem 3.39 *Suppose that P is a frame. Then there is a complete Boolean algebra B, and a topos morphism $i : \mathbf{Shv}(B) \to \mathbf{Shv}(P)$ such that the inverse image functor $i^* : \mathbf{Shv}(P) \to \mathbf{Shv}(B)$ is faithful.*

A geometric morphism i as in the statement of the theorem is called a *Boolean cover* of $\mathbf{Shv}(P)$.

Suppose that $\mathcal{C}$ is a (small) Grothendieck site. Write $\mathbf{St}(\mathcal{C})$ for the poset of all finite strings
$$\sigma : x_n \to \cdots \to x_0,$$
where $\tau \leq \sigma$ if τ extends σ to the left in the sense that τ is of the form
$$y_k \to \cdots \to y_{m+1} \to x_n \to \cdots \to x_0.$$
There is a functor $\pi : \mathbf{St}(\mathcal{C}) \to \mathcal{C}$ that is defined by $\pi(\sigma) = x_n$ for σ as above.

If $R \subset \hom(\ , \sigma)$ is a sieve of $\mathbf{St}(\mathcal{C})$, then $\pi(R) \subset \hom(\ , x_n)$ is a sieve of $\mathcal{C}$. In effect, if $\tau \leq \sigma$ is in R and $z \to y_k$ is morphism of $\mathcal{C}$, then the string
$$\tau_* : z \to y_k \to \cdots \to y_{m+1} \to x_n \to \ldots x_0$$
refines τ and the relation $\tau_* \leq \tau$ maps to $z \to y_k$.

Say that a sieve $R \subset \hom(\ , \sigma)$ is covering if $\pi(R)$ is a covering sieve of $\mathcal{C}$. Then $\mathbf{St}(\mathcal{C})$ acquires the structure of a Grothendieck site.

Lemma 3.40 *Suppose that F is a sheaf of sets on $\mathcal{C}$. Then $\pi^*(F) = F \cdot \pi$ is a sheaf on $\mathbf{St}(\mathcal{C})$.*

The proof of this result is an exercise.

Lemma 3.41 *The functor $F \mapsto \pi^*(F)$ is faithful.*

Proof For $x \in \mathcal{C}$ let $\{x\}$ denote the corresponding string of length 0. Then we have $\pi^*F(\{x\}) = F(x)$. If sheaf morphisms $f, g : F \to G$ on $\mathcal{C}$ induce maps $f_*, g_* : \pi^*(F) \to \pi^*(G)$ such that $f_* = g_*$, then $f_* = g_* : \pi^*F(\{x\}) \to \pi^*G(\{x\})$ for all $x \in \mathcal{C}$. This means, then, that $f = g$.

Lemma 3.42 *The functor π^* preserves local epimorphisms and local monomorphisms of presheaves.*

Proof Suppose $m : P \to Q$ is a local monomorphism of presheaves on $\mathcal{C}$. This means that if $m(\alpha) = m(\beta)$ for $\alpha, \beta \in P(x)$ there is a covering $\phi_i : y_i \to x$ such that $\phi_i^*(\alpha) = \phi_i^*(\beta)$ for all ϕ_i.

Suppose that $\alpha, \beta \in \pi^*P(\sigma)$ such that $m_*(\alpha) = m_*(\beta)$ in $\pi^*Q(\sigma)$. Then $\alpha, \beta \in P(x_n)$ and $m(\alpha) = m(\beta) \in Q(x_n)$. There is a covering $\phi_i : y_i \to x_n$ such that $\phi_i^*(\alpha) = \phi_i^*(\beta)$ for all ϕ_i. But then α, β map to the same element of
$$\pi^*P(y_i \to x_n \to \cdots \to x_0)$$

for all members of a cover of σ.

Suppose that $p : P \to Q$ is a local epimorphism of presheaves on $\mathcal{C}$. Then for all $\alpha \in Q(x)$ there is a covering $\phi_i : y_i \to x$ such that $\phi_i^*(\alpha)$ lifts to an element of $P(y_i)$ for all i. Given $\alpha \in \pi^* Q(\sigma)$, $\alpha \in Q(x_n)$, and there is a cover $\phi_i : y_i \to x_n$ such that $\phi_i^*(\alpha)$ lifts to $P(y_i)$. It follows that there is a cover of σ such that the image of α in

$$\pi^* Q(y_i \to x_n \to \cdots \to x_0)$$

lifts to

$$\pi^* P(y_i \to x_n \to \cdots \to x_0)$$

for all members of the cover.

Lemma 3.43 *The functor*

$$\pi^* : \mathbf{Shv}(\mathcal{C}) \to \mathbf{Shv}(\mathbf{St}(\mathcal{C}))$$

preserves all small colimits.

Proof Suppose that $A : I \to \mathbf{Shv}(\mathcal{C})$ is a small diagram of sheaves. Write $\varinjlim_i A$ for the presheaf colimit, and let

$$\eta : \varinjlim A \to L^2(\varinjlim A)$$

be the natural associated sheaf map. The map η is a local epimorphism and a local monomorphism. The functor π^* plainly preserves presheaf colimits, and there is a diagram

$$\begin{array}{ccc} \pi^*(\varinjlim A) & \xrightarrow{\pi^*(\eta)} & \pi^*(L^2(\varinjlim A)) \\ \cong \uparrow & & \uparrow \\ \varinjlim \pi^* A & \xrightarrow{\eta} & L^2(\varinjlim \pi^* A) \end{array}$$

Then $\pi^*(\eta)$ is a local epimorphism and a local monomorphism by Lemma 3.42. It follows that the map

$$L^2(\varinjlim \pi^* A) \to \pi^*(L^2(\varinjlim A))$$

is a local epimorphism and a local monomorphism of sheaves (use Lemma 3.40) and is therefore an isomorphism.

It is a consequence of the following result that any Grothendieck topos has a localic cover (see also [77, IX.9]). The topos $\mathbf{Shv}(\mathbf{St}(\mathcal{C}))$ is also called the *Diaconescu cover* of $\mathbf{Shv}(\mathcal{C})$.

3.5 Boolean Localization

Theorem 3.44 (Diaconescu) *The right adjoint* $\pi_* : \mathbf{Pre}(\mathbf{St}(\mathcal{C})) \to \mathbf{Pre}(\mathcal{C})$ *of precomposition with π restricts to a functor*

$$\pi_* : \mathbf{Shv}(\mathbf{St}(\mathcal{C})) \to \mathbf{Shv}(\mathcal{C}),$$

which is right adjoint to π^*. The functors π^* and π_* determine a geometric morphism

$$\pi : \mathbf{Shv}(\mathbf{St}(\mathcal{C})) \to \mathbf{Shv}(\mathcal{C}).$$

The functor π^* is faithful.

Proof A covering sieve $R \subset \hom(\ , x)$ in $\mathcal{C}$ determines an isomorphism of sheaves

$$\varinjlim_{y \to x} L^2\hom(\ , y) \cong L^2\hom(\ , x).$$

The functor π^* preserves colimits of sheaves, and so $\pi_* G$ is a sheaf if G is a sheaf. The functor π^* plainly preserves finite limits, so that the functors π^* and π_* form a geometric morphism. The last statement is Lemma 3.41.

We have thus assembled a proof of Barr's theorem (Theorem 3.31), which can be restated as follows:

Theorem 3.45 *Suppose that $\mathcal{C}$ is a small Grothendieck site. Then there are geometric morphisms*

$$\mathbf{Shv}(B) \xrightarrow{f} \mathbf{Shv}(\mathbf{St}(\mathcal{C})) \xrightarrow{\pi} \mathbf{Shv}(\mathcal{C})$$

such that the inverse image functors f^* and π^* are faithful, and such that B is a complete Boolean algebra.

Part II
Simplicial Presheaves and Simplicial Sheaves

Chapter 4
Local Weak Equivalences

Local weak equivalences have been with us, in one form or another, for a very long time.

Quasi-isomorphisms, or maps of chain complexes that induce isomorphisms of all homology sheaves, were already present in protean form in Grothendieck's Tôhoku paper [36] of 1957, and the concept was well entrenched in algebraic geometry by the time that Illusie's thesis appeared in 1971. Illusie expanded the definition to maps of simplicial sheaves in [47], and what he said was a quasi-isomorphism of simplicial sheaves is what we call a local weak equivalence today.

The concept was further expanded to maps of simplicial presheaves in [50] in 1987, following the form introduced by Illusie: a map $X \to Y$ of simplicial presheaves is a local weak equivalence if it induces an isomorphism in sheaves of path components and in all possible sheaves of homotopy groups with locally defined base points.

Local weak equivalences of simplicial sheaves were given a simple and very useful topos-theoretic description by Joyal in his letter to Grothendieck [70] in 1984, and Joyal's definition is easily adapted to simplicial presheaves. All of these descriptions of local weak equivalences are equivalent and important, and are presented in the first section of this chapter.

The concept of local fibration of simplicial sheaves, together with the local left lifting property that describes it, was introduced in the 1986 paper [49]. The object of that paper was to set up a homotopy theoretic machinery for simplicial sheaves, as a framework for the study of some problems of algebraic K-theory of the day, including the Lichtenbaum–Quillen conjecture and the Friedlander–Milnor conjecture for the cohomology of algebraic groups. The machinery of [49] is a bit limited, in that it is an artifact of a category of fibrant objects structure for locally fibrant simplicial sheaves (defined in Brown's thesis [16]), but it is a departure from stalkwise arguments. It is also sufficiently strong computationally to give a sheaf-theoretic approach to Suslin's calculation of the K-theory of algebraically closed fields, starting from Gabber rigidity. The category of fibrant objects result for simplicial sheaves from [49] is proved here in Corollary 4.34, while Proposition 4.33 is the simplicial presheaves version. The theory of local fibrations is the subject of the second section of this chapter.

A category of fibrant objects structure is essentially half of a Quillen model structure. A full model structure for simplicial sheaves, with the same (local) weak equivalences and a very different class of fibrations, was the subject of Joyal's letter to Grothendieck [70]. The analogue of Joyal's structure for simplicial presheaves was introduced in [50]. These model structures for simplicial sheaves and simplicial presheaves are now tools of choice for the study of local homotopy theory, and all theories based on it.

These model structures and their close cousins form the subject of Chap. 5. The present chapter is taken up with a description of the interplay between local fibrations and local weak equivalences for simplicial presheaves and simplicial sheaves, much of which is introduced with a view to proving the model structure results that come later.

Local fibrations have their own intrinsic technical interest. It is a rather serious result (Theorem 4.32 below) that maps of simplicial presheaves, which are local fibrations and local weak equivalences, can be completely described by a local right lifting property in arbitrary simplicial presheaf categories. From this point of view, such morphisms are the natural generalizations of the hypercovers that one finds in the Artin–Mazur book [4] and throughout étale homotopy theory. Other highlights include the properness results of Corollary 4.36 and Lemma 4.37. The latter, in particular, is a right properness assertion for local weak equivalences and local fibrations which is quite useful.

We use Boolean localization methods from Chap. 3 thoughout this discussion. The interested reader can replace all Boolean localization arguments by stalkwise arguments, in situations where there are enough stalks. That said, such situations are usually classical, while stalkwise arguments do not address major settings of interest, such as the flat site of a scheme or a variety.

4.1 Local Weak Equivalences

Suppose that $\mathcal{C}$ is a small Grothendieck site. The notations $s\mathbf{Pre}(\mathcal{C})$ and $s\mathbf{Shv}(\mathcal{C})$ denote the categories of simplicial presheaves and simplicial sheaves on $\mathcal{C}$, respectively.

Recall that a simplicial set map $f : X \to Y$ is a weak equivalence if and only if the induced map $|X| \to |Y|$ is a weak equivalence of topological spaces in the classical sense. This is equivalent to the assertion that all induced morphisms

a) $\pi_0 X \to \pi_0 Y$, and
b) $\pi_n(X, x) \to \pi_n(Y, f(x))$, $x \in X_0, x \in X_0, n \geq 1$

of homotopy "groups" are bijections.

Define

$$\pi_n(X, x) = \pi_n(|X|, x)$$

for now. This is standard, but this definition will be revised later.

4.1 Local Weak Equivalences

We have an identification
$$\pi_n(X, x) = [(S^n, *), (X, x)]$$
with pointed homotopy classes of maps simplicial sets, where $S^n = \Delta^n/\partial\Delta^n$ is the simplicial n-sphere. If X is a Kan complex, then we have
$$\pi_n(X, x) = \pi((S^n, *), (X, x))$$
by the Milnor Theorem (Theorem 2.18), where $\pi((S^n, *), (X, x))$ is pointed (naive) simplicial homotopy classes of maps.

Write
$$\pi_n X = \bigsqcup_{x \in X_0} \pi_n(X, x)$$
for a simplicial set X, where $n \geq 1$. Then the canonical function $\pi_n X \to X_0$ gives $\pi_n X$ the structure of a group object over X_0, which is abelian if $n \geq 2$.

One verifies easily that a map $f : X \to Y$ of simplicial sets is a weak equivalence if and only if the following conditions are satisfied:

a) the function $\pi_0 X \to \pi_0 Y$ is a bijection, and
b) the diagrams

$$\begin{array}{ccc} \pi_n X & \longrightarrow & \pi_n Y \\ \downarrow & & \downarrow \\ X_0 & \longrightarrow & Y_0 \end{array}$$

are pullbacks for $n \geq 1$.

If X is a Kan complex, then the object $\pi_n(X, x)$ can also be defined as a set by setting
$$\pi_n(X, x) = \pi_0 F_n(X)_x,$$
where the simplicial set $F_n(X)_x$ is defined by the pullback diagram

$$\begin{array}{ccc} F_n(X)_x & \longrightarrow & \mathbf{hom}(\Delta^n, X) \\ \downarrow & & \downarrow i^* \\ \Delta^0 & \xrightarrow{x} & \mathbf{hom}(\partial\Delta^n, X) \end{array}$$

in which i^* is the Kan fibration between function spaces that is induced by the inclusion $i : \partial\Delta^n \subset \Delta^n$. Define the space $F_n(X)$ by the pullback diagram

$$\begin{array}{ccc} F_n(X) & \longrightarrow & \mathbf{hom}(\Delta^n, X) \\ \downarrow & & \downarrow i^* \\ X_0 & \xrightarrow{\alpha} & \mathbf{hom}(\partial\Delta^n, X) \end{array}$$

where $\alpha(x)$ is the constant map $\partial \Delta^n \to \Delta^0 \xrightarrow{x} X$ at the vertex x. Then

$$F_n(X) = \bigsqcup_{x \in X_0} F_n(X)_x,$$

so that

$$\pi_0 F_n(X) = \bigsqcup_{x \in X_0} \pi_0 F_n(X)_x = \bigsqcup_{x \in X_0} \pi_n(X, x) = \pi_n X,$$

as object fibred over X_0.

It follows that if X and Y are Kan complexes, then the map $f : X \to Y$ is a weak equivalence if and only if

a) the induced function $\pi_0 X \to \pi_0 Y$ is a bijection, and
b) all diagrams

$$\begin{array}{ccc} \pi_0 F_n(X) & \longrightarrow & \pi_0 F_n(Y) \\ \downarrow & & \downarrow \\ X_0 & \longrightarrow & Y_0 \end{array}$$

are pullbacks for $n \geq 1$.

Kan's Ex^∞ *construction* gives a natural combinatorial method of replacing a simplicial set by a Kan complex up to weak equivalence.

The functor $\mathrm{Ex} : s\mathbf{Set} \to s\mathbf{Set}$ is defined by

$$\mathrm{Ex}(X)_n = \hom(\mathrm{sd}\, \Delta^n, X).$$

The subdivision $\mathrm{sd}\, \Delta^n$ is the nerve $BN\Delta^n$, where $N\Delta^n$ is the poset of non-degenerate simplices of Δ^n (subsets of $\{0, 1, \ldots, n\}$). Any ordinal number map $\theta : \mathbf{m} \to \mathbf{n}$ induces a functor $N\Delta^m \to N\Delta^n$, and hence induces a simplicial set map $\mathrm{sd}\, \Delta^m \to \mathrm{sd}\, \Delta^n$. Precomposition with this map gives the simplicial structure map θ^* of $\mathrm{Ex}(X)$. There is a last vertex functor $N\Delta^n \to \mathbf{n}$, which is natural in ordinal numbers $\mathbf{n}$; the collection of such functors determines a natural simplicial set map

$$\eta : X \to \mathrm{Ex}(X).$$

Observe that $\mathrm{Ex}(X)_0 = X_0$, and that η induces a bijection on vertices.

Iterating gives a simplicial set $\mathrm{Ex}^\infty(X)$ that is defined by the assignment

$$\mathrm{Ex}^\infty(X) = \varinjlim \mathrm{Ex}^n(X),$$

and a natural map $j : X \to \mathrm{Ex}^\infty(X)$.

The salient features of the construction are the following (see [32, III.4]):

1) the map $\eta : X \to \mathrm{Ex}(X)$ is a weak equivalence,

4.1 Local Weak Equivalences

2) the functor $X \mapsto \text{Ex}(X)$ preserves Kan fibrations
3) the simplicial set $\text{Ex}^\infty(X)$ is a Kan complex, and the natural map $j : X \to \text{Ex}^\infty(X)$ is a weak equivalence.

The weak equivalence claimed in statement 3) is a consequence of statement 1).

It follows that a simplicial set map $f : X \to Y$ is a weak equivalence if and only if the induced map of Kan complexes $f_* : \text{Ex}^\infty(X) \to \text{Ex}^\infty(Y)$ is a weak equivalence, so that f is a weak equivalence if and only if

a) the function $\pi_0 X \to \pi_0(Y)$ is a bijection, and
b) the diagram

$$\begin{array}{ccc} \pi_0 F_n(\text{Ex}^\infty(X)) & \longrightarrow & \pi_0 F_n(\text{Ex}^\infty(Y)) \\ \downarrow & & \downarrow \\ X_0 & \longrightarrow & Y_0 \end{array}$$

is a pullback for $n \geq 1$.

The sets of vertices X_0 and $\text{Ex}^\infty X_0$ coincide, and the set $\pi_0 F_n(\text{Ex}^\infty(X))$ is a disjoint union of simplicial homotopy groups $\pi_n(\text{Ex}^\infty X, x)$ for $x \in X_0$.

We shall define the nth homotopy group object $\pi_n X$ of a simplicial set X by setting

$$\pi_n X = \pi_0 F_n(\text{Ex}^\infty(X))$$

in all that follows.

The fundamental idea of local homotopy theory is that the topology of the underlying site $\mathcal{C}$ should create weak equivalences.

It is relatively easy to see what the local weak equivalences should look like for simplicial presheaves on a topological space: a map $f : X \to Y$ of simplicial presheaves on $op|_T$ for a topological space T should be a local weak equivalence if and only if it induces a weak equivalence of simplicial sets $X_x \to Y_x$ in stalks for all $x \in T$. In particular the map f should induce isomorphisms

$$\pi_n(X_x, y) \to \pi_n(Y_x, f(y))$$

for all $n \geq 1$ *and all choices of base point* $y \in X_x$ and $x \in T$, as well as bijections

$$\pi_0 X_x \xrightarrow{\cong} \pi_0 Y_x$$

for all $x \in T$.

Recall that the stalk

$$X_x = \varinjlim_{x \in U} X(U)$$

is a filtered colimit, and so each base point y of X_x comes from somewhere, namely some $z \in X(U)$ for some U. The point z determines a global section of $X|_U$, where the restriction $X|_U$ is the composite

$$((op|_T)/U)^{op} \to (op|_T)^{op} \xrightarrow{X} s\mathbf{Set}.$$

The map f restricts to a simplicial presheaf map $f|_U : X|_U \to Y|_U$. Then one can show that f is a weak equivalence in all stalks if and only if all induced maps

a) $\tilde{\pi}_0 X \to \tilde{\pi}_0 Y$, and
b) $\tilde{\pi}_n(X|_U, z) \to \tilde{\pi}_n(Y|_U, f(z))$, for all $n \geq 1$, $U \in \mathcal{C}$, and $z \in X_0(U)$

in associated sheaves.

This is equivalent to the following: the map $f : X \to Y$ of simplicial presheaves on the topological space T is a local weak equivalence if and only if

a) $\tilde{\pi}_0 X \to \tilde{\pi}_0 Y$ is an isomorphism
b) the presheaf diagrams

$$\begin{array}{ccc} \pi_n X & \longrightarrow & \pi_n Y \\ \downarrow & & \downarrow \\ X_0 & \longrightarrow & Y_0 \end{array}$$

induce pullback diagrams of associated sheaves for $n \geq 1$.

These last two descriptions generalize to equivalent sets of conditions for maps of simplicial presheaves on an arbitrary site $\mathcal{C}$, but the equivalence of conditions requires proof.

We begin by saying that a map $X \to Y$ of simplicial presheaves on a site $\mathcal{C}$ is a *local weak equivalence* if and only if

a) the map $\tilde{\pi}_0 X \to \tilde{\pi}_0 Y$ is an isomorphism of sheaves, and
b) the diagrams of presheaf maps

$$\begin{array}{ccc} \pi_n X & \longrightarrow & \pi_n Y \\ \downarrow & & \downarrow \\ X_0 & \longrightarrow & Y_0 \end{array}$$

induce pullback diagrams of associated sheaves for $n \geq 1$.

The following result gives a first example. A *sectionwise weak equivalence* is a simplicial presheaf map $X \to Y$ such that all maps $X(U) \to Y(U)$ in sections are weak equivalences of simplicial sets.

Lemma 4.1 *Suppose that the map $f : X \to Y$ is a sectionwise weak equivalence of simplicial presheaves. Then f is a local weak equivalence.*

4.1 Local Weak Equivalences

Proof The map $\pi_0 X \to \pi_0 Y$ is an isomorphism of presheaves, and all diagrams

$$\begin{array}{ccc} \pi_n X & \longrightarrow & \pi_n Y \\ \downarrow & & \downarrow \\ X_0 & \longrightarrow & Y_0 \end{array}$$

are pullbacks of presheaves. Apply the associated sheaf functor, which is exact by Lemma 3.14.

The Ex^∞ construction extends to a construction for simplicial presheaves, which construction preserves and reflects local weak equivalences:

Lemma 4.2 *A map $f : X \to Y$ of simplicial presheaves is a local weak equivalence if and only if the induced map $\mathrm{Ex}^\infty X \to \mathrm{Ex}^\infty Y$ is a local weak equivalence.*

Proof The natural simplicial set map $j : X \to \mathrm{Ex}^\infty X$ consists, in part, of a natural bijection

$$X_0 \xrightarrow{\cong} \mathrm{Ex}^\infty X_0$$

of vertices for all simplicial sets X, and the horizontal arrows in the natural pullback diagrams

$$\begin{array}{ccc} \pi_n X & \longrightarrow & \pi_n \mathrm{Ex}^\infty X \\ \downarrow & & \downarrow \\ X_0 & \longrightarrow & \mathrm{Ex}^\infty X_0 \end{array}$$

are isomorphisms. It follows that the diagram

$$\begin{array}{ccc} \tilde{\pi}_n X & \longrightarrow & \tilde{\pi}_n Y \\ \downarrow & & \downarrow \\ \tilde{X}_0 & \longrightarrow & \tilde{Y}_0 \end{array}$$

is a pullback if and only if the diagram

$$\begin{array}{ccc} \tilde{\pi}_n \mathrm{Ex}^\infty X & \longrightarrow & \tilde{\pi}_n \mathrm{Ex}^\infty Y \\ \downarrow & & \downarrow \\ \tilde{X}_0 & \longrightarrow & \tilde{Y}_0 \end{array}$$

is a pullback.

Remark 4.3 The map $j : X \to \mathrm{Ex}^\infty X$ is a sectionwise equivalence, and is therefore a local weak equivalence by Lemma 4.2. We do not yet have a 2 out of 3 lemma for local weak equivalences so the apparent trick in the proof of Lemma 4.2 is required to show that the Ex^∞ functor preserves local weak equivalences. This situation is repaired in Lemma 4.30 below.

We close this section by showing that the local weak equivalences that are defined here coincide with the topological weak equivalences of [50]. A map $f : X \to Y$ of simplicial presheaves is a *topological weak equivalence* if

1) the map $\tilde{\pi}_0 X \to \tilde{\pi}_0 Y$ is an isomorphism of sheaves, and
2) all induced maps $\tilde{\pi}_n(X|_U, x) \to \tilde{\pi}_n(Y|_U, f(x))$ are isomorphisms of sheaves on $\mathcal{C}/U$ for all $n \geq 1$, all $U \in \mathcal{C}$, and all $x \in X_0(U)$.

The claim that local and topological weak equivalences coincide is an immediate consequence of Lemma 4.4 below. To state and prove that result we need some notation:

1) Suppose that X' is a presheaf and that the presheaf map $x : * \to X'$ is a global section of X'. Suppose that $X \to X'$ is a presheaf morphism, and define a presheaf $X(x)$ by the pullback diagram

$$\begin{array}{ccc} X(x) & \longrightarrow & X \\ \downarrow & & \downarrow \\ * & \xrightarrow{x} & X' \end{array}$$

2) The *restriction* $X|_U$ of a presheaf X to the site $\mathcal{C}/U$ is the composite

$$(\mathcal{C}/U)^{op} \to \mathcal{C}^{op} \xrightarrow{X} s\mathbf{Set}.$$

Recall that the notations

$$L^2(X) = \tilde{X}$$

denote the sheaf that is associated with a presheaf X.

Lemma 4.4 *Suppose given a commutative diagram of presheaves*

$$\begin{array}{ccc} Z & \longrightarrow & W \\ \downarrow & & \downarrow \\ Z' & \xrightarrow{f} & W' \end{array} \qquad (4.1)$$

Then the induced diagram of associated sheaves is a pullback if and only if the maps

$$L^2(Z|_U)(x)) \to L^2(W|_U)(f(x))$$

4.1 Local Weak Equivalences

are isomorphisms of sheaves for all $x \in Z'(U)$ and $U \in \mathcal{C}$.

Applying Lemma 4.4 to the diagrams

$$\begin{array}{ccc} \pi_n X & \longrightarrow & \pi_n Y \\ \downarrow & & \downarrow \\ X_0 & \longrightarrow & Y_0 \end{array}$$

associated to a simplicial presheaf map $f : X \to Y$ then gives the following:

Corollary 4.5 *A map $f : X \to Y$ of simplicial presheaves on $\mathcal{C}$ is a local weak equivalence if and only if it is a topological weak equivalence.*

Proof [Proof of Lemma 4.1] This proof is a formality—it is displayed to fix ideas.

Suppose that $x \in Z'(U)$ is a section of Z' and form the pullback diagram

$$\begin{array}{ccc} Z_{U,x} & \longrightarrow & Z \\ \downarrow & & \downarrow \\ U & \xrightarrow{x} & Z' \end{array}$$

in presheaves, where $U = \hom(\ ,U)$ is the presheaf represented by $U \in \mathcal{C}$. Then there is an isomorphism

$$\varinjlim_{U \xrightarrow{x} Z'} Z_{U,x} \xrightarrow{\cong} Z$$

that is natural in presheaves over Z', and hence there is an isomorphism

$$\varinjlim_{U \xrightarrow{x} Z'} \tilde{Z}_{U,x} \xrightarrow{\cong} \tilde{Z}$$

in sheaves over $\tilde{Z}'$. The diagrams

$$\begin{array}{ccc} \tilde{Z}_{U,x} & \longrightarrow & \tilde{Z} \\ \downarrow & & \downarrow \\ \tilde{U} & \xrightarrow{x} & \tilde{Z}' \end{array}$$

of associated sheaves are pullbacks.

It follows that the diagram (4.1) induces a pullback diagram of sheaves if and only if all sheaf maps

$$\tilde{Z}_{U,x} \to \tilde{W}_{U,f(x)}$$

are isomorphisms.

Write r^* for the left adjoint of the restriction functor $X \mapsto X|_U$, for both presheaves and sheaves. There is a natural isomorphism of presheaves

$$r^*(X|_U(x)) \cong X_{U,x},$$

and hence a natural isomorphism of sheaves

$$r^* L^2(X|_U(x)) \cong \tilde{X}_{U,x}. \tag{4.2}$$

Restriction commutes with formation of the associated sheaf, and preserves pullbacks. Thus, if the diagram of sheaves

$$\begin{array}{ccc} \tilde{Z} & \longrightarrow & \tilde{W} \\ \downarrow & & \downarrow \\ \tilde{Z}' & \xrightarrow{f} & \tilde{W}' \end{array}$$

is a pullback, then the diagram

$$\begin{array}{ccc} L^2(Z|_U) & \longrightarrow & L^2(W|_U) \\ \downarrow & & \downarrow \\ L^2(Z'|_U) & \xrightarrow{f} & L^2(W'|_U) \end{array}$$

is a pullback, and it follows that the map

$$L^2(Z|_U)(x)) \to L^2(W|_U)(f(x)) \tag{4.3}$$

is an isomorphism of sheaves.

If the map (4.3) is an isomorphism, then the map

$$\tilde{Z}_{U,x} \to \tilde{W}_{U,f(x)}$$

is an isomorphism on account of the identification (4.2).

The sheaf $\tilde{\pi}_0 X$ is the *path component sheaf* of a simplicial presheaf X, and it is standard to say (as in [50]) that the sheaves

$$\tilde{\pi}_n(X|_U, z),$$

which are defined for all $n \geq 1$, $U \in \mathcal{C}$, and $z \in X_0(U)$, are the *sheaves of homotopy groups* of X.

The dependence on local choices of base points $z \in X_0(U)$ for the homotopy group sheaves $\tilde{\pi}_n(X|_U, z)$ is important and cannot be suppressed.

Local base points are, for example, the source of the base points for the stalks X_x of X, where one has stalks. More generally, the homotopy group sheaf $\tilde{\pi}_n(X, z)$ is the fibre over the vertex $z \in X_0(U)$ of the sheaf map

$$\tilde{\pi}_n(X) \to \tilde{X}_0.$$

This sheaf map is a group object in sheaves over $\tilde{X}_0$, and contains all information about the higher homotopy group sheaves in degree n. One often says that *this map* is the nth sheaf of homotopy groups of the simplicial presheaf X.

4.2 Local Fibrations

Suppose that the map $i : K \subset L$ is a cofibration of finite simplicial sets and that $f : X \to Y$ is a map of simplicial presheaves. We say that f has the *local right lifting property* with respect to i if for every commutative diagram

$$\begin{array}{ccc} K & \longrightarrow & X(U) \\ i \downarrow & & \downarrow f \\ L & \longrightarrow & Y(U) \end{array}$$

in simplicial sets there is a covering sieve $R \subset \hom(\ , U)$ such that the lift exists in the diagram

$$\begin{array}{ccccc} K & \longrightarrow & X(U) & \xrightarrow{\phi^*} & X(V) \\ i \downarrow & & & & \downarrow f \\ L & \longrightarrow & Y(U) & \xrightarrow{\phi^*} & Y(V) \end{array}$$

for every $\phi : V \to U$ in R.

Remark 4.6 There is no requirement for consistency between the lifts along the various members of the sieve R. Thus, if R is generated by a covering family $\phi_i : V_i \to U$, we just require liftings

$$\begin{array}{ccccc} K & \longrightarrow & X(U) & \xrightarrow{\phi_i^*} & X(V_i) \\ i \downarrow & & & & \downarrow f \\ L & \longrightarrow & Y(U) & \xrightarrow{\phi_i^*} & Y(V_i) \end{array}$$

The proof of the following result is an exercise.

Lemma 4.7

1) Suppose given simplicial presheaf maps

$$X \xrightarrow{f} Y \xrightarrow{g} Z$$

such that f and g have the local right lifting property with respect to $i : K \subset L$. Then the composite $g \cdot f$ has the local right lifting property with respect to the map i.

2) Suppose given a pullback diagram

$$\begin{array}{ccc} Z \times_Y X & \longrightarrow & X \\ {\scriptstyle f_*} \downarrow & & \downarrow {\scriptstyle f} \\ Z & \longrightarrow & Y \end{array}$$

such that f has the local right lifting property with respect to $i : K \subset L$. Then the map f_* has the local right lifting property with respect to i.

Lemma 4.7 says, in summary, that the class of simplicial presheaf maps having the local right lifting property with respect to $i : K \subset L$ is closed under composition and base change.

Write X^K for the presheaf that is defined by the function complexes

$$X^K(U) = \mathbf{hom}(K, X(U))$$

Lemma 4.8 *A map $f : X \to Y$ of simplicial presheaves has the local right lifting property with respect to the simplicial set map $i : K \to L$ if and only if the simplicial presheaf map*

$$X^L \xrightarrow{(i^*, f_*)} X^K \times_{Y^K} Y^L$$

is a local epimorphism in degree 0.

Proof The proof is again an exercise.

The condition on the map $f : X \to Y$ of Lemma 4.8 is the requirement that the presheaf map

$$\hom(L, X) \xrightarrow{(i^*, f_*)} \hom(K, X) \times_{\hom(K, Y)} \hom(L, Y) \qquad (4.4)$$

is a local epimorphism, where $\hom(K, X)$ is the presheaf that is specified in sections by

$$\hom(K, X)(U) = \hom(K, X(U)),$$

or the set of simplicial set morphisms $K \to X(U)$.

4.2 Local Fibrations

Lemma 4.9 *Suppose that $\pi : X \to Y$ is a map of simplicial sheaves on $\mathcal{C}$ that has the local right lifting property with respect to an inclusion $i : K \subset L$ of finite simplicial sets, and suppose that $p : \mathbf{Shv}(\mathcal{D}) \to \mathbf{Shv}(\mathcal{C})$ is a geometric morphism. Then the induced map $p^*(\pi) : p^*X \to p^*Y$ has the local right lifting property with respect to $i : K \subset L$.*

Proof The identifications
$$p^* \hom(\Delta^n, X) \cong p^*X_n \cong \hom(\Delta^n, p^*X)$$
are natural in simplices Δ^n and simplicial sheaves X, and therefore induce a natural map
$$p^* \hom(K, X) \to \hom(K, p^*X).$$
This map is an isomorphism for all simplicial sheaves X and all finite simplicial sets K, since the inverse image functor p^* preserves finite limits.

The map (4.4) is a sheaf epimorphism, since $\pi : X \to Y$ has the local right lifting property with respect to i. The inverse image functor p^* preserves sheaf epimorphisms, so applying p^* to the map (4.4) gives a sheaf epimorphism which is isomorphic to the map
$$\hom(L, p^*X) \xrightarrow{(i^*, p^* f_*)} \hom(K, p^*X) \times_{\hom(K, p^*Y)} \hom(L, p^*Y).$$

Lemma 4.10 *A simplicial presheaf map $g : X \to Y$ has the local right lifting property with respect to an inclusion $i : K \subset L$ of finite simplicial sets if and only if the induced map $g_* : \tilde{X} \to \tilde{Y}$ of associated simplicial sheaves has the local right lifting property with respect to i.*

Proof The presheaf map (4.4) is a local epimorphism if and only if the induced map
$$\hom(L, \tilde{X}) \to \hom(K, \tilde{X}) \times_{\hom(K, \tilde{Y})} \hom(L, \tilde{Y})$$
of associated sheaves is a local epimorphism (i.e. an epimorphism of sheaves), by Lemma 3.16.

A *local fibration* is a map $X \to Y$ of simplicial presheaves that has the local right lifting property with respect to all inclusions $\Lambda^n_k \subset \Delta^n$ of horns in simplices, i.e. where $n \geq 1$ and $0 \leq k \leq n$. A simplicial presheaf X is *locally fibrant* if the map $X \to *$ is a local fibration.

Example 4.11 Every sectionwise fibration is a local fibration.

Corollary 4.12

1) Suppose that $p : \mathbf{Shv}(\mathcal{D}) \to \mathbf{Shv}(\mathcal{C})$ is a geometric morphism. Then the inverse image functor p^* preserves local fibrations $X \to Y$ of simplicial sheaves.
2) The associated sheaf functor preserves and reflects local fibrations of simplicial presheaves.

Say that a map $p : X \to Y$ that has the local right lifting property with respect to the simplicial set maps $\partial \Delta^n \subset \Delta^n$, $n \geq 0$, is a *local trivial fibration*. Such a map is also sometimes called a *hypercover*.

This is the natural generalization, to simplicial presheaves, of the concept of hypercover of a scheme for the étale topology which was introduced by Artin and Mazur [4].

To see this, suppose that X is a simplicial sheaf. Then the map $X \to *$ is a hypercover (or local trivial fibration) if the maps

$$X_0 \to *,$$
$$\hom(\Delta^n, X) \to \hom(\partial \Delta^n, X), \; n \geq 1, \tag{4.5}$$

are sheaf epimorphisms. There is a standard definition

$$\cosk_m(X)_n = \hom(\sk_m \Delta^n, X)$$

and $\sk_{n-1} \Delta^n = \partial \Delta^n$, so that the second map of (4.5) can be written as

$$X_n \to \cosk_{n-1}(X)_n,$$

which is the way that it is displayed in [4].

Example 4.13 Every map which is a sectionwise fibration and a sectionwise weak equivalence is a local trivial fibration.

Remark 4.14 It is an exercise to show that a map $p : X \to Y$ is a local trivial fibration if and only if it has the local right lifting property with respect to all inclusions $K \subset L$ of finite simplicial sets.

Corollary 4.15

1) Suppose that $p : \mathbf{Shv}(\mathcal{D}) \to \mathbf{Shv}(\mathcal{C})$ is a geometric morphism. Then the inverse image functor p^* preserves local trivial fibrations $X \to Y$ of simplicial sheaves.
2) The associated sheaf functor preserves and reflects local trivial fibrations of simplicial presheaves.

Corollary 4.16 *The maps $\nu : X \to LX$ and $\eta : X \to L^2X$ are local trivial fibrations.*

Proof Both maps induce isomorphisms of associated simplicial sheaves, and every isomorphism is a local trivial fibration.

Example 4.17 Suppose that $f : X \to Y$ is a function. There is a groupoid $C(f)$, whose objects are the elements x of X, and whose morphisms are the pairs (x_1, x_2) such that $f(x_1) = f(x_2)$. The set of path components

$$\pi_0 C(f) = \pi_0 BC(f)$$

of the groupoid (and of its associated nerve) is isomorphic to the image $f(X)$ of f, and there is a trivial Kan fibration $BC(f) \to f(X)$ that is natural in functions f.

4.2 Local Fibrations 73

The nerve $BC(f)$ of the groupoid $C(f)$ and constant simplicial set $f(X)$ are both Kan complexes.

Thus, if $f : X \to Y$ is a map of presheaves, then there is a sectionwise trivial fibration $BC(f) \to f(X)$, where $B(C(f))(U)$ is the nerve of the groupoid associated to the function $f : X(U) \to Y(U)$. If the map f is a local epimorphism, then the inclusion $f(X) \subset Y$ induces an isomorphism of associated sheaves, and is therefore a local trivial fibration of constant simplicial presheaves. The groupoid $C(f)$ is the Čech groupoid for the map f.

It follows that the canonical map $BC(f) \to Y$ is a local trivial fibration if the presheaf map f is a local epimorphism.

Write $\check{C}(U) = BC(U)$ for the copy of $BC(t)$ associated to the presheaf map $t : U \to *$, where $*$ is the terminal presheaf. If the presheaf map t is a local epimorphism, then the map $\check{C}(U) \to *$ is a local trivial fibration (and therefore a hypercover). This is the Čech resolution of the terminal object that is associated to the covering $U \to *$.

If $\mathcal{C}$ is the site op $|_T$ of open subsets of a topological space T, and $U_\alpha \subset T$ is an open cover, then the subspaces U_α represent sheaves having the same names, and the map $U = \sqcup_\alpha U_\alpha \to *$ is a sheaf epimorphism. The n-fold product $U^{\times n}$ has the form

$$U^{\times n} = \bigsqcup_{(\alpha_1,\ldots,\alpha_n)} U_{\alpha_1} \cap \cdots \cap U_{\alpha_n},$$

and so the simplicial sheaf $\check{C}(U)$ is represented by a simplicial space, which is the classical Čech resolution associated to the covering $U_\alpha \subset T$.

If $\mathcal{C}$ is the étale site $et|_k$ of a field k, and L/k is a finite Galois extension with Galois group G, then the scheme homomorphism $\mathrm{Sp}(L) \to \mathrm{Sp}(k)$ represents a covering $\mathrm{Sp}(L) \to *$ on $et|_k$. There is a canonical sheaf isomorphism

$$G \times \mathrm{Sp}(L) \xrightarrow{\cong} \mathrm{Sp}(L) \times \mathrm{Sp}(L)$$

by Galois theory, and the sheaf of groupoids underlying $C(\mathrm{Sp}(L))$ is isomorphic to the translation groupoid $E_G \mathrm{Sp}(L)$ for the action of G on $\mathrm{Sp}(L)$. It follows that the Čech resolution $\check{C}(L) := \check{C}(\mathrm{Sp}(L))$ is isomorphic in simplicial sheaves to the Borel construction $EG \times_G \mathrm{Sp}(L)$ for the action of the Galois group G on $\mathrm{Sp}(L)$.

Such an observation holds, more generally, for all principal bundles (torsors) in sheaf categories. This phenomenon is discussed in much more detail in Chap. 9.

Lemma 4.18 *Suppose that X and Y are presheaves of Kan complexes. Then a map $p : X \to Y$ is a local fibration and a local weak equivalence if and only if p is a local trivial fibration.*

We show in Theorem 4.32 that an arbitrary map $p : X \to Y$ of simplicial presheaves is a local weak equivalence and a local fibration if and only if it is a local trivial fibration.

Example 4.19 If $f : X \to Y$ is a local epimorphism of presheaves, then the local trivial fibration $BC(f) \to Y$ of Example 4.17 is a local weak equivalence. In particular, every Čech resolution $\check{C}(U) \to *$ associated to a covering $U \to *$ is a local weak equivalence.

Proof (Proof of Lemma 4.18) Suppose that p is a local fibration and a local weak equivalence, and that we have a commutative diagram

$$\begin{array}{ccc} \partial\Delta^n & \longrightarrow & X(U) \\ \downarrow & & \downarrow p \\ \Delta^n & \longrightarrow & Y(U) \end{array} \qquad (4.6)$$

of simplicial set maps. The idea is to show that this diagram is locally homotopic to diagrams

$$\begin{array}{ccc} \partial\Delta^n & \longrightarrow & X(V) \\ \downarrow & \nearrow & \downarrow p \\ \Delta^n & \longrightarrow & Y(V) \end{array}$$

for which the lift exists. This means that there are homotopies

$$\begin{array}{ccc} \partial\Delta^n \times \Delta^1 & \longrightarrow & X(V) \\ \downarrow & & \downarrow p \\ \Delta^n \times \Delta^1 & \longrightarrow & Y(V) \end{array}$$

from the diagrams

$$\begin{array}{ccccc} \partial\Delta^n & \longrightarrow & X(U) & \xrightarrow{\phi^*} & X(V) \\ \downarrow & & & & \downarrow p \\ \Delta^n & \longrightarrow & Y(U) & \xrightarrow{\phi^*} & Y(V) \end{array}$$

to the corresponding diagrams above for all $\phi : V \to U$ in a covering for U. If such local homotopies exist, then solutions to the lifting problems

$$\begin{array}{ccc} (\partial\Delta^n \times \Delta^1) \cup (\Delta^n \times \{0\}) & \longrightarrow & X(V) \\ \downarrow & & \downarrow p \\ \Delta^n \times \Delta^1 & \longrightarrow & Y(V) \end{array}$$

4.2 Local Fibrations

have local solutions for each V, and so the original lifting problem is solved on a refined covering of U.

The required local homotopies are created by arguments similar to the proof of the corresponding result in the simplicial set case [32, I.7.10]. Here are the steps in the construction:

a) The diagram (4.6) is homotopic to a diagram

$$\begin{array}{ccc} \partial\Delta^n & \xrightarrow{(\alpha_0,x,\ldots,x)} & X(U) \\ \downarrow & & \downarrow p \\ \Delta^n & \longrightarrow & Y(U) \end{array} \qquad (4.7)$$

for some choice of base point $x \in X(U)$, since X and Y are presheaves of Kan complexes.

b) The element $[\alpha_0] \in \pi_{n-1}(X(U), x)$ vanishes locally in X since its image vanishes in $\pi_{n-1}(Y(U), p(x))$, so that the diagram (4.7) is locally homotopic to a diagram

$$\begin{array}{ccc} \partial\Delta^n & \xrightarrow{x} & X(V) \\ \downarrow & & \downarrow p \\ \Delta^n & \xrightarrow{\beta} & Y(V). \end{array} \qquad (4.8)$$

c) The element $[\beta] \in \pi_n(Y(V), p(x))$ lifts locally to X, and so the diagram (4.8) is locally homotopic to a diagram

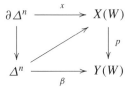

for which the indicated lifting exists.

For the converse, show that the induced presheaf maps

$$\pi_0 X \to \pi_0 Y,$$
$$\pi_i(X|_U, x) \to \pi_i(Y|_U, p(x))$$

are local epimorphisms and local monomorphisms—use presheaves of simplicial homotopy groups for this.

Lemma 4.20 *Suppose that the simplicial presheaf map $f : X \to Y$ is a local trivial fibration. Then f is a local fibration and a local weak equivalence.*

Proof The local fibration part of the claim is easy, since the map f has the right lifting property with respect to all inclusions of finite simplicial sets.

The induced map
$$f : \mathrm{Ex}(X) \to \mathrm{Ex}(Y)$$
has the local right lifting property with respect to all $\partial \Delta^n \subset \Delta^n$, since f has the local right lifting property with respect to all $\mathrm{sd}\,\partial \Delta^n \to \mathrm{sd}\,\Delta^n$. It follows that the map
$$f : \mathrm{Ex}^\infty(X) \to \mathrm{Ex}^\infty(Y)$$
has the local right lifting property with respect to all $\partial \Delta^n \subset \Delta^n$ and is a map of presheaves of Kan complexes. Finish by using Lemmas 4.2 and 4.18.

Corollary 4.21 *The maps $\nu : X \to LX$ and $\eta : X \to L^2 X$ are local fibrations and local weak equivalences.*

Proof This result is a consequence of Corollary 4.16 and Lemma 4.20.

Corollary 4.22

1) A map $f : X \to Y$ of simplicial presheaves is a local weak equivalence if and only if the induced map $f_* : LX \to LY$ is a local weak equivalence.
2) A map $f : X \to Y$ of simplicial presheaves is a local weak equivalence if and only if the induced map $f_* : \tilde{X} \to \tilde{Y}$ of associated sheaves is a local weak equivalence.
3) A map $f : X \to Y$ of simplicial presheaves is a local weak equivalence if and only if the induced map $f_* : L^2 \mathrm{Ex}^\infty(X) \to L^2 \mathrm{Ex}^\infty(Y)$ is a local weak equivalence.

Proof For statement 1), the map $\eta : X \to LX$ is a local weak equivalence, so that the induced diagram of sheaves

$$\begin{array}{ccc} \tilde{\pi}_n X & \longrightarrow & \tilde{\pi}_n LX \\ \downarrow & & \downarrow \\ \tilde{X}_0 & \longrightarrow & \widetilde{LX}_0 \end{array}$$

is a pullback. The map $\tilde{X}_0 \to \widetilde{LX}_0$ is an isomorphism, so that both horizontal arrows in the diagram are isomorphisms of sheaves. Finish with the argument for Lemma 4.2.

For statement 2), recall that $\tilde{X} = L^2 X$, and use statement 1). Statement 3) is a consequence of statement 2) and Lemma 4.2.

4.3 First Applications of Boolean Localization

The local weak equivalence and local fibration concepts for simplicial presheaves have very special interpretations for simplicial sheaves on a complete Boolean algebra $\mathcal{B}$.

The key point is that the sheaf category $\mathbf{Shv}(\mathcal{B})$ satisfies the Axiom of Choice (Lemma 3.29), which says that every epimorphism $F \to F'$ of such sheaves has a section. In particular, every sheaf epimorphism must be a sectionwise epimorphism.

It follows that a map $F \to F'$ of sheaves on $\mathcal{B}$ is an epimorphism of sheaves if and only if it is a sectionwise epimorphism. We use this observation somewhat relentlessly.

Lemma 4.23 *Suppose that $\mathcal{B}$ is a complete Boolean algebra.*

1) *A map $p : X \to Y$ of simplicial sheaves on $\mathcal{B}$ is a local (respectively local trivial) fibration if and only if all maps $p : X(b) \to Y(b)$ are Kan fibrations (respectively trivial Kan fibrations) for all $b \in \mathcal{B}$.*
2) *A map $f : X \to Y$ of locally fibrant simplicial sheaves on $\mathcal{B}$ is a local weak equivalence if and only if all maps $f : X(b) \to Y(b)$ are weak equivalences of simplicial sets for all $b \in \mathcal{B}$.*

Proof The induced map

$$X^{\Delta^n} \to Y^{\Delta^n} \times_{Y^{\partial\Delta^n}} X^{\partial\Delta^n}$$

is a sheaf epimorphism in degree 0 if and only if it is a sectionwise epimorphism in degree 0, since the sheaf category $\mathbf{Shv}(\mathcal{B})$ satisfies the Axiom of Choice. The local fibration statement is similar.

For part 2), suppose that f is a local weak equivalence. The map f has a factorization

$$\begin{array}{ccc} X & \xrightarrow{j} & X \times_Y Y^{\Delta^1} \\ & {\scriptstyle f} \searrow & \downarrow p \\ & & Y \end{array}$$

where p is a sectionwise Kan fibration and j is right inverse to a sectionwise trivial Kan fibration, by a standard construction (see also Sect. 6.1 below). All objects in the diagram are sheaves of Kan complexes. The map p is a local weak equivalence and a local fibration, and is therefore a sectionwise weak equivalence by Lemma 4.18 and part 1). It follows that f is a sectionwise weak equivalence.

The converse follows from Lemma 4.1.

Recall that a *Boolean localization* (or Boolean cover) is a geometric morphism

$$p : \mathbf{Shv}(\mathcal{B}) \to \mathbf{Shv}(\mathcal{C})$$

such that $\mathcal{B}$ is a complete Boolean algebra, and such that the inverse image functor
$$p^* : \mathbf{Shv}(\mathcal{C}) \to \mathbf{Shv}(\mathcal{B})$$
is faithful. Recall also that Barr's theorem (Theorem 3.31) says that Boolean localizations exist for every Grothendieck topos $\mathbf{Shv}(\mathcal{C})$.

Lemma 4.24 *Suppose that the geometric morphism $p : \mathbf{Shv}(\mathcal{B}) \to \mathbf{Shv}(\mathcal{C})$ is a Boolean localization. A map $f : X \to Y$ of simplicial sheaves on $\mathcal{C}$ is a local trivial fibration (respectively local fibration) if and only if the induced map*
$$p^* f : p^* X \to p^* Y$$
is a sectionwise trivial Kan fibration (respectively sectionwise Kan fibration) in $s\mathbf{Shv}(\mathcal{B})$.

Proof The simplicial sheaf map
$$X^{\Delta^n} \to X^{\partial \Delta^n} \times_{Y^{\partial \Delta^n}} Y^{\Delta^n}$$
is a sheaf epimorphism in degree zero if and only if the induced map
$$p^* X^{\Delta^n} \to p^* X^{\partial \Delta^n} \times_{p^* Y^{\partial \Delta^n}} p^* Y^{\Delta^n}$$
is a sheaf epimorphism in degree 0, by Lemma 3.26. Now use Lemma 4.23.

The argument for the local fibration statement is similar.

Lemma 4.25 *Suppose that the map $f : X \to Y$ of simplicial presheaves on $\mathcal{C}$ is a local fibration (respectively local trivial fibration). Then the induced map $f_* : \mathrm{Ex}^\infty X \to \mathrm{Ex}^\infty Y$ is a local fibration (respectively local trivial fibration).*

Proof We have already seen a proof of the local trivial fibration statement in the proof of Lemma 4.20.

In more detail, if $f : X \to Y$ is a local trivial fibration, then it has the local right lifting property with respect to all inclusions $\mathrm{sd}\,(\partial \Delta^n) \to \mathrm{sd}\,(\Delta^n)$, and so the induced map $f_* : \mathrm{Ex}\,(X) \to \mathrm{Ex}\,(Y)$ is a local trivial fibration. It follows inductively that all maps $f_* : \mathrm{Ex}^n\,(X) \to \mathrm{Ex}^n\,(Y)$ are local trivial fibrations, and so the map $f_* : \mathrm{Ex}^\infty(X) \to \mathrm{Ex}^\infty(Y)$ of filtered colimits is a local trivial fibration.

For the local fibration statement, suppose that $p : \mathbf{Shv}(\mathcal{B}) \to \mathbf{Shv}(\mathcal{C})$ is a Boolean localization.

The map f_* is a local fibration of simplicial presheaves if and only if the induced map $p^* L^2(f_*) : p^* L^2 \mathrm{Ex}^\infty X \to p^* L^2 \mathrm{Ex}^\infty Y$ is a local fibration of simplicial sheaves on $\mathcal{B}$, by Corollary 4.12 and Lemma 4.24. This last map coincides up to isomorphism with the map
$$L^2 \mathrm{Ex}^\infty p^* \tilde{X} \to L^2 \mathrm{Ex}^\infty p^* \tilde{Y}. \tag{4.9}$$

The map $\tilde{f} : \tilde{X} \to \tilde{Y}$ of associated simplicial sheaves on $\mathcal{C}$ is a local fibration, as is the induced map $p^* \tilde{f} : p^* \tilde{X} \to p^* \tilde{Y}$ of simplicial sheaves on $\mathcal{B}$. This map $p^* \tilde{f}$ is a sectionwise Kan fibration, as is the induced simplicial presheaf map
$$\mathrm{Ex}^\infty p^* \tilde{f} : \mathrm{Ex}^\infty p^* \tilde{X} \to \mathrm{Ex}^\infty p^* \tilde{Y},$$

4.3 First Applications of Boolean Localization

since the Ex^∞ functor preserves Kan fibrations of simplicial sets [32, III.4.5]. But then the map (4.9) of associated simplicial sheaves on $\mathcal{B}$ is a local fibration, as required.

Remark 4.26 One can give an alternate proof of the local fibration statement in Lemma 4.25, by using the fact that the subdivision functor takes the inclusions $\Lambda^n_k \subset \Delta^n$ to inclusions $\mathrm{sd}\,(\Lambda^n_k) \subset \mathrm{sd}\,(\Delta^n)$, which are anodyne extensions in the sense that they occur as finite iterated pushouts of maps of the form $\Lambda^r_j \subset \Delta^r$. This is proved in [62] and [28].

Every local fibration $f : X \to Y$ therefore has the local right lifting property with respect to all maps $\mathrm{sd}\,(\Lambda^n_k) \subset \mathrm{sd}\,(\Delta^n)$, and thus induces a local fibration $f_* : \mathrm{Ex}\,(X) \to \mathrm{Ex}\,(Y)$.

Lemma 4.27 *Suppose that $p : \mathbf{Shv}(\mathcal{B}) \to \mathbf{Shv}(\mathcal{C})$ is a Boolean localization. Then the simplicial sheaf map $f : X \to Y$ is a local weak equivalence if and only if the map*

$$f_* : p^*X \to p^*Y$$

is a local weak equivalence of $s\mathbf{Shv}(\mathcal{B})$.

Proof The map $f : X \to Y$ is a local weak equivalence if and only if the induced map $f_* : L^2 \mathrm{Ex}^\infty(X) \to L^2 \mathrm{Ex}^\infty(Y)$ is a local weak equivalence.

The simplicial presheaf map $\mathrm{Ex}^\infty(X) \to \mathrm{Ex}^\infty(Y)$ has a factorization

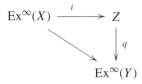

such that q is a sectionwise Kan fibration and the map i is a section of a sectionwise trivial Kan fibration $Z \to \mathrm{Ex}^\infty(X)$, by Proposition 2.22. There is an induced diagram

$$
\begin{array}{ccc}
p^*L^2\,\mathrm{Ex}^\infty(X) & \xrightarrow{i_*} & p^*L^2 Z \\
& \searrow{\scriptstyle f_*} & \downarrow{\scriptstyle q_*} \\
& & p^*L^2\,\mathrm{Ex}^\infty(Y)
\end{array}
$$

in simplicial sheaves on $\mathcal{B}$, in which the map i_* is a local weak equivalence and the map q_* is a local fibration.

One uses the natural isomorphism

$$p^*L^2\,\mathrm{Ex}^\infty X \cong L^2\,\mathrm{Ex}^\infty p^*X$$

to show that the map $p^*X \to p^*Y$ is a local weak equivalence of $s\mathbf{Shv}(\mathcal{B})$ if and only if the map $f_* : p^*L^2\,\mathrm{Ex}^\infty X \to p^*L^2\,\mathrm{Ex}^\infty Y$ is a local weak equivalence. This map f_* is a local weak equivalence if and only if the map q_* is a local trivial fibration, and this is so if and only if the map q is a local trivial fibration of simplicial sheaves on $\mathcal{C}$ by Lemma 4.24.

The following result is now a corollary of Lemma 4.27:

Proposition 4.28 *Suppose that* $p : \mathbf{Shv}(\mathcal{B}) \to \mathbf{Shv}(\mathcal{C})$ *is a Boolean localization. Then the simplicial presheaf map* $f : X \to Y$ *is a local weak equivalence if and only if the map*

$$f_* : p^*\tilde{X} \to p^*\tilde{Y}$$

is a local weak equivalence of $s\mathbf{Shv}(\mathcal{B})$.

Proof The associated sheaf functor preserves and reflects local weak equivalences by Corollary 4.16, and the inverse image functor p^* preserves and reflects local weak equivalences of simplicial sheaves by Lemma 4.25.

The following result is a further consequence of this line of argument:

Corollary 4.29 *Suppose that* $p : \mathbf{Shv}(\mathcal{B}) \to \mathbf{Shv}(\mathcal{C})$ *is a Boolean localization. Then a map* $f : X \to Y$ *of simplicial presheaves on* $\mathcal{C}$ *is a local weak equivalence if and only if the induced map*

$$f_* : p^*L^2 \operatorname{Ex}^\infty(X) \to p^*L^2 \operatorname{Ex}^\infty(Y)$$

is a sectionwise equivalence of simplicial sheaves on $\mathcal{B}$.

Proof The map f is a local weak equivalence if and only if the map f_* is a local weak equivalence, by Lemma 4.2, Corollary 4.12 and Lemma 4.25. The simplicial sheaves $p^*L^2 \operatorname{Ex}^\infty(X)$ and $p^*L^2 \operatorname{Ex}^\infty(Y)$ are locally fibrant by Corollary 4.12. Finish the proof by using Lemma 4.23.

We can now prove the two out of three axiom for local weak equivalences.

Lemma 4.30 *Suppose given a commutative diagram of simplicial presheaf maps*

$$\begin{array}{ccc} X & \xrightarrow{f} & Y \\ & \searrow_{h} & \downarrow_{g} \\ & & Z \end{array} \quad (4.10)$$

on a Grothendieck site $\mathcal{C}$. *If any two of* f, g *or* h *are local weak equivalences then so is the third.*

Proof Suppose that $p : \mathbf{Shv}(\mathcal{B}) \to \mathbf{Shv}(\mathcal{C})$ is a Boolean localization. Then a simplicial presheaf map $f : X \to Y$ is a local weak equivalence if and only if the induced map

$$f_* : p^*L^2 \operatorname{Ex}^\infty X \to p^*L^2 \operatorname{Ex}^\infty Y$$

4.3 First Applications of Boolean Localization

is a sectionwise equivalence of sheaves of Kan complexes on $\mathcal{B}$ by Corollary 4.29. Apply the functor $p^*L^2\operatorname{Ex}^\infty$ to the triangle (4.10) to prove the result.

Generally, if U is an object of a Grothendieck site $\mathcal{C}$, then the left adjoint L_U of the U-sections functor $X \mapsto X(U)$ can be defined for simplicial sets K by

$$L_U(K) = K \times \hom(\ ,U).$$

It is an exercise to show that L_U preserves cofibrations, takes weak equivalences (respectively fibrations) to sectionwise weak equivalences (respectively sectionwise fibrations).

Lemma 4.31 *Suppose that $\mathcal{B}$ is a complete Boolean algebra. Suppose that the map $p : X \to Y$ is a local fibration and a local weak equivalence of simplicial sheaves on $\mathcal{B}$. Then p is a sectionwise trivial fibration.*

Proof The map p is a sectionwise Kan fibration by Lemma 4.23.

The functor $X \mapsto L^2\operatorname{Ex}^\infty X$ preserves sectionwise (or local) fibrations of simplicial sheaves on $\mathcal{B}$ (Corollary 4.12), and it preserves pullbacks. A sectionwise fibration $p : X \to Y$ is local weak equivalence if and only if the induced map $p_* : L^2\operatorname{Ex}^\infty X \to L^2\operatorname{Ex}^\infty Y$ is a sectionwise weak equivalence, by Lemma 4.23. It follows that the family of all maps of simplicial sheaves on $\mathcal{B}$, which are simultaneously local fibrations and local weak equivalences, is closed under pullback.

Suppose given a diagram

$$\begin{array}{ccc} \partial\Delta^n & \xrightarrow{\alpha} & X(b) \\ {\scriptstyle i}\downarrow & & \downarrow{\scriptstyle p} \\ \Delta^n & \xrightarrow{\beta} & Y(b) \end{array}$$

The simplex Δ^n contracts onto the vertex 0; write $h : \Delta^n \times \Delta^1 \to \Delta^n$ for the contracting homotopy. Let $h' : \partial\Delta^n \times \Delta^1 \to X(b)$ be a choice of lifting in the diagram

$$\begin{array}{ccc} \partial\Delta^n & \xrightarrow{\alpha} & X(b) \\ \downarrow & \nearrow{\scriptstyle h'} & \downarrow{\scriptstyle p} \\ \partial\Delta^n \times \Delta^1 & \xrightarrow[\beta\cdot h\cdot(i\times 1)]{} & Y(b) \end{array}$$

Then the original diagram is homotopic to a diagram of the form

$$\begin{array}{ccc} \partial\Delta^n & \xrightarrow{\alpha'} & X(b) \\ {\scriptstyle i}\downarrow & & \downarrow{\scriptstyle p} \\ \Delta^n & \xrightarrow{x} & Y(b) \end{array}$$

where $x : \Delta^n \to Y(b)$ is constant at the vertex $x \in Y(b)$. Consider the induced diagram

$$\begin{array}{ccc} \partial \Delta^n & \longrightarrow & (L_b\Delta^0 \times_Y X)(b) \\ {\scriptstyle i} \downarrow & \nearrow & \downarrow {\scriptstyle p_*} \\ \Delta^n & \longrightarrow & L_b\Delta^0(b), \end{array}$$

where L_b is the left adjoint of the b-sections functor $X \mapsto X(b)$ in sheaves. The object $L_b\Delta^0$ is the sheaf associated to a diagram of points and is therefore locally fibrant, and is thus a sheaf of Kan complexes. The map $p_* : L_b\Delta^0 \times_Y X \to L_b\Delta^0$ is a local fibration and a local weak equivalence between sheaves of Kan complexes and is therefore a sectionwise trivial fibration by Lemma 4.24, so the indicated lift exists.

We can now prove the following result. It is essentially a corollary of Lemma 4.31.

Theorem 4.32 *A map $q : X \to Y$ of simplicial presheaves is a local weak equivalence and a local fibration if and only if it has the local right lifting property with respect to all inclusions $\partial \Delta^n \subset \Delta^n$, $n \geq 0$.*

To paraphrase, this result says that a map is a local fibration and a local weak equivalence if and only if it is a local trivial fibration.

Proof If q has the local right lifting property with respect to all $\partial \Delta^n \subset \Delta^n$ then q is a local fibration and a local weak equivalence by Lemma 4.20. We prove the converse statement here.

Suppose that the map q is a local weak equivalence and a local fibration. Suppose that $p : \mathbf{Shv}(\mathcal{B}) \to \mathbf{Shv}(\mathcal{C})$ is a Boolean localization. Then p^*L^2q is a local weak equivalence by Proposition 4.28, and is a local fibration by Corollary 4.12, and is therefore a sectionwise trivial fibration by Lemma 4.31. The functor p^*L^2 reflects local epimorphisms by Lemma 3.16 and 3.26, so that the map

$$X^{\Delta^n} \to Y^{\Delta^n} \times_{Y^{\partial \Delta^n}} X^{\partial \Delta^n}$$

is a local epimorphism in degree 0.

Suppose that X is a locally fibrant simplicial presheaf, and form the standard diagram

$$\begin{array}{ccc} & & X^I \\ & {\scriptstyle s} \nearrow & \downarrow {\scriptstyle (d_0,d_1)} \\ X & \xrightarrow{\Delta} & X \times X, \end{array} \quad (4.11)$$

where $X^I = \mathbf{hom}(\Delta^1, X)$ is the path space function complex, which is defined in sections by

$$\mathbf{hom}(\Delta^1, X)(U) = \mathbf{hom}(\Delta^1, X(U)).$$

4.3 First Applications of Boolean Localization

In other words, $X^I(U)$ is the ordinary unbased path space $X(U)^I$. The maps d_0 and d_1 are defined by restriction to end points, and s is the constant path map. It is an exercise to show that the map (d_0, d_1) is a local fibration and the maps d_i are local trivial fibrations, since X is locally fibrant. It follows from Lemma 4.20 that the maps d_i and s are local weak equivalences. The simplicial presheaf X^I is also locally fibrant.

Let $s\mathbf{Pre}(\mathcal{C})_f$ denote the full subcategory of the simplicial presheaf category $s\mathbf{Pre}(\mathcal{C})$, whose objects are the locally fibrant simplicial presheaves.

Proposition 4.33 *The category $s\mathbf{Pre}(\mathcal{C})_f$ of locally fibrant simplicial presheaves, with the classes of local weak equivalences and local fibrations within that category, together satisfy the following:*

A) Given a commutative diagram

of morphisms in $s\mathbf{Pre}(\mathcal{C})_f$, if any two of the maps f, g or h is a local weak equivalence, then so is the third.

B) Local fibrations are closed under composition. Any isomorphism is a local fibration.

C) The classes of local fibrations and of those maps that are simultaneously local fibrations and local weak equivalences are closed under pullback.

D) For every object X of $s\mathbf{Pre}(\mathcal{C})_f$, there is a factoriztion

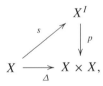

where Δ is the diagonal map, s is a local weak equivalence and p is a local fibration.

E) For all objects X of $s\mathbf{Pre}(\mathcal{C})_f$, the map $X \to *$ is a local fibration.

Proposition 4.33 says that the category of locally fibrant objects, and local fibrations and local weak equivalences in that category satisfies the axioms for a *category of fibrant objects* in the sense of K. Brown's thesis [16].

Proof [Proof of Proposition 4.33] Statement A) follows from Lemma 4.30. Statement B) follows from Lemma 4.7. The requisite construction for statement C) is given in the diagram (4.11) above. Statement E) is a tautology.

We know from Theorem 4.32 that a map of simplicial presheaves is a local fibration and a local weak equivalence if and only if it has the right lifting property with respect to all inclusions $\partial\Delta^n \subset \Delta^n$, $n \geq 0$. Statement D) therefore follows from Lemma 4.7.

Let $s\mathbf{Shv}(\mathcal{C})_f$ be the full subcategory of the category of simplicial sheaves, whose objects are the locally fibrant simplicial sheaves.

Corollary 4.34 *The category $s\mathbf{Shv}(\mathcal{C})_f$ of locally fibrant simplicial sheaves and the classes of local weak equivalences and local fibrations within that category, together satisfy the axioms for a category of fibrant objects.*

Proof All statements for Proposition 4.33 restrict to simplicial sheaves. For statement D), observe that the path object $\mathbf{hom}(\Delta^1, X)$ is a simplicial sheaf if X is a simplicial sheaf.

Corollary 4.34 first appeared in [49], with a very different proof. This was the first published method of constructing a local homotopy theory within an arbitrary Grothendieck topos.

Say that a simplicial presheaf map $i : A \to B$ is a *cofibration* if it is a monomorphism in all sections. It follows from the proof of Proposition 2.22 that every projective cofibration is a cofibration.

Lemma 4.35 *Suppose given a pushout diagram*

$$\begin{array}{ccc} A & \xrightarrow{\alpha} & C \\ {\scriptstyle i}\downarrow & & \downarrow{\scriptstyle i_*} \\ B & \xrightarrow[\alpha_*]{} & D \end{array} \qquad (4.12)$$

of simplicial sheaves on a complete Boolean algebra $\mathcal{B}$ such that i is a cofibration and a local weak equivalence. Then the map i_* is a cofibration and

1) the map i_* is a local weak equivalence if i is a local weak equivalence, and
2) the map α_* is a local weak equivalence if α is a local weak equivalence.

Proof Use the Ex^∞ functor and write A' for $\mathrm{Ex}^\infty(A)$ to form a diagram

$$\begin{array}{ccccc} B & \xleftarrow{i} & A & \xrightarrow{\alpha} & C \\ {\scriptstyle \simeq}\downarrow & & {\scriptstyle \simeq}\downarrow & & \downarrow{\scriptstyle \simeq} \\ B' & \xleftarrow[i']{} & A' & \xrightarrow[\alpha']{} & C' \end{array}$$

in simplicial presheaves on $\mathcal{B}$, in which the vertical maps are sectionwise weak equivalences, i' is a cofibration, and the objects A', B' and C' are sectionwise fibrant.

4.3 First Applications of Boolean Localization

Form the pushout

$$\begin{array}{ccc} A' & \xrightarrow{\alpha'} & C' \\ {\scriptstyle i'}\downarrow & & \downarrow{\scriptstyle i'_*} \\ B' & \xrightarrow{\alpha'_*} & D' \end{array}$$

all in the simplicial presheaf category on $\mathcal{B}$. The map i' is a cofibration since the Ex^∞ functor preserves cofibrations, so the induced map $D \to D'$ is a sectionwise weak equivalence by left properness of the standard model structure for simplicial sets. It follows that

1) the map i_* is a local weak equivalence if and only if i'_* is a local weak equivalence, and
2) the map α_* is a local weak equivalence if and only if α'_* is a local weak equivalence.

Sheafifying gives a pushout diagram of simplicial sheaves

$$\begin{array}{ccc} \tilde{A}' & \xrightarrow{\tilde{\alpha}'} & \tilde{C}' \\ {\scriptstyle \tilde{i}'}\downarrow & & \downarrow \\ \tilde{B}' & \longrightarrow & \tilde{D}', \end{array}$$

which is locally equivalent to the original, and for which $\tilde{i}'$ is a cofibration. We can therefore assume that the objects A, B and C in the diagram (4.12) are locally fibrant simplicial sheaves.

If the map i is a local weak equivalence, it is a sectionwise equivalence by Lemma 4.23. Sectionwise trivial cofibrations are closed under pushout in the simplicial presheaf category, and since D is the associated sheaf of the presheaf pushout, the map $i_* : C \to D$ must then be a local weak equivalence by Lemma 4.30.

Similarly, if α is a local weak equivalence, then it is a sectionwise equivalence, and so the map α_* must be a sectionwise weak equivalence.

Corollary 4.36 *Suppose given a pushout diagram*

$$\begin{array}{ccc} A & \xrightarrow{\alpha} & C \\ {\scriptstyle i}\downarrow & & \downarrow{\scriptstyle i_*} \\ B & \xrightarrow{\alpha_*} & D \end{array}$$

of simplicial presheaves on a Grothendieck site $\mathcal{C}$, and suppose that i is a cofibration. If the map i is a local weak equivalence then i_* is a local weak equivalence. Also, if α is a local weak equivalence then α_* is a local weak equivalence.

Proof Suppose that $p : \mathbf{Shv}(\mathcal{B}) \to \mathbf{Shv}(\mathcal{C})$ is a Boolean localization. The functor p^*L^2 preserves cofibrations and pushouts, and preserves and reflects local weak equivalences by Proposition 4.28.

The map $p^*\tilde{A} \to p^*\tilde{B}$ induced by i is a local weak equivalence and a cofibration, so the map $p^*\tilde{C} \to p^*\tilde{D}$ induced by i_* is a local weak equivalence by Lemma 4.35. It follows that the map i_* is a local weak equivalence.

One uses the same argument to show that α_* is a local weak equivalence if α is a local weak equivalence.

Corollary 4.36 incorporates a left properness assertion for local weak equivalences and cofibrations of simplicial presheaves. The following result is right properness for local fibrations and local weak equivalences.

Lemma 4.37 *Suppose given a pullback diagram*

$$\begin{array}{ccc} W & \xrightarrow{g_*} & X \\ \downarrow & & \downarrow p \\ Z & \xrightarrow{g} & Y \end{array}$$

in $s\mathbf{Pre}(\mathcal{C})$ such that p is a local fibration and g is a local weak equivalence. Then the map g_* is a local weak equivalence.

Proof One can either use a direct Boolean localization argument to prove this result (this is an exercise), or argue as follows.

Kan's functor $X \mapsto \mathrm{Ex}^\infty(X)$ takes values in the category of locally fibrant simplicial presheaves. It preserves local fibrations (Lemma 4.25) and pullbacks. It preserves and reflects local weak equivalences by Lemma 4.2.

We can therefore assume that the diagram consists of locally fibrant simplicial presheaves, and hence in a category of fibrant objects by Proposition 4.33. The map g_* is then a local weak equivalence by a formal property of categories of fibrant objects—see, for example, [32, II.8.5].

Corollary 4.38 *Suppose given a commutative diagram*

$$\begin{array}{ccccc} Z & \longrightarrow & Y & \xleftarrow{p} & X \\ {\scriptstyle f_Z}\downarrow & & {\scriptstyle f_Y}\downarrow & & \downarrow {\scriptstyle f_X} \\ Z' & \longrightarrow & Y' & \xleftarrow[p']{} & X' \end{array} \qquad (4.13)$$

of simplicial presheaves on a site $\mathcal{C}$ such that the maps p and p' are local fibrations, and such that f_X, f_Y and f_Z are local weak equivalences. Then the induced map

$$Z \times_Y X \to Z' \times_{Y'} X'$$

is a local weak equivalence.

4.3 First Applications of Boolean Localization

Proof The right properness assertion for local fibrations of Lemma 4.35 implies this result in a standard way—see [32, II.8.8].

Alternatively, we can assume that the diagram (4.13) lives in simplicial sheaves on $\mathcal{B}$, and that all objects in the diagram are locally fibrant. This is done in the usual way, by applying the functor $X \mapsto p^* L^2 \operatorname{Ex}^\infty X$, where $p : \mathbf{Shv}(\mathcal{B}) \to \mathbf{Shv}(\mathcal{C})$ is a Boolean localization. We also use Lemma 4.25.

But then the maps f_X, f_Y and f_Z are sectionwise weak equivalences and the maps p and p' are sectionwise Kan fibrations, so the result follows from the coglueing lemma for simplicial sets [32, II.8.10].

The following result is a strong form of right properness for simplicial presheaves fibred over presheaves which is useful in applications.

Lemma 4.39 *Suppose given a commutative diagram of simplicial presheaf maps*

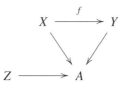

such that A is simplicially discrete and f is a local weak equivalence. Then the induced map

$$f_* : Z \times_A X \to Z \times_A Y$$

is a local weak equivalence.

Proof The result holds in the category of simplicial sets. In this case, all objects are coproducts of fibres over $x \in A$, so that $X = \bigsqcup_{x \in A} X_x$ for example. Then $f : X \to Y$ is a weak equivalence if and only if all maps of fibres $X_x \to Y_x$ are weak equivalences. Then all maps $Z_x \times X_x \to Z_x \times Y_x$ are weak equivalences, so that the map $f_* : Z \times_A X \to Z \times_A Y$ is a weak equivalence.

More generally, form a diagram

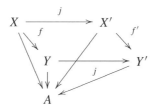

such that the maps j are sectionwise equivalences and the maps $X' \to A$ and $Y' \to A$ are sectionwise Kan fibrations. This can be done in the projective model structure for simplicial presheaves.

It suffices, by the observation about simplicial sets in the first paragraph, to show that the map $f'_* : Z \times_A X' \to Z \times_A Y'$ is a local weak equivalence.

The maps $X' \to A$ and and $Y' \to A$ are local fibrations and the map f' is a local weak equivalence. It follows from Lemma 4.38 that the induced map f'_* is a local weak equivalence.

Corollary 4.40 *Suppose that $f : X \to Y$ is a local weak equivalence of simplicial presheaves and that Z is a simplicial presheaf. Then the map*

$$f \times 1 : X \times Z \to Y \times Z$$

is a local weak equivalence.

The following is an enriched version of the simplicial model axiom **SM7** for simplicial presheaves:

Corollary 4.41 *Suppose that $i : A \to B$ and $j : C \to D$ are cofibrations of simplicial presheaves. Then the induced cofibration*

$$(i, j) : (B \times C) \cup (A \times D) \to B \times D$$

is a local weak equivalence if either i or j is a local weak equivalence.

Proof The map (i, j) is plainly a cofibration.

Suppose that the map $i : A \to B$ is a local weak equivalence. Then the diagram

$$\begin{array}{ccc} A \times D & & \\ {\scriptstyle (i \times 1)_*} \downarrow & \searrow^{i \times 1} & \\ (B \times C) \cup (A \times D) & \xrightarrow[(i,j)]{} & B \times D, \end{array}$$

the maps $i \times 1$ and $(i \times 1)_*$ are trivial cofibrations, so (i, j) is a local weak equivalence.

One could, alternatively, prove Corollary 4.41 directly with a Boolean localization argument.

We close this chapter with a first result about local weak equivalences between homotopy colimits.

Lemma 4.42

1) *Suppose that the simplicial presheaf maps $X_i \to Y_i$, $i \in I$, are local weak equivalences, where I is a set. Then the induced map*

$$\bigsqcup_{i \in I} X_i \to \bigsqcup_{i \in I} Y_i$$

is a local weak equivalence.

2) *Suppose that $f : X \to Y$ is a natural transformation of J-diagrams of simplicial presheaves, where J is a right filtered category. Then the induced map*

$$\varinjlim_{j \in J} X_j \to \varinjlim_{j \in J} Y_j$$

4.3 First Applications of Boolean Localization

of filtered colimits is a local weak equivalence of simplicial presheaves.

3) Suppose that

$$X_0 \to X_1 \to \cdots \to X_s \to X_{s+1} \to \cdots$$

is an inductive system of simplicial presheaf maps, indexed by $s < \gamma$, where γ is some limit ordinal. Suppose that all maps $X_s \to X_{s+1}$ and the maps

$$\varinjlim_{s<t} X_s \to X_t$$

corresponding to limit ordinals $t < \gamma$ are local weak equivalences. Then the map

$$X_0 \to \varinjlim_{s<\gamma} X_s$$

is a local weak equivalence.

Proof For all statements, by applying a functor $p^* L^2 \operatorname{Ex}^\infty$ for some Boolean localization p, it is enough to assume that all objects are locally fibrant simplicial sheaves on a complete Boolean algebra $\mathcal{B}$. But then the weak equivalences are sectionwise weak equivalences, and the statements follow from the corresponding results for simplicial sets.

Chapter 5
Local Model Structures

This chapter presents constructions of the basic model structures for simplicial presheaves and simplicial sheaves. These include the injective model structures for both categories—these structures, and the Quillen equivalence between them, are the subject of Sect. 5.1. Precise statements appear in Theorems 5.8 and 5.9.

The construction of the injective model structures depends on the results for local weak equivalences and local fibrations from Chap. 3, and uses Boolean localization techniques. The other main technical device that is used is a bounded monomorphism statement, which appears in Lemma 5.2. Variants of the bounded monomorphism concept appear repeatedly in subsequent chapters.

The injective model structure for simplicial sheaves first appeared in a letter of Joyal to Grothendieck [70], while the injective model structure for simplicial presheaves had its first appearance in [50]. There are now many related structures with the same (local) weak equivalences: the local projective structure was introduced by Blander in [9], and the intermediate model structures appeared in [64].

The basic properties of injective fibrations and injective fibrant objects are discussed in Sect. 5.2. Injective fibrant objects, meaning objects which are acyclic and fibrant for the injective model structure, behave like injective objects in an abelian category, hence the name.

The overall message of Sect. 5.3 is that a geometric morphism of Grothendieck sites induces a Quillen adjunction between the injective model structures on the respective categories of simplicial presheaves. This means, in particular, that direct image functors preserve injective fibrations—this observation leads to calculations in the stye of the Leray spectral sequence, as in [52].

Section 5.3 closes with a discussion of restrictions to sites fibred over a simplicial object, such as one encounters when discussing the cohomology of a simplicial scheme. One outcome is that the cohomology of a simplicial scheme X, for whatever topology, can be calculated either on the fibred site, as is traditional, or by interpreting X as a representable simplicial presheaf on some bigger site. The homotopy theoretic approach to cohomology calculations is discussed later in Chap. 8.

An injective fibrant model of a simplicial presheaf X is a local weak equivalence $X \to Z$ with Z injective fibrant. Such morphisms exist for all objects X by a standard argument, but can be mysterious. At the same time, one can start to calculate the homotopy groups in the sections of an injective fibrant model Z with sheaf cohomology (see Chap. 8), starting from the sheaves of homotopy groups of the original object X. Thus, the closer that an object X is to being injective fibrant, the more tools we have to compute its homotopy groups in sections.

Most coarsely, we say that a simplicial presheaf X *satisfies descent* if we can find a local weak equivalence $f : X \to Z$ with Z injective fibrant, such that all induced maps $X(U) \to Z(U)$ in sections are weak equivalences of simplicial sets.

The descent concept first arose in discussions of patching data for sheaves and sheaf cohomology theory. The homotopy theoretic variant that is discussed here, which is a great strengthening of the cohomological concept, had its origins in the study of some of the leading open questions of algebraic K-theory of the 1970s and 1980s.

Variants of this concept are described in Sect. 5.4, and we prove two of the main descent theorems from algebraic K-theory as examples, namely, the Brown–Gersten descent theorem for the Zariski topology (Theorem 5.33) and the Morel–Voevodsky descent theorem for the Nisnevich topology (Theorem 5.37). The arguments for both results are quite geometric, and one generally expects that the proof of a descent statement will be anything but purely homotopy theoretic. Descent theorems of this kind are rare and important, and typically have interesting proofs.

Injective model structures for presheaf objects are the traditional platform for homotopy theoretic descent questions and theorems. In particular, these model structures and their descent theories form the basis for the description of the theory of stacks which appears in Chap. 9.

The local projective structure is constructed in Sect. 5.5, together with the intermediate model structures for simplicial presheaves. All of these variants of the injective structure have the same weak equivalences, namely, the local weak equivalences, and the respective model structures are found by varying the class of cofibrations. The injective structure has the largest class of cofibrations, namely, all monomorphisms, while the local projective structure has the smallest class: these are the monomorphisms which are projective cofibrations, as in Proposition 2.22. The intermediate structures have classes of cofibrations which lie somewhere between the projective cofibrations and the class of all monomorphisms, hence the name.

One says that a simplicial presheaf X is an n-type if the homotopy group sheaves $\tilde{\pi}_k X$ are trivial for $k > n$. In particular, X has only finitely many nontrivial sheaves of homotopy groups. Such objects can be extremely important for the study of descent questions, as in [56] and [99].

A model theoretic approach to the study of n-types of simplicial presheaves is presented in Sect. 5.6, in which derived Postnikov sections $\mathbf{P}_n X$ of simplicial presheaves X are used to describe weak equivalences, here called n-equivalences. The resulting model structure appears in Theorem 5.49. The proof of this result, which is due to Biedermann [8], is an interesting variant of the Bousfield–Friedlander construction of classical stable homotopy theory [13].

5.1 The Injective Model Structure

We begin by reviewing the fibration replacement construction from classical simplicial homotopy theory.

Suppose that $f : X \to Y$ is a map of Kan complexes, and form the diagram

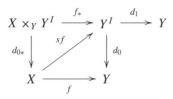

in which the square is a pullback. Then d_0 is a trivial fibration since Y is a Kan complex, so d_{0*} is a trivial fibration. The section s of d_0 (and d_1) induces a section s_* of d_{0*}, and

$$(d_1 f_*)s_* = d_1(sf) = f$$

Finally, there is a pullback diagram

$$\begin{array}{ccc} X \times_Y Y^I & \xrightarrow{f_*} & Y^I \\ {\scriptstyle (d_{0*}, d_1 f_*)} \downarrow & & \downarrow {\scriptstyle (d_0, d_1)} \\ X \times Y & \xrightarrow{f \times 1} & Y \times Y \end{array}$$

and the projection map $pr_R : X \times Y \to Y$ is a fibration since X is a Kan complex, so that $pr_R(d_{0*}, d_1 f_*) = d_1 f_*$ is a fibration.

Write $Z_f = X \times_Y Y^I$ and $\pi = d_1 f_*$. Then we have functorial replacement

$$\begin{array}{ccc} X & \xrightarrow{s_*} & Z_f & \xrightarrow{d_{0*}} & X \\ & {\scriptstyle f} \searrow & \downarrow {\scriptstyle \pi} & \\ & & Y & \end{array} \qquad (5.1)$$

of f by a fibration π, where d_{0*} is a trivial fibration such that $d_{0*}s_* = 1$.

The same argument can be repeated exactly within the theory of local fibrations, giving the following:

Lemma 5.1 *Suppose that $f : X \to Y$ is a map between locally fibrant simplicial presheaves. Then we have the following:*

1) The map f has a natural factorization (5.1) for which π is a local fibration, d_{0*} is a local trivial fibration, and $d_{0*}s_* = 1_X$.

2) *The map f is a local weak equivalence if and only if the map π in the factorization (5.1) is a local trivial fibration.*

The second statement of the Lemma follows from Theorem 4.32.

Suppose again that $f : X \to Y$ is a simplicial set map, and form the diagram

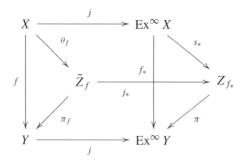

in which the front face is a pullback. Then π_f is a fibration, and θ_f is a weak equivalence since j_* is a weak equivalence by properness of the model structure for simplicial sets. The construction taking a map f to the factorization

$$X \xrightarrow{\theta_f} \tilde{Z}_f \xrightarrow{\pi_f} Y \qquad (5.2)$$

(with f as the diagonal)

is natural and preserves filtered colimits in f.

Say that a simplicial set X is α-*bounded* if $|X_n| < \alpha$ for all $n \geq 0$, or in other words if α is an upper bound for the cardinality of all sets of simplices of X. A simplicial presheaf Y is α-bounded if all of the simplicial sets $Y(U)$, $U \in \mathcal{C}$, are α-bounded.

This construction (5.2) carries over to simplicial presheaves, giving a natural factorization

of a simplicial presheaf map $f : X \to Y$ such that θ_f is a sectionwise weak equivalence and π_f is a sectionwise fibration. Here are some further properties of this factorization:

5.1 The Injective Model Structure

a) the factorization preserves filtered colimits in f
b) if X and Y are α-bounded where α is some infinite cardinal, then so is $\tilde{Z}_f$
c) the map f is a local weak equivalence if and only if π_f has the local right lifting property with respect to all $\partial\Delta^n \subset \Delta^n$.

Statement c) is a consequence of Theorem 4.32, since π_f is a local fibration.

Suppose that $\mathcal{C}$ is a Grothendieck site, and recall that we assume that such a category is small. Suppose that α is a regular cardinal such that $\alpha > |\operatorname{Mor}(\mathcal{C})|$.

Regular cardinals are used throughout this book, in order that the size of filtered colimits works out correctly.

Specifically, if α is a regular cardinal and $F = \varinjlim_{i \in I} F_i$ is a filtered colimit of sets F_i such that $|I| < \alpha$ and all $|F_i| < \alpha$, then $|F| < \alpha$. One could take this condition to be the definition of a regular cardinal.

It is easy to see that if β is an infinite cardinal, then the successor cardinal $\beta + 1$ is regular, so that regular cardinals abound in nature. There are well known examples of limit cardinals that are not regular.

The following result is the *bounded monomorphism property* for simplicial presheaves.

Lemma 5.2 *Suppose that* $i : X \to Y$ *is a monomorphism and a local weak equivalence of* $s\mathbf{Pre}(\mathcal{C})$. *Suppose that* $A \subset Y$ *is an* α-*bounded subobject of* Y. *Then, there is an* α-*bounded subobject* B *of* Y *such that* $A \subset B$ *and such that the map* $B \cap X \to B$ *is a local weak equivalence.*

Proof Write $\pi_B : Z_B \to B$ for the natural pointwise Kan fibration replacement for the monomorphism $B \cap X \to B$. The map $\pi_Y : Z_Y \to Y$ has the local right lifting property with respect to all $\partial\Delta^n \subset \Delta^n$.

Suppose given a lifting problem

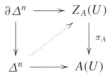

where A is α-bounded. The lifting problem can be solved locally over Y along some covering sieve for U having at most α elements. We have an identification

$$Z_Y = \varinjlim_{|B| < \alpha} Z_B$$

since Y is a filtered colimit of its α-bounded subobjects. It follows from the regularity assumption on α that there is an α-bounded subobject $A' \subset Y$ with $A \subset A'$ such that the original lifting problem can be solved over A'. The list of all such lifting problems is α-bounded, so there is an α-bounded subobject $B_1 \subset Y$ with $A \subset B_1$ so that all lifting problems as above over A can be solved locally over B_1. Repeat this procedure countably many times to produce an ascending family

$$A = B_0 \subset B_1 \subset B_2 \subset \dots$$

of α-bounded subobjects of Y such that all lifting local lifting problems

$$
\begin{array}{ccc}
\partial \Delta^n & \longrightarrow & Z_{B_i}(U) \\
\downarrow & \nearrow & \downarrow \pi_{B_i} \\
\Delta^n & \longrightarrow & B_i(U)
\end{array}
$$

over B_i can be solved locally over B_{i+1}. Set $B = \cup_i B_i$.

Say that a map $p : X \to Y$ of $s\mathbf{Pre}(\mathcal{C})$ is an *injective fibration* if p has the right lifting property with respect to all maps $A \to B$ which are cofibrations and local weak equivalences.

Remark 5.3 Injective fibrations are also called *global fibrations* in the literature, for example in [50]. The use of the term "global fibration" originated in early work of Brown and Gersten [17], but has declined with the introduction of the various model structures associated with motivic homotopy theory.

Say that a map $A \to B$ of simplicial presheaves is an *α-bounded cofibration* if it is a cofibration and the object B is α-bounded. It follows that the subobject A is α-bounded as well.

A *trivial cofibration* is a map of simplicial presheaves that is both a cofibration and a local weak equivalence. This is standard terminology, and is consistent with the injective model structure which appears in Theorem 5.8 below. Similarly, for that model structure, a *trivial fibration* is a map which is an injective fibration and a local weak equivalence.

Lemma 5.4 *The map $p : X \to Y$ is an injective fibration if and only if it has the right lifting property with respect to all α-bounded trivial cofibrations.*

Proof Suppose that $p : X \to Y$ has the right lifting property with respect to all α-bounded trivial cofibrations, and suppose given a diagram

where i is a trivial cofibration. Consider the poset of partial lifts

in which the maps $A \to A' \to B$ are trivial cofibrations. This poset is non-trivial: given $x \in B(U) - A(U)$ there is an α-bounded subcomplex $C \subset B$ with $x \in C(U)$

5.1 The Injective Model Structure

(let C be the image of the map $L_U \Delta^n \to B$ which is adjoint to the simplex $x : \Delta^n \to B(U)$), and there is an α-bounded subcomplex $C' \subset B$ with $C \subset C'$ and $i_* : C' \cap A \to C'$ a trivial cofibration. Then $x \in C' \cup A$, and there is a diagram

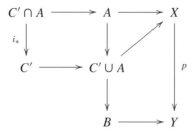

where the indicated lift exists because p has the right lifting property with respect to the α-bounded trivial cofibration i_*. The map $A \to C' \cup A$ is a trivial cofibration by Corollary 4.36.

The poset of partial lifts has maximal elements by Zorn's Lemma, and the maximal elements of the poset must have the form

$$\begin{array}{ccc} A & \longrightarrow & X \\ {\scriptstyle i}\downarrow & \nearrow & \downarrow {\scriptstyle p} \\ B & \longrightarrow & Y \end{array}$$

Recall that one defines

$$L_U K = \hom(\ , U) \times K$$

for $U \in \mathcal{C}$ and simplicial sets K, and that the functor $K \mapsto L_U K$ is left adjoint to the U-sections functor $X \mapsto X(U)$.

Lemma 5.5 *Suppose that $q : Z \to W$ has the right lifting property with respect to all cofibrations. Then q is an injective fibration and a local weak equivalence.*

Proof The map q is obviously an injective fibration, and it has the right lifting property with respect to all cofibrations $L_U \partial \Delta^n \to L_U \Delta^n$, so that all maps $q : Z(U) \to W(U)$ are trivial Kan fibrations. It follows from Lemma 4.1 that q is a local weak equivalence.

Lemma 5.6 *A map $q : Z \to W$ has the right lifting property with respect to all cofibrations if and only if it has the right lifting property with respect to all α-bounded cofibrations.*

Proof The proof of this result is an exercise.

Lemma 5.7 *Any simplicial presheaf map* $f : X \to Y$ *has factorizations*

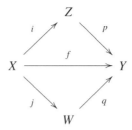

where

1) *the map i is a cofibration and a local weak equivalence, and p is an injective fibration,*
2) *the map j is a cofibration and p has the right lifting property with respect to all cofibrations (and is therefore an injective fibration and a local weak equivalence)*

Proof For the first factorization, choose a cardinal $\lambda > 2^\alpha$ and do a transfinite small object argument of size λ to solve all lifting problems

arising from locally trivial cofibrations i which are α-bounded. We need to know that locally trivial cofibrations are closed under pushout, but this is proved in Corollary 4.36. The small object argument stops on account of the condition on the size of the cardinal λ.

The second factorization is similar, and uses Lemma 5.6.

The main results of this section say that the respective categories of simplicial presheaves and simplicial sheaves on a Grothendieck site admit well-behaved model structures, and that these model structures are Quillen equivalent.

Theorem 5.8 *Suppose that $\mathcal{C}$ is a small Grothendieck site. Then, the category $s\mathbf{Pre}(\mathcal{C})$, with the classes of local weak equivalences, cofibrations and injective fibrations, satisfies the axioms for a proper closed simplicial model category. This model structure is cofibrantly generated.*

Proof The simplicial presheaf category $s\mathbf{Pre}(\mathcal{C})$ has all small limits and colimits, giving **CM1**. The weak equivalence axiom **CM2** was proved in Lemma 4.30 with a Boolean localization argument. The retract axiom **CM3** is trivial to verify—use the pullback description of local weak equivalences to see the weak equivalence part. The factorization axiom **CM5** is Lemma 5.7.

5.1 The Injective Model Structure

Suppose that $\pi : X \to Y$ is an injective fibration and a local weak equivalence. Then by Lemma 5.7, the map π has a factorization

where p has the right lifting property with respect to all cofibrations and is therefore a local weak equivalence. Then j is a local weak equivalence, and so π is a retract of p. It follows that π has the right lifting property with respect to all cofibrations, giving **CM4**.

The simplicial model structure comes from the *function complex* $\mathbf{hom}(X, Y)$. This is the simplicial set with n-simplices given by

$$\mathbf{hom}(X, Y)_n = \hom_{s\mathbf{Pre}(\mathcal{C})}(X \times \Delta^n, Y).$$

One uses Corollary 4.41 to show that, if $i : A \to B$ is a cofibration of simplicial presheaves and $j : K \to L$ is a cofibration of simplicial sets, then the induced map

$$(B \times K) \cup (A \times L) \to B \times L$$

is a cofibration which is a local weak equivalence if either i is a local weak equivalence of simplicial presheaves or j is a weak equivalence of simplicial sets.

The properness of the model structure follows from Corollary 4.36 and Lemma 4.37.

It is a consequence of the proof of the model axioms that a generating set J for the class of trivial cofibrations is given by the set of all α-bounded trivial cofibrations, while the set I of α-bounded cofibrations generates the class of cofibrations.

Write $s\mathbf{Shv}(\mathcal{C})$ for the category of simplicial sheaves on $\mathcal{C}$. Say that a map $f : X \to Y$ is a *local weak equivalence* of simplicial sheaves if it is a local weak equivalence of simplicial presheaves. A *cofibration* of simplicial sheaves is a monomorphism, and an *injective fibration* is a map which has the right lifting property with respect to all trivial cofibrations.

Theorem 5.9 *Let $\mathcal{C}$ be a small Grothendieck site.*

1) *The category $s\mathbf{Shv}(\mathcal{C})$ with the classes of local weak equivalences, cofibrations and injective fibrations, satisfies the axioms for a proper closed simplicial model category. This model structure is cofibrantly generated.*
2) *The inclusion i of sheaves in presheaves and the associated sheaf functor L^2 together induce a Quillen equivalence*

$$L^2 : s\mathbf{Pre}(\mathcal{C}) \leftrightarrows s\mathbf{Shv}(\mathcal{C}) : i.$$

Proof The associated sheaf functor L^2 preserves cofibrations, and it preserves and reflects local weak equivalences (Corollary 4.22). The inclusion functor i therefore preserves injective fibrations. The associated sheaf map $\eta : X \to L^2 X$ is a local weak equivalence, while the counit of the adjunction is an isomorphism. Thus, we have statement 2) if we can prove statement 1).

The axiom **CM1** follows from completeness and cocompleteness for the sheaf category $s\mathbf{Shv}(\mathcal{C})$. The axioms **CM2**, **CM3** and **CM4** follow from the corresponding statements for simplicial presheaves.

A map $p : X \to Y$ is an injective fibration (respectively trivial injective fibration) of $s\mathbf{Shv}(\mathcal{C})$ if and only if it is an injective fibration (respectively trivial injective fibration) of $s\mathbf{Pre}(\mathcal{C})$.

Choose a regular cardinal β such that $\beta > |\tilde{B}|$ for all α-bounded simplicial presheaves B. Then a simplicial sheaf map p is an injective fibration if and only if it has the right lifting property with respect to all monomorphisms $A \subset B$ of β-bounded objects of $s\mathbf{Shv}(\mathcal{C})$ which are local weak equivalences.

The factorization axiom **CM5** is proved by a transfinite small object argument of size λ, where λ is a regular cardinal such that $\lambda > 2^\beta$.

The simplicial model structure is inherited from simplicial presheaves, as is properness.

Example 5.10 The category $s\mathbf{Pre}(\mathcal{C})$ of simplicial presheaves is also the category of simplicial sheaves for the "chaotic" Grothendieck topology on $\mathcal{C}$ whose covering sieves are the representable functors $\hom(\ ,U)$, $U \in \mathcal{C}$ (Example 3.8). The injective model structures, for simplicial presheaves or simplicial sheaves, specialize to the injective model structure for diagrams of simplicial sets. The injective model structure for diagrams is the good setting for describing homotopy inverse limits—see [32, VIII.2]. The existence of this model structure is attributed to Heller [37], but it is also a consequence of Joyal's theorem (Theorem 5.9.1) for simplicial sheaves [70].

5.2 Injective Fibrations

Injective fibrant simplicial presheaves are usually only formally defined, but here is a first simple example:

Lemma 5.11 *Suppose that F is a sheaf of sets on $\mathcal{C}$. Then, the associated constant simplicial sheaf $K(F, 0)$ is injective fibrant.*

The object $K(F, 0)$ has n-simplices

$$K(F, 0)_n = F,$$

and all simplicial structure maps are the identity on F.

Proof There is a natural bijection

$$\hom(X, K(F, 0)) \cong \hom(\tilde{\pi}_0(X), F)$$

5.2 Injective Fibrations

for all simplicial presheaves X and sheaves F. Any local weak equivalence $f : X \to Y$ induces an isomorphims $\tilde{\pi}_0(X) \cong \tilde{\pi}_0(Y)$, and so f induces a bijection

$$f^* : \hom(Y, K(F, 0)) \xrightarrow{\cong} \hom(X, K(F, 0)).$$

Thus, all lifting problems

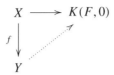

have unique solutions.

Many of the applications of local homotopy theory are based on the sectionwise properties of injective fibrations and injective fibrant objects.

Lemma 5.12

1) *Every injective fibration $p : X \to Y$ is a sectionwise Kan fibration.*
2) *Every injective fibration is a local fibration.*
3) *Every trivial injective fibration is a sectionwise trivial Kan fibration.*

Proof An injective fibration $p : X \to Y$ has the right lifting property with respect to the trivial cofibrations $L_U \Lambda^n_k \to L_U \Delta^n$. If p is a trivial injective fibration then it has the right lifting property with respect to the cofibrations $L_U \partial \Delta^n \to L_U \Delta^n$. All sectionwise Kan fibrations are local fibrations.

Corollary 5.13 *Suppose that the map $f : X \to Y$ is a local weak equivalence, and that X and Y are injective fibrant simplicial presheaves. Then f is a sectionwise weak equivalence.*

Proof According to the method of proof of Lemma 5.1, the map f has a factorization

where π is an injective fibration and s_* is a section of a trivial injective fibration. The map π is then a trivial injective fibration, and it follows from Lemma 5.12 that the maps s_* and π are both sectionwise weak equivalences.

An *injective fibrant model* of a simplicial presheaf X is a local weak equivalence $j : X \to Z$ such that Z is injective fibrant.

Remark 5.14 One can make a functorial choice $j : X \to GX$ of injective fibrant models, since the injective model structure on simplicial presheaves (or on simplicial sheaves) is cofibrantly generated. This functoriality means that there is a functor $X \mapsto GX$ such that GX is injective fibrant, together with a natural transformation $j : X \to GX$ which consists of local weak equivalences.

The notation is consistent with the usage of [50], where one speaks of global fibrations (see Remark 5.3)—in that language, one says that the map $j : X \to GX$ is a global fibrant model.

In isolated circumstances, one can use Godement resolutions to construct functorial injective fibrant models—see [50, Prop. 3.3], [56, Th. 5.8], [99, Def. 1.33]. This is another historical origin for the notation $X \to GX$.

A simplicial presheaf X on a site $\mathcal{C}$ is said to satisfy *descent* (or have the *descent property*) if some injective fibrant model $j : X \to Z$ is a sectionwise weak equivalence in the sense that the simplicial set maps $j : X(U) \to Z(U)$ are weak equivalences for all objects U of $\mathcal{C}$.

All injective fibrant objects Z satisfy descent, since any injective fibrant model $Z \to Z'$ is a local weak equivalence between injective fibrant objects, and is therefore a sectionwise weak equivalence by Corollary 5.13.

We also have the following:

Corollary 5.15 *Suppose that the simplicial presheaf X satisfies descent, and suppose that the map $f : X \to W$ is a local weak equivalence such that W is injective fibrant. Then the map f is a sectionwise weak equivalence.*

In other words, a simplicial presheaf X satisfies descent if and only if all of its injective fibrant models $X \to Z$ are sectionwise weak equivalences.

Proof Suppose that the injective fibrant model $j : X \to Z$ is a sectionwise weak equivalence. We can suppose that the map j is a cofibration by a factorization argument and Corollary 5.13. Let $i : Z \cup_X W \to W'$ be an injective fibrant model for the pushout $Z \cup_X W$. Then by left properness of the injective model structure for simplicial presheaves, all maps in the resulting commutative diagram

are local weak equivalences, and the objects Z, W and W' are injective fibrant. It follows that the map f is a sectionwise weak equivalence.

Corollary 5.16 *Any two injective fibrant models of a simplicial presheaf X are sectionwise weakly equivalent.*

The injective model structure of Theorem 5.8 is a closed simplicial model structure in which all objects are cofibrant. Suppose that the simplicial presheaf Z is injective fibrant and that $f : X \to Y$ is a morphism of simplicial presheaves. It follows that

5.2 Injective Fibrations

the simplicial set map

$$f^* : \hom(Y, Z) \to \hom(X, Z)$$

is a weak equivalence if the map f is a local weak equivalence of simplicial presheaves.

Example 5.17 Suppose that U is an object of the underlying site $\mathcal{C}$ and that there is a simplicial object $V_\bullet$ in $\mathcal{C}$, together with a map $V_\bullet \to U$ which represents a local weak equivalence of simplicial presheaves.

Such things arise, for example, as Čech resolutions associated to covering morphisms $V \to U$ in sites $\mathcal{C}$ which have all finite products. More generally the map $V_\bullet \to U$ could be a representable hypercover.

If the simplicial presheaf Z is injective fibrant, then the induced simplicial set map

$$\hom(U, Z) \to \hom(V_\bullet, Z)$$

is a weak equivalence.

In general, if X is a simplicial presheaf, then there is an isomorphism of simplicial sets

$$\hom(U, X) \cong X(U)$$

since U is simplicially discrete and representable.

Suppose that X is a presheaf of Kan complexes, and let $j : X \to Z$ be an injective fibrant model. By methods of Bousfield and Kan [14], the function complex $\hom(V_\bullet, X)$ is naturally the homotopy inverse limit of the cosimplicial space

$$\mathbf{n} \mapsto \hom(V_n, X) = X(V_n).$$

Injective fibrant objects are also presheaves of Kan complexes, and it follows that there is a commutative diagram of simplicial set maps

$$\begin{array}{ccccc} X(U) & \longrightarrow & \hom(V_\bullet, X) & \xleftarrow{\simeq} & \operatorname{holim}_n X(V_n) \\ {\scriptstyle j}\downarrow & & \downarrow{\scriptstyle j_*} & & \downarrow{\scriptstyle j_*} \\ Z(U) & \underset{\simeq}{\longrightarrow} & \hom(V_\bullet, Z) & \underset{\simeq}{\longleftrightarrow} & \operatorname{holim}_n Z(V_n) \end{array}$$

The vertical maps in the diagram are weak equivalences if X satisfies descent, and so there is a weak equivalence

$$X(U) \simeq \operatorname{holim}_n X(V_n)$$

in that case. We would then have a means of computing homotopy groups or the homotopy type of $X(U)$ from patching data, using the spaces $X(V_n)$. The device which is commonly used for computing homotopy groups is the Bousfield–Kan spectral sequence for a cosimplicial space.

The existence of a weak equivalence

$$X(U) \simeq \varprojlim_n X(V_n)$$

in the presence of a representable hypercover is an abstraction of the original meaning of cohomological descent.

The assertion that X satisfies descent in the sense that any injective fibrant model $X \to Z$ is a sectionwise equivalence is much stronger than such cohomological descent statements.

Example 5.18 Suppose that L/k is a finite Galois extension of fields with Galois group G. The Čech resolution associated to the covering $\mathrm{Sp}(L) \to \mathrm{Sp}(k)$ on the étale site $et|_k$ for the field k has the form

$$EG \times_G \mathrm{Sp}(L) \to *$$

as a map of simplicial simplicial sheaves, as in Example 4.17. Recall that the Borel construction $EG \times_G \mathrm{Sp}(L)$ is specified in simplicial degree n by the sheaf

$$(EG \times_G \mathrm{Sp}(L))_n = \bigsqcup_{(g_1, g_2, \dots, g_n)} \mathrm{Sp}(L).$$

Suppose that X is a presheaf of Kan complexes X on $et|_k$. The space

$$\mathbf{hom}(EG \times_G \mathrm{Sp}(L), X)$$

is the homotopy fixed point space for the action of the group G on the space $X(L)$, in the sense that there is an isomorphism

$$\varprojlim_G X(L) \cong \mathbf{hom}(EG \times_G \mathrm{Sp}(L), X).$$

The Bousfield–Kan spectral sequence for the cosimplicial space

$$\mathbf{n} \mapsto \mathbf{hom}((EG \times_G \mathrm{Sp}(L))_n X) \cong \prod_{g_1, g_2, \dots, g_n} X(L)$$

is the homotopy fixed points spectral sequence.

If the simplicial presheaf X satsifies descent for the étale topology on the field k, then the canonical map

$$X(k) \to \varprojlim_G X(L)$$

is a weak equivalence of simplicial sets, because this is certainly true if X is injective fibrant.

The problem of determining whether or not the map

$$X(k) \to \varprojlim_G X(L)$$

5.2 Injective Fibrations

is a weak equivalence is called, variously, the finite Galois descent problem or the homotopy fixed points problem for the simplicial presheaf X.

The existence of solutions to finite descent problems for a simplicial presheaf X over all finite separable extensions L/k *does not* imply that X satisfies étale descent over the field k [69].

Étale descent for fields is often called Galois descent, or Galois cohomological descent, for the usual reason that the étale and Galois cohomology theories coincide for all fields.

If U is an object of the Grothendieck site $\mathcal{C}$, recall that the category $\mathcal{C}/U$ of inherits a topology for which a collection of morphisms $V_i \to V \to U$ is covering for the object $V \to U$ if and only if the morphisms $V_i \to V$ cover the object V of $\mathcal{C}$.

For a presheaf F on the site $\mathcal{C}$, recall that the restriction $F|_U$ is the composite

$$(\mathcal{C}/U)^{op} \xrightarrow{q^{op}} (\mathcal{C})^{op} \xrightarrow{F} \mathbf{Set},$$

where $q : \mathcal{C}/U \to \mathcal{C}$ is the canonical functor which takes an object $V \to U$ to V. Observe that $F|_U$ is a sheaf if F is a sheaf.

If $\phi : U \to U'$ is a morphism of $\mathcal{C}$, then the diagram of functors

commutes, and so a morphism $E|_{U'} \to F|_{U'}$ restricts to a morphism $E|_U \to F|_U$ by composition with ϕ_*.

It follows that there is a presheaf $\mathbf{Hom}(E, F)$ on the site $\mathcal{C}$ with

$$\mathbf{Hom}(E, F)(U) = \hom(E|_U, F|_U).$$

The presheaf $\mathbf{Hom}(E, F)$ is a sheaf if E and F are sheaves.

The standard exponential law, applied sectionwise, implies that there is an adjunction isomorphism

$$\hom(A, \mathbf{Hom}(E, F)) \cong \hom(A \times E, F) \tag{5.3}$$

for all presheaves A. It follows that a map $E|_U \to F|_U$ can be identified with a presheaf map $E \times U \to F$, where U is identified notationally with the representable presheaf $U = \hom(\ , U)$. We can therefore write

$$\mathbf{Hom}(E, F)(U) = \hom(E \times U, F) \tag{5.4}$$

for all objects U of $\mathcal{C}$ and all presheaves E and F.

If X and Y are simplicial presheaves on $\mathcal{C}$, then the *internal function complex* $\mathbf{Hom}(X, Y)$ is the simplicial presheaf whose U-sections are defined in terms of the function complex on $\mathcal{C}/U$ by the assignment

$$\mathbf{Hom}(X, Y)(U) = \mathbf{hom}(X \times U, Y).$$

There is a corresponding exponential law, meaning an isomorphism

$$\hom(A, \mathbf{Hom}(X, Y)) \cong \hom(X \times A, Y)$$

which is natural in simplicial presheaves A, X and Y. This is a consequence of the identifications

$$\hom(U \times \Delta^n, \mathbf{Hom}(X, Y)) \cong \mathbf{Hom}(X, Y)(U)_n = \hom(X \times U \times \Delta^n, Y),$$

and the fact that every simplicial presheaf A is a colimit of objects $U \times \Delta^n$.

The statement of Corollary 4.41 amounts to the existence of an enriched simplicial model structure on the category $s\mathbf{Pre}(\mathcal{C})$. The following is an equivalent formulation:

Corollary 5.19 *Suppose that $p : X \to Y$ is an injective fibration and that $i : A \to B$ is a cofibration of simplicial presheaves. Then the induced map of simplicial presheaves*

$$\mathbf{Hom}(B, X) \to \mathbf{Hom}(A, X) \times_{\mathbf{Hom}(A,Y)} \mathbf{Hom}(A, X)$$

is an injective fibration which is a local weak equivalence if either i or p is a local weak equivalence.

We close this section by observing that right properness for local fibrations and local weak equivalences (Lemma 4.37) has the following useful interpretation within the injective model structure:

Lemma 5.20 *Suppose given a pullback diagram*

$$\begin{array}{ccc} Z \times_Y X & \longrightarrow & X \\ \downarrow & & \downarrow \pi \\ Z & \longrightarrow & Y \end{array}$$

in which the map π is a local fibration of simplicial presheaves. Then, the diagram is homotopy cartesian for the injective model structure on the simplicial presheaf category $s\mathbf{Pre}(\mathcal{C})$.

Proof Choose a factorization

in the category of simplicial presheaves, such that q is an injective fibration, and j is a cofibration and a local weak equivalence. The map q is a local fibration by Lemma 5.12, and then j defines a local weak equivalence between local fibrations. The induced map $j_* : Z \times_Y X \to Z \times_Y W$ is a local weak equivalence by Corollary 4.38.

5.3 Geometric and Site Morphisms

Suppose that $\pi : \mathbf{Shv}(\mathcal{C}) \to \mathbf{Shv}(\mathcal{D})$ is a geometric morphism. Then, the inverse image and direct image functors for π induce adjoint functors

$$\pi^* : s\mathbf{Shv}(\mathcal{D}) \leftrightarrows s\mathbf{Shv}(\mathcal{C}) : \pi_*$$

between the respective categories of simplicial sheaves.

Lemma 5.21 *Suppose that $\pi : \mathbf{Shv}(\mathcal{C}) \to \mathbf{Shv}(\mathcal{D})$ is a geometric morphism. Then the inverse image functor*

$$\pi^* : s\mathbf{Shv}(\mathcal{D}) \to s\mathbf{Shv}(\mathcal{C})$$

preserves cofibrations and local weak equivalences.

Proof The functor π^* is exact, and therefore preserves cofibrations since every monomorphism is an equalizer (Lemma 3.14).

The functor π^* commutes with the sheaf theoretic Ex^∞-functor, up to natural isomorphism. It therefore suffices to show that π^* preserves local weak equivalences between locally fibrant objects. If $g : X \to Y$ is a local weak equivalence between locally fibrant simplicial sheaves, then g has a factorization

such that p is a local trivial fibration and the map j is a section of a local trivial fibration, by Lemma 5.1. The inverse image functor π^* preserves local trivial fibrations, by exactness, so that $\pi^*(g)$ is a local weak equivalence of $s\mathbf{Shv}(\mathcal{C})$.

Corollary 5.22 *Suppose that $\pi : \mathbf{Shv}(\mathcal{C}) \to \mathbf{Shv}(\mathcal{D})$ is a geometric morphism. Then the adjoint functors*

$$\pi^* : s\mathbf{Shv}(\mathcal{D}) \leftrightarrows s\mathbf{Shv}(\mathcal{C}) : \pi_*$$

form a Quillen adjunction for the injective model structures on the respective categories of simplicial sheaves. In particular if X is an injective fibrant simplicial sheaf on $\mathcal{C}$, then its direct image $\pi_* X$ is injective fibrant.

Suppose that $\mathcal{C}$ and $\mathcal{D}$ are Grothendieck sites. Any functor $f : \mathcal{C} \to \mathcal{D}$ induces a functor

$$f_* : \mathbf{Pre}(\mathcal{D}) \to \mathbf{Pre}(\mathcal{C})$$

of presheaf categories, by precomposition with the induced functor $f : \mathcal{C}^{op} \to \mathcal{D}^{op}$. The functor f_* has a left adjoint

$$f^p : \mathbf{Pre}(\mathcal{C}) \to \mathbf{Pre}(\mathcal{D})$$

which is the left Kan extension of f_*. Explicitly,

$$f^p(F)(d) = \varinjlim_{d \to f(c)} F(c),$$

where the colimit is defined on the slice category d/f.

Following SGA4 [2, IV.5.9], a *site morphism* is a functor $f : \mathcal{C} \to \mathcal{D}$ such that

1) the functor f_* is continuous in the sense that it preserves sheaves, and
2) the functor f^p is exact in the sense that it preserves finite colimits.

Every site morphism $f : \mathcal{C} \to \mathcal{D}$ induces a geometric morphism

$$f : \mathbf{Shv}(\mathcal{D}) \to \mathbf{Shv}(\mathcal{C})$$

with direct image f_* defined by precomposition with f as above, and with inverse image f^* defined by

$$f^*(F) = L^2(f^p(F))$$

for sheaves F.

Lemma 5.23 *Suppose that the functor $f : \mathcal{C} \to \mathcal{D}$ is a site morphism. Then the inverse image functor*

$$f^p : s\mathbf{Pre}(\mathcal{C}) \to s\mathbf{Pre}(\mathcal{D})$$

preserves cofibrations and local weak equivalences.

Proof The proof is similar to that of Lemma 5.21. Every monomorphism of $s\mathbf{Pre}(\mathcal{C})$ is an equalizer and f^p preserves equalizers, so that f^p preserves monomorphisms. The functor f^p preserves local weak equivalences, since the inverse image functor f^* preserves local weak equivalences of simplicial sheaves by Lemma 5.21.

Corollary 5.24 *Suppose that the functor $f : \mathcal{C} \to \mathcal{D}$ is a site morphism. Then the adjoint functors*

$$f^p : s\mathbf{Pre}(\mathcal{C}) \leftrightarrows s\mathbf{Pre}(\mathcal{D}) : f_*$$

5.3 Geometric and Site Morphisms

form a Quillen adjunction for the respective injective model structures. In particular, the functor f_* preserves injective fibrant objects.

The assertion that the direct image functor f_* preserves injective fibrant objects first appeared in [52], with essentially the same proof.

Recall that the *forgetful functor* $q : \mathcal{C}/U \to \mathcal{C}$ is defined on objects by

$$q(V \xrightarrow{\phi} U) = V.$$

This functor is continuous for the topology on $\mathcal{C}/U$ which is inherited from the site $\mathcal{C}$, but it is not necessarily a site morphism. We nevertheless have the following useful result:

Lemma 5.25 *Suppose that $\mathcal{C}$ is a Grothendieck site and that U is an object of $\mathcal{C}$. Then the functor*

$$q^p : s\mathbf{Pre}(\mathcal{C}/U) \to s\mathbf{Pre}(\mathcal{C})$$

preserves cofibrations and local weak equivalences.

Proof The functor q^p is defined, for a simplicial presheaf X on $\mathcal{C}/U$, by

$$q^p(X)(V) = \bigsqcup_{\phi : V \to U} X(\phi)$$

for $V \in \mathcal{C}$. This functor plainly preserves cofibrations.

Suppose that $p : X \to Y$ is a locally trivial fibration on $\mathcal{C}/U$ and that there is a commutative diagram

$$\begin{array}{ccc} \partial\Delta^n & \longrightarrow & q^p X(V) \\ \downarrow & & \downarrow{\scriptstyle p_*} \\ \Delta^n & \longrightarrow & q^p Y(V) \end{array}$$

The Δ^n is connected for all $n \geq 0$, so that there is a factorization of this diagram

$$\begin{array}{ccccc} \partial\Delta^n & \longrightarrow & X(\phi) & \xrightarrow{in_\phi} & \sqcup_{\phi:V\to U} X(\phi) \\ \downarrow & & \downarrow{\scriptstyle p} & & \downarrow{\scriptstyle p_*} \\ \Delta^n & \longrightarrow & Y(\phi) & \xrightarrow{in_\phi} & \sqcup_{\phi:V\to U} Y(\phi) \end{array}$$

for some map $\phi : V \to U$, where in_ϕ is the inclusion of the summand corresponding to the map ϕ. There is a covering of ϕ such that the liftings exist in the diagrams

$$\begin{array}{ccc} \partial\Delta^n & \longrightarrow X(\phi) \longrightarrow & X(\phi_i) \\ \downarrow & & \downarrow p \\ \Delta^n & \longrightarrow Y(\phi) \longrightarrow & Y(\phi_i). \end{array}$$

It follows that the liftings exist in the diagrams

$$\begin{array}{ccc} \partial\Delta^n & \longrightarrow q^p X(V) \longrightarrow & q^p X(V_i) \\ \downarrow & & \downarrow p_* \\ \Delta^n & \longrightarrow q^p Y(V) \longrightarrow & q^p Y(V_i) \end{array}$$

after refinement along the covering $V_i \to V$.

The functor q^p therefore preserves local trivial fibrations. It also commutes up to isomorphism with the Ex^∞ functor. It follows from Lemma 5.1 that q^p preserves local weak equivalences.

Corollary 5.26 *Suppose that $\mathcal{C}$ is a Grothendieck site, U is an object of $\mathcal{C}$ and that $q : \mathcal{C}/U \to \mathcal{C}$ is the forgetful functor. Then the adjoint functors*

$$q^p : s\mathbf{Pre}(\mathcal{C}/U) \leftrightarrows s\mathbf{Pre}(\mathcal{C}) : q_*$$

define a Quillen adjunction for the respective injective model structures. In particular, the restriction functor

$$X \mapsto q_*(X) = X|_U$$

preserves injective fibrant objects.

The presheaf-level restriction functor

$$q_* : \mathbf{Pre}(\mathcal{C}) \to \mathbf{Pre}(\mathcal{C}/U)$$

is exact (preserves limits and colimits) and preserves local epimorphisms. It follows that the functor

$$q_* : s\mathbf{Pre}(\mathcal{C}) \to s\mathbf{Pre}(\mathcal{C}/U)$$

commutes with the Ex^∞ functor up to natural isomorphism and preserves local fibrations and local trivial fibrations. In particular, the restriction functor q_* preserves local weak equivalences, by Lemma 5.1. We also have the following:

Lemma 5.27 *A map $f : X \to Y$ is a local fibration (respectively local trivial fibration, respectively, local weak equivalence) if and only if the restrictions $f|_U : X|_U \to Y|_U$ are local fibrations (respectively, local trivial fibrations, respectively, local weak equivalences) for all objects $U \in \mathcal{C}$.*

5.3 Geometric and Site Morphisms

Proof A map $F \to G$ of presheaves on $\mathcal{C}$ is a local epimorphism if and only if the restrictions $F|_U \to G|_U$ are local epimorphisms on $\mathcal{C}/U$ for all $U \in \mathcal{C}$. The claims about local fibrations and local trivial fibrations follow immediately. One proves the claim about local weak equivalences with another appeal to Lemma 5.1.

Example 5.28 The ideas of this section occur frequently in examples.

Suppose that $\phi : T \to S$ is a morphism of schemes which is locally of finite type. Then ϕ occurs as an object of the big étale site $(Sch|_S)_{et}$. Pullback along the scheme homomorphism ϕ determines a site morphism

$$\phi : (Sch|_S)_{et} \to (Sch|_T)_{et}.$$

One can identify the site $(Sch|_T)_{et}$ with the slice category $(Sch|_S)_{et}/\phi$, and the presheaf-level inverse image functor

$$\phi^p : \mathbf{Pre}((Sch|_S)_{et}) \to \mathbf{Pre}((Sch|_T)_{et})$$

(i.e. the left adjoint of composition with the pullback functor) is isomorphic to the restriction functor which is induced by composition with ϕ. It follows from Corollary 5.26 that the functor

$$\phi^p : s\mathbf{Pre}((Sch|_S)_{et}) \to s\mathbf{Pre}((Sch|_T)_{et})$$

preserves injective fibrations. This functor also preserves local weak equivalences since it is an inverse image functor for a site morphism. The restriction functor ϕ^p therefore preserves injective fibrant models.

The inclusion $i : et|_T \subset (Sch|_T)_{et}$ of the étale site in the big étale site is a site morphism for each S-scheme T. Restriction to $et|_T$ is exact and preserves local epimorphisms for presheaves on $(Sch|_T)_{et}$, and it therefore preserves local weak equivalences. This restriction functor also preserves injective fibrations, by Corollary 5.24, and therefore preserves injective fibrant models.

It follows that composite restriction functors

$$s\mathbf{Pre}((Sch|_S)_{et}) \xrightarrow{\phi^p} s\mathbf{Pre}((Sch|_T)_{et}) \xrightarrow{i_*} s\mathbf{Pre}((et|_T))$$

preserve local weak equivalences and injective fibrations for all S-schemes $\phi : T \to S$. These functors are exact and commute with the Ex^∞ construction. Taken together, these functors reflect local epimorphisms. Thus, a simplicial presheaf map $X \to Y$ on the big étale site for S is a local weak equivalence if and only if the induced map $i_*\phi^p(X) \to i_*\phi^p(Y)$ is a local weak equivalence on the ordinary étale site $et|_T$ for each S-scheme $\phi : T \to S$.

The foregoing is only a paradigm. Similar arguments and results are available for the flat, Zariski and Nisnevich topologies (for example), and for variations of the big site such as the smooth site. These results are very useful for cohomology calculations.

We close this section with a general result (Proposition 5.29) about simplicial objects S in a site $\mathcal{C}$; this result is effectively a nonabelian version of a cohomology isomorphism which is given by Lemma 8.34 below. In colloquial terms, this result asserts that cohomological invariants for such an object S can be computed either in simplicial presheaves on $\mathcal{C}$, or in simplicial presheaves on a site $\mathcal{C}/S$ which is fibred over S. The latter is the usual setting for the classical approach to the cohomology of simplicial schemes [26].

Suppose that S is a simplicial object in the site $\mathcal{C}$. The site $\mathcal{C}/S$ *fibred over* S has for objects all morphisms $U \to S_n$, and for morphisms all commutative diagrams

$$\begin{array}{ccc} U & \xrightarrow{\phi} & V \\ \downarrow & & \downarrow \\ S_n & \xrightarrow{\theta^*} & S_m \end{array} \tag{5.5}$$

where θ^* is a simplicial structure map. The covering families of the site $\mathcal{C}$ are the families

$$\begin{array}{ccc} U_i & \xrightarrow{\phi_i} & U \\ \downarrow & & \downarrow \\ S_n & \xrightarrow{1} & S_n \end{array}$$

where the family $U_i \to U$ is covering for U in $\mathcal{C}$.

There is a simplicial object 1_S in $\mathcal{C}/Y$, with n-simplices given by the identity $1 : S_n \to S_n$, and with the diagrams

$$\begin{array}{ccc} S_n & \xrightarrow{\theta^*} & S_m \\ 1 \downarrow & & \downarrow 1 \\ S_n & \xrightarrow{\theta^*} & S_m \end{array}$$

as structure maps. This simplicial object represents a simplicial presheaf on $\mathcal{C}/S$, which will also be denoted by 1_S.

There is a functor $q : \mathcal{C}/S \to \mathcal{C}$ which takes the morphism (5.5) to the morphism $\phi : U \to V$ of $\mathcal{C}$. Composition with q defines a restriction functor

$$q_* : s\mathbf{Pre}(\mathcal{C}) \to s\mathbf{Pre}(\mathcal{C}/S),$$

and we write

$$X|_S = q_*(X)$$

5.3 Geometric and Site Morphisms

for simplicial presheaves X on $\mathcal{C}$.

There are obvious inclusions $j_n : \mathcal{C}/S_n \to \mathcal{C}/S$ which induce restriction functors

$$j_{n*} : s\mathbf{Pre}(\mathcal{C}/S) \to s\mathbf{Pre}(\mathcal{C}/S_n)$$

by precomposition. The composite

$$\mathcal{C}/S_n \xrightarrow{j_n} \mathcal{C}/S \xrightarrow{q} \mathcal{C}$$

is an instance of the forgetful functor $\mathcal{C}/S_n \to \mathcal{C}$ (see Lemma 5.25). We shall denote this composite functor by q_n.

Precomposition with j_n defines a restriction functor

$$j_{n*} : s\mathbf{Pre}(\mathcal{C}/S) \to s\mathbf{Pre}(\mathcal{C}/S_n).$$

The restriction functor j_{n*} has a left adjoint

$$j_n^p : s\mathbf{Pre}(\mathcal{C}/S_n) \to s\mathbf{Pre}(\mathcal{C}/S)$$

which is defined for a simplicial presheaf X by

$$j_n^p(X)(U \xrightarrow{\phi} S_m) = \bigsqcup_{\mathbf{n} \xrightarrow{\theta} \mathbf{m}} X(U \xrightarrow{\phi} S_m \xrightarrow{\theta^*} S_n).$$

This functor j_n^p preserves cofibrations. The functors j_n^p and j_{n*} both commute with the Ex^∞ functor and preserve local trivial fibrations, and therefore both preserve local weak equivalences. It follows that the functors j_{n*} preserve injective fibrant models.

We also conclude that a map $f : X \to Y$ of simplicial presheaves on $\mathcal{C}/S$ is a local weak equivalence if and only if the restrictions $j_{n*}f : j_{n*}X \to j_{n*}Y$ are local weak equivalences on $\mathcal{C}/S_n$ for all n.

Proposition 5.29 *Suppose that S is a simplicial object in a site $\mathcal{C}$ and that Z is an injective fibrant simplicial presheaf on $\mathcal{C}$. Choose an injective fibrant model $j : Z|_S \to W$ on $\mathcal{C}/S$. Then there is a weak equivalence of simplicial sets*

$$\mathbf{hom}(S, Z) \simeq \mathbf{hom}(*, W).$$

This weak equivalence is natural in the map $j : Z|_S \to W$.

Proof There is an isomorphism

$$1_S(U \to S_m) \cong \Delta^m$$

of simplicial sets. It follows that the map $1_S \to *$ is a sectionwise weak equivalence on $\mathcal{C}/S$.

The restricted object $Z|_S$ satisfies descent. In effect, the restricted map

$$j_* : j_{n*}(Z|_S) \to j_{n*}(W)$$

is a local weak equivalence of simplicial presheaves on $\mathcal{C}/S_n$ for all $n \geq 0$. The restriction $j_{n*}(Z|_S) = q_{n*}(Z)$ is injective fibrant on $\mathcal{C}/S_n$ for all $n \geq 0$ by Lemma 5.25 and the discussion above. Local weak equivalences of injective fibrant objects are sectionwise weak equivalences, and it follows that the maps

$$Z|_S(\phi) \to W(\phi)$$

are weak equivalences of simplicial sets for all objects $\phi : U \to S_n$ of $\mathcal{C}/S$.

It follows that the maps

$$Z|_S(1_{S_n}) \to W(1_{S_n})$$

are weak equivalences for all $n \geq 0$. There is an isomorphism

$$Z|_S(1_{S_n}) \cong Z(S_n)$$

of cosimplicial spaces. It follows from Lemma 5.30 below that there is a weak equivalence

$$\operatorname*{holim}_n Z|_S(1_{S_n}) \simeq \mathbf{hom}(S, Z).$$

There are also weak equivalences

$$\operatorname*{holim}_n Z|_S(1_{S_n}) \xrightarrow{\simeq} \operatorname*{holim}_n W(1_{S_n}) \simeq \mathbf{hom}(1_S, W) \xleftarrow{\simeq} \mathbf{hom}(*, W)$$

since W is injective fibrant on $\mathcal{C}/S$ and the map $1_S \to *$ is a local weak equivalence.

If the following result looks familiar, it should. Lemma 5.30 generalizes the well-known theorem of Bousfield–Kan [14] which asserts that there is a natural weak equivalence of simplicial sets

$$\operatorname{Tot} X \simeq \operatorname*{holim}_n X_n$$

for all cosimplicial spaces X which are Bousfield–Kan fibrant. The main idea in the proof of Lemma 5.30 is that, for a cosimplicial object S in the site $\mathcal{C}$, the cosimplicial space $\mathbf{n} \mapsto Z(S_n)$ is Bousfield–Kan fibrant if the simplicial presheaf Z is injective fibrant.

Lemma 5.30 *Suppose that the simplicial presheaf S is represented by a simplicial object in the site $\mathcal{C}$, and suppose that Z is an injective fibrant simplicial presheaf. Then there is a weak equivalence*

$$\mathbf{hom}(S, Z) \simeq \operatorname*{holim}_n Z(S_n).$$

Proof Let $Z(S)$ be the cosimplicial space with

$$Z(S)^n = Z(S_n)$$

for $n \geq 0$.

5.3 Geometric and Site Morphisms

We begin by carefully establishing an identification ((5.6) below) which is well-known [14, X.3.2]. There is a natural bijection

$$\hom(A, Z(S)) \cong \hom(A \otimes S, Z)$$

relating morphisms of cosimplicial spaces to morphisms of simplicial presheaves.

Here, $A \otimes S$ is a coend in the sense that it is described by the coequalizer

$$\bigsqcup_{\theta: \mathbf{m} \to \mathbf{n}} A^m \times S_n \rightrightarrows \bigsqcup_n A^n \times S_n \to A \otimes S$$

in simplicial presheaves. Observe that the simplicial presheaf S also is a coend, in that there is a coequalizer

$$\bigsqcup_{\theta: \mathbf{m} \to \mathbf{n}} \Delta^m \times S_n \rightrightarrows \bigsqcup_n \Delta^n \times S_n \to S,$$

so there is an isomorphism $S \cong \Delta \otimes S$.

A cosimplicial space map $\Delta \times \Delta^n \to Z(S)$ therefore corresponds uniquely to a simplicial presheaf map

$$S \times \Delta^n \cong (\Delta \otimes S) \times \Delta^n \to Z,$$

and it follows that there is a natural isomorphism

$$\operatorname{Tot} Z(S) \cong \mathbf{hom}(S, Z). \tag{5.6}$$

The degenerate part DS_n of the presheaf S_n is a subobject of S_n, and is defined by a coequalizer

$$\bigsqcup_{i<j} S_{n-2} \rightrightarrows \bigsqcup_i S_{n-1} \xrightarrow{s} DS_n,$$

where the map s is induced by the degeneracy $s_i : S_{n-1} \to S_n$ on the summand corresponding to i. The cofibration $DS_n \subset S_n$ induces a Kan fibration

$$Z(S)^n \cong \mathbf{hom}(S_n, Z) \to \mathbf{hom}(DS_n, Z) = M^{n-1} Z(S)$$

since Z is injective fibrant. The cosimplicial space $Z(S)$ is therefore Bousfield–Kan fibrant [14, X.4.6], and so the canonical map

$$\operatorname{Tot} Z(S) \to \underleftarrow{\operatorname{holim}}_n Z(S_n)$$

is a weak equivalence of simplicial sets [14, XI.4.4].

Remark 5.31 The homotopy inverse limit for a cosimplicial space X can be defined by

$$\underleftarrow{\operatorname{holim}}_n X = \underleftarrow{\lim}_n Z,$$

where $j : X \to Z$ is an injective fibrant model for X in the category of cosimplicial spaces, or diagrams in spaces indexed on the ordinal number category. This follows from the fact that every injective fibrant cosimplicial space is Bousfield–Kan fibrant.

5.4 Descent Theorems

Recall from Sect. 5.2 that a simplicial presheaf X satisfies descent if some (hence any) injective fibrant model $j : X \to Z$ for X is a sectionwise weak equivalence.

One can, more concretely, ask for a set of criteria for objects S of the underlying site such that the induced simplicial set map

$$j : X(S) \to Z(S) \tag{5.7}$$

is a weak equivalence for all such objects S.

Injective fibrant models do not have explicit constructions. Thus, if one can find a set of geometric criteria for a simplicial presheaf X and a set of objects S (particularly schemes, in applications) such that maps of the form (5.7) are weak equivalences, then one has probably found a major result. This result would then be called a descent theorem. The outcome of a descent theorem is that one then has available the tools of sheaf cohomology theory for calculating the homotopy groups of the space $X(S)$.

Variants on this theme are possible, and do occur: one could, for example, ask for criteria on S such that the induced homomorphisms

$$j_* : \pi_p X(S) \to \pi_p Z(S)$$

in homotopy groups are isomorphisms for p sufficiently large.

Two fundamental descent theorems are proved in this section. These are the Brown–Gersten descent theorem (Theorem 5.33) for the Zariski topology [17], and the Morel–Voevodsky descent theorem (Theorem 5.39) for the Nisnevich topology [82]. The latter depends on the former, and is often called "Nisnevich descent". A special case of this result was first proved, with a different method, by Nisnevich for algebraic K-theory with torsion coefficients [83].

The other major early descent statement from K-theory is Thomason's descent theorem for Bott periodic algebraic K-theory [99]. The proof of this result involved the first application of Nisnevich descent outside of Nisnevich's original paper. Nisnevich descent has since become a central feature of motivic homotopy theory [82, 57]. See also [56].

Before the advent of motivic homotopy theory, most attempts to compute algebraic K-groups by cohomological methods focused on descent questions for the étale topology, and usually involved trying to prove special cases of the Lichtenbaum–Quillen conjecture.

The Lichtenbaum–Quillen conjecture, subject to suitable niceness hypotheses on a scheme S (including finite dimension d), says that if ℓ is a prime which is distinct from the residue characteristics of S and $K/\ell \to Z$ is a stably fibrant model for the mod ℓ K-theory presheaf of spectra (see Sect. 10.1), then the induced map

$$\pi_p K/\ell(S) \to \pi_p Z(S)$$

in stable homotopy groups is an isomorphism for $p \geq d-1$. The stably fibrant model Z is alternatively called the étale K-theory presheaf of spectra, and is often denoted

5.4 Descent Theorems

by K/ℓ^{et}. This conjecture, which was the main open problem of its time in algebraic K-theory, has been discussed at length in the literature—see, for example, [56] and [99]. The Lichtenbaum–Quillen conjecture is now known to be a consequence of the Bloch–Kato conjecture [96].

All topological versions of the traditional algebraic K-theory presheaf of spectra are constructed in this way: start with a variant of the K-theory presheaf of spectra on a category of schemes or varieties [68], call it K, and then find a stably fibrant model $K \to Z$ for the presheaf of spectra K relative to some topology **T**, as in Chap. 10 below. The stably fibrant object

$$K_{\mathbf{T}} := Z$$

is the "**T**" K-theory presheaf of spectra. Thus, one speaks of Zariski K-theory, Nisnevich K-theory, and étale K-theory, according to the topology.

More generally, one can take any simplicial presheaf or presheaf of spectra E, and take a fibrant model $E \to E_{\mathbf{T}}$ of E with respect to the topology **T**, to obtain the **T**-topologized version $E_{\mathbf{T}}$ of the presheaf E.

We begin with the results of Brown and Gersten for the Zariski topology.

Theorem 5.32 *Suppose that S is a Noetherian scheme of finite dimension. Suppose that X is a simplicial presheaf on the small Zariski site $Zar|_S$ such that*

1) the space $X(\emptyset)$ is contractible
2) all stalks X_x of X are contractible
3) the diagram

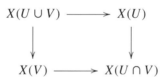

is homotopy cartesian for each pair of open subsets U, V of S.

Then all spaces $X(U)$ are contractible.

Proof We first show that $X(U)$ is nonempty for all open $U \subset S$, under the assumption that $X(U)$ is contractible if $X(U) \neq \emptyset$.

Suppose that $X(U) = \emptyset$ and pick a maximal open subset V such that $X(V) \neq \emptyset$. Such a V exists because all stalks are contractible, hence nonempty, and S is Noetherian. Pick an element $x \in U - V$. Then x has an open neighbourhood $W \subset U$ such that $X(W) \neq \emptyset$. But then the diagram

$$\begin{array}{ccc} X(V \cup W) & \longrightarrow & X(W) \\ \downarrow & & \downarrow \\ X(V) & \longrightarrow & X(V \cap W) \end{array}$$

is homotopy cartesian, and it follows that $X(V \cup W)$ is nonempty. In effect, either $V \cap W$ is empty or this is a contradiction to the maximality of V, and so $X(U) \ne \emptyset$.

Given $\alpha \in \pi_k X(U)$ there is a maximal open subset $V \subset U$ such that $\alpha \mapsto 0 \in \pi_k X(V)$. If $U - V \ne \emptyset$ there is a point $x \in U - V$. Pick a data set (α, U, V, x) which meets these conditions and such that the point x has maximal dimension.

There is an open neighbourhood $x \in W$ such that $\alpha \mapsto 0 \in \pi_k X(W)$ since all stalks are contractible. From the exact sequence

$$\pi_{k+1} X(V \cap W) \xrightarrow{\partial} \pi_k X(V \cup W) \to \pi_k X(V) \oplus \pi_k X(W),$$

there is a $\beta \in \pi_{k+1}(V \cap W)$ such that $\partial(\beta)$ is the restriction of α to $V \cup W$.

There is a maximal subset $V' \subset V \cap W$ such that $\beta \mapsto 0 \in \pi_{k+1} X(V')$.

The closure $\bar{x}$ of the point x is an irreducible component of $U - V'$, for otherwise there is a point $y \in U - V'$ with corresponding data set $(\beta, U \cap V, V', y)$ such that $\dim(y) > \dim(x)$, contradicting the maximality of the dimension of x.

There is a decomposition

$$U - V' = \bar{x} \cup C_1 \cup \cdots \cup C_k$$

as irreducible components. Write

$$F = C_1 \cup \cdots \cup C_k.$$

Then $\bar{x} \cap V = \emptyset$ by construction, so that

$$(V \cap W) - V' \subset (U - V') \cap V \subset F,$$

and it follows that

$$V' \subset V \cap (W - F) \subset V \cap (U - F) \subset V',$$

and so $V' = V \cap (W - F)$.

Finally, $x \in W - F \ne \emptyset$, for otherwise $x \in F$ and $\bar{x} = C_i$ for some i. Then

$$\alpha \mapsto 0 \in \pi_k X(V \cup (W - F)),$$

by comparing exact sequences, and this contradicts the maximality of V if $U - V \ne \emptyset$. Thus, $U = V$ and $\alpha = 0 \in \pi_k X(U)$.

The following result is the Brown–Gersten descent theorem:

Theorem 5.33 *Suppose that X is a simplicial presheaf on the Zariski site $Zar|_S$ such that*

1) the map $X(\emptyset) \to *$ is a weak equivalence, and
2) the diagram

$$\begin{array}{ccc} X(U \cup V) & \longrightarrow & X(U) \\ \downarrow & & \downarrow \\ X(V) & \longrightarrow & X(U \cap V) \end{array}$$

5.4 Descent Theorems

which is associated to each pair of open subsets U, V of S is homotopy cartesian.

Let $j : X \to Z$ be an injective fibrant model for the Zariski topology. Then j is a sectionwise equivalence.

Proof It suffices to show that the induced map $j : X(S) \to Z(S)$ is a weak equivalence. The map $X(U) \to Z(U)$ is global sections of the restriction of $j|_U$ to the Zariski site $Zar|_U$, for all open subschemes $U \subset S$, and the restricted map $j|_U$ is an injective fibrant model—see Example 5.28.

We shall also assume that the simplicial set $Z(S)$ is nonempty. Otherwise, if $Z(S)$ is empty, then so is $X(S)$ and the map $X(S) \to Z(S)$ is a weak equivalence.

Following Proposition 2.22, choose a factorization

such that i is a sectionwise equivalence and p is a sectionwise Kan fibration. Then, the simplicial presheaf Y satisfies conditions 1) and 2) of the statement of the Theorem, and the local weak equivalence $p : Y \to Z$ is an injective fibrant model for the simplicial presheaf Y.

Suppose that $x \in Z(S)$ is a vertex of $Z(S)$, and form the pullback diagram

in simplicial presheaves. Then, the simplicial presheaf F_x satisfies the conditions of Theorem 5.32, and is therefore sectionwise contractible.

The map $F_x(\emptyset)$ is contractible since $Y(\emptyset)$ is contractible and $Z(\emptyset)$ is contractible. For the latter claim, use the fact that Z is sectionwise equivalent to an injective fibrant simplicial sheaf (Theorem 5.9, Corollary 5.13).

The map $F_x(S) \to *$ is a weak equivalence, so that the simplicial set $F_x(S)$ is nonempty, and the vertex x lifts to $Y(S)$. This is true for all vertices of $Z(S)$, so the induced map $\pi_0 Y(S) \to \pi_0 Z(S)$ is surjective.

All fibres $F_{p(y)}$ associated to all vertices $y \in Y(S)$ are sectionwise contractible. It follows that the map $\pi_0 Y(S) \to \pi_0 Z(S)$ is injective, and that all homomorphisms

$$\pi_n(Y(S), y) \to \pi_n(Z(S), p(y))$$

are isomorphisms.

Suppose that $T \to S$ is an étale morphism. An *elementary distinguished square* is a pullback diagram in $(et|_S)_{Nis}$

$$\begin{array}{ccc} \phi^{-1}(U) & \longrightarrow & V \\ \downarrow & & \downarrow \phi \\ U & \xrightarrow{j} & T \end{array} \qquad (5.8)$$

such that j is an open immersion, the morphism ϕ is is étale, and such that the induced morphism

$$\phi^{-1}(T - U) \to T - U$$

of closed subschemes (with reduced structure) is an isomorphism.

Example 5.34 If U and V are open subschemes of T, then the diagram of inclusions

$$\begin{array}{ccc} U \cap V & \longrightarrow & V \\ \downarrow & & \downarrow \\ U & \longrightarrow & U \cup V \end{array}$$

is an elementary distinguished square.

Example 5.35 Suppose that $x \in T$ is a closed point of T, and suppose that the map $\phi : U \to T$ is an étale morphism which has finite fibres. Suppose also that there is a section

$$\begin{array}{ccc} & & U \\ & \nearrow^{y} & \downarrow \phi \\ \mathrm{Sp}(k(x)) & \xrightarrow{x} & T \end{array}$$

over the residue field $k(x)$ of x. The set-theoretic fibre $\phi^{-1}(x)$ is a finite set of closed points, of the form

$$\phi^{-1}(x) = \{y, y_1, \ldots, y_k\}.$$

Let V be the open subset $U - \{y_1, \ldots, y_k\}$ of U, and let ϕ also denote the restriction of ϕ to V. Then there is a diagram

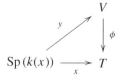

5.4 Descent Theorems

The map ϕ induces an isomorphism

$$\mathrm{Sp}\,(k(y)) \cong \mathrm{Sp}\,(k(x)),$$

and $\mathrm{Sp}\,(k(y))$ is the reduced closed fibre of $\phi : V \to T$ over the closed subscheme $\mathrm{Sp}\,(k(x))$ of T. Let U be the open subscheme $T - \{x\}$ of T, with inclusion $j : U \subset T$. It follows that the pullback diagram

$$\begin{array}{ccc} \phi^{-1}(U) & \longrightarrow & V \\ \downarrow & & \downarrow \phi \\ U & \xrightarrow{j} & T \end{array}$$

is an elementary distinguished square.

Every elementary distinguished square defines a Nisnevich cover $\{j : U \subset T, \phi : V \to T\}$ of T, because every residue field map $\mathrm{Sp}\,(k(x)) \to T$ for T factors through one of the two maps ϕ and j.

Following [82], say that a simplicial presheaf X on the Nisnevich site $(et|_S)_{Nis}$ has the *BG-property* if

1) the space $X(\emptyset)$ is contractible, and
2) the simplicial presheaf X takes elementary distinguished squares (5.8) to homotopy cartesian diagrams

$$\begin{array}{ccc} X(T) & \xrightarrow{j^*} & X(U) \\ \phi^* \downarrow & & \downarrow \\ X(V) & \longrightarrow & X(\phi^{-1}(U)) \end{array}$$

of simplicial sets.

If the simplicial presheaf X has the BG-property and U, V are open subschemes of a scheme T is étale over S, then the diagram

$$\begin{array}{ccc} X(U \cup V) & \longrightarrow & X(V) \\ \downarrow & & \downarrow \\ X(U) & \longrightarrow & X(U \cap V) \end{array}$$

is homotopy cartesian, so that the restriction of X to the Zariski site $Zar|_T$ satisfies the conditions of for Brown-Gersten descent (Theorem 5.33).

Lemma 5.36 *Suppose that Z is an injective fibrant simplicial presheaf on the Nisnevich site $(et|_S)_{Nis}$. Then Z has the BG-property.*

Proof Every open immersion $j : U \to T$ is a cofibration of simplicial presheaves, and all induced inclusions

$$(U \times \Delta^n) \cup (T \times \Lambda^n_k) \subset T \times \Delta^n$$

are trivial cofibrations. It follows that the map $j^* : Z(T) \to Z(U)$ is a Kan fibration.

The square (5.8) is a pushout in the category of sheaves (and simplicial sheaves) on $et|_S)_{Nis}$. Thus, if Z' is an injective fibrant simplicial sheaf, then the diagram of simplicial set maps

$$\begin{array}{ccc} Z'(T) & \xrightarrow{j^*} & Z'(U) \\ \phi^* \downarrow & & \downarrow \\ Z'(V) & \longrightarrow & Z'(\phi^{-1}(U)) \end{array}$$

is a pullback in which both horizontal maps are Kan fibrations, and is therefore homotopy cartesian.

There is a local weak equivalence $\eta : Z \to Z'$ such that Z' is an injective fibrant simplicial sheaf (Theorem 5.9). The map η is a sectionwise weak equivalence by Corollary 5.13, and the property of taking elementary distinguished squares to homotopy cartesian diagrams is an invariant of sectionwise equivalence.

The map η induces a weak equivalence

$$Z(\emptyset) \to Z'(\emptyset) \cong *$$

of simplicial sets, by assumption.

Here is the analogue of Theorem 5.32 for the Nisnevich topology:

Theorem 5.37 *Suppose that S is a Noetherian scheme of finite dimension. Suppose that X is a simplicial presheaf on the Nisnevich site $(et|_S)_{Nis}$ such that*

1) X has the BG-property, and
2) the map $X \to *$ is a local weak equivalence for the Nisnevich topology.

Then the map $X(S) \to *$ is a weak equivalence of simplicial sets.

Proof We prove this result by showing that all Zariski stalks of X on S are contractible. We then invoke Theorem 5.33.

Write $\mathcal{O}_x$ for the local ring $\mathcal{O}_{x,S}$ of $x \in S$, and let $x : \text{Sp}(\mathcal{O}_x) \to S$ be the canonical map. Let x^p be the left adjoint of the direct image functor

$$x_* : \mathbf{Pre}(et|_{\text{Sp}(\mathcal{O}_x)})_{Nis} \to \mathbf{Pre}(et|_S)_{Nis}.$$

The simplicial set

$$X(\mathcal{O}_x) := x^p X(\mathcal{O}_x)$$

5.4 Descent Theorems

is the Zariski stalk of X at the point x. The functor x^p preserves local weak equivalences for the Nisnevich topology, since it is defined by a site morphism (Corollary 5.24).

It is a consequence of Lemma 5.38 below that the simplicial presheaf $x^p X$ satisfies the BG-property on the site $(et|_{Sp(\mathcal{O}_x)})_{Nis}$.

Suppose that the point x has dimension 0, so that $\mathcal{O}_x$ is an Artinian local ring. The functor

$$U \mapsto U \times_{Sp(\mathcal{O}_x)} Sp(k(x))$$

defines an equivalence of categories

$$et|_{Sp(\mathcal{O}_x)} \to et|_{Sp(k)}$$

(see [79, I.3.23]). Every diagram

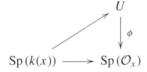

with ϕ étale therefore determines a section $\sigma : Sp(\mathcal{O}_x) \to U$ of the map ϕ. It follows that the global sections functor $Y \mapsto Y(\mathcal{O}_x)$ for simplicial presheaves Y on $et|_{Sp(\mathcal{O}_x)}$ takes sheaf epimorphisms for the Nisnevich topology to surjections, and hence is exact. The functor $Y \mapsto Y(\mathcal{O}_x)$ therefore takes local weak equivalences for the Nisnevich topology on $et|_{Sp(\mathcal{O}_x)}$ to weak equivalences of simplicial sets.

In particular, the simplicial set $X(\mathcal{O}_x) = x^p(\mathcal{O}_x)$ is contractible for points x of dimension 0.

We show by induction on the dimension of $x \in S$ that $X(\mathcal{O}_x)$ is contractible for all points x of S. Take an element $x \in S$ and assume that $X(\mathcal{O}_y)$ is contractible for all points y (in all étale S-schemes) of smaller dimension.

Write x for the closed point of $Sp(\mathcal{O}_x)$, and suppose given an element $\alpha \in \pi_k X(\mathcal{O}_x)$. Then α is 0 locally for the Nisnevich topology, so that, following the prescription of Example 5.35, there is an étale morphism $\phi : V \to Sp(\mathcal{O}_x)$ with a diagram

$$\begin{array}{ccc}
V \times_{Sp(\mathcal{O}_x)} Sp(k(x)) & \longrightarrow & V \\
\cong \downarrow & & \downarrow \phi \\
Sp(k(x)) & \xrightarrow{x} & Sp(\mathcal{O}_x)
\end{array}$$

such that $\phi^*(\alpha) = 0$ in $\pi_k x^p X(V)$. The simplicial set $x^p X(V)$ is nonempty, since the simplicial presheaf $x^p X$ has contractible (hence nonempty) stalks for the Nisnevich topology.

Write $U = \mathrm{Sp}(\mathcal{O}_x) - \{x\}$. Then all points of U and all points of $\phi^{-1}(U)$ have dimension smaller than that of x, and the simplicial presheaf $x^p X$ has the BG-property. Brown–Gersten descent (Theorem 5.32) and the inductive assumption together imply that the spaces $x^p X(U)$ and $x^p X(\phi^{-1}(U))$ are contractible. It follows that the map

$$\phi^* : X(\mathcal{O}_x) = x^p X(\mathcal{O}_x) \to x^p X(V)$$

is a weak equivalence, but then $\alpha = 0$ in $\pi_k X(\mathcal{O}_x)$.

All homotopy groups and the set of path components of $X(\mathcal{O}_x)$ are therefore trivial if the space $X(\mathcal{O}_x)$. The simplicial set $X(\mathcal{O}_x)$ is also nonempty, and it follows that $X(\mathcal{O}_x)$ is weakly equivalent to a point.

Lemma 5.38 *Suppose that the simplicial presheaf X on $(et|_S)_{Nis}$ has the BG-property, and let $\mathcal{O}_x$ be the local ring of $x \in S$ with canonical map $x : \mathrm{Sp}(\mathcal{O}_x) \to S$. Then, the inverse image $x^p X$ on the site $(et|_{\mathrm{Sp}(\mathcal{O}_x)})_{Nis}$ has the BG-property.*

Proof Suppose that the morphism $f : T \to \mathrm{Sp}(\mathcal{O}_x)$ is étale. Then, there is an open affine neighbourhood U of x in S and a U-scheme $f' : T' \to U$ which is étale, along with an isomorphism of $\mathcal{O}_x$-schemes

$$T \cong \mathrm{Sp}(\mathcal{O}_x) \times_U T'$$

If f is an open immersion (respectively closed immersion), then the "thickening" f' can be chosen to have the same property. In particular, if $\phi : V \to \mathrm{Sp}(\mathcal{O}_x)$ is étale and has étale thickening $\phi' : V' \to U$ over an open neighbourhood U, then there is an isomorphism

$$x^p X(V) = \varinjlim_{x \in W \subset U} X(W \times_U V'),$$

where W varies over the open neighbourhoods of x which are contained in U.

It follows that every elementary distinguished square

$$\begin{array}{ccc} \phi^{-1}(U) & \longrightarrow & V \\ \downarrow & & \downarrow \phi \\ U & \xrightarrow{j} & T \end{array}$$

over $\mathrm{Sp}(\mathcal{O}_x)$ has a thickening

$$\begin{array}{ccc} (\phi')^{-1}(U') & \longrightarrow & V' \\ \downarrow & & \downarrow \phi' \\ U' & \xrightarrow{j'} & T' \end{array}$$

5.4 Descent Theorems

which is defined over some affine open neighbourhood of x in S, and that the diagram

$$\begin{array}{ccc} x^p X(T) & \longrightarrow & x^p X(U) \\ \downarrow & & \downarrow \\ x^p X(V) & \longrightarrow & x^p X(\phi^{-1}(U)) \end{array} \qquad (5.9)$$

is a filtered colimit of homotopy cartesian diagrams

$$\begin{array}{ccc} X(T') & \longrightarrow & X(U') \\ \downarrow & & \downarrow \\ X(V') & \longrightarrow & X(\phi'^{-1}(U')) \end{array}$$

The diagram (5.9) is therefore homotopy cartesian.

The space $x^p X(\emptyset)$ is isomorphic to the space $X(\emptyset)$, and is therefore contractible.

The following result is the Morel–Voevodsky descent theorem [82, Prop. 1.16]. It is the analogue for the Nisnevich topology of Theorem 5.33 and has the same proof, in the presence of Theorem 5.37. It is also, commonly, called the Nisnevich descent theorem.

Theorem 5.39 *Suppose that S is a Noetherian scheme of finite dimension. Suppose that X is a simplicial presheaf on the Nisnevich site $(et|_S)_{Nis}$ which satisfies the BG-property, and let $j : X \to Z$ be an injective fibrant model for the Nisnevich topology. Then, the induced map $X(S) \to Z(S)$ is a weak equivalence of simplicial sets.*

Proof The map j can be replaced up to sectionwise weak equivalence by a sectionwise Kan fibration $p : Y \to Z$. It suffices to show that the induced map $p : Y(S) \to Z(S)$ in global sections is a weak equivalence of simplicial sets. We can assume that the simplicial set $Z(S)$ is nonempty.

The fibres F_x of the map p over all global vertices $x \in Z(S)$ satisfy the conditions of Theorem 5.37, and are therefore sectionwise equivalent to a point. It follows that the map $\pi_0 Y(S) \to \pi_0 Z(S)$ is a bijection, and all homomorphisms $\pi_n(Y(S), y) \to \pi_n(Z(S), p(y))$ are isomorphisms.

Corollary 5.40 *Suppose that S is a Noetherian scheme of finite dimension. Suppose that $f : X \to Y$ is a local weak equivalence of simplicial presheaves on the Nisnevich site $(et|_S)_{Nis}$), and that X and Y have the BG-property. Then, the induced map $f : X(S) \to Y(S)$ is a weak equivalence of simplicial sets.*

Corollary 5.40 is equivalent to Theorem 5.39, via Corollary 5.13 and Lemma 5.36.

5.5 Intermediate Model Structures

Throughout this section, suppose that **M** is a model structure on the category $s\mathbf{Pre}(\mathcal{C})$ of simplicial presheaves on $\mathcal{C}$ for which the cofibrations are the monomorphisms. Suppose that every local weak equivalence is a weak equivalence of **M**. Every object is cofibrant in the model structure **M**, so that **M** is a left proper model structure.

The class of weak equivalences of **M** is closed under products with simplicial sets. In effect, if $f : X \to Y$ is a weak equivalence of **M** then each induced map

$$f \times 1 : X \times \Delta^n \to Y \times \Delta^n$$

is locally equivalent and hence weakly equivalent to f in **M**, and is therefore a weak equivalence of **M**. It follows by an induction on skeleta (which involves the left properness of **M**) that the map

$$f \times 1 : X \times K \to Y \times K$$

is a weak equivalence of **M** for each simplicial set K.

The standard function complex construction $\mathbf{hom}(X, Y)$ therefore gives **M** the structure of a simplicial model category.

Recall, from Proposition 2.22, that there is a *projective model structure* on the category $s\mathbf{Pre}(\mathcal{C})$ of simplicial presheaves, for which the fibrations are sectionwise Kan fibrations and the weak equivalences are sectionwise weak equivalences.

The cofibrations for this theory are the *projective cofibrations*, and this class of maps has a generating set S_0 consisting of all maps $L_U(\partial \Delta^n) \to L_U(\Delta^n)$, $n \geq 0$, $U \in \mathcal{C}$. The functor L_U is the left adjoint of the U-sections functor $X \mapsto X(U)$ from simplicial presheaves to simplicial sets.

Write $\mathbf{C}_P$ for the class of projective cofibrations, and write $\mathbf{C}$ for the full class of cofibrations, which are the simplicial presheaf monomorphisms. Every projective cofibration is a cofibration, so there is a relation $\mathbf{C}_P \subset \mathbf{C}$.

Let S be any set of cofibrations which contains S_0. Let $\mathbf{C}_S$ be the saturation of the set of all cofibrations of the form

$$(B \times \partial \Delta^n) \cup (A \times \Delta^n) \subset B \times \Delta^n$$

which are induced by members $A \to B$ of the set S. Say that $\mathbf{C}_S$ is the class of *S-cofibrations*.

An *S-fibration* is a map $p : X \to Y$ of simplicial presheaves which has the right lifting property with respect to all S-cofibrations which are weak equivalences of **M**. Observe that every fibration of **M** is an S-fibration.

Theorem 5.41 *Let **M** be a model structure on the category $s\mathbf{Pre}(\mathcal{C})$ of simplicial presheaves for which the cofibrations are the monomorphisms, and suppose that every local weak equivalence is a weak equivalence of **M**. Then, the category $s\mathbf{Pre}(\mathcal{C})$, together with the classes of S-cofibrations, weak equivalences of **M**, and S-fibrations, satisfies the axioms for a left proper closed simplicial model category.*

5.5 Intermediate Model Structures

Proof The axioms **CM1**, **CM2** and **CM3** are easily verified.

Any map $f : X \to Y$ has a factorization

where $j \in \mathbf{C}_S$ and p has the right lifting property with respect to all members of $\mathbf{C}_S$. Then p is an S-fibration and is a sectionwise weak equivalence. The map p is therefore a weak equivalence of **M**.

The map f also has a factorization

for which q is a fibration of **M** and i is a trivial cofibration of **M**. Then q is an S-fibration. Factorize the map i as $i = p \cdot j$ where $j \in \mathbf{C}_S$ and p is an S-fibration and a weak equivalence of **M** (as above). Then j is a weak equivalence of **M**, so $f = (qp) \cdot j$ factorizes f as an S-fibration following a map which is an S-cofibration and a weak equivalence of **M**.

We have therefore proved the factorization axiom **CM5**.

It is an exercise to prove **CM4**. One shows that if $p : X \to Y$ is an S-fibration and a weak equivalence of **M**, then p is a retract of a map which has the right lifting property with respect to all S-cofibrations.

Suppose that $j : K \to L$ is a cofibration of simplicial sets. The collection of all cofibrations $i : C \to D$ of simplicial presheaves such that the induced map

$$(D \times K) \cup (C \times L) \to D \times L \tag{5.10}$$

is an S-cofibration is saturated, and contains all generators

$$(B \times \partial \Delta^n) \cup (A \times \Delta^n) \subset B \times \Delta^n$$

of the class $\mathbf{C}_S$. It follows that the map (5.10) is an S-cofibration if $i : C \to D$ is an S-cofibration. This map is a weak equivalence of **M** if either i is a weak equivalence of **M** or j is a weak equivalence of simplicial sets. The model structure of the statement of the theorem is therefore a simplicial model structure, with the standard function complex.

The left properness of this model structure is a consequence of the left properness for the ambient model category **M**.

Example 5.42 The case $S = S_0$ for Theorem 5.41, and where **M** is the injective model structure on $s\mathbf{Pre}(\mathcal{C})$, gives the *projective local model structure* of Blander [9] for simplicial presheaves on $\mathcal{C}$.

If **M** is still the injective model structure on $s\mathbf{Pre}(\mathcal{C})$, but the set of cofibrations S is allowed to vary, then Theorem 5.41 gives the *intermediate model structures* for simplicial presheaves of [64].

All intermediate model structures are right proper, because all S-fibrations are local fibrations and pullbacks of local weak equivalences along local fibrations are local weak equivalences by Lemma 4.37.

Suppose that a simplicial presheaf X is fibrant (ie. S-fibrant) for one of the intermediate model structures, and suppose that $j : X \to Z$ is an injective fibrant model for X. Then Z is S-fibrant, and j is a weak equivalence of S-fibrant objects. The closed simplicial model structure and the properness of the S-structure together guarantee (via the classical replacement of a map by a fibration) that the map j can be replaced up to sectionwise weak equivalence by a trivial S-fibration. Trivial S-fibrations are sectionwise equivalences, so that the map j is a sectionwise weak equivalence.

In other words, all simplicial presheaves X which are fibrant for one of the intermediate model structures satisfy descent in the sense of Sect. 5.2.

Example 5.43 Let $(Sm/T)_{Nis}$ be the Nisnevich site of smooth schemes over a scheme T, and let **M** be the motivic model structure on the category of simplicial presheaves on this site (see Example 7.20 below). The case $S = S_0$ of Theorem 5.41 for the motivic model structure gives the *projective motivic model structure* for $s\mathbf{Pre}((Sm|_T)_{Nis})$—see also [89], [105] and [106]. Theorem 5.41 also gives a large collection of other motivic model structures which are intermediate between the projective and standard motivic model structures.

Similar considerations apply to all localizations of injective model structures on all categories of simplicial presheaves, as in Theorem 7.18 below. There is always a projective version of such a model structure, and a class of intermediate structures between the given localized structure and its corresponding projective structure.

The model structure of Theorem 5.41 is cofibrantly generated, under an extra assumption on the model structure **M** that is satisfied in the usual examples. This was proved for the original intermediate model structures of Example 5.42 by Beke [6], whose method was to verify a solution set condition. Beke's argument can be deconstructed (as in [64]) to give a basic and useful trick for verifying cofibrant generation in the presence of some kind of cardinality calculus. That trick is reprised here, in the proof of Lemma 5.44 below.

The proof of Lemma 5.44 requires the assumption that the model structure **M** satisfies a *bounded monomorphism condition*. This means that there is a set **A** of objects of **M** (which is closed under subobjects and quotients) such that the following condition holds:

5.5 Intermediate Model Structures

Given a diagram

such that i and j are monomorphisms, i is an **M**-trivial cofibration and A is in **A**, there is a subobject B of Y which contains A, such that B is in **A** and the induced map $B \cap X \to B$ is a weak equivalence of **M**.

The bounded monomorphism property for the injective model structure is verified in Lemma 5.2. In that case, **A** is the set of α-bounded objects, where α is a regular cardinal which is bigger than $|\operatorname{Mor}(\mathcal{C})|$.

The use of a cardinality bound to define a set **A** which appears in a bounded monomorphism statement is standard practice.

Choose a regular cardinal α such that $\alpha > |\operatorname{Mor}(\mathcal{C})|$, and that $|D| < \alpha$ for all members $C \to D$ of the set of cofibrations generating $\mathbf{C}_S$. Let **A** be the set of α-bounded simplicial presheaves.

Suppose that $u : A \to B$ is a trivial cofibration of **M** with B in the set **A**. Then u has a factorization

such that j_u is an S-cofibration, p_u is an S-fibration and both maps are weak equivalences of **M**. Write I for the set of all S-cofibrations j_u which are constructed in this way.

Lemma 5.44 *Suppose that, in addition to the assumption of Theorem 5.41, the model structure* **M** *on the category* $s\mathbf{Pre}(\mathcal{C})$ *has the bounded monomorphism property. Then, the members of the set I generate the class of trivial S-cofibrations, and the S-model structure is cofibrantly generated.*

Proof Suppose given a commutative diagram

such that i is a member of $\mathbf{C}_S$ with B in $\mathbf{A}$, and such that f is a weak equivalence of $\mathbf{M}$. Then, since B is in $\mathbf{A}$, this diagram has a factorization

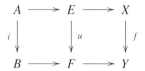

where j is a member of the set of S-cofibrations I.

In effect, by factorizing $f = p \cdot u$, where u is an trivial cofibration and q is a trivial fibration of $\mathbf{M}$, we can assume that f is an $\mathbf{M}$-trivial cofibration. The bounded monomorphism property then implies that there is a factorization

as above with u a trivial cofibration with B in $\mathbf{A}$. The object F is an extension of the image of B in Y, which image is in $\mathbf{A}$ since $\mathbf{A}$ is closed under quotients. Use the factorization $u = p_u j_u$ displayed above. Then p_u is a trivial S-fibration and therefore has the right lifting property with respect to i, and j_u is the desired member of the set I.

Every trivial S-cofibration $j : A' \to B'$ has a factorization

$$\begin{array}{c} A' \xrightarrow{\beta} C' \\ \searrow_{j} \downarrow q \\ B' \end{array}$$

such that β is an S-cofibration in the saturation of the set I and q has the right lifting property with respect to all members of I. Then q is also a weak equivalence of $\mathbf{M}$, and therefore has the right lifting property with respect to all members of the class $\mathbf{C}_S$ of S-cofibrations by the previous paragraph, since all generators of $\mathbf{C}_S$ have targets in the set $\mathbf{A}$. It follows that the lifting problem

has a solution, so that j is a retract of β.

The argument for Lemma 5.44 is useful in multiple contexts. It reappears, for example, in the proof of Lemma 7.3 below.

5.6 Postnikov Sections and n-Types

Recall [32, p. 172] that the nth *Postnikov section* $P_n X$ for a simplicial set X is defined by $P_n X = X/\sim$, where two simplices $\alpha, \beta : \Delta^k \to X$ are related if they restrict to the same map on the subcomplex $\text{sk}_n \Delta^k$. In more categorical terms, the set $P_n X_m$ is the coequalizer of the kernel pair defined by the function $X_m \to \text{cosk}_n X_m$.

It is an exercise to verify the following statements, under the assumption that X is a Kan complex:

1) the canonical map $\pi : X \to P_n X$ is a Kan fibration and $P_n X$ is a Kan complex
2) the induced maps $\pi_0 X \to \pi_0(P_n X)$ and all maps $\pi_k(X, x) \to \pi_k(P_n X, x)$ $k \leq n$, are isomorphisms
3) there are isomorphisms $\pi_k(P_n X, x) = 0$ for $k > n$

The functor $X \mapsto P_n X$ preserves weak equivalences of Kan complexes X, but must be derived in general. For an arbitrary simplicial set Y, define $\mathbf{P}_n Y$ by setting

$$\mathbf{P}_n Y := P_n \text{Ex}^\infty Y.$$

The composite

$$Y \xrightarrow{\simeq} \text{Ex}^\infty Y \xrightarrow{\pi} P_n \text{Ex}^\infty Y = \mathbf{P}_n Y$$

defines a natural map $\eta : Y \to \mathbf{P}_n Y$. The object $\mathbf{P}_n Y$ is the nth *derived Postnikov section* of the simplicial set Y.

The definitions of the functor $Y \to \mathbf{P}_n Y$ and the natural map $\eta : Y \to \mathbf{P}_n Y$ extend to a functor $X \mapsto \mathbf{P}_n X$ and a natural map $\eta : X \to \mathbf{P}_n X$, respectively, for simplicial presheaves X on a site $\mathcal{C}$: the object $\mathbf{P}_n X$ is defined in sections by

$$(\mathbf{P}_n X)(U) := \mathbf{P}_n(X(U))$$

for objects U of the underlying site, and the simplicial presheaf map $\eta : X \to \mathbf{P}_n X$ is the simplicial set map η in sections.

A map $f : X \to Y$ of simplicial presheaves is said to be an *local n-equivalence* if the map $\mathbf{P}_n X \to \mathbf{P}_n Y$ is a local weak equivalence. A simplicial presheaf X is an *n-type* if the canonical map $\eta : X \to \mathbf{P}_n X$ is a local weak equivalence.

Remark 5.45 The definition of n-type given here specializes to the usual notion of n-type in classical homotopy theory, but the description of n-equivalence is nonstandard: the usual definition says that a map $f : X \to Y$ is an n-equivalence if all homotopy fibres of f are n-connected. We shall depart from tradition and use the above definition of n-equivalence in all contexts, even for simplicial sets. Explicitly, a map $f : X \to Y$ of simplicial sets is an n-*equivalence* if the map $\mathbf{P}_n X \to \mathbf{P}_n Y$ is a weak equivalence.

Lemma 5.46

1) If X is a locally fibrant simplicial presheaf on $\mathcal{C}$, then the natural weak equivalence $X \to \operatorname{Ex}^\infty X$ induces a local weak equivalence
$$P_n X \xrightarrow{\simeq} \mathbf{P}_n X.$$

2) The functor $X \mapsto \mathbf{P}_n X$ preserves local weak equivalences of simplicial presheaves X.

Proof Suppose that $p : \mathbf{Shv}(\mathcal{B}) \to \mathbf{Shv}(\mathcal{C})$ is a Boolean localization.

The associated sheaf map $X \to L^2 X$ induces a map $P_n X \to P_n L^2 X$ of simplicial presheaves which induces an isomorphism of simplicial sheaves
$$L^2 P_n X \xrightarrow{\cong} L^2 P_n L^2 X \tag{5.11}$$

on account of the kernel pair description of $P_n X$. Similarly, there is a natural isomorphism
$$L^2 P_n p^* Y \cong p^* L^2 P_n Y \tag{5.12}$$

of simplicial sheaves on $\mathcal{B}$, for all simplicial sheaves Y on $\mathcal{C}$.

To prove statement 1), first suppose that Z is a locally fibrant simplicial sheaf, and observe that the composite local weak equivalence
$$Z \to \operatorname{Ex}^\infty Z \to L^2 \operatorname{Ex}^\infty Z$$

of locally fibrant simplicial sheaves induces a sectionwise weak equivalence
$$L^2 P_n p^* Z \xrightarrow{\simeq} L^2 P_n p^* L^2 \operatorname{Ex}^\infty Z,$$

which map is isomorphic to the map
$$p^* L^2 P_n Z \to p^* L^2 P_n \operatorname{Ex}^\infty Z$$

via the natural isomorphisms (5.11) and (5.12). It follows that the simplicial presheaf map
$$P_n Z \to P_n \operatorname{Ex}^\infty Z$$

is a local weak equivalence.

Suppose that X is a locally fibrant simplicial presheaf. The vertical maps in the diagram

$$\begin{array}{ccc} P_n X & \longrightarrow & P_n \operatorname{Ex}^\infty X \\ \downarrow & & \downarrow \\ P_n L^2 X & \xrightarrow{\simeq} & P_n \operatorname{Ex}^\infty L^2 X \end{array}$$

5.6 Postnikov Sections and n-Types

induce isomorphisms on associated sheaves by (5.11), and the indicated map is a local weak equivalence since $L^2 X$ is a locally fibrant simplicial sheaf. It follows that the map

$$P_n X \to P_n \operatorname{Ex}^\infty X$$

is a local weak equivalence

For statement 2), it is enough to assume, by statement 1), that $f : X \to Y$ is a local weak equivalence of locally fibrant simplicial presheaves and show that the induced map $P_n X \to P_n Y$ is a local weak equivalence. The associated sheaf map $X \to L^2 X$ induces an isomorphism $L^2 P_n X \cong L^2 P_n L^2 X$, by (5.11), so it is enough to assume that X and Y are locally fibrant simplicial sheaves, and then show that the induced map $L^2 P_n X \to L^2 P_n Y$ is a local weak equivalence. But then the induced map $p^* L^2 P_n X \to p^* L^2 P_n Y$ is isomorphic to the map $L^2 P_n p^* X \to L^2 P_n p^* Y$ by (5.12), and the map $p^* X \to p^* Y$ is a sectionwise weak equivalence of sheaves of Kan complexes, so that $P_n p^* X \to P_n p^* Y$ is a sectionwise equivalence of presheaves of Kan complexes, and the desired result follows.

Lemma 5.47 *Suppose given a pullback diagram*

$$\begin{array}{ccc} A & \xrightarrow{\alpha} & X \\ \downarrow & & \downarrow p \\ B & \xrightarrow{\beta} & Y \end{array} \quad (5.13)$$

in the simplicial presheaf category $s\mathbf{Pre}(\mathcal{C})$ such that p is a local fibration. Suppose that Y is an n-type and that the map β is a local n-equivalence. Then the map α is a local n-equivalence.

Proof First of all, suppose that $s\mathbf{Pre}(\mathcal{C})$ is the category of simplicial sets. We can assume that all objects in the diagram (5.13) are Kan complexes and that p is a Kan fibration. One shows that the function $\pi_0 A \to \pi_0 X$ is surjective. The homotopy fibres of the map α are homotopy fibres of the map β, and are therefore n-connected since Y is an n-type. It follows that α is an n-equivalence.

In the more general setting of simplicial presheaves, we can assume that all objects in the diagram are locally fibrant simplicial sheaves by Lemma 5.46. In that case, the map $\alpha : A \to X$ is a local n-equivalence if and only if the map $\alpha_* : L^2 P_n A \to L^2 P_n X$ is a local weak equivalence, and this holds if and only if $p^* A \to p^* B$ is an n-equivalence, where $p : \mathbf{Shv}(\mathcal{B}) \to \mathbf{Shv}(\mathcal{C})$ is a Boolean localization. Further, Y is an n-type if and only if $p^* Y$ is an n-type. We can therefore assume that the diagram (5.12) is in the category of sheaves of Kan complexes on a complete Boolean algebra $\mathcal{B}$. But then α is a sectionwise n-equivalence by the simplicial sets case.

Lemma 5.48 *The simplicial presheaf $\mathbf{P}_n X$ is an n-type and the map $\eta : X \to \mathbf{P}_n X$ is a local n-equivalence, for all simplicial presheaves X.*

Proof If Y is a Kan complex, it is well known that $P_n Y$ is an n-type and that the map $Y \to P_n Y$ is an n-equivalence (in the sense of Remark 5.45). These two statements mean, respectively, that the natural maps $\eta, P_n\eta : P_n Y \to P_n P_n Y$ are weak equivalences.

In the diagram

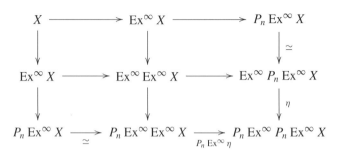

the indicated maps are sectionwise weak equivalences of presheaves of Kan complexes. The map η is sectionwise weakly equivalent to the map $\eta : P_n \operatorname{Ex}^\infty X \to P_n P_n \operatorname{Ex}^\infty X$, and is therefore a sectionwise weak equivalence. The map $P_n \operatorname{Ex}^\infty \eta$ is sectionwise weakly equivalent to the map $P^n \eta : P_n \operatorname{Ex}^\infty X \to P_n P_n \operatorname{Ex}^\infty X$, and is therefore a sectionwise weak equivalence. The vertical composite on the right is $\eta : P_n X \to P_n P_n X$ and the bottom horizontal composite is $P_n \eta$. Both maps are sectionwise hence local weak equivalences, and so the Lemma is proved.

To summarize, the functor $P_n : s\mathbf{Pre}(\mathcal{C}) \to s\mathbf{Pre}(\mathcal{C})$ satisfies the following axioms:

A4 The functor $\mathbf{P}_n$ preserves local weak equivalences.
A5 The maps $\eta, \mathbf{P}_n(\eta) : \mathbf{P}_n Y \to \mathbf{P}_n \mathbf{P}_n Y$ are local weak equivalences.
A6 Suppose given a pullback diagram

in simplicial presheaves with p an injective fibration. Suppose that the maps $\eta : Y \to \mathbf{P}_n Y$, $\eta : X \to \mathbf{P}_n X$ and $\beta_* : \mathbf{P}_n B \to \mathbf{P}_n Y$ are local weak equivalences. Then the map $\alpha_* : \mathbf{P}_n A \to \mathbf{P}_n X$ is a local weak equivalence.

The statement **A4** is part of Lemma 5.46, **A6** is a consequence of Lemma 5.47, and **A5** is a restatement of Lemma 5.48.

An *n-fibration* in $s\mathbf{Pre}(\mathcal{C})$ is a map which has the right lifting property with respect to all cofibrations which are local n-equivalences. Then we have the following:

5.6 Postnikov Sections and n-Types

Theorem 5.49

1) *The category $s\mathbf{Pre}(\mathcal{C})$ of simplicial presheaves, together with the classes of cofibrations, local n-equivalences and n-fibrations, satisfies the axioms for a proper closed simplicial model category.*
2) *A map $p : X \to Y$ is an n-fibration if and only if it is an injective fibration and the diagram*

$$\begin{array}{ccc} X & \xrightarrow{\eta} & \mathbf{P}_n X \\ p \downarrow & & \downarrow p_* \\ Y & \xrightarrow{\eta} & \mathbf{P}_n Y \end{array}$$

is homotopy cartesian for the injective model structure on simplicial presheaves.

This result is due to Biedermann [8]. With the statements **A4**, **A5** and **A6** in place, it is a formal consequence of a result of Bousfield [12]. Bousfield's proof uses a refinement of the Bousfield–Friedlander localization technique [13], which was used in their description of the stable homotopy category. The proof of Theorem 5.49 is outlined here.

Proof [Proof of Theorem 5.49] We verify the existence of the model structure, prove statement 2), and then verify right properness. The model structure is easily seen to be left proper, because every object is cofibrant.

One shows first that $p : X \to Y$ is an n-fibration and a local n-equivalence if and only if p is a trivial injective fibration of simplicial presheaves. In effect, every map which is an n-fibration and a local n-equivalence is a retract of a trivial injective fibration.

The next step is to show that if $p : X \to Y$ is an injective fibration such that the maps $\eta : X \to \mathbf{P}_n X$ and $\eta : Y \to \mathbf{P}_n Y$ are local weak equivalences, then p is an n-fibration. Suppose given a lifting problem

$$\begin{array}{ccc} A & \longrightarrow & X \\ j \downarrow & \nearrow & \downarrow p \\ B & \longrightarrow & Y \end{array} \qquad (5.14)$$

where j is a cofibration and a local n-equivalence. Form the diagram

$$\begin{array}{ccccc} & & X_* & \longrightarrow & X \\ & & \eta_* \downarrow & & \downarrow \eta \\ \mathbf{P}_n A & \xrightarrow{i} & V_X & \xrightarrow{p} & \mathbf{P}_n X \end{array}$$

in which p is an injective fibration, i is a trivial cofibration, $p \cdot i$ is the map $\mathbf{P}_n A \to \mathbf{P}_n X$, and the square is a pullback. Then, the map η_* is a local weak equivalence by properness for the injective model structure. This construction can be made functorially, to produce a local weak equivalence $X_* \to Y_*$ which gives a factorization

$$\begin{array}{ccccc} A & \longrightarrow & X_* & \longrightarrow & X \\ {\scriptstyle j}\downarrow & & \downarrow & & \downarrow{\scriptstyle p} \\ B & \longrightarrow & Y_* & \longrightarrow & Y \end{array}$$

of the diagram (5.14). Then, the desired lift exists, by factoring the map $X_* \to Y_*$ as a trivial cofibration followed by a trivial injective fibration.

If the map $f : \mathbf{P}_n X \to \mathbf{P}_n Y$ has a factorization

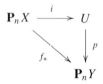

where i is a trivial cofibration and p is an injective fibration, then the map $\eta : U \to \mathbf{P}_n U$ is a weak equivalence, so that p is an n-fibration by the previous paragraph. Pull such a factorization of f_* back along the diagram

$$\begin{array}{ccc} X & \xrightarrow{\eta} & \mathbf{P}_n X \\ {\scriptstyle f}\downarrow & & \downarrow{\scriptstyle f_*} \\ Y & \xrightarrow{\eta} & \mathbf{P}_n Y \end{array}$$

to show that f has a factorization $f = q \cdot j$ where j is a cofibration and an n-equivalence and q is an n-fibration.

This gives the model structure. Explicitly, $q = p_* \cdot \theta$, where p_* is the pullback of p, and the map $\theta : X \to Y \times_{\mathbf{P}_n Y} \mathbf{P}_n X$ is factored $\theta = \pi \cdot j$ where π is a trivial injective fibration. The map θ is an n-equivalence by **A6**, so the cofibration j is an n-equivalence.

For statement 2), if p is an injective fibration and the indicated diagram is homotopy cartesian, then p is a retract of an n-fibration which is constructed by the methods above.

5.6 Postnikov Sections and n-Types

For the converse, one shows that every n-fibration p is a retract of an n-fibration q for which the diagram is homotopy cartesian.

To verify right properness, suppose given a pullback diagram

$$\begin{array}{ccc} A \times_Y X & \xrightarrow{f_*} & X \\ \downarrow & & \downarrow p \\ A & \xrightarrow{f} & Y \end{array}$$

such that p is an n-fibration and f is a local n-equivalence. Then the composite square

$$\begin{array}{ccccc} A \times_Y X & \xrightarrow{f_*} & X & \xrightarrow{\eta} & \mathbf{P}_n X \\ \downarrow & & \downarrow p & & \downarrow p_* \\ A & \xrightarrow{f} & Y & \xrightarrow{\eta} & \mathbf{P}_n Y \end{array}$$

is homotopy cartesian, and so the composite

$$A \times_Y X \xrightarrow{f_*} X \xrightarrow{\eta} \mathbf{P}_n X$$

is a local n-equivalence by Lemma 5.47.

The simplicial structure is given by the standard function complex construction. One shows that if $i : A \to B$ is a cofibration of simplicial presheaves and $j : C \to D$ are cofibrations of simplicial presheaves, then the cofibration

$$(B \times C) \cup (A \times D) \to B \times D$$

is a local n-equivalence if either i or j is a local n-equivalence. To prove this, observe that the map $i \times 1 : A \times C \to B \times C$ is a local n-equivalence for all simplicial presheaves C if i is a local n-equivalence, since the functor $\mathbf{P}_n$ preserves finite products.

The model structure of Theorem 5.49 is the *n-equivalence model structure* for simplicial presheaves.

Lemma 5.50 *The n-equivalence model structure on the category $s\mathbf{Pre}(\mathcal{C})$ of simplicial presheaves is cofibrantly generated.*

Proof We begin by proving a bounded cofibration statement for local n-equivalences. To this end, observe that the functor $\mathbf{P}_n$ preserves cofibrations, filtered colimits and regular cardinal bounds. Suppose that α is a regular cardinal such that $\alpha > |\operatorname{Mor}(\mathcal{C})|$.

Suppose given a cofibration $i : X \to Y$ which is a local n-equivalence and an α-bounded subobject A of Y. Then, in the induced diagram

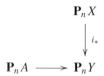

all maps are cofibrations, the object $\mathbf{P}_n A$ is α-bounded, and the map i_* is a local weak equivalence. By Lemma 5.2, there is an α-bounded subobject $B_1 \subset \mathbf{P}_n Y$ with $\mathbf{P}_n A \subset B_1$ and such that the map $B_1 \cap \mathbf{P}_n X \to B_1$ is a local weak equivalence. Since $\mathbf{P}_n$ preserves filtered colimits, there is an α-bounded subobject A_1 of Y such that $A \subset A_1$, $B_1 \subset \mathbf{P}_n A_1$, and $B_1 \cap \mathbf{P}_n X \subset \mathbf{P}_n (A_1 \cap X)$.

Continue inductively to find a sequence of α-bounded subobjects $A_1 \subset A_2 \subset \ldots$ of Y and $B_1 \subset B_2 \subset \ldots$ of $\mathbf{P}_n Y$, such that $B_i \subset \mathbf{P}_n A_i$, $B_i \cap \mathbf{P}_n X \subset \mathbf{P}_n (A_i \cap X)$, $\mathbf{P}_n A_i \subset B_{i+1}$, and the maps $B_i \cap \mathbf{P}_n X \to B_i$ are local weak equivalences. Then the map $\mathbf{P}_n(\varinjlim A_i \cap X) \to \mathbf{P}_n(\varinjlim A_i)$ is isomorphic to the map $\varinjlim (B_i \cap \mathbf{P}_n X) \to \varinjlim B_i$, which is a local weak equivalence.

Let $C = \varinjlim A_i$. Then C is α-bounded, $A \subset C \subset Y$ and the induced map $C \cap X \to Y$ is an n-equivalence.

It follows that the α-bounded cofibrations which are local n-equivalences generate the class of all trivial cofibrations for the n-equivalence model structure.

We already know that the α-bounded cofibrations generate the class of all cofibrations, from the proof of Theorem 5.8.

Chapter 6
Cocycles

Classically, cocycles come in two major flavours:

Example 6.1 Suppose that T is a topological space and G is a topological group. Then G represents a sheaf of groups G on the site $op|_T$, by defining $G(U)$ for an open subset $U \subset T$ to be the set hom (U, G) of continuous maps $U \to G$.

Suppose that $U_\alpha \subset T$ is an open cover of T, and let

$$U = \sqcup_\alpha U_\alpha \to T$$

be the resulting covering map. The corresponding Čech resolution $C(U)$ is represented by the simplicial space

$$\sqcup_\alpha U_\alpha \leftleftarrows \sqcup_{\alpha,\beta} U_\alpha \cap U_\beta \lllarrow \sqcup_{\alpha,\beta,\gamma} U_\alpha \cap U_\beta \cap U_\gamma \quad \ldots,$$

where the displayed face maps correspond to inclusions $U_\alpha \cap U_\beta \subset U_\alpha, U_\beta$ and the inclusions of $U_\alpha \cap U_\beta \cap U_\gamma$ into three double intersections $U_\alpha \cap U_\beta, U_\alpha \cap U_\gamma$ and $U_\beta \cap U_\gamma$.

A map $C(U) \to BG$ of simplicial sheaves on the site $op|_T$, by the representability of $C(U)$, is therefore defined by a set of elements $g_{\alpha,\beta} \in G(U_\alpha \cap U_\beta)$ such that

a) $g_{\alpha,\alpha} = e$ in the group $G(U_\alpha)$ for all α, and
b) the restrictions of the elements $g_{\alpha,\beta}, g_{\beta,\gamma}$ and $g_{\alpha,\gamma}$ to the group $G(U_\alpha \cap U_\beta \cap U_\gamma)$ satisfy

$$g_{\alpha,\gamma} = g_{\alpha,\beta} \cdot g_{\beta,\gamma}.$$

In other words, a simplicial sheaf map $C(U) \to BG$ can be identified with a normalized cocycle in G, which is defined with respect to the covering $U_\alpha \subset T$.

There is nothing special about the fact that G is a topological group—G could be an arbitrary sheaf of groups on the site $op|_T$ in the above.

A normalized cocycle in a sheaf of groups H that is defined with respect to a Zariski open covering $U_\alpha \subset S$ of a scheme S can be similarly represented as morphism of simplicial sheaves $C(U) \to BH$ on the Zariski site $Zar|_S$ of the scheme S, by the same argument.

Example 6.2 Suppose that L/k is a finite Galois extension of the field k, with Galois group G. Suppose that H is a sheaf of groups on the étale site $et|_k$ of the field. Of course, the sheaf of groups H could be represented by an algebraic group.

The Čech resolution corresponding to the étale covering $\mathrm{Sp}\,(L) \to \mathrm{Sp}\,(k)$ can be rewritten as a Borel construction $EG \times_G \mathrm{Sp}\,(L)$ in the category of simplicial sheaves on $et|_k$ (see Example 4.17), which is representable by a simplicial scheme

$$\mathrm{Sp}\,(L) \rightleftarrows \sqcup_{g \in G} \mathrm{Sp}\,(L) \rightleftarrows \sqcup_{(g_1,g_2) \in G \times G} \mathrm{Sp}\,(L).$$

In this case, the face map $d_0 : \sqcup_g \mathrm{Sp}\,(L) \to \mathrm{Sp}\,(L)$ is the map $g : \mathrm{Sp}\,(L) \to \mathrm{Sp}\,(L)$ defined by the action of g on the summand corresponding to g, and the face map $d_1 : \sqcup_g \mathrm{Sp}\,(L) \to \mathrm{Sp}\,(L)$ is the fold map ∇, meaning the map which restricts to the identity on $\mathrm{Sp}\,(L)$ on all summands. The face maps $d_0, d_1, d_2 : \sqcup_{(g_1,g_2)} \mathrm{Sp}\,(L) \to \sqcup_g \mathrm{Sp}\,(L)$ are defined, respectively, on the summand corresponding to the pair (g_1, g_2) by the composite

$$\mathrm{Sp}\,(L) \xrightarrow{g_1} \mathrm{Sp}\,(L) \xrightarrow{in_{g_2}} \sqcup_g \mathrm{Sp}\,(L),$$

the inclusion

$$in_{g_2 \cdot g_1} : \mathrm{Sp}\,(L) \to \sqcup_g \mathrm{Sp}\,(L),$$

of the summand corresponding to the product $g_2 \cdot g_1$, and the inclusion

$$in_{g_1} : \mathrm{Sp}\,(L) \to \sqcup_g \mathrm{Sp}\,(L),$$

on the summand corresponding to the pair (g_1, g_2).

It follows that a map

$$EG \times_G \mathrm{Sp}\,(L) \to BH$$

of simplicial sheaves on $et|_k$ can be identified with a function $\sigma : G \to H(L)$ such that

a) $\sigma(e) = e$ in the group $H(L)$, and
b) $\sigma(g_2 \cdot g_1) = g_1^*(\sigma(g_2)) \cdot \sigma(g_1)$.

The function σ is a normalized cocycle for the Galois group G with coefficients in the sheaf of groups H.

The cocycles that are described in Examples 6.1 and 6.2 are special cases of a diagram in simplicial sheaves (or simplicial presheaves) having the form

$$* \xleftarrow{\simeq} U \to Y, \qquad (6.1)$$

where the object U is acyclic in the sense that the terminal map $U \to *$ is a local weak equivalence. In effect, both cocycle constructions involve Čech resolutions associated to covers $V \to *$, and we know that all such covers determine local weak equivalences $C(V) \to *$ of simplicial presheaves—see Example 4.17. We can say more: the maps $C(V) \to *$ which are associated to covers $V \to *$ are hypercovers.

These examples, while standard, are a bit misleading for the present purposes, since the point of departure for the theory of cocycles which is presented in this chapter is that the map $U \to *$ in the picture (6.1) is only required to be a local weak equivalence.

More generally, suppose that X and Y are simplicial presheaves. We define a cocycle from X to Y as a picture

$$X \xleftarrow[\simeq]{f} U \xrightarrow{g} Y \qquad (6.2)$$

in the simplicial presheaf category such that the map f is a local weak equivalence. These are the objects of a cocycle category $h(X, Y)$, with morphisms defined by refinement.

There is a function

$$\phi : \pi_0 h(X, Y) \to [X, Y]$$

that is defined on the set of path components of the category $h(X, Y)$ and takes values in the set of morphisms from X to Y in the homotopy category Ho $(s\mathbf{Pre}(\mathcal{C}))$. This function is defined by sending the cocycle (6.2) to the induced morphism $g \cdot f^{-1}$ in the homotopy category. Then the category of simplicial presheaves (or simplicial sheaves) is sufficiently well behaved that we can show that the function ϕ is a *bijection*. This statement is a consequence of Theorem 6.5.

Theorem 6.5 gives a different picture of the construction of morphisms in the homotopy categories for many model structures of practical interest and should be viewed as a generalization of the Verdier hypercovering theorem.

The Verdier hypercovering theorem appears here as Theorem 6.12. This theorem is a little stronger than the classical version of the theorem [16], which is a formal result that holds for all categories of fibrant objects, and hence for locally fibrant simplicial presheaves by Proposition 4.33. The classical result therefore requires the use of pictures of the form (6.2) as input data, in which all objects are locally fibrant and the map g is a hypercover or local trivial fibration. For Theorem 6.12, the only requirement on the objects of (6.2) is that the target object Y should be locally fibrant.

Much of the utility of cocycle categories results from the fact that we do not require assumptions about the objects and maps and maps that form a cocycle. There is a model category theoretic version of Theorem 6.5 that says that the nerve of the cocycle category $h(X, Y)$ is a model for the function space $\mathbf{hom}(X, Y)$ if Y is fibrant—this is a consequence of the Dwyer–Kan theory of hammock localizations [22]—but that result has limited utility for injective model structures since injective fibrant objects have such strong descent properties. One can, however, prove a hammock localization statement for the cocycle category $h(X, Y)$ if Y is only assumed to be locally fibrant, and this result appears as Corollary 6.19.

The Verdier hypercovering theorem is important historically, because the cohomology theory of locally fibrant simplicial sheaves and its relation with sheaf cohomology can then be developed using only this result, as in [49]. One of the outcomes is a theory of cup products in cohomology theories for arbitrary Grothendieck toposes (such as sheaves on the flat site) that may not have enough points. There is, however, a technical cost, in that serious use of the classical form of the Verdier hypercovering theorem often requires rather delicate formal manipulations of hypercovers and their homotopy classes.

The generalized Verdier hypercovering statement given by Theorem 6.5 and the theory of cocycle categories together form a much more flexible and even ubiquitous device. This theory is a fundamental tool in the proofs of the results comprising the cohomology theories for simplicial presheaves, abelian and non-abelian, which appear in Chaps. 8 and 9 below. Cocycle categories are of particular use, in Chap. 9, in the discussion of torsors, stacks and higher stacks, and in classification results for gerbes.

6.1 Cocycle Categories

Let **M** be a closed model category such that

1) **M** is right proper in the sense that weak equivalences pull back to weak equivalences along fibrations, and
2) the class of weak equivalences is closed under finite products: if $f : X \to Y$ is a weak equivalence, so is any map $f \times 1 : X \times Z \to Y \times Z$.

Examples include all of the model structures on simplicial presheaves and simplicial sheaves that we have seen so far, where the weak equivalences are local weak equivalences. In effect, these model structures are proper by Theorems 5.8 and 5.9, and the observations of Example 5.42. The class of local weak equivalences is closed under finite products by Corollary 4.40.

Suppose that X and Y are objects of **M**, and write $h(X, Y)$ for the category whose objects are all pairs of maps (f, g)

$$X \xleftarrow[\simeq]{f} Z \xrightarrow{g} Y,$$

where f is a weak equivalence. A morphism

$$\gamma : (f, g) \to (f', g')$$

of $h(X, Y)$ is a map $\gamma : Z \to Z'$ which makes the diagram

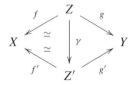

6.1 Cocycle Categories

commute. The category $h(X, Y)$ is the *category of cocycles*, or *cocycle category*, from X to Y. The objects of $h(X, Y)$ are called *cocycles*.

In general, write $\pi_0 h(X, Y)$ for the path components of the cocycle category $h(X, Y)$, which is defined in the model category **M**. There is a function

$$\phi : \pi_0 h(X, Y) \to [X, Y],$$

which is defined by the assignment $(f, g) \mapsto g \cdot f^{-1}$, and where $[X, Y]$ denotes morphisms from X to Y in the homotopy category Ho(**M**).

Lemma 6.3 *Suppose that $\gamma : X \to X'$ and $\omega : Y \to Y'$ are weak equivalences. Then the function*

$$(\gamma, \omega)_* : \pi_0 h(X, Y) \to \pi_0 h(X', Y')$$

is a bijection.

Proof An object (f, g) of $h(X', Y')$ is a map $(f, g) : Z \to X' \times Y'$ such that f is a weak equivalence. There is a factorization

$$\begin{array}{ccc} Z & \xrightarrow{j} & W \\ & \searrow{\scriptstyle (f,g)} & \downarrow{\scriptstyle (p_{X'},p_{Y'})} \\ & & X' \times Y' \end{array}$$

such that j is a trivial cofibration and $(p_{X'}, p_{Y'})$ is a fibration. The map $p_{X'}$ is a weak equivalence. Form the pullback

$$\begin{array}{ccc} W_* & \xrightarrow{(\gamma \times \omega)_*} & W \\ {\scriptstyle (p_X^*, p_Y^*)}\downarrow & & \downarrow{\scriptstyle (p_{X'},p_{Y'})} \\ X \times Y & \xrightarrow{\gamma \times \omega} & X' \times Y' \end{array}$$

Then the map (p_X^*, p_Y^*) is a fibration and $(\gamma \times \omega)_*$ is a local weak equivalence since $\gamma \times \omega$ is a weak equivalence, by right properness. The map p_X^* is also a weak equivalence.

The assignment $(f, g) \mapsto (p_X^*, p_Y^*)$ defines a function

$$\pi_0 h(X', Y') \to \pi_0 h(X, Y),$$

which is inverse to $(\gamma, \omega)_*$.

Lemma 6.4 *Suppose that the object X is cofibrant and Y is fibrant. Then the function*

$$\phi : \pi_0 h(X, Y) \to [X, Y]$$

144   6 Cocycles

is a bijection.

Proof The function $\pi(X,Y) \to [X,Y]$ relating homotopy classes of maps $X \to Y$ to morphisms in the homotopy category is a bijection since X is cofibrant and Y is fibrant.

If $f, g : X \to Y$ are homotopic, there is a diagram

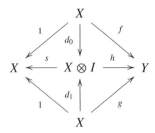

where h is the homotopy. Thus, sending $f : X \to Y$ to the class of $(1_X, f)$ defines a function

$$\psi : \pi(X,Y) \to \pi_0 h(X,Y)$$

and there is a diagram

$$\pi(X,Y) \xrightarrow{\psi} \pi_0 h(X,Y)$$
$$\searrow_{\cong} \qquad \downarrow \phi$$
$$[X,Y]$$

It suffices to show that ψ is surjective, or that any cocycle

$$X \xleftarrow{f}_{\simeq} Z \xrightarrow{g} Y$$

is in the path component of a cocycle $X \xleftarrow{1} X \xrightarrow{\alpha} Y$ for some map α.

The weak equivalence f has a factorization

where j is a trivial cofibration and p is a trivial fibration. The object Y is fibrant, so the dotted arrow θ exists in the diagram

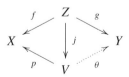

6.1 Cocycle Categories

The object X is cofibrant, so the trivial fibration p has a section σ, and there is a commutative diagram

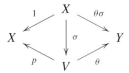

The composite $\theta\sigma$ is the required map α.

We now have the tools to prove the following result.

Theorem 6.5 *Suppose that the model category* **M** *is right proper and that its class of weak equivalences is closed under finite products. Suppose that X and Y are objects of* **M**. *Then the function*

$$\phi : \pi_0 h(X, Y) \to [X, Y]$$

is a bijection.

Proof There are weak equivalences $\pi : X' \to X$ and $j : Y \to Y'$ such that X' and Y' are cofibrant and fibrant, respectively, and there is a commutative diagram

$$\begin{array}{ccc}
\pi_0 h(X, Y) & \xrightarrow{\phi} & [X, Y] \\
{\scriptstyle (1,j)_*} \downarrow \cong & & \cong \downarrow {\scriptstyle j_*} \\
\pi_0 h(X, Y') & \xrightarrow{\phi} & [X, Y'] \\
{\scriptstyle (\pi,1)_*} \uparrow \cong & & \cong \downarrow {\scriptstyle \pi^*} \\
\pi_0 h(X', Y') & \xrightarrow[\phi]{\cong} & [X', Y']
\end{array}$$

The functions $(1, j)_*$ and $(\pi, 1)_*$ are bijections by Lemma 6.3, and the bottom map ϕ is a bijection by Lemma 6.4.

Remark 6.6 Cocycle categories have appeared before, in the context of Dwyer–Kan hammock localizations [20, 22]. One of the main results of the theory, which holds for arbitrary model categories **M**, says roughly that the nerve $Bh(X, Y)$ is a model for the function space of maps from X to Y if Y is fibrant. This result implies Theorem 6.5 if the target object Y is fibrant.

The statement of Theorem 6.5 must be interpreted with some care because the cocycle category $h(X, Y)$ may not be small. The theorem says that two cocycles are in the same path component in the sense that they are connected by a finite string of morphisms of $h(X, Y)$ if and only if they represent the same morphism in the homotopy category, and that every morphism in the homotopy category can be represented by a cocycle. Similar care is required for the interpretation of Lemmas 6.3 and 6.4 in general.

For simplicial presheaves and simplicial sheaves, we have the following alternative:

Proposition 6.7 *Suppose that a simplicial presheaf X is α-bounded, where α is a regular cardinal such that $\alpha > |\operatorname{Mor}(\mathcal{C})|$ and $\mathcal{C}$ is the underlying site. Let $h(X,Y)_\alpha$ be the full subcategory of $h(X,Y)$ on those cocycles*

$$X \xleftarrow{\simeq} U \to Y$$

such that U is α-bounded. Then the induced function

$$\pi_0 h(X,Y)_\alpha \to \pi_0 h(X,Y)$$

is a bijection.

The category $h(X,Y)_\alpha$ in the statement of Proposition 6.7 is small. We can therefore the category of all cocycles by a small category of bounded cocycles, suitably defined, for simplicial presheaf categories.

The proof of Proposition 6.7 uses the following technical lemmas:

Lemma 6.8 *Suppose that $i : A \to B$ is a cofibration such that A is α-bounded. Then there is an α-bounded subobject $C \subset B$ with $A \subset B$ such that all presheaf maps $\pi_* C \to \pi_* B$ are monomorphisms.*

Proof The simplicial presheaf $\operatorname{Ex}^\infty B$ is a filtered colimit of simplicial presheaves $\operatorname{Ex}^\infty D$, where D varies through the α-bounded subcomplexes of B. Any commutative diagram

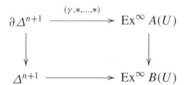

therefore factors through $\operatorname{Ex}^\infty D(U)$, where D is an α-bounded subobject of B. Such lifting problems are indexed on simplices γ of A which represent homotopy group elements, so there is an α-bounded subcomplex $A_1 \subset B$ such that all diagrams as above factor through A_1. Repeat this process inductively to produce a string of inclusions

$$A = A_0 \subset A_1 \subset A_2 \subset \dots$$

of α-bounded subcomplexes of B. Then the subcomplex $C = \bigcup_i A_i$ is α-bounded, and the presheaf maps $\pi_* C(U) \to \pi_* B(U)$ are monomorphisms.

6.1 Cocycle Categories

Lemma 6.9 *Suppose given a diagram*

of simplicial presheaf maps such that h is a local weak equivalence and the induced maps $f_* : \tilde{\pi}_n Y \to \tilde{\pi}_n Z$ are monomorphisms of sheaves for $n \geq 0$. Then the map f is a local weak equivalence.

Proof The analogous claim for morphisms of Kan complexes is true. In that case, we can suppose that f is a Kan fibration, and then we show that f has the right lifting property with respect to all inclusions $\partial \Delta^n \subset \Delta^n$.

In general, we can assume that X, Y and Z are locally fibrant simplicial sheaves and that f is a local fibration. Take a Boolean localization $p : \mathbf{Shv}(\mathcal{B}) \to \mathbf{Shv}(\mathcal{C})$, and observe that the induced diagram

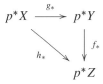

of simplicial sheaf maps on $\mathcal{B}$ is a map of diagrams of Kan complexes that satisfies the conditions of the lemma in all sections.

Proof (*Proof of Proposition 6.7*) Suppose that

$$X \xleftarrow[\simeq]{g} U \to Y$$

is a cocycle, and that X is α-bounded. The map $g : U \to Y$ has a factorization

where i is a trivial cofibration and p is a trivial injective fibration. The map p has a section $\sigma : X \to Z$ that is a trivial cofibration.

There is an α-bounded subobject X_1 of Z that contains X such that the induced map $X_1 \cap U \to X_1$ is a local weak equivalence, by Lemma 5.2. There is an α-bounded subobject X_1' of Z which contains X_1, and such that the cofibration $X_1' \to Z$ is a weak equivalence, by Lemmas 6.8 and 6.9.

Repeat these constructions inductively, to form a sequence of cofibrations

$$X \subset X_1 \subset X_1' \subset X_2 \subset X_2' \subset \dots$$

between α-bounded subobjects of X, and let A be the union of these subobjects. Then, A is α-bounded, and the map $A \subset Z$ is a weak equivalence, as is the map $A \cap U \to A$. The induced map $A \cap U \to U$ is also a weak equivalence, and is therefore a trivial cofibration.

We have therefore found an α-bounded object $B = A \cap U$ together with a trivial cofibration $i : B \to U$. It follows that the function

$$\pi_0 h(X, Y)_\alpha \to \pi_0 h(X, Y)$$

is surjective.

Suppose given a diagram

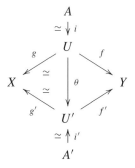

where the maps i and i' are trivial cofibrations and the objects A and A' are α-bounded. Then the subobject $\theta(A) \subset U'$ is α-bounded, as is the union $A' \cup \theta(A)$. By Lemma 6.8 (and Boolean localization), there is an α-bounded subobject B of U' with $A' \cup \theta(A) \subset B$ and such that $B \subset U'$ is a weak equivalence. The cocycles (gi, fi) and $(g'i', f'i')$ are therefore in the same path component of $h(X, Y)_\alpha$.

It follows that the function $\pi_0 h(X, Y)_\alpha \to \pi_0 h(X, Y)$ is injective.

We shall also need the following result:

Proposition 6.10 *Suppose that the simplicial presheaf X is α-bounded, where α is a regular cardinal such that $\alpha > |\operatorname{Mor}(\mathcal{C})|$. Suppose that β is a regular cardinal such that $\beta > \alpha$. Then the inclusion functor $j : h(X,Y)_\alpha \subset h(X,Y)_\beta$ induces a weak equivalence of simplicial sets*

$$j_* : Bh(X,Y)_\alpha \simeq Bh(X,Y)_\beta.$$

Proof Suppose that

$$X \xleftarrow[\simeq]{f} V \xrightarrow{g} Y$$

6.1 Cocycle Categories   149

is a cocycle such that V is β-bounded. We show that the slice category $j/(f,g)$ has a contractible nerve. The lemma then follows from Quillen's Theorem B (or Theorem A) [32, IV.4.6], [87].

The category $j/(f,g)$ is isomorphic to the category we_α/V whose objects are the local weak equivalences $\theta : U \to V$ with U α-bounded, and whose morphisms $\theta \to \theta'$ are commutative diagrams

of simplicial presheaf morphisms. Write cof_α/V for the full subcategory of we_α/V whose objects are the cofibrations, and let $i : cof_\alpha/V \subset we_\alpha/V$ be the inclusion functor.

The slice category θ/i is non-empty. In effect, the image $\theta(U)$ of the weak equivalence θ is an α-bounded subobject of V, $\theta(U)$ and is contained in an α-bounded subobject A of V such that $\pi_*(A) \to \pi_*(V)$ is a monomorphism of presheaves by Lemma 6.8, and then the inclusion $A \subset V$ is a weak equivalence by Lemma 6.9. The category θ/i is also filtered, again by Lemmas 6.8 and 6.9.

This is true for all $\theta : U \to V$ in the category we_α/V, so that the induced map

$$i_* : B(cof_\alpha/V) \subset B(we_\alpha/V)$$

is a weak equivalence.

Finally, the category cof_α/V is non-empty filtered by Lemmas 6.8 and 6.9, and it follows that the simplicial set $B(we_\alpha/V)$ is contractible.

Corollary 6.11 *Suppose that $f : X \to X'$ is a local weak equivalence of α-bounded simplicial presheaves, where α is a regular cardinal such that $\alpha > |\operatorname{Mor}(\mathcal{C})|$. Suppose that $g : Y \to Y'$ is a local weak equivalence. Then the induced simplicial set map*

$$(f,g)_* : Bh(X,Y)_\alpha \to Bh(X',Y')_\alpha$$

is a weak equivalence.

Proof Following the proof of Lemma 6.3, suppose that $(f,g) : Z \to X' \times Y'$ is an α-bounded cocycle, and take the functorial factorization

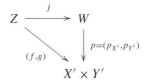

such that j is a trivial cofibration and p is an injective fibration. Form the pullback diagram

$$\begin{CD}
W_* @>(\gamma \times \omega)_*>> W \\
@V(p_X^*, p_Y^*)VV @VV(p_{X'}, p_{Y'})V \\
X \times Y @>>\gamma \times \omega> X' \times Y'
\end{CD}$$

as before. Then there is a regular cardinal $\beta > \alpha$ such that all objects in this diagram are β-bounded, and the map (p_X^*, p_Y^*) is a cocycle since the map $(\gamma \times \omega)_*$ is a local weak equivalence. It follows that there is a homotopy commutative diagram

$$\begin{CD}
Bh(X,Y)_\alpha @>(\gamma, \omega)_*>> B(X', Y')_\alpha \\
@V\simeq VV @VV\simeq V \\
Bh(X,Y)_\beta @>>(\gamma, \omega)_*> B(X', Y')_\beta
\end{CD}$$

of simplicial set maps in which the vertical maps are weak equivalences by Proposition 6.10. The statement of the corollary follows.

6.2 The Verdier Hypercovering Theorem

The discussion that follows will be confined to simplicial presheaves. All statements of this section have exact analogues for simplicial sheaves, which the reader can state and prove.

Recall that a hypercover $p : Z \to X$ is a locally trivial fibration. This means, equivalently (Theorem 4.32), that p is a local fibration and a local weak equivalence, or that p has the local right lifting property with respect to all inclusions $\partial \Delta^n \subset \Delta^n$, $n \geq 0$.

The objects of the category $Triv/X$ are the naive simplicial homotopy classes of maps $[p] : Z \to X$ which are represented by hypercovers $p : Z \to X$. The morphisms of this category are commutative triangles of simplicial homotopy classes of maps in the obvious sense.

To be completely explicit, suppose that $f, g : Z \to X$ are simplicial presheaf morphisms. A *simplicial homotopy* from f to g is a commutative diagram of simplicial

6.2 The Verdier Hypercovering Theorem

presheaf maps

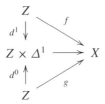

In the presence of such a diagram, we say that f and g are *simplicially homotopic* and write $f \sim g$. The set $\pi(Z, X)$ is the effect of collapsing the morphism set $\hom(Z, X)$ by the equivalence relation which is generated by the simplicial homotopy relation. The set $\pi(Z, X)$ is the set of *simplicial homotopy classes* of maps from Z to X.

Observe that the simplicial homotopy relation respects local weak equivalences, in the sense that if $f : Z \to Y$ is a local weak equivalence and $f \sim g$, then g is a local weak equivalence.

There is a contravariant set-valued functor that takes an object $[p] : Z \to X$ of $Triv/X$ to the set $\pi(Z, Y)$ of simplicial homotopy classes of maps between Z and Y. There is a function

$$\phi_h : \varinjlim_{[p]:Z \to X} \pi(Z, Y) \to [X, Y]$$

that is defined by sending the diagram of homotopy classes

$$X \xleftarrow{[p]} Z \xrightarrow{[f]} Y$$

to the morphism $f \cdot p^{-1}$ in the homotopy category.

The colimit

$$\varinjlim_{[p]:Z \to X} \pi(Z, Y)$$

is the set of path components of a category $H_h(X, Y)$ whose objects are the pictures of simplicial homotopy classes

$$X \xleftarrow{[p]} Z \xrightarrow{[f]} Y,$$

such that $p : Z \to X$ is a hypercover, and whose morphisms are the commutative diagrams

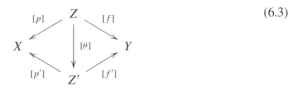

(6.3)

in homotopy classes of maps. The map ϕ_h therefore has the form

$$\phi_h : \pi_0 H_h(X,Y) \to [X,Y]$$

The following result is the *Verdier hypercovering theorem*:

Theorem 6.12 *The function*

$$\phi_h : \pi_0 H_h(X,Y) \to [X,Y]$$

is a bijection if the simplicial presheaf Y is locally fibrant.

Remark 6.13 Theorem 6.12 is a generalization of the Verdier hypercovering theorem of [16, p. 425] and [49], in which X is required to be locally fibrant. The statement of Theorem 6.12 first appeared in [82].

There are multiple variants of the category $H_h(X,Y)$:
1) Write $H'_h(X,Y)$ for the category whose objects are pictures

$$X \xleftarrow{p} Z \xrightarrow{[f]} Y$$

where p is a hypercover and $[f]$ is a homotopy class of maps. The morphisms of $H'_h(X,Y)$ are diagrams

$$\begin{array}{c} & Z & \\ {}^p\swarrow & \downarrow{[\theta]} & \searrow{[f]} \\ X & & Y \\ {}_{p'}\nwarrow & & \nearrow{[f']} \\ & Z' & \end{array} \qquad (6.4)$$

such that $[\theta]$ is a fibrewise homotopy class of maps over x, and $[f'][\theta] = [f]$ as simplicial homotopy classes. There is a functor

$$\omega : H'_h(X,Y) \to H_h(X,Y)$$

that is defined by the assignment $(p,[f]) \mapsto ([p],[f])$, and which sends the morphism (6.4) to the morphism (6.3).

2) Write $H''_h(X,Y)$ for the category whose objects are the pictures

$$X \xleftarrow{p} Z \xrightarrow{[f]} Y$$

where p is a hypercover and $[f]$ is a simplicial homotopy class of maps. The morphisms of $H''_h(X,Z)$ are commutative diagrams such that $[f' \cdot \theta] = [f]$. There is a canonical functor

$$H_h'''(X,Y) \xrightarrow{\omega'} H_h'(X,Y)$$

which is the identity on objects, and takes morphisms θ to their associated fibrewise homotopy classes.

3) Let $h_{hyp}(X,Y)$ be the full subcategory of $h(X,Y)$ whose objects are the cocycles

$$X \xleftarrow{p} Z \xrightarrow{f} Y$$

with p a hypercover. There is a functor

$$\omega'' : h_{hyp}(X,Y) \to H_h'''(X,Y)$$

that takes a cocycle (p, f) to the object $(p, [f])$.

Lemma 6.14 *Suppose that the simplicial presheaf Y is locally fibrant. Then the inclusion functor*

$$i : h_{hyp}(X,Y) \subset h(X,Y)$$

is a homotopy equivalence of categories.

Proof Objects of the cocycle category $h(X,Y)$ can be identified with maps (g, f) : $Z \to X \times Y$ such that the morphism g is a local weak equivalence, and morphisms of $h(X,Y)$ are commutative triangles in the obvious way. Maps of the form (g, f) have functorial factorizations

$$\begin{array}{ccc} Z & \xrightarrow{j} & V \\ & \searrow_{(g,f)} & \downarrow_{(p,g')} \\ & & X \times Y \end{array} \qquad (6.5)$$

such that j is a sectionwise trivial cofibration and (p, g') is a sectionwise Kan fibration. It follows that (p, g') is a local fibration and the map p, or rather the composite

$$Z \xrightarrow{(p,g')} X \times Y \xrightarrow{pr} X,$$

is a local weak equivalence. The projection map pr is a local fibration since Y is locally fibrant, so the map p is also a local fibration, and hence a hypercover.

It follows that the assignment $(u, g) \mapsto (p, g')$ defines a functor

$$\psi' : h(X,Y) \to h_{hyp}(X,Y).$$

The local weak equivalences j of (6.5) define natural maps $1 \to \psi' \cdot i$ and $1 \to i \cdot \psi'$.

Proof (Proof of Theorem 6.12) Write ψ for the composite functor

$$h(X,Y) \xrightarrow{\psi'} h_{hyp}(X,Y) \xrightarrow{\omega''} H_h''(X,Y) \xrightarrow{\omega'} H_h'(X,Y) \xrightarrow{\omega} H_h(X,Y).$$

The composite function

$$\pi_0 h(X,Y) \xrightarrow{\psi'_*} \pi_0 h_{hyp}(X,Y) \xrightarrow{\omega''_*} \pi_0 H_h''(X,Y) \xrightarrow{\omega'_*} \pi_0 H_h'(X,Y) \xrightarrow{\omega_*} \pi_0 H_h(X,Y) \xrightarrow{\phi_h} [X,Y] \tag{6.6}$$

is the bijection ϕ of Theorem 6.5. The function ψ'_* is a bijection by Lemma 6.14, and the functions ω''_*, ω'_* and ω_* are surjective, as is the function ϕ_h. It follows that all of the functions that make up the string (6.6) are bijections.

The following corollary of the proof of Theorem 6.12 deserves independent mention:

Corollary 6.15 *Suppose that the simplicial presheaf Y is locally fibrant. Then the induced functions*

$$\pi_0 h_{hyp}(X,Y) \xrightarrow{\omega''_*} \pi_0 H_h''(X,Y) \xrightarrow{\omega'_*} \pi_0 H_h'(X,Y) \xrightarrow{\omega_*} \pi_0 H_h(X,Y)$$

are bijections, and all of these sets are isomorphic to the set $[X,Y]$ of morphisms $X \to Y$ in the homotopy category $\text{Ho}(s/\mathbf{Pre}(\mathcal{C}))$.

The bijections of the path component objects in the statement of Corollary 6.15 with the set $[X,Y]$ all represent specific variants of the Verdier hypercovering theorem.

Remark 6.16 There is a relative version of Theorem 6.12 which holds for the model structures on slice category $A/s\mathbf{Pre}(\mathcal{C})$ which is induced from the injective model structure. Recall that the objects of this category are the simplicial presheaf maps $x : A \to X$, and the morphisms $f : x \to y$ are the commutative diagrams

In the induced model structure, the morphism $f : x \to y$ is a weak equivalence (respectively, cofibration, fibration) if and only if the underlying map $f : X \to Y$ is a local weak equivalence (respectively, cofibration, injective fibration) of simplicial presheaves.

In general, if $\mathbf{M}$ is a closed model category and A is an object of $\mathbf{M}$, then the slice category $A/\mathbf{M}$ inherits a model structure from $\mathbf{M}$ with the same definitions of weak equivalence, fibration and cofibration as above.

6.2 The Verdier Hypercovering Theorem

The slice category $A/s\mathbf{Pre}(\mathcal{C})$ has a theory of cocycles by Theorem 6.5, and then the argument for Theorem 6.12 goes through as displayed above in the case where the target Y of the object $y : A \to Y$ is locally fibrant.

These observations apply, in particular, to give a Verdier hypercovering theorem for pointed simplicial presheaves. More detail can be found in [67].

A *pointed simplicial presheaf* is a map $x : * \to X$ in simplicial presheaves, or alternatively a choice of global section x of a simplicial presheaf X. The category $*/s\mathbf{Pre}(\mathcal{C})$ of pointed simplicial presheaves is usually denoted by $s\mathbf{Pre}_*(\mathcal{C})$. See also Sect. 8.4.

In some respects, Lemma 6.14 is the best statement of the Verdier hypercovering theorem, although the result is again a little awkward to interpret because the cocycle categories in the statement might not be small. This situation is easily remedied by introducing cardinality bounds.

Suppose that α is a regular cardinal such that the simplicial presheaves X and Y are α-bounded, and write $h_{hyp}(X,Y)_\alpha$ for full subcategory of the cocycle category $h(X,Y)$ on the cocycles

$$X \xleftarrow{p} Z \xrightarrow{f} Y$$

for which p is a hypercover and Z is α-bounded. Then $h_{hyp}(X,Y)_\alpha$ is a full subcategory of the category $h_\alpha(X,Y)$ of Proposition 6.7, and we have the following:

Theorem 6.17 *Suppose that α is a regular cardinal such that $\alpha > |\mathrm{Mor}(\mathcal{C})|$. Suppose that the simplicial presheaves X and Y on the site $\mathcal{C}$ are α-bounded and that Y is locally fibrant. Then the inclusion $h_{hyp}(X,Y)_\alpha \subset h(X,Y)_\alpha$ induces a weak equivalence of simplicial sets*

$$Bh_{hyp}(X,Y)_\alpha \xrightarrow{\simeq} Bh(X,Y)_\alpha.$$

Proof The inclusion $h_{hyp}(X,Y)_\alpha \subset h(X,Y)_\alpha$ is a homotopy equivalence of small categories, by the same argument as for Lemma 6.14. In particular, the construction of the homotopy inverse functor

$$h(X,Y)_\alpha \to h_{hyp}(X,Y)_\alpha$$

respects cardinality bounds, by the assumptions on the size of the cardinal α.

Theorem 6.17 also leads to "hammock localization" results for simplicial presheaves. In the following, suppose that X is an α-bounded simplicial presheaf.

As in the proof of Proposition 6.10, write we_α/X for the category whose objects are all local weak equivalences $U \to X$ such that U is α-bounded. The morphisms of we_α/X are the commutative diagrams

Suppose that the simplicial presheaf Z is injective fibrant, and consider the functor

$$\mathbf{hom}(\ ,Z) : (we_\alpha/X)^{op} \to s\mathbf{Set}$$

which is defined by the assignment

$$U \xrightarrow{\simeq} X \mapsto \mathbf{hom}(U,Z).$$

Since X is α-bounded, the category we_α/X has a terminal object, namely 1_X, so $B(we_\alpha/X)^{op}$ is contractible, while the diagram $\mathbf{hom}(\ ,Z)$ is a diagram of weak equivalences since Z is injective fibrant. It follows [32, IV.5.7] that the canonical map

$$\mathbf{hom}(X,Z) \to \varinjlim\text{holim}_{U \xrightarrow{\simeq} X} \mathbf{hom}(U,Z)$$

is a weak equivalence. At the same time, the horizontal simplicial set

$$\varinjlim\text{holim}_{U \xrightarrow{\simeq} X} \mathbf{hom}(U,Z)_n$$

is the nerve of the cocycle category $h(X, Z^{\Delta^n})_\alpha$ and is therefore weakly equivalent to $Bh(X,Z)_\alpha$, for all n, by Corollary 6.11. It follows that there is a weak equivalence

$$Bh(X,Z)_\alpha \to \varinjlim\text{holim}_{U \xrightarrow{\simeq} X} \mathbf{hom}(U,Z).$$

We have proved the following variant of the hammock localization theorem:

Theorem 6.18 *Suppose that the simplicial presheaf Z is injective fibrant and that X is α-bounded, where α is a regular cardinal such that $\alpha > |\operatorname{Mor}(\mathcal{C})|$. Then there are weak equivalences of simplicial sets*

$$Bh(X,Z)_\alpha \xrightarrow{\simeq} \varinjlim\text{holim}_{U \xrightarrow{\simeq} X} \mathbf{hom}(U,Z) \xleftarrow{\simeq} \mathbf{hom}(X,Z),$$

where the homotopy colimit is indexed over all local weak equivalences $U \to X$ such that U is α-bounded.

The assertion that $Y \to Y^{\Delta^n}$ is a local weak equivalence holds for any locally fibrant object Y, and so we have the following:

Corollary 6.19 *Suppose that Y is locally fibrant and that $j : Y \to Z$ is an injective fibrant model in simplicial presheaves. Suppose that X is α-bounded. Then the simplicial set maps*

$$\begin{array}{ccc} Bh(X,Y)_\alpha & \longrightarrow & \varinjlim\text{holim}_{U \xrightarrow{\simeq} X} \mathbf{hom}(U,Y) \\ & & \downarrow \\ & & \varinjlim\text{holim}_{U \xrightarrow{\simeq} X} \mathbf{hom}(U,Z) \longleftarrow \mathbf{hom}(X,Z) \end{array}$$

6.2 The Verdier Hypercovering Theorem

are weak equivalences.

Proof The maps $Y \to Y^{\Delta^n} \to Z^{\Delta^n}$ are local weak equivalences since Y and Z are locally fibrant. The maps

$$Bh(X,Y)_\alpha \to Bh(X,Y^{\Delta^n})_\alpha \to Bh(X,Z^{\Delta^n})_\alpha$$

are therefore weak equivalences by Corollary 6.11.

Chapter 7
Localization Theories

This chapter establishes the formal basis for all localized homotopy theories that are derived in this monograph.

The method is to start with a model category **M** which satisfies a list of assumptions met in all examples of interest, together with a functor $L : \mathbf{M} \to \mathbf{M}$ that satisfies a shorter list of assumptions, including preservation of weak equivalences. Then it is shown, in Theorem 7.5, that there is a model structure on the category underlying **M** whose cofibrations are those of the original model structure, and whose weak equivalences are the L-equivalences. The L-equivalences are those maps $X \to Y$ with induced maps $LX \to LY$ that are weak equivalences of **M**.

Typically, one constructs a localized model structure by formally inverting a set of cofibrations $\mathcal{F}$ which has certain closure properties in the model structure **M**. The method for doing so is initiated in the first section, and the result is the production of an $\mathcal{F}$-injective model functor

$$L = L_{\mathcal{F}} : \mathbf{M} \to \mathbf{M}.$$

The main point of Sect. 7.1 is that the model category **M** and the $\mathcal{F}$-injective model functor L together satisfy all of the assumptions on **M** (which have not varied), and all assumptions on the functor **L** that guarantee Theorem 7.5, with the possible exception of a "bounded monomorphism condition" for the functor **L**.

This general setup for formally inverting a set of cofibrations $\mathcal{F}$ in a nice model category **M**, together with verification of the bounded monomorphism condition in particular cases, is the method that is used to generate all localized homotopy theories in this book. These include the traditional f-localization for a cofibration $f : A \to B$, f-localizations of presheaves of chain complexes (Sect. 8.5), and all stable homotopy theories of Chaps. 10 and 11.

The bounded monomorphism condition is new. It generalizes the bounded cofibration property for simplicial presheaves of Lemma 5.2, and it is particularly well suited to contexts where not all monomorphisms are cofibrations, or where the class of cofibrations might not be stable under pullback. For example, it is certainly true that not all monomorphisms are cofibrations in the base "strict" model structures for spectrum objects. The same is true for presheaves of chain complexes, where pullbacks of cofibrations can also fail to be cofibrations.

The first concrete application appears in Sect. 7.2. The line of argument which is established in the first section is used to give an explicit construction of the $\mathcal{F}$-local model structure for the category of simplicial presheaves for a set of cofibrations $\mathcal{F}$ in Theorem 7.18. This is a localization of the injective model structure on $s\mathbf{Pre}(\mathcal{C})$ in the traditional sense, whose class of trivial cofibrations is effectively the smallest class of maps, which contains the trivial cofibrations of the injective model structure and the cofibrations of $\mathcal{F}$. In the case where the set $\mathcal{F}$ is generated by a single cofibration $f : A \to B$, we obtain the f-local model structure for simplicial presheaves. See Example 7.19.

Examples of f-localizations appear at the end of the section. These include, in Example 7.20, the motivic homotopy theory of Morel and Voevodsky [82]. Example 7.21 gives a construction of Blander's local projective model structure (Example 5.42) for the simplicial presheaf category as a localization of the Bousfield–Kan model structure on the underlying diagram category of Proposition 2.22. The local projective model structure can itself be localized with the same techniques—the motivic projective model structure of [89] is an example of such a construction.

The question of when an f-local model structure is right proper can be important in applications—this was especially so in the original construction of the motivic stable category [57, 102]—and this is the subject of the third section. Theorem 7.27 says that the f-local model structure is right proper if f has the form $* \to I$ of a global section of a simplicial presheaf I.

Motivic homotopy theory is constructed by formally collapsing the affine line $\mathbb{A}^1$ over a scheme S to a global section $* \to \mathbb{A}^1$ in the Nisnevich topology, so that the motivic model structure is right as well as left proper. Left properness is a formal statement for all of these theories, and is proved in Lemma 7.8.

The localization method which is presented here is not the most general, and has a handicrafted flavour by comparison to standard references such as [41], and more recent discussions of combinatorial model structures [5, 75]. It does, however, have the benefit of being relatively simple to describe and apply in presheaf-theoretic contexts.

One of the more interesting and useful variants of the theory appears in Cisinski's thesis [19]. It is a fundamental insight of that work that one can construct homotopy theories in the style of localizations without reference to an underlying model structure. Examples include Joyal's theory of quasi-categories, which is a model structure for simplicial sets and has fewer trivial cofibrations than the standard structure. More generally, Cisinski shows how to construct a panoply of model structures for presheaves on a Grothendieck test category for which the cofibrations are monomorphisms, with or without an ambient Grothendieck topology. These methods apply, for example, to cubical sets and presheaves of cubical sets: examples include a cubical presheaves model for motivic homotopy theory [63].

7.1 General Theory

The purpose of this section is to set up a general framework for the localization constructions that are used for model categories **M**, which are based on simplicial presheaves.

We shall also require that the category **M** and its model structure satisfy the following conditions:

M1: The category **M** has a left proper, closed simplicial model structure. This model structure is cofibrantly generated, with generating sets I and J for the classes of cofibrations, respectively, trivial cofibrations.

M2: The category **M** has epi-monic factorizations, and every cofibration of **M** is a monomorphism.

M3: We have a set **A** of objects of **M** which contains the set of all targets of morphisms of I and J, and is closed under subobjects and quotients. This means that if $A \subset B$ and $B \in \mathbf{A}$ then $A \in \mathbf{A}$, and if $A \to C$ is an epimorphism with $A \in \mathbf{A}$ then $C \in \mathbf{A}$.

M4: The source objects A are cofibrant for all morphisms $A \to B$ in the generating sets I and J.

M5: The class of weak equivalences is closed under inductive colimits in **M**. This means that, if κ is a limit ordinal and $\alpha : X \to Y$ is a natural transformation of diagrams $X, Y : \kappa \to \mathbf{M}$ such that all component maps $\alpha : X_s \to Y_s$, $s < \kappa$, are weak equivalences of **M**, then the induced map

$$\alpha_* : \varinjlim_{s<\kappa} X_s \to \varinjlim_{s<\kappa} Y_s$$

is a weak equivalence of **M**.

M6: Suppose that the map $A \to B$ is a weak equivalence of **M** and that K is a simplicial set. Then the map

$$A \otimes K \to B \otimes K$$

is a weak equivalence of **M**.

We use tensor product notation for the simplicial model structure: in particular, the function complex $\mathbf{hom}(X, Y)$ is the simplicial set with n-simplices given by the set of morphisms

$$X \otimes \Delta^n \to Y$$

in **M**.

We *do not* assume that every monomorphism of **M** is a cofibration.

Example 7.1 This seems at first like a formidable list of requirements for the model category **M**, but the injective model structure on the category of simplicial presheaves of Theorem 5.8 has all of these properties.

The fact that inductive colimits preserve local weak equivalences is proved in Lemma 4.42. All simplicial presheaves are cofibrant for the injective model structure, and condition **M6** follows.

These conditions are also met by the projective local structure and all intermediate model structures for simplicial presheaves. These model structures all have local weak equivalences as their weak equivalences, so that conditions **M5** and **M6** are inherited from the injective model structure. Recall that there are simplicial presheaves, which are not cofibrant for these structures.

Among the listed conditions for the simplicial model structure **M**, the statement **M6** is perhaps the most brutal categorically, but it is satisfied in simplicial model structures that are built from simplicial presheaves and local weak equivalences. In particular, the conditions **M1**–**M6** hold for all model structures of simplicial modules, spectrum objects and symmetric spectrum objects which are discussed in that are encountered in later chapters.

Suppose that $L : \mathbf{M} \to \mathbf{M}$ is a functor. Say that a map $X \to Y$ of **M** is an *L-equivalence* if it induces a weak equivalence $LX \to LY$.

Here are the additional conditions that we shall require for the functor L:

L1: The functor $L : \mathbf{M} \to \mathbf{M}$ preserves weak equivalences.

L2: The class of cofibrations which are L-equivalences is saturated, meaning that the class is closed under pushouts, (transfinite) composition and coproducts,

L3: Suppose given a diagram of monomorphisms

such that i is an L-equivalence, and the subobject A of Y is a member of **A**. Then there is a subobject B of Y with $A \subset B$, B is a member of **A**, and such that the map $B \times_Y X \to B$ is an L-equivalence.

We shall say that statement **L3** is the *bounded monomorphism condition*. It is a variant of the bounded monomorphism property that one finds, for example, in Lemma 5.2.

Say that a map $p : X \to Y$ is an *L-fibration* if it has the right lifting property with respect to all maps which are cofibrations and L-equivalences. An object Z is *L-fibrant* if the map $Z \to *$ is an L-fibration.

Weak equivalences are L-equivalences by condition **L1**, so that every L-fibration is a fibration of **M**.

Lemma 7.2 *A map $p : X \to Y$ is an L-fibration and an L-equivalence if and only if it is a fibration and a weak equivalence.*

Proof If p is a trivial fibration then it has the right lifting property with respect to all cofibrations and is therefore an L-fibration. It is also an L-equivalence since it is a weak equivalence.

7.1 General Theory

Suppose that the morphism $p : X \to Y$ is an L-fibration and an L-equivalence. Then p is a retract of a trivial fibration, by a standard argument.

The collection of monomorphisms $m : C \to D$ of **M** with $D \in \mathbf{A}$ and such that m is an L-equivalence forms a set, since **A** is closed under subobjects. For each such map m, make a fixed choice of factorization

$$\begin{array}{ccc} C & \xrightarrow{j_m} & E \\ & \searrow{\scriptstyle m} & \downarrow{\scriptstyle p_m} \\ & & D \end{array}$$

with p_m a trivial fibration and j_m a cofibration. Write J_L for the set of all resulting cofibrations j_m. Each map j_m is an L-equivalence.

Lemma 7.3 *Suppose that $q : X \to Y$ is an L-equivalence which has the right lifting property with respect to all cofibrations j_m in the set J_L. Then the map q is a trivial fibration.*

Proof Suppose given a commutative diagram

$$\begin{array}{ccc} A & \longrightarrow & X \\ \downarrow{\scriptstyle i} & & \downarrow{\scriptstyle q} \\ B & \xrightarrow{\beta} & Y \end{array}$$

where i is a cofibration with $B \in \mathbf{A}$.

The map q has a factorization

such that p is a trivial fibration and j is a cofibration. The cofibration j is an L-equivalence.

The lifting β_* exists in the diagram

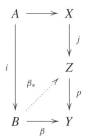

The image $\beta_*(B)$ is in **A** by **M3**, so there is a subobject D of Z which is in **A** such that the induced monomorphism $\alpha : D \times_Z X \to D$ is an L-equivalence by **L3**. It also follows that the original diagram has a factorization

$$\begin{array}{ccccc} A & \longrightarrow & D \times_Z X & \longrightarrow & X \\ \downarrow i & & \downarrow m & & \downarrow q \\ B & \longrightarrow & D & \longrightarrow & Y \end{array}$$

where m is a monomorphism and an L-equivalence, and D is a member of **A**.

The map m has the factorization $m = p_m \cdot j_m$, which is described above, and there is a commutative diagram

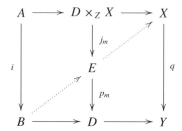

The dotted arrow lifts exist because p_m is a trivial fibration and we assume that q has the right lifting property with respect to all j_m.

It follows that the map q has the right lifting property with respect to the generating set I for the class of cofibrations, and is therefore a trivial fibration.

Lemma 7.4 *A map $X \to Y$ is an L-fibration if and only if it has the right lifting property with respect to all cofibrations in the set J_L.*

Proof Suppose that the morphism $i : A \to B$ is a cofibration and an L-equivalence. Then i has a factorization

where j is the saturation of J_L and p has the right lifting property with respect to all members of J_L. The cofibration j is an L-equivalence by the assumption (**L5**) that the class of L-trivial cofibrations is saturated, so that the map p is an L-equivalence.

7.1 General Theory

It follows from Lemma 7.3 that the map p is a trivial fibration, and so the lift exists in the diagram

The map i is therefore a retract of j.

Thus, if a morphism p has the right lifting property with respect to all cofibrations in the set J_L, then it has the right lifting property with respect to all maps, which are cofibrations and L-equivalences, and is therefore an L-fibration.

Theorem 7.5 *Suppose that the model category* **M** *and the functor* $L : \mathbf{M} \to \mathbf{M}$ *satisfy the conditions above. Then the category* **M**, *together with the cofibrations, L-equivalences and L-fibrations, satisfies the conditions for a cofibrantly generated closed model category.*

Proof The model axioms **CM1**, **CM2** and **CM3** are easily verified.

A map $p : X \to Y$ is an L-fibration (respectively trivial L-fibration) if and only if it has the right lifting property with respect to the set of cofibrations J_L (respectively the set I) by Lemma 7.2 and Lemma 7.4 above. It follows that every map $f : X \to Y$ has factorizations

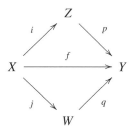

where i is a cofibration and an L-equivalence and p is an L-fibration, and such that j is as cofibration and q is an L-fibration and an L-equivalence. This gives the factorization axiom **CM5**.

The lifting axiom **CM4** follows from the identification of trivial L-fibrations with trivial fibrations of **M** in Lemma 7.2.

The sets I and J_L generate the classes of cofibrations and L-trivial cofibrations, respectively.

The functor $L : \mathbf{M} \to \mathbf{M}$ and the set of objects **A** in **M** arise, in practice, from a very specific construction.

Start with a set $\mathcal{F}$ of cofibrations $A \to B$ of **M**, and suppose the following:

F1: the set $\mathcal{F}$ includes all members of the generating set J of trivial cofibrations

F2: all cofibrations $A \to B$ in $\mathcal{F}$ have cofibrant source objects A

F3: if the map $A \to B$ is a member of $\mathcal{F}$, then so are all induced cofibrations

$$(B \otimes \partial \Delta^n) \cup (A \otimes \Delta^n) \to B \otimes \Delta^n \tag{7.1}$$

The $\otimes$ construction is part of the simplicial model structure for **M**. All maps (7.1) have cofibrant source objects, since the objects A and B are cofibrant.

The set **A** contains the objects appearing as targets of members of $\mathcal{F}$ and of the members of the set I, and is closed under subobjects and quotients. Typically, the set **A** is specified by a cardinality condition, and so closure with respect to subobjects and quotients is automatic.

Say that a map $p : X \to Y$ of **M** is $\mathcal{F}$-*injective* if it has the right lifting property with respect to all members of $\mathcal{F}$, and hence with respect to the saturation of $\mathcal{F}$. An object Z is $\mathcal{F}$-injective if the map $Z \to *$ is $\mathcal{F}$-injective.

The $\mathcal{F}$-injective maps are fibrations of **M** since $\mathcal{F}$ contains the generators for the class of trivial cofibrations of **M**, and so the $\mathcal{F}$-injective objects are fibrant. All trivial fibrations of **M** are $\mathcal{F}$-injective maps.

Say that a map $g : X \to Y$ is an $\mathcal{F}$-*equivalence* if some (and hence any) cofibrant replacement $X_c \to Y_c$ of g induces a weak equivalence of simplicial sets

$$\mathbf{hom}(Y_c, Z) \to \mathbf{hom}(X_c, Z)$$

for all $\mathcal{F}$-injective objects Z.

A *cofibrant replacement* of a map $g : X \to Y$ is a commutative diagram

$$\begin{array}{ccc} X_c & \longrightarrow & Y_c \\ \simeq \downarrow & & \downarrow \simeq \\ X & \xrightarrow{g} & Y \end{array}$$

in **M** such that the vertical maps are weak equivalences and the objects X_c and Y_c are cofibrant. One can further assume that the map $X_c \to Y_c$ is a cofibration.

All weak equivalences of **M** are $\mathcal{F}$-equivalences—the proof of this claim is an exercise.

Lemma 7.6 *A map $g : Z \to W$ of $\mathcal{F}$-injective objects is an $\mathcal{F}$-equivalence if and only if it is a weak equivalence of **M**.*

Proof Suppose that $g : Z \to W$ is an $\mathcal{F}$-equivalence. We show that g is a weak equivalence.

By replacing by cofibrant models $Z_c \to Z$ and $W_c \to W$ which are trivial fibrations of **M**, it is enough to assume that the objects Z and W are cofibrant as well as $\mathcal{F}$-injective.

The map g induces a weak equivalence

$$g^* : \mathbf{hom}(W, F) \xrightarrow{\simeq} \mathbf{hom}(Z, F)$$

7.1 General Theory

for each $\mathcal{F}$-injective object F, and thus induces a bijection

$$g^* : [W, F] \xrightarrow{\cong} [Z, F]$$

of morphisms in the homotopy category for **M** for each such object F.

The map g induces a bijection

$$g^* : [W, Z] \xrightarrow{\cong} [Z, Z],$$

so that g has a homotopy left inverse $\sigma : W \to Z$. The composite $g \cdot \sigma : W \to W$ and the identity on W restrict to $g : Z \to W$ along g up to homotopy, and are therefore homotopic.

We have shown that the map g is a homotopy equivalence of **M**. It is therefore a weak equivalence of **M**, [33, II.1.14].

All members $A \to B$ of the set $\mathcal{F}$ induce trivial fibrations

$$\mathbf{hom}(B, Z) \to \mathbf{hom}(A, Z)$$

for all $\mathcal{F}$-injective objects Z, by the closure property for $\mathcal{F}$ which is specified by condition **L3**. The objects A and B are cofibrant, by assumption. It follows that all members of the set $\mathcal{F}$ are $\mathcal{F}$-equivalences.

Lemma 7.7 *Suppose that κ is a limit ordinal, and that $A, B : \kappa \to$ **M** are inductive systems defined on κ. Suppose that $f : A \to B$ is a natural transformation such that each map $f : A_s \to B_s$ is an $\mathcal{F}$-equivalence. Then the induced map*

$$f_* : \varinjlim_{s < \kappa} A_s \to \varinjlim_{s < \kappa} B_s$$

is an $\mathcal{F}$-equivalence.

Proof Weak equivalences are closed under inductive colimits in **M**, by assumption **M5**, so we can assume that the diagram $\{A_s\}$ is cofibrant in the sense that A_0 is cofibrant, all maps $A_s \to A_{s+1}$ are cofibrations, and the morphisms

$$\varinjlim_{s < t} A_s \to A_t$$

are cofibrations for all limit ordinals $t < \kappa$. The objects

$$\varinjlim_{s < t} A_s$$

are cofibrant for all limit ordinals $t < \kappa$.

By the same argument, we can assume that the map $A_0 \to B_0$ is a cofibration, that all maps

$$B_s \cup A_{s+1} \to B_{s+1}$$

are cofibrations, and all maps

$$(\varinjlim_{s<t} B_s) \cup A_t \to B_t$$

are cofibrations for all limit ordinals $t < \kappa$. Then the map

$$\varinjlim_{s<t} A_s \to \varinjlim_{s<t} B_s$$

is a cofibration for each limit ordinal $t < \kappa$.

It follows that all maps $A_s \to B_s$, $B_s \cup A_{s+1} \to B_{t+1}$ and $(\varinjlim_{s<t} B_s) \cup A_t \to B_t$ are cofibrations and $\mathcal{F}$-equivalences. The claim that the map $A_t \to B_t$ is an $\mathcal{F}$-equivalence at each limit ordinal $t < \gamma$ is proved by showing inductively that the cofibration

$$\varinjlim_{s<t} A_s \to \varinjlim_{s<t} B_s$$

(of cofibrant objects) has the left lifting property with respect to all maps

$$\mathbf{hom}(\Delta^n, Z) \to \mathbf{hom}(\partial \Delta^n, Z),\ n \geq 0,$$

which are associated to $\mathcal{F}$-injective objects Z.

Lemma 7.8

1) The class of cofibrations which are $\mathcal{F}$-equivalences is saturated.
2) The class of $\mathcal{F}$-equivalences is stable under pushout along cofibrations.

Proof Suppose given a pushout diagram

$$\begin{array}{ccc} C & \xrightarrow{g} & E \\ j \downarrow & & \downarrow \tilde{j} \\ D & \xrightarrow{\tilde{g}} & F \end{array} \qquad (7.2)$$

in which the map j is a cofibration. Form the diagram

$$\begin{array}{ccccc} D_c & \xleftarrow{i} & C_c & \xrightarrow{i'} & E_c \\ \pi \downarrow & & \pi \downarrow & & \pi \downarrow \\ D & \xleftarrow{j} & C & \xrightarrow{g} & E \end{array}$$

7.1 General Theory

in which the object C_c is cofibrant, the vertical maps π are trivial fibrations, and the maps i and i' are cofibrations. Form the pushout

$$\begin{array}{ccc} C_c & \longrightarrow & E_c \\ i \downarrow & & \downarrow \\ D_c & \longrightarrow & F_c \end{array}$$

Then the induced map $F_c \to F$ is a weak equivalence since the model structure on **M** is left proper. We can therefore assume that all objects in the diagram (7.2) are cofibrant.

The induced diagram of simplicial set maps

$$\begin{array}{ccc} \mathbf{hom}(F, Z) & \xrightarrow{\tilde{g}^*} & \mathbf{hom}(D, Z) \\ \tilde{j}^* \downarrow & & \downarrow j^* \\ \mathbf{hom}(E, Z) & \xrightarrow{g^*} & \mathbf{hom}(C, Z) \end{array}$$

is a pullback in which the map j^* is a Kan fibration for each $\mathcal{F}$-injective object Z.

Thus, if the cofibration j is an $\mathcal{F}$-equivalence then $\tilde{j}$ is an $\mathcal{F}$-equivalence, and so the class consisting of cofibrations which are $\mathcal{F}$-equivalences is closed under pushout.

If g is an $\mathcal{F}$-equivalence then g^* is a weak equivalence and so $\tilde{g}^*$ is a weak equivalence by properness for simplicial sets. Again, this is true for all $\mathcal{F}$-injective objects Z, so that $\tilde{g}$ is an $\mathcal{F}$-equivalence if g is an f-equivalence, giving statement 2).

Suppose given an inductive system

$$A_0 \to A_1 \to \cdots \to A_s \to A_{s+1} \to \cdots$$

of morphisms of **M** with $s < \kappa$, where κ is some limit ordinal. Suppose that all maps $A_s \to A_{s+1}$ and all maps

$$\varinjlim_{s < \gamma} A_s \to A_\gamma$$

defined by limit ordinals $\gamma < \kappa$ are $\mathcal{F}$-equivalences. Then the canonical map

$$A_0 \to \varinjlim_{s < \kappa} A_s$$

is an $\mathcal{F}$-equivalence, by Lemma 7.7.

We use the assumption **M6** to prove the following result:

Lemma 7.9 *Suppose that the map $A \to B$ is a cofibration of **M** and that $K \to L$ is a cofibration of simplicial sets. Then the induced cofibration*

$$(B \otimes K) \cup (A \otimes L) \to B \otimes L \qquad (7.3)$$

is an $\mathcal{F}$-equivalence if either $A \to B$ is an $\mathcal{F}$-equivalence or the map $K \to L$ is a weak equivalence of simplicial sets.

Proof If $A \to B$ is an arbitrary cofibration and the map $K \to L$ is a trivial cofibration of simplicial sets, then the map (7.3) is a weak equivalence, from the simplicial model structure of **M**.

Suppose that the map $A \to B$ is a cofibration of **M** with A cofibrant. Then $A \to B$ is an $\mathcal{F}$-equivalence if and only if the dotted arrow lift exists in all diagrams

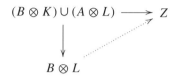

for all $\mathcal{F}$-injective objects Z and all cofibrations $K \to L$ of simplicial sets. This is equivalent to the requirement that the map $\mathbf{hom}(B, Z) \to \mathbf{hom}(A, Z)$ is a trivial fibration of simplicial sets for all such objects Z.

The map (7.3) is a cofibration with cofibrant source since $A \to B$ is a cofibration with cofibrant source. If $K' \to L'$ is any second choice of cofibration of simplicial sets, then the $\mathcal{F}$-injective object Z has the right lifting property with respect to the map

$$B \otimes ((K \times L') \cup (L \times K')) \cup (A \otimes (L \times L')) \to B \otimes (L \times L').$$

This holds for all cofibrations $K' \to L'$ and all $\mathcal{F}$-injective objects Z, and so the map (7.3) is an $\mathcal{F}$-equivalence.

More generally we can find a cofibrant replacement

$$\begin{array}{ccc} A_c & \xrightarrow{i_c} & B_c \\ \simeq \downarrow & & \downarrow \simeq \\ A & \xrightarrow{i} & B \end{array}$$

for the map i in which the vertical maps are weak equivalences of **M**, the object A_c is cofibrant, and the map i_c is a cofibration. Then the cofibration i_c is an $\mathcal{F}$-equivalence, and there is a commutative diagram

$$\begin{array}{ccc} (B_c \otimes K) \cup (A_c \otimes L) & \longrightarrow & B_c \otimes L \\ \downarrow & & \downarrow \\ (B \otimes K) \cup (A \otimes L) & \longrightarrow & B \otimes L \end{array}$$

in which the vertical maps are weak equivalences of **M** by condition **M6** and left properness of **M**. The top horizontal map is an $\mathcal{F}$-equivalence by the previous paragraph, and so the bottom horizontal map is an $\mathcal{F}$-equivalence.

7.1 General Theory

Every map $g : X \to Y$ of **M** has a functorial factorization

where p is $\mathcal{F}$-injective and η is in the saturation of the set $\mathcal{F}$, by a small object argument. In particular, there is a natural map

$$\eta : X \to L_{\mathcal{F}} X$$

where $L_{\mathcal{F}} X$ is $\mathcal{F}$-injective and the map η is a cofibration and an $\mathcal{F}$-equivalence. It is a consequence of Lemma 7.6 that a map $g : X \to Y$ is an $\mathcal{F}$-equivalence if and only if the induced map $L_{\mathcal{F}} X \to L_{\mathcal{F}} Y$ is a weak equivalence of **M**.

We have our functor $L = L_{\mathcal{F}}$.

Of the list of assumptions which is displayed at the beginning of the section, we have verified statements **L1** and **L2** for the functor $L_{\mathcal{F}}$, in Lemmas 7.8 and 7.9, respectively. It only remains to verify the bounded monomorphism condition **L3** for the functor $L_{\mathcal{F}}$, in order to show that Theorem 7.5 holds for this construction. We do this in multiple contexts below.

Say that a map $X \to Y$ is an $\mathcal{F}$-*fibration* if it has the right lifting property with respect to all cofibrations which are $\mathcal{F}$-equivalences. In the language of Theorem 7.5, the $\mathcal{F}$-fibrations coincide with the $L_{\mathcal{F}}$-fibrations.

In summary, we have the following:

Theorem 7.10 *Suppose that* **M** *is a closed simplicial model category which satisfies conditions* **M1** – **M6**. *Suppose that* $\mathcal{F}$ *is a set of cofibrations* $A \to B$ *of* **M** *having cofibrant source objects* A. *Suppose that the bounded monomorphism condition* **L3** *holds for the class of* $\mathcal{F}$-*equivalences.*

Then the category **M** *and the classes of cofibrations (of* **M***),* $\mathcal{F}$-*equivalences and* $\mathcal{F}$-*fibrations, together satisfy the axioms for a left proper closed simplicial model structure. This model structure is cofibrantly generated.*

Proof The indicated model structure exists by Theorem 7.5. The left properness for this model structure is proved in Lemma 7.8. The function complex construction for the underlying model structure on **M** gives the simplicial model structure, and the simplicial model axiom **SM7** is verified in Lemma 7.9.

The model structure of Theorem 7.10 is called the $\mathcal{F}$-*local model structure*, when it exists. See, for example, Theorem 7.18.

We close this section with some formal consequences of the general construction.

Lemma 7.11 *Suppose that the map* $p : Z \to Y$ *is* $\mathcal{F}$-*injective, and that the object* Y *is* $\mathcal{F}$-*injective. Then the map* p *is an* $\mathcal{F}$-*fibration.*

Corollary 7.12 *An object* Z *is* $\mathcal{F}$-*injective if and only if it is* $\mathcal{F}$-*fibrant.*

Proof [Proof of Lemma 7.11] Suppose given a commutative diagram

$$
\begin{array}{ccc}
A & \xrightarrow{\alpha} & Z \\
{\scriptstyle i}\downarrow & & \downarrow{\scriptstyle p} \\
B & \xrightarrow{\beta} & Y
\end{array}
$$

such that i is a cofibration and an $\mathcal{F}$-equivalence. The object Y is $\mathcal{F}$-injective, so the map β extends along the canonical map $\eta : B \to L_{\mathcal{F}}B$ to a morphism $\beta' : L_{\mathcal{F}}B \to Y$. The map p has the right lifting property with respect to the cofibration $\eta : A \to L_{\mathcal{F}}A$, so the original diagram has a factorization

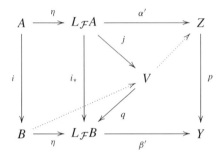

where the map i_* is a weak equivalence, j is a trivial cofibration and q is a trivial fibration. Thus, the dotted arrow liftings exist.

Lemma 7.13 *Suppose that the map $p : X \to Y$ is a fibration of* **M**, *and suppose that the diagram*

$$
\begin{array}{ccc}
X & \xrightarrow{\eta} & L_{\mathcal{F}}X \\
{\scriptstyle p}\downarrow & & \downarrow{\scriptstyle L_{\mathcal{F}}p} \\
Y & \xrightarrow{\eta} & L_{\mathcal{F}}Y
\end{array}
$$

is homotopy cartesian in **M**. *Then p is an $\mathcal{F}$-fibration.*

Proof There is a factorization

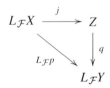

7.1 General Theory

of the map $L_{\mathcal{F}}p$ such that j is an $L_{\mathcal{F}}$-equivalence and the map q is is an $\mathcal{F}$-fibration. The object Z is $\mathcal{F}$-fibrant, so that j is a weak equivalence of $\mathbf{M}$ (Lemma 7.6). By pulling back q along i, we see from the hypothesis that the induced map

$$\theta : X \to Y \times_{L_{\mathcal{F}}Y} Z$$

is a weak equivalence of $\mathbf{M}$.

Factorize $\theta = \pi \cdot i$, where π is a trivial fibration and i is a trivial cofibration of $\mathbf{M}$. Then π is an $\mathcal{F}$-fibration, and there is a commutative diagram

$$\begin{array}{ccc} X & \xrightarrow{1} & X \\ i \downarrow & \nearrow & \downarrow p \\ V & \xrightarrow{q_* \cdot \pi} & Y \end{array}$$

in which the indicated lift exists, since, p is a fibration and i is a trivial cofibration of $\mathbf{M}$. Here,

$$q_* : Y \times_{L_{\mathcal{F}}Y} Z \to Y$$

is the pullback of the $\mathcal{F}$-fibration q. The map π is an $\mathcal{F}$-fibration, and it follows that p is a retract of the $\mathcal{F}$-fibration $q_* \cdot \pi$.

In an important class of examples, where the $\mathcal{F}$-local model structure is right proper, the statement of Lemma 7.13 has a converse:

Lemma 7.14 *Suppose that the $\mathcal{F}$-local model structure on $\mathbf{M}$ is right proper and that the map $p : X \to Y$ is a fibration of $\mathbf{M}$. Then p is an $\mathcal{F}$-fibration if and only if the diagram*

$$\begin{array}{ccc} X & \xrightarrow{\eta} & L_{\mathcal{F}}X \\ p \downarrow & & \downarrow L_{\mathcal{F}}p \\ Y & \xrightarrow{\eta} & L_{\mathcal{F}}Y \end{array}$$

is homotopy cartesian in $\mathbf{M}$.

Proof We suppose that the map p is an $\mathcal{F}$-fibration. Take a factorization

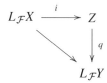

such that q is an $\mathcal{F}$-fibration and i is an $\mathcal{F}$-equivalence of **M**. Then the induced map

$$\eta_* : Y \times_{L_\mathcal{F} Y} Z \to Z$$

is an $\mathcal{F}$-equivalence by right properness of the $\mathcal{F}$-local structure, and so the map

$$\theta : X \to Y \times_{L_\mathcal{F} Y} Z$$

defines an $\mathcal{F}$-equivalence of $\mathcal{F}$-fibrations.

This map θ is a weak equivalence of **M**. One proves this with the standard factorization of a map between fibrant objects as a fibration following a section of a trivial fibration, together with the observation (Lemma 7.2) that the trivial fibrations of **M** coincide with the trivial $\mathcal{F}$-fibrations.

Examples of the phenomenon which is described by Lemma 7.14 include the standard description of stable fibrations that occurs in stable homotopy theory—for example, see Lemma 10.9 and Corollary 10.37.

7.2 Localization Theorems for Simplicial Presheaves

Suppose that $\mathcal{C}$ is a small Grothendieck site. Let α be an infinite cardinal such that $\alpha > |\operatorname{Mor}(\mathcal{C})|$. Recall that a cofibration $C \to D$ is said to be α-bounded if $\alpha > |D|$.

The injective model structure on the category $s\mathbf{Pre}(\mathcal{C})$ of simplicial presheaves given by Theorem 5.8 is an example of the sort of category **M** which is described in the previous section. See Example 7.1.

Recall that its weak equivalences are the local weak equivalences, its cofibrations are the monomorphisms, and its fibrations (injective fibrations) are defined by a right lifting property. This is a proper, cofibrantly generated, closed simplicial model structure. The generating sets I and J are, respectively, the α-bounded cofibrations and α bounded trivial cofibrations for this theory.

Suppose that $\mathcal{F}$ is a set of cofibrations of simplicial presheaves, such that

C1: the set $\mathcal{F}$ contains all members of the generating set J of trivial cofibrations for the injective model structure,

C2: if the map $i : A \to B$ is a member of $\mathcal{F}$, and $j : C \to D$ is an α-bounded cofibration, then the cofibration

$$(B \times C) \cup (A \times D) \to B \times D$$

is a member of $\mathcal{F}$.

Condition **C2** is a closure property. We can construct any such set of cofibrations $\mathcal{F}$ by starting with a set of cofibrations S, adding on the generating set J, and then let $\mathcal{F}$ be the smallest set of cofibrations which contains S and J, subject to satisfying condition **C2**. One writes $\mathcal{F} = \langle S \rangle$ in this case.

7.2 Localization Theorems for Simplicial Presheaves

It is an exercise to show that the set $\mathcal{F}$ consists of the cofibrations

$$(F \times C) \cup (E \times D) \to F \times D$$

which are induced by cofibrations $E \to F$ and $C \to D$, where the map $E \to F$ is either in S or in J, and $C \to D$ is an α-bounded cofibration. In particular, if the cofibration $C \to D$ is in $\mathcal{F}$, then so are all cofibrations

$$(D \times \partial\Delta^n) \cup (C \times \Delta^n) \to D \times \Delta^n.$$

All simplicial presheaves are cofibrant for the injective model structure. The set $\mathcal{F}$ therefore satisfies conditions **L1**, **L2** and **L3** of Sect. 7.1.

Example 7.15 Suppose that $f : A \to B$ is a cofibration. The set of cofibrations that is used to construct the f-local theory (see Example 7.19 below) is the set $\mathcal{F} = \langle f \rangle$, which is generated by the singleton set $\{f\}$ and the set J, in the sense described above.

The set $\langle f \rangle$ consists of the set of α-bounded trivial cofibrations for the injective model structure, together with the set of cofibrations

$$(f, i) : (B \times C) \cup (A \times D) \to B \times D$$

which are induced by the α-bounded cofibrations $i : C \to D$ of $s\mathbf{Pre}(\mathcal{C})$.

A morphism of simplicial presheaves $p : Z \to W$ is said to be $\mathcal{F}$-*injective* if it has the right lifting property with respect to all members of the set $\mathcal{F}$. Every $\mathcal{F}$-injective map is a fibration for the injective model structure on $s\mathbf{Pre}(\mathcal{C})$. A simplicial presheaf Z is said to be $\mathcal{F}$-*injective* if the map $Z \to *$ is $\mathcal{F}$-injective.

The condition **C2** in the description of the set of maps $\mathcal{F}$ reflects the enriched simplicial model structure of the simplicial presheaf category. It follows from the definition of the set $\mathcal{F}$ that an object Z is $\mathcal{F}$-injective if and only if Z is injective fibrant, and the induced map

$$i^* : \mathbf{Hom}(B, Z) \to \mathbf{Hom}(A, Z)$$

of internal function complexes is an injective fibration and a local weak equivalence for each member $i : A \to B$ of a generating set S for $\mathcal{F}$ in this context.

Say that a map $g : X \to Y$ of simplicial presheaves is an $\mathcal{F}$-*equivalence* if the induced map

$$g^* : \mathbf{hom}(Y, Z) \to \mathbf{hom}(X, Z)$$

of ordinary function complexes is a weak equivalence of simplicial sets for all $\mathcal{F}$-injective objects Z.

All simplicial presheaves are cofibrant for the injective model structure, so there is no need to use a cofibrant replacement for the map $g : X \to Y$ in the definition of $\mathcal{F}$-equivalence.

Suppose that β is a regular cardinal such that $\beta > |\mathcal{F}|$, that $\beta > |D|$ for all $C \to D$ in $\mathcal{F}$, and that $\beta > |\operatorname{Mor}(\mathcal{C})|$.

Suppose that λ is a cardinal such that $\lambda > 2^\beta$.

Every map $f : X \to Y$ of simplicial presheaves has a functorial system of factorizations

$$X \xrightarrow{i_s} E_s(f) \xrightarrow{f_s} Y$$
(with f the diagonal composite)

for $s < \lambda$ defined by the lifting property for maps in $\mathcal{F}$, and which form the stages of a transfinite small object argument.

Specifically, given the factorization $f = f_s i_s$ form the pushout diagram

$$\begin{array}{ccc} \bigsqcup_{\mathbf{D}} C & \longrightarrow & E_s(f) \\ \downarrow & & \downarrow \\ \bigsqcup_{\mathbf{D}} D & \longrightarrow & E_{s+1}(f) \end{array}$$

where **D** is the set of diagrams

$$\begin{array}{ccc} C & \longrightarrow & E_s(f) \\ i \downarrow & & \downarrow \\ D & \longrightarrow & Y \end{array}$$

with i in $\mathcal{F}$. Then $f_{s+1} : E_{s+1}(f) \to Y$ is the obvious induced map. Set $E_t(f) = \varinjlim_{s<t} E_s(f)$ at limit ordinals $t < \lambda$.

There is an induced natural factorization

$$X \xrightarrow{i_\lambda} E_\lambda(f) \xrightarrow{f_\lambda} Y$$

with

$$E_\lambda(f) = \varinjlim_{s<\lambda} E_s(f).$$

The map f_λ has the right lifting property with respect to all maps $C \to D$ in $\mathcal{F}$, and the cofibration i_λ is in the saturation of $\mathcal{F}$.

Write

$$L_{\mathcal{F}}(X) = E_\lambda(X \to *).$$

7.2 Localization Theorems for Simplicial Presheaves

The assignment $X \mapsto L_{\mathcal{F}}(X)$ is functorial in simplicial presheaves X.

Lemma 7.16

1) Suppose that the assignment $t \mapsto X_t$ defines a diagram of simplicial presheaves, indexed by $\gamma > 2^\beta$. Then the map

$$\varinjlim_{t < \gamma} L_{\mathcal{F}}(X_t) \to L_{\mathcal{F}}(\varinjlim_{t < \gamma} X_t)$$

is an isomorphism.

2) Suppose that ζ is a cardinal with $\zeta > \beta$, and let $B_\zeta(X)$ be the family of subobjects of X having cardinality less than ζ. Then the map

$$\varinjlim_{Y \in B_\zeta(X)} L_{\mathcal{F}}(Y) \to L_{\mathcal{F}}(X)$$

is an isomorphism.

3) The functor $X \mapsto L_{\mathcal{F}}(X)$ preserves monomorphisms.

4) Suppose that A and B are subobjects of X. Then the natural map

$$L_{\mathcal{F}}(A \cap B) \to L_{\mathcal{F}}(A) \cap L_{\mathcal{F}}(B)$$

is an isomorphism.

5) If $|X| \leq 2^\mu$ where $\mu \geq \lambda$ then $|L_{\mathcal{F}}(X)| \leq 2^\mu$.

Proof It suffices to prove statements 1)–4) with $L_{\mathcal{F}}(X)$ replaced by $E_1(X)$. There is a pushout diagram

$$\bigsqcup_{\mathcal{F}} (C \times \hom(C, X)) \longrightarrow X$$
$$\downarrow \qquad\qquad\qquad\qquad \downarrow$$
$$\bigsqcup_{\mathcal{F}} (D \times \hom(C, X)) \longrightarrow E_1 X$$

Then, in sections and in a fixed simplicial degree,

$$E_1 X(U) = (\bigsqcup_{\mathcal{F}} (D(U) - C(U))_n \times \hom(C, X)) \sqcup X(U)_n.$$

To prove statement 1), one shows that the map

$$\varinjlim_{t < \gamma} \hom(C, X_t) \to \hom(C, \varinjlim_{t < \gamma} X_t)$$

is an isomorphism.

For statement 2), show that the map

$$\varinjlim_{Y \in B_\zeta(X)} \hom(C, Y) \to \hom(C, X)$$

is an isomorphism by using the fact that the image of a map $C \to X$ is contained in a subobject $Y \subset X$ with $|Y| < \beta < \zeta$. The object X is also a colimit of subobjects Y with $|Y| < \zeta$, since $\zeta > |\operatorname{Mor}(\mathcal{C})|$.

Statement (3) is a consequence of the fact a monomorphism $X \to Y$ induces injective functions $\hom(C, X) \to \hom(C, Y)$.

Statement (4) follows from the observation that the functor $X \mapsto \hom(C, X)$ preserves pullbacks.

For statement 5), one shows inductively that all objects $E_s X$ have $|E_s X| \le 2^\mu$ for $s < \gamma$. This follows from the fact that the set $\hom(C, X)$ has cardinality bounded by $(2^\mu)^\beta = 2^{\mu \cdot \beta} = 2^\mu$. Then we have

$$|L_{\mathcal{F}}(X)| \le \lambda \cdot 2^\mu \le 2^\mu \cdot 2^\mu = 2^\mu.$$

Say that a simplicial presheaf map $X \to Y$ is an $L_{\mathcal{F}}$-*equivalence* if the induced map $L_{\mathcal{F}} X \to L_{\mathcal{F}} Y$ is a local weak equivalence.

Let κ be the successor cardinal for 2^μ, where μ is cardinal of statement 5) of Lemma 7.16. Then κ is a regular cardinal, and Lemma 7.16 implies that if a simplicial presheaf X is κ-bounded then $L_{\mathcal{F}}(X)$ is κ-bounded. The following result is the bounded monomorphism property for $L_{\mathcal{F}}$-equivalences:

Lemma 7.17 *Suppose given a diagram of monomorphisms*

such that i is an $L_{\mathcal{F}}$-equivalence and A is κ-bounded. Then there is a factorization $A \subset B \subset Y$ of j by monomorphisms such that B is κ-bounded and the map $B \cap X \to B$ is an $L_{\mathcal{F}}$-equivalence.

Proof There is an induced diagram of monomorphisms

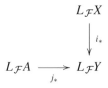

(statements 3) and 5) of Lemma 7.16) in which the map i_* is a local weak equivalence and $L_{\mathcal{F}} A$ is κ-bounded. Then, by Lemma 5.2 (the bounded cofibration property for the injective model structure), there is a κ-bounded subobject $A_0 \subset L_{\mathcal{F}} Y$ such that the map $A_0 \cap L_{\mathcal{F}} X \to A_0$ is a local weak equivalence.

There is a κ-bounded subobject $B_0 \subset Y$ such that $A \subset B_0$ and $A_0 \subset L_{\mathcal{F}} B_0$. In effect, $L_{\mathcal{F}} Y$ is a union of subobjects $L_{\mathcal{F}} F$ associated to the κ-bounded subobjects

7.2 Localization Theorems for Simplicial Presheaves

$F \subset Y$, by statement 2) of Lemma 7.16, and every section of the κ-bounded object B_0 is in some $L_\mathcal{F} F$.

Continue inductively. Find a κ-bounded object A_1 such that the map $L_\mathcal{F} B_0 \subset A_1$ and $A_1 \cap L_\mathcal{F} X \to L_\mathcal{F} Y$ is a local weak equivalence, and find a κ-bounded subobject B_1 of Y with $B_0 \subset B_1$ and $A_1 \subset L_\mathcal{F} B_1$. Repeat the construction λ times, and set $B = \varinjlim_{s < \lambda} B_s$.
Then the map

$$L_\mathcal{F}(B \cap X) \cong L_\mathcal{F} B \cap L_\mathcal{F} X \to L_\mathcal{F} B$$

(statement 4) of Lemma 7.16) coincides with the map

$$\varinjlim_s (A_s \cap L_\mathcal{F} X) \to \varinjlim_s A_s,$$

which is a local weak equivalence.

Now let **A** be the set of κ-bounded simplicial presheaves, where κ is the cardinal that is chosen above.

All objects $L_\mathcal{F} X$ are $\mathcal{F}$-injective by construction, and a map $X \to Y$ is an $L_\mathcal{F}$-equivalence if and only if it is an $\mathcal{F}$-equivalence.

All $\mathcal{F}$-injective objects are injective fibrant. Corollary 5.13 and Lemma 7.6 therefore imply that a map $X \to Y$ of simplicial presheaves is an $\mathcal{F}$-equivalence if and only if the induced map $L_\mathcal{F} X \to L_\mathcal{F} Y$ is a sectionwise weak equivalence.

In the language of Theorem 7.10, a simplicial presheaf map $p : X \to Y$ is an $\mathcal{F}$-fibration if it has the right lifting property with respect to all cofibrations which are $\mathcal{F}$-equivalences.

We have now assembled a proof of the following theorem, which gives the $\mathcal{F}$-*local model structure* for the category of simplicial presheaves. This result is a special case of Theorem 7.10, in the presence of Lemma 7.17.

Theorem 7.18 *Suppose that $\mathcal{F}$ is a set of cofibrations of simplicial presheaves on a small site $\mathcal{C}$ which satisfies conditions 1) and 2) above. Then the category s**Pre**($\mathcal{C}$) of simplicial presheaves on $\mathcal{C}$, together with the classes of cofibrations, $\mathcal{F}$-equivalences and $\mathcal{F}$-fibrations, satisfies the axioms for a left proper closed simplicial model category. This model structure is cofibrantly generated.*

If $\mathcal{F} = \langle S \rangle$ is generated by a set of cofibrations S, together with the generating set J of the trivial cofibrations in the sense described above, one also says that the $\mathcal{F}$-local model structure of Theorem 7.18 is the S-*local model structure*.

Example 7.19 Suppose that $f : A \to B$ is a cofibration of simplicial presheaves. Theorem 7.18 specializes, for the case $\mathcal{F} = \langle f \rangle$ to the f-*local model structure* for the category of simplicial presheaves.

The $\langle f \rangle$-equivalences are usually called f-*equivalences*, and the $\langle f \rangle$-fibrations are called f-*fibrations*. The $\langle f \rangle$-fibrant objects are said to be f-*fibrant*. One also writes $L_f X = L_{\langle f \rangle} X$, so that a map $X \to Y$ is an f-equivalence if and only if the induced map $L_f X \to L_f Y$ is a local (or even sectionwise) weak equivalence.

All $\mathcal{F}$-local theories are f-local theories for some f. In effect, if S is a set of cofibrations $A_i \to B_i$, then one can add up these cofibrations to form the cofibration

$$f = \sqcup f_i : \sqcup A_i \to \sqcup B_i.$$

If the set $\mathcal{F} = \langle S \rangle$ is generated by the set S, then the classes $\mathcal{F}$-equivalences and f-equivalences coincide, so that the f-local and $\mathcal{F}$-local model structures coincide. To see this, one shows that a simplicial presheaf Z is f-fibrant if and only if it is $\mathcal{F}$-fibrant, so that there is a natural weak equivalence $L_f X \simeq L_{\mathcal{F}} X$ for all simplicial presheaves X.

Aficionados of cellular model structures [41] will recognize that the injective model structure on the simplicial presheaf category $s\mathbf{Pre}(\mathcal{C})$ is cellular, for set theoretic reasons.

In particular, all simplicial presheaves A are compact in a strong sense. Suppose that β and γ are cardinals such that $\gamma > 2^\beta$ and $|A| < \beta$. Then if $s \to X_s$ is a diagram of monomorphisms indexed by $s < \gamma$, it follows that the function

$$\varinjlim_{s<\gamma} \hom(A, X_s) \to \hom(A, \varinjlim_{s<\gamma} X_s)$$

is a bijection. This observation is the basis for transfinite small object arguments in categories related to simplicial presheaves.

Theorem 7.18 is a consequence of Hirschhorn's Theorem 4.1.1, which asserts that each left proper cellular model structure admits left Bousfield localization at a set of maps.

The proof of Theorem 7.18 which is displayed here has the same ingredients as the proof of Hirschhorn's result, albeit in a rearranged and compressed form. The formal part of this proof (along with the proofs of all of our other localization results) amounts to a reduction to the bounded monomorphism condition, which condition appears as Proposition 4.5.15 of [41].

Example 7.20 (Motivic homotopy theory) Suppose that S is a scheme which is Noetherian and of finite dimension (typically a field, in practice), and let $(Sm|_S)_{Nis}$ be the category of smooth schemes of finite type over S, equipped with the Nisnevich topology. The motivic model structure on $s\mathbf{Pre}(Sm|_S)_{Nis}$ is the f-local theory for the 0-section $f : * \to \mathbb{A}^1$ of the affine line, in which the affine line is formally contracted to a point.

The motivic model structure is called the $\mathbb{A}^1$-model structure in [82], where also motivic weak equivalences are $\mathbb{A}^1$-weak equivalences and motivic fibrations are $\mathbb{A}^1$-fibrations. Strictly speaking, the Morel–Voevodsky model structure is defined on the category of simplicial sheaves on the smooth Nisnevich site, but the model structures for simplicial sheaves and simplicial presheaves are Quillen equivalent by the usual argument [57]. There are many other models for motivic homotopy theory, including model structures on presheaves and sheaves (not simplicial!) on the smooth Nisnevich site [57], and all of the models arising from test categories [63], including motivic cubical presheaves.

7.2 Localization Theorems for Simplicial Presheaves

Corollary 7.12 implies that a simplicial presheaf Z is motivic fibrant if and only if Z is injective fibrant for the Nisnevich topology, and all projection maps $U \times \mathbb{A}^1 \to U$ of S-schemes induce weak equivalences of simplicial sets

$$Z(U) \xrightarrow{\simeq} Z(U \times \mathbb{A}^1).$$

It follows from Theorem 5.39 (Nisnevich descent) that a simplicial presheaf X satisfies *motivic descent* in the sense that any motivic fibrant model $j : X \to Z$ is a sectionwise weak equivalence if and only if the following criteria are satisfied:

1) The simplicial presheaf X has the BG-property, meaning that $X(\phi)$ is contractible and X takes elementary distinguished squares

$$\begin{array}{ccc} \phi^{-1}(U) & \longrightarrow & V \\ \downarrow & & \downarrow \phi \\ U & \xrightarrow{j} & T \end{array} \qquad (7.4)$$

over S to homotopy cartesian diagrams of simplicial sets

$$\begin{array}{ccc} X(T) & \longrightarrow & X(U) \\ \downarrow & & \downarrow \\ X(V) & \longrightarrow & X(\phi^{-1}(U)) \end{array}$$

and

2) The simplicial presheaf X takes all projections $U \times \mathbb{A}^1 \to U$ over S to weak equivalences of simplicial sets $X(U) \xrightarrow{\simeq} X(U \times \mathbb{A}^1)$.

Properties 1 and 2 are preserved by filtered colimits in simplicial presheaves X and it follows that the collections of simplicial presheaves which satisfy, respectively, Nisnevich descent and motivic descent, are closed under filtered colimits. This observation is very important in applications.

Example 7.21 (Projective model structures) The projective model structure for the category of simplicial presheaves of Proposition 2.22 has weak equivalences and fibrations defined sectionwise. This is a closed simplicial model structure which is cofibrantly generated and proper. In particular, the generating set J for the class of trivial cofibrations consists of the maps of the form

$$\Lambda^n_k \times \hom(\,,U) \to \Delta^n \times \hom(\,,U),$$

all of which have cofibrant source objects. The projective model structure is a candidate for a model structure **M** in the notation of Sect. 7.1, and we use that notation here.

The cofibrations of **M** are the projective cofibrations. These maps are monomorphisms, but the class of projective cofibrations does not include all monomorphisms.

Now consider the set J of all α-bounded monomorphisms $f: A \to B$ of simplicial presheaves, which are local weak equivalences, and take projective cofibrant replacements

$$\begin{array}{ccc} A_c & \xrightarrow{i_f} & B_c \\ {\scriptstyle \simeq}\downarrow & & \downarrow{\scriptstyle \simeq} \\ A & \xrightarrow{f} & B \end{array}$$

for all such maps f, so that i_f is a projective cofibration between projective cofibrant objects.

Let $\mathcal{F}$ be the set containing all such projective cofibrations i_f and the members of the set J, and which satisfies the closure property that if $A \to B$ is in $\mathcal{F}$ then all induced projective cofibrations

$$(B \times \partial \Delta^n) \cup (A \times \Delta^n) \to B \times \Delta^n$$

are in $\mathcal{F}$.

Define a functor $X \mapsto L_{\mathcal{F}}X$ according to the method displayed above, so that each object $L_{\mathcal{F}}X$ is $\mathcal{F}$-injective. Then Lemma 7.17 and Theorem 7.10 together imply that there is a model structure on the category of simplicial presheaves, for which the cofibrations are the projective cofibrations and the weak equivalences are the $L_{\mathcal{F}}$-equivalences.

One uses a cofibrant replacement technique, as in the proof of Lemma 7.8, to show that the class of monomorphisms of simplicial presheaves which are $L_{\mathcal{F}}$-equivalences is saturated. It follows that every trivial cofibration for the injective model stucture is an $L_{\mathcal{F}}$-equivalence.

Every injective fibrant object Z is $\mathcal{F}$-injective, since all maps j_f are trivial cofibrations for the injective model structure. Thus, if the map $X \to Z$ is an injective fibrant model for an $\mathcal{F}$-injective (or $\mathcal{F}$-fibrant) simplicial presheaf X, then it is an $L_{\mathcal{F}}$-equivalence of $\mathcal{F}$-injective objects, and is therefore a sectionwise weak equivalence by Lemma 7.6. All $\mathcal{F}$-injective objects X, therefore, satisfy descent in the sense of Sect. 5.2.

It also follows that a map $X \to Y$ of simplicial presheaves is an $L_{\mathcal{F}}$-equivalence if and only if it is a local weak equivalence. This $\mathcal{F}$-local model structure therefore coincides with the local projective model structure of Example 5.42.

We have shown that the local projective structure can be obtained from the projective structure on the category of simplicial presheaves by formally inverting a set of local trivial cofibrations.

The f-local projective model structures (eg. the projective motivic model structure of Example 5.43) can be constructed in the same way, by formally inverting a generating set of trivial cofibrations for the f-local structure of Theorem 7.18 in the projective model structure. An alternative is to use Theorem 5.41 directly, starting from the f-local structure.

7.2 Localization Theorems for Simplicial Presheaves

Example 7.22 Suppose that $(Sm|_S)_{Nis}$ is the smooth Nisnevich site for a nice scheme S, as in Example 7.20.

Suppose that $\mathcal{F}$ is a set of projective cofibrations of simplicial presheaves on $(Sm|_S)_{Nis}$ which contains the following:

1) The members of the generating set

$$\Lambda^n_k \times \hom(\ , U) \to \Delta^n \times \hom(\ , U)$$

for the trivial cofibrations of the projective model structure
2) Projective cofibrant replacements of the maps

$$U \cup_{\phi^{-1}(U)} V \to T,$$

which are associated to all elementary distinguished squares (7.4)
3) The map

$$\emptyset \to \hom(\ , \emptyset).$$

from the empty simplicial presheaf to the presheaf which is represented by the empty scheme

We also state that $\mathcal{F}$ is the smallest set which contains the cofibrations listed above and satisfies the closure property that if $A \to B$ is a member of $\mathcal{F}$, then so are all morphisms

$$(B \times \partial \Delta^n) \cup (A \times \Delta^n) \to B \times \Delta^n.$$

Then, as in Example 7.21, we can localize the projective model structure on $s\mathbf{Pre}(Sm|_S)_{Nis}$ with respect to the set of projective cofibrations $\mathcal{F}$. In this case, an $\mathcal{F}$-injective (or $\mathcal{F}$-fibrant) object X is precisely a presheaf of Kan complexes on $(Sm|_S)_{Nis}$ which has the BG-property. The object X therefore satisfies Nisnevich descent in the sense that any injective fibrant model $j : X \to Z$ is a sectionwise equivalence (Theorem 5.39).

The maps $A \to B$ of $\mathcal{F}$ are local weak equivalences for the Nisnevich topology, since any such map induces a weak equivalence of simplicial sets

$$\mathbf{hom}(B, Z) \to \mathbf{hom}(A, Z)$$

for each injective fibrant simplicial sheaf Z. It follows that the classes of local weak equivalences and $\mathcal{F}$-equivalences coincide.

We therefore have another construction of the projective local model structure for the Nisnevich topology—see also Example 5.42.

If one adds the set of 0-sections

$$U \to \mathbb{A}^1 \times U$$

of the affine line over all smooth S-schemes U to the set $\mathcal{F}$ above, then the corresponding $\mathcal{F}$-injective objects are precisely the presheaves of Kan complexes which satisfy motivic descent, and the $\mathcal{F}$-equivalences coincide with the motivic weak equivalences of simplicial presheaves on the smooth site $(Sm|_S)_{Nis}$. In that case, the corresponding $\mathcal{F}$-local structure is the *projective motivic model structure* for the smooth Nisnevich site. See Example 5.43

We close this section with a return to the setting in which we formally invert a set $\mathcal{F}$ of cofibrations of simplicial presheaves. The following result is the $\mathcal{F}$-local analogue of Corollary 4.41.

Lemma 7.23 *Suppose that the maps $i : C \to D$ and $j : E \to F$ are cofibrations of simplicial presheaves. Then the induced cofibration*

$$(i, j) : (D \times E) \cup (C \times F) \to D \times F$$

is an $\mathcal{F}$-equivalence if either i of j is an $\mathcal{F}$-equivalence.

Proof The map (i, j) is plainly a cofibration.

The set of cofibrations $\mathcal{F}$ satisfies the closure property **C2**, which asserts that if $i : A \to B$ is a member of $\mathcal{F}$ and $j : C \to D$ is an α-bounded cofibration, then the cofibration

$$(i, j) : (B \times C) \cup (A \times D) \to B \times D$$

is a member of $\mathcal{F}$, and is therefore an $\mathcal{F}$-equivalence.

It follows that the map (i, j) is an $\mathcal{F}$-equivalence for all cofibrations $j : C \to D$. In effect, the class of cofibrations j for which the maps (i, j) are f-equivalences is closed under composition, pushouts and retractions, while the class of all cofibrations is generated by the α-bounded cofibrations under these operations.

It also follows, more generally, that the map (i, j) is an $\mathcal{F}$-equivalence for any cofibration j and any map i in the saturation of the set $\mathcal{F}$.

In particular, if $j : X \to L_{\mathcal{F}} X$ is the standard $\mathcal{F}$-fibrant model and E is any simplicial presheaf, then the induced cofibration

$$1 \times j : E \times X \to E \times L_{\mathcal{F}} X$$

is an $\mathcal{F}$-equivalence.

It follows that if the map $g : X \to Y$ is an $\mathcal{F}$-equivalence, then so is the map

$$1 \times g : E \times X \to E \times Y.$$

To show this, one uses the commutative diagram

$$\begin{array}{ccc} E \times X & \xrightarrow{1 \times g} & E \times Y \\ {\scriptstyle 1 \times j} \downarrow & & \downarrow {\scriptstyle 1 \times j} \\ E \times L_{\mathcal{F}} X & \xrightarrow{1 \times g_*} & E \times L_{\mathcal{F}} Y. \end{array}$$

The vertical maps $1 \times j$ are $\mathcal{F}$-equivalences by the previous paragraph, and the map $1 \times g_*$ is a sectionwise equivalence since g is an $\mathcal{F}$-equivalence of $\mathcal{F}$-fibrant objects.

7.3 Properness

Finally, suppose that the cofibration $i : C \to D$ is an $\mathcal{F}$-equivalence. Then the cofibrations $i \times 1$ in the diagram

$$\begin{array}{ccc} C \times E & \xrightarrow{1 \times j} & C \times F \\ {\scriptstyle i \times 1} \downarrow & & \downarrow {\scriptstyle i \times 1} \\ D \times E & \xrightarrow[1 \times j]{} & D \times F \end{array}$$

are $\mathcal{F}$-equivalences. The desired result follows, since there is a commutative diagram

$$\begin{array}{c} C \times F \\ {\scriptstyle (i \times 1)_*} \downarrow \quad \searrow {\scriptstyle i \times 1} \\ (D \times E) \cup (C \times F) \xrightarrow[(i,j)]{} D \times F \end{array}$$

and the class of maps which are cofibrations and $\mathcal{F}$-equivalences is closed under pushout.

We shall need the following consequence of Lemma 7.23 later.

Corollary 7.24 *Suppose that the maps $i : C \to D$ and $j : E \to F$ are cofibrations of pointed simplicial presheaves. Then the induced cofibration*

$$(i, j) : (D \wedge E) \cup (C \wedge F) \to D \wedge F$$

is an $\mathcal{F}$-equivalence if either i of j is an $\mathcal{F}$-equivalence.

The proof of this result is an exercise.

7.3 Properness

In this section, we establish a condition on the cofibration $f : A \to B$ which guarantees properness of the f-local model structure. Theorem 7.27 says that we have this property if the map f is a global section $* \to I$ of a simplicial presheaf I.

Lemma 7.25 *Suppose that every morphism of $\langle f \rangle$ pulls back to an f-equivalence along all f-fibrations $p : X \to Y$ with Y f-fibrant. Then the f-local model structure on $s\mathbf{Pre}(\mathcal{C})$ is right proper.*

Say that a map $g : A \to Y$ pulls back to an f-equivalence along $q : Z \to W$ if for every diagram

$$\begin{array}{ccc} & & Z \\ & & \downarrow {\scriptstyle q} \\ A & \xrightarrow{g} Y & \longrightarrow W \end{array}$$

the induced map $g_* : A \times_W Z \to Y \times_W Z$ is an f-equivalence.

Proof [Proof of Lemma 7.25] Suppose given a pullback diagram

$$\begin{array}{ccc} A \times_Y X & \xrightarrow{g_*} & X \\ \downarrow & & \downarrow p \\ A & \xrightarrow{g} & Y \end{array}$$

with g an f-equivalence and p an f-fibration. We want to show, under the conditions of the Lemma, that the map g_* is an f-equivalence.

Form the diagram

$$\begin{array}{ccccc} A \times_Y X & \xrightarrow{g_*} & X & \xrightarrow{j} & X' \\ \downarrow & & \downarrow p & & \downarrow q \\ A & \xrightarrow{g} & Y & \xrightarrow{\eta} & LY \end{array}$$

where the cofibration j is in the saturation of $\langle f \rangle$ (and is therefore an f-equivalence) and the map q is f-injective. The map q is an f-fibration by Lemma 7.11. All cofibrations in the saturation of $\langle f \rangle$ pull back to f-equivalences along f-fibrations with f-fibrant targets by assumption and exactness of pullback, so the map $\eta_* : Y \times_{LY} X' \to X'$ is an f-equivalence. The diagram

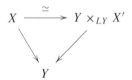

therefore defines an f-equivalence of f-fibrations, which is a weak equivalence of f-fibrations. This weak equivalence pulls back along g to a weak equivalence of f-fibrations

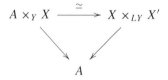

over A, by properness of the injective model structure (Lemma 4.37). It follows that g pulls back to an f-equivalence along p if and only if it pulls back to an f-equivalence along q.

7.3 Properness

Use Lemma 7.11 again to construct a diagram

$$
\begin{array}{ccc}
A & \xrightarrow{g} & Y \\
{\scriptstyle j}\downarrow & & \downarrow{\scriptstyle \eta} \\
V & \xrightarrow{q'} & LY
\end{array}
$$

where q' is an f-fibration, and j is in the saturation of $\langle f \rangle$. Then the map q' is a trivial f-fibration, and it pulls back to an f-equivalence along q. The maps j and η pull back to a f-equivalences along q by assumption. It follows that g pulls back to an f-equivalence along q.

Lemma 7.26 *Suppose that the cofibration f has the form $f : * \to I$ for some simplicial presheaf I, and that $p : X \to Y$ is an f-fibration of simplicial presheaves such that Y is f-fibrant. Then every map $f_* : A \to A \times I$ pulls back to an f-equivalence along p.*

Proof Suppose given the iterated pullback diagram

$$
\begin{array}{ccccc}
A \times_Y X & \xrightarrow{f'} & (A \times I) \times_Y X & \longrightarrow & X \\
\downarrow & & \downarrow & & \downarrow{\scriptstyle p} \\
A & \xrightarrow{f_*} & A \times I & \xrightarrow{g} & Y
\end{array}
$$

We want to show that the map f' is an f-equivalence.

The map g is simplicially homotopic to a composite

$$ A \times I \xrightarrow{pr} A \xrightarrow{g'} Y $$

since Y is f-fibrant, so it suffices to assume that g has this form since the injective model structure is proper. There is an iterated pullback diagram

$$
\begin{array}{ccccccc}
V & \xrightarrow{1 \times f} & V \times I & \xrightarrow{pr} & V & \longrightarrow & X \\
\downarrow & & \downarrow & & \downarrow & & \downarrow{\scriptstyle p} \\
A & \xrightarrow{1 \times f} & A \times I & \xrightarrow{pr} & A & \xrightarrow{g'} & Y
\end{array}
$$

where $f_* = 1 \times f : A \to A \times I$. The map $1 \times f : V \to V \times I$ is an f-equivalence by Lemma 7.23.

Theorem 7.27 *The f-local model structure for the category $s\mathbf{Pre}(\mathcal{C})$ of simplicial presheaves is proper if the cofibration f has the form $f : * \to I$.*

Proof We only need to verify right properness.

On account of Lemma 7.25, it suffices to assume that all maps

$$B \cup_A (A \times I) \xrightarrow{(i,f)} B \times I$$

pull back to f-equivalences along all f-fibrations $p : X \to Y$ with Y f-fibrant.

Lemma 7.26 says that the maps $A \to A \times I$ and $B \to B \times I$ pull back to cofibrations which are f-equivalences along all such p. Pullback preserves pushouts, and the result follows.

Example 7.28 It is an immediate consequence of Theorem 7.27 that the motivic model structure of Example 7.21 on the category $s\mathbf{Pre}(Sm|_S)_{Nis}$ of simplicial presheaves on the smooth Nisnevich site $(Sm|_S)_{Nis}$ is proper.

The motivic weak equivalences are the $\mathcal{F}$-equivalences for the $\mathcal{F}$-local model structure (the motivic projective model structure) which is constructed in Example 7.22.

The projective motivic model structure is right proper. In effect, suppose given a pullback diagram

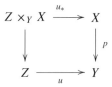

in which p is an $\mathcal{F}$-fibration and u is an $\mathcal{F}$-local (hence motivic) equivalence. Take a factorization

in which the map q is a motivic fibration and i is a motivic equivalence. Then i is a $\mathcal{F}$-equivalence between $\mathcal{F}$-fibrant objects of the slice category $s\mathbf{Pre}((Sm|_S)_{Nis})/Y$, and so i is a sectionwise weak equivalence, by Lemma 7.2 and a standard argument. The maps p and q are sectionwise Kan fibrations, so that the map

$$Z \times_Y X \to Z \times_Y U$$

is a sectionwise weak equivalence. The map $Z \times_Y U \to U$ is a motivic weak equivalence since the motivic model structure is proper. It follows that the map

$$u_* : Z \times_Y X \to X$$

is a motivic weak equivalence.

Part III
Sheaf Cohomology Theory

Chapter 8
Homology Sheaves and Cohomology Groups

The concept of local weak equivalence has its origins in, and restricts to, the classical notion of quasi-isomorphism of sheaves of chain complexes.

In particular a morphism $C \to D$ of ordinary chain complexes of sheaves is a quasi-isomorphism if and only if the associated map $\Gamma(C) \to \Gamma(D)$ of simplicial sheaves of abelian groups is a local weak equivalence of simplicial sheaves when one forgets the group structure (see Lemma 8.4). Here, Γ is one of the functors of Dold–Kan correspondence which gives the equivalence of ordinary chain complexes with simplicial abelian groups. This relation between quasi-isomorphisms and local weak equivalences is a straightforward generalization of the standard relation between homology isomorphisms of chain complexes and weak equivalences of their associated simplicial abelian groups [32].

The classical relation between chain complexes and simplicial abelian groups is somewhat deceptive, however, in that the Dold–Kan correspondence induces a Quillen equivalence between the model structure on simplicial abelian groups, which is induced from the standard model structure on simplicial sets, and the naive model structure on ordinary chain complexes from basic homological algebra. That naive model structure for chain complexes has no analogue for sheaves or presheaves of chain complexes: if one has a map $C \to D$ of sheaves of chain complexes which consists of sheaf epimorphisms $C_n \to D_n$ for $n > 0$, then the corresponding map $\Gamma(C) \to \Gamma(D)$ of simplicial abelian sheaves is certainly a local fibration, but it is usually not an injective fibration.

It is, however, relatively straightforward to show that the injective model structure on simplicial presheaves induces a model structure on sheaves (or presheaves) of simplicial abelian groups. For this structure, a map $A \to B$ is a local weak equivalence (respectively injective fibration) if the underlying map of simplicial presheaves is a local weak equivalence (respectively injective fibration). This is the injective model structure for simplicial abelian presheaves—its existence is a special case of Theorem 8.6, which deals with simplicial R-modules for a presheaf of commutative unitary rings R.

The most interesting detail in the proof of the simplicial abelian groups variant of Theorem 8.6 is that the free abelian group functor takes local weak equivalences

$X \to Y$ of simplicial presheaves to local weak equivalences $\mathbb{Z}(X) \to \mathbb{Z}(Y)$ of simplicial abelian presheaves. This is proved in Lemma 8.2, with a Boolean localization argument. The corresponding statement for simplicial sheaves is a form of a conjecture of Illusie that was first proved by van Osdol [101]. It is, and always was, a simple exercise to prove this statement for toposes having enough points.

The Dold–Kan correspondence induces an injective model structure for sheaves of ordinary chain complexes in which the cofibrant objects and cofibrant models (or resolutions) take the place of chain complexes of projectives and projective resolutions, respectively. This model structure extends to several model structures for categories based on unbounded complexes, all of which model the full derived category. These are effectively stable homotopy theories—the first of these structures appears in Proposition 8.16, which gives a model structure for spectrum objects in chain complexes, suitably defined.

The injective model structures for simplicial abelian presheaves and simplicial abelian sheaves are Quillen equivalent, by analogy with the pattern established for simplicial presheaves and simplicial sheaves in Theorem 5.9. Lemma 8.2 also implies that the forgetful and free abelian functor together determine a Quillen adjunction

$$\mathbb{Z} : s\mathbf{Pre} \leftrightarrows s\mathbf{Pre}_\mathbb{Z} : u$$

between the respective injective model structures. There is a corresponding result for simplicial sheaves and simplicial abelian sheaves. It follows in particular that there is a relation

$$[X, K(A, n)] = [X, u(K(A, n))] \cong [\mathbb{Z}(X), K(A, n)],$$

between morphisms in the respective homotopy categories, where

$$K(A, n) = \Gamma(A[-n])$$

is the Eilenberg–Mac Lane object associated to an abelian presheaf A, given by the standard method of applying the Dold–Kan correspondence functor to the chain complex $A[-n]$ which consists of a copy of A concentrated in degree n. This relation makes possible various descriptions of the cohomology $H^*(X, A)$ of a simplicial presheaf X with coefficients in the abelian presheaf A. Most succinctly, one defines

$$H^n(X.A) := [X, K(A, n)].$$

One can replace A by an injective resolution J up to local weak equivalence in the chain complex category, and one shows in Lemma 8.24 that the good truncations of the shifts of the resolution J satisfy descent. We can, moreover, identify $H^*(X, A)$ with the cohomology groups of bicomplex $\hom(X, J)$ which is associated to the simplicial presheaf X and the cochain complex J, as is done in Theorem 8.25. It follows, in particular, that sheaf cohomology is representable in the homotopy category in the sense that there is an isomorphism

$$H^n(\mathcal{C}, \tilde{A}) \cong [*, K(A, n)],$$

where $\tilde{A}$ is the sheaf associated to A and $*$ is the terminal sheaf or presheaf—this statement appears as Theorem 8.26.

The bicomplex hom (X, J) is a fundamental calculational device. It leads directly to the universal coefficients spectral sequence

$$E_2^{p,q} = \text{Ext}^q(\tilde{H}_p(X, \mathbb{Z}), \tilde{A}) \Rightarrow H^{p+q}(X, A)$$

of Corollary 8.28. There is a corresponding spectral sequence for R-module categories, in Lemma 8.30.

The universal coefficients spectral sequence was central to the original method for showing that a map $X \to Y$ of simplicial presheaves which induces homology sheaf isomorphisms

$$\tilde{H}_*(X, R) \xrightarrow{\cong} \tilde{H}_*(Y, R)$$

must also induce cohomology isomorphisms

$$H^*(Y, A) \xrightarrow{\cong} H^*(X, A).$$

for any R-module A. One now uses a trivial model theoretic argument.

This observation about the relation between homology sheaves and cohomology groups is the heart of the sheaf-theoretic approach to proving Suslin's rigidity theorem for the K-theory of algebraically closed fields, which first appeared in [49].

At one time, cup products in sheaf cohomology were defined by using Godement resolutions, as in [79]. Godement resolutions are defined only in toposes having enough points, and this means, for example, that there is no way to use this technique to construct cup products in flat cohomology. We define cup products for the cohomology of simplicial presheaves on arbitrary Grothendieck sites in Sect. 8.4. It has been known since [49] that one can construct such cup products for arbitrary sites, but the method presented in that paper depends on the Verdier hypercovering theorem. The cup product construction which is displayed here uses cocycle category methods, and it is straightforward to define and use.

The localization methods of Chap. 7 apply equally well to categories of chain complex objects, provided that one starts from a simplicial presheaf map, or set of maps, that is to be formally inverted in the chain complex category. This is the subject of Sect. 8.5.

The model structure for simplicial R-modules that is obtained by formally inverting a set of maps $\mathcal{F}$ in the injective model structure is constructed in Theorem 8.39. The existence of a motivic model structure for presheaves of chain complexes on the smooth Nisnevich site is one of the consequences.

One can, alternatively, formally invert a set of maps $\mathcal{F}$ in the projective model structure for simplicial R-modules, and in a projective model structure for a linear variant of the category of simplicial R-modules. This is done in the last section of this chapter, with the main result being Theorem 8.48.

Voevodsky's category of simplicial presheaves with transfers on the smooth Nisnevich site for a perfect field k is such a linear variant. These objects are contravariant additive functors on the category Cor_k of finite correspondences over k. Formally inverting a set of cofibrant replacements for the generating trivial cofibrations for the injective model structure gives a Nisnevich local model structure for the category of simplicial presheaves with transfers. Going further, in the sense that one also formally contracts the affine line to a point, gives a model structure whose associated homotopy category is Voevodsky's category of effective motives over the field k. See Example 8.49.

8.1 Chain Complexes

Suppose that $\mathcal{C}$ is a fixed Grothendieck site, and suppose that R is a presheaf of commutative rings with unit on $\mathcal{C}$.

Write $\mathbf{Pre}_R = \mathbf{Pre}_R(\mathcal{C})$ for the category of R-modules, or abelian presheaves which have an R-module structure. Then $s\mathbf{Pre}_R$ is the category of simplicial R-modules, $\mathrm{Ch}_+(\mathbf{Pre}_R)$ is the category of positively graded (i.e. ordinary) chain complexes in $\mathbf{Pre}_R$, and $\mathrm{Ch}(\mathbf{Pre}_R)$ is the category of unbounded chain complexes in $\mathbf{Pre}_R$.

Much of the time in applications, R is a constant presheaf of rings such as $\mathbb{Z}$ or $\mathbb{Z}/n$. In particular, $\mathbf{Pre}_\mathbb{Z}$ is the category of presheaves of abelian groups, $s\mathbf{Pre}_\mathbb{Z}$ is presheaves of simplicial abelian groups, and $\mathrm{Ch}(\mathbb{Z})$ and $\mathrm{Ch}_+(\mathbb{Z})$ are categories of presheaves of chain complexes. The category $\mathbf{Pre}_{\mathbb{Z}/n}$ is the category of n-torsion abelian presheaves, and so on.

All of these presheaf categories have corresponding sheaf categories, based on the category $\mathbf{Shv}_R$ of sheaves in R-modules. Explicitly, an object E of the category $\mathbf{Shv}_R$ is an R-module which happens to be a sheaf. Thus, $s\mathbf{Shv}_R$ is the category of simplicial sheaves in R-modules, $\mathrm{Ch}_+(\mathbf{Shv}_R)$ is the category of positively graded chain complexes of sheaves in R-modules, and $\mathrm{Ch}(\mathbf{Shv}_R)$ is the category of unbounded complexes of sheaves in R-modules.

There is a free R-module functor

$$R : s\mathbf{Pre}(\mathcal{C}) \to s\mathbf{Pre}_R,$$

written $X \mapsto R(X)$ for simplicial presheaves X, where $R(X)_n$ is the free R-module on the presheaf X_n. This functor is left adjoint to the forgetful functor

$$u : s\mathbf{Pre}_R \to s\mathbf{Pre}(\mathcal{C}).$$

The simplicial sheaf associated to $R(X)$ is denoted by $\tilde{R}(X)$.

One also writes $R(X)$ for the presheaf of Moore chains on X, which is the complex with $R(X)_n$ in degree n, with boundary maps

$$\partial = \sum_{i=0}^{n} (-1)^i d_i : R(X)_n \to R(X)_{n-1}.$$

The *homology sheaf* $\tilde{H}_n(X, R)$ is the sheaf associated to the presheaf $H_n(R(X))$, where the latter is the nth homology presheaf of the Moore complex $R(X)$. More

8.1 Chain Complexes

generally, if A is an R-module, then $\tilde{H}_n(X, A)$ is the sheaf associated to the presheaf $H_n(R(X) \otimes A)$.

Remark 8.1 Despite the notation, the homology sheaves $\tilde{H}_n(X, R)$ have no relation to the classical reduced homology of a pointed space. The homology sheaf $\tilde{H}_n(X, R)$ has a reduced variant, which can be defined as the nth homology sheaf of the kernel $R_\bullet(X)$ of the map $R(X) \to R(*)$. The simplicial R-module $R_\bullet(X)$ is the *reduced free simplicial R-module* for the pointed simplicial presheaf X. See also Sects. 8.4 and 10.7.

The normalized chains functor induces a functor

$$N : s\mathbf{Pre}_R \to \mathrm{Ch}_+(\mathbf{Pre}_R),$$

which is part of an equivalence of categories (the Dold–Kan correspondence [32, III.2.3])

$$N : s\mathbf{Pre}_R \simeq \mathrm{Ch}_+(\mathbf{Pre}_R) : \Gamma.$$

The normalized chain complex NA is the complex with

$$NA_n = \cap_{i=0}^{n-1} \ker(d_i)$$

and boundary

$$\partial = (-1)^n d_n : NA_n \to NA_{n-1}.$$

It is well known [32, III.2.4] that the obvious natural inclusion $NA \subset A$ of NA in the Moore chains is split by killing degeneracies, and is a homology isomorphism. This map induces a natural isomorphism

$$H_*(NA) \cong H_*(A)$$

of homology presheaves, and hence an isomorphism

$$\tilde{H}_*(NA) \cong \tilde{H}_*(A)$$

of homology sheaves.

Lemma 8.2 *Suppose that $f : X \to Y$ is a local weak equivalence of simplicial presheaves. Then the induced map $f_* : R(X) \to R(Y)$ of simplicial R-modules is a local weak equivalence.*

Proof It is enough to assume that R is a sheaf of rings, and show that if $f : X \to Y$ is a local equivalence of locally fibrant simplicial sheaves, then the induced map $f_* : \tilde{R}(X) \to \tilde{R}(Y)$ is a local equivalence of simplicial abelian sheaves.

We further assume that the map $f : X \to Y$ is a morphism of locally fibrant simplicial sheaves on a complete Boolean algebra $\mathcal{B}$, since the inverse image functor p^* for a Boolean localization $p : \mathbf{Shv}(\mathcal{B}) \to \mathbf{Shv}(\mathcal{C})$ commutes with the free R-module functor.

In this case, the map $f : X \to Y$ is a sectionwise weak equivalence, so that $f_* : R(X) \to R(Y)$ is a sectionwise weak equivalence, and so the map $f_* : \tilde{R}(X) \to \tilde{R}(Y)$ of associated sheaves is a local weak equivalence.

Remark 8.3 At one time, Lemma 8.2 and its variants were forms of the *Illusie conjecture*. There are various proofs of this result in the literature: the earliest, by van Osdol [101], is one of the first applications of Boolean localization. See also [49].

Suppose that A is a simplicial abelian group. Then A is a Kan complex, and we know [32, III.2.7] that there are natural isomorphisms

$$\pi_n(A, 0) \cong H_n(NA)$$

for $n \geq 0$. There is a canonical isomorphism

$$\pi_n(A, 0) \xrightarrow{\cong} \pi_n(A, a)$$

which is defined for any $a \in A_0$ by $[\alpha] \mapsto [\alpha + a]$ where we have written a for the composite

$$\Delta^n \to \Delta^0 \xrightarrow{a} A$$

The collection of these isomorphisms, taken together, define isomorphisms

$$\pi_n(A, 0) \times A_0 \xrightarrow{\cong} \pi_n A$$
$$\searrow_{pr} \quad \swarrow$$
$$A_0$$

of abelian groups fibred over A_0, and these isomorphisms are natural in simplicial abelian group homomorphisms.

Lemma 8.4 *A map $A \to B$ of simplicial R-modules induces a local weak equivalence $u(A) \to u(B)$ of simplicial presheaves if and only if the induced map $NA \to NB$ induces an isomorphism in all homology sheaves.*

Proof The map $NA \to NB$ induces an isomorphism in all homology sheaves if and only if the map $\tilde{\pi}_0(A) \to \tilde{\pi}_0(B)$ and all maps $\tilde{\pi}_n(A, 0) \to \tilde{\pi}_n(B, 0)$ are isomorphisms of sheaves. The diagram of sheaves associated to the presheaf diagram

$$\begin{array}{ccc} \pi_n(A, 0) \times A_0 & \longrightarrow & \pi_n(B, 0) \times B_0 \\ \downarrow & & \downarrow \\ A_0 & \longrightarrow & B_0 \end{array}$$

is a pullback if and only if the map $\tilde{\pi}_n(A, 0) \to \tilde{\pi}_n(B, 0)$ is an isomorphism of sheaves.

8.1 Chain Complexes

Corollary 8.5 *Suppose given a pushout diagram*

$$\begin{array}{ccc} A & \longrightarrow & C \\ i \downarrow & & \downarrow i_* \\ B & \longrightarrow & D \end{array}$$

in simplicial R-modules, such that the map i is a monomorphism and a homology sheaf isomorphism. Then the induced map i_* is a homology sheaf isomorphism.

Proof The cokernel of the monomorphism i_* is B/A, which is acyclic in the sense that $\tilde{H}_*(B/A) = 0$. The Moore chains functor is exact, and the short exact sequence

$$0 \to C \xrightarrow{i_*} D \to B/A \to 0$$

of simplicial R-modules induces a long exact sequence

$$\ldots \to \tilde{H}_n(C) \xrightarrow{i_*} \tilde{H}_n(D) \to \tilde{H}_n(B/A) \xrightarrow{\partial} \tilde{H}_{n-1}(A) \to \ldots$$

$$\xrightarrow{\partial} \tilde{H}_0(C) \xrightarrow{i_*} \tilde{H}_0(D) \to \tilde{H}_0(B/A) \to 0$$

in homology sheaves. It follows that all maps

$$\tilde{H}_n(C) \xrightarrow{i_*} \tilde{H}_n(D)$$

are isomorphisms.

Say that a map $A \to B$ of simplicial R-modules is a *local weak equivalence* (respectively *injective fibration*) if the simplicial presheaf map $u(A) \to u(B)$ is a local weak equivalence (respectively injective fibration).

A *cofibration* of simplicial R-modules is a map which has the left lifting property with respect to all trivial injective fibrations.

In view of Lemma 8.4, the morphism $A \to B$ is a local weak equivalence if and only if the induced maps $NA \to NB$ and $A \to B$ of normalized and Moore chains, respectively, are homology sheaf isomorphisms. Homology sheaf isomorphisms are often called *quasi-isomorphisms*.

Analogous definitions are available for morphisms of sheaves of simplicial R-modules. Say that a map $E \to F$ in $s\mathbf{Shv}_R$ is a *local weak equivalence* (respectively *injective fibration*) if the underlying simplicial sheaf map $u(E) \to u(F)$ is a local weak equivalence (respectively injective fibration). Cofibrations are defined by a left lifting property with respect to trivial fibrations.

If the simplicial presheaf map $i : A \to B$ is a cofibration, then the induced map $i_* : R(A) \to R(B)$ of simplicial R-modules is a cofibration. The map i_* is a monomorphism, because the free R-module functor preserves monomorphisms. The corresponding map $i_* : \tilde{R}(A) \to \tilde{R}(B)$ of associated simplicial sheaves is a cofibration and a monomorphism of simplicial sheaves in R-modules.

If A and B are simplicial R-modules, their *tensor product* $A \otimes B$ is the simplicial R-module which is defined in sections (and in all simplicial degrees) by

$$(A \otimes B)(U) = A(U) \otimes_{R(U)} B(U)$$

for U in the underlying site $\mathcal{C}$.

Theorem 8.6

1) With the definitions of local weak equivalence, injective fibration and cofibration given above, the category $s\mathbf{Pre}_R$ of simplicial R-modules satisfies the axioms for a proper closed simplicial model category. This model structure is cofibrantly generated. Every cofibration is a monomorphism.

2) With the corresponding definitions of local weak equivalence, injective fibration and cofibration, the category $s\mathbf{Shv}_R$ of simplicial sheaves in R-modules satisfies the axioms for a proper closed simplicial model category. This model structure is cofibrantly generated. Every cofibration is a monomorphism.

3) The inclusion and associated sheaf functors define a Quillen equivalence

$$L^2 : s\mathbf{Pre}_R \leftrightarrows s\mathbf{Shv}_R : i$$

between the model structures of parts 1) and 2).

Proof The injective model structure on $s\mathbf{Pre}$ is cofibrantly generated. It follows from this, together with Lemma 8.2 and Corollary 8.5, that every map $f : A \to B$ of $s\mathbf{Pre}_R$ has factorizations

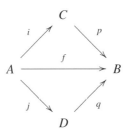

such that p is an injective fibration, i is a trivial cofibration which has the left lifting property with respect to all injective fibrations, q is a trivial injective fibration, j is a cofibration, and both i and j are monomorphisms. This proves the factorization axiom **CM5**. It also follows that every trivial cofibration is a retract of a map of the form i and therefore has the left lifting property with respect to all injective fibrations, giving **CM4**. The remaining closed model axioms for the category $s\mathbf{Pre}_R$ of simplicial R-modules are easy to verify.

Every cofibration is a retract of a map of the form j above, so that every cofibration is a monomorphism.

The generating set $A \to B$ of cofibrations (respectively trivial cofibrations) for simplicial presheaves induces a generating set $R(A) \to R(B)$ of cofibrations (respectively trivial cofibrations) for the category of simplicial R-modules.

8.1 Chain Complexes

The simplicial structure is given by the function complexes $\mathbf{hom}(A, B)$, where $\mathbf{hom}(A, B)_n$ is the abelian group of homomorphisms

$$A \otimes R(\Delta^n) \to B.$$

If $A \to B$ is a cofibration of simplicial presheaves and $j : K \to L$ is a cofibration of simplicial sets, then the cofibration

$$(B \times K) \cup (A \times L) \subset B \times L$$

induces a cofibration

$$(R(B) \otimes R(K)) \cup (R(A) \otimes R(L)) \subset R(B) \otimes R(L)$$

which is a local weak equivalence if either $A \to B$ is a local weak equivalence or $K \to L$ is a weak equivalence of simplicial sets, by Lemma 8.2 and Corollary 8.5. It follows that if $C \to D$ is a cofibration of $s\mathbf{Pre}_R$, then the map

$$(D \otimes R(K)) \cup (C \otimes R(L)) \to D \otimes R(L)$$

is a cofibration, which is a local weak equivalence if either $C \to D$ is a local weak equivalence or $K \to L$ is a weak equivalence of simplicial sets.

Left properness is proved with a comparison of long exact sequences in homology sheaves, which starts with the observation that every cofibration is a monomorphism. Right properness is automatic, from the corresponding property for the injective model structure for simplicial presheaves.

The proof of statement 2), for simplicial sheaves in R-modules is completely analogous, and the verification of 3) follows the pattern established in the proof of Theorem 5.9.

The model structures of Theorem 8.6 are the *injective model structures* for the categories of presheaves and sheaves of simplicial R-modules. The Quillen equivalence between the two structures is analogous to and is a consequence of the Quillen equivalence of Theorem 5.9 between the injective model structures for simplicial presheaves and simplicial sheaves.

It is an immediate consequence of the definitions that the adjunction

$$R : s\mathbf{Pre}(\mathcal{C}) \leftrightarrows s\mathbf{Pre}_R : u$$

defines a Quillen adjunction between the respective injective model structures. In particular, if X is a simplicial presheaf and A is a simplicial R-module, then there is an isomorphism

$$[R(X), A] \cong [X, u(A)]$$

relating morphisms in the respective homotopy categories, which is natural in X and A.

If A is a simplicial R-module and K is a simplicial presheaf on $\mathcal{C}$, we shall write

$$A \otimes K := A \otimes R(K). \tag{8.1}$$

Then, for example, the n-simplices of the function complex $\mathbf{hom}(A, B)$ can be written as simplicial R-module maps $A \otimes \Delta^n \to B$.

The Dold–Kan correspondence

$$N : s\mathbf{Pre}_R \simeq \mathrm{Ch}_+(\mathbf{Pre}_R) : \Gamma.$$

induces an injective model structure on the category $\mathrm{Ch}_+(\mathbf{Pre}_R)$ of presheaves of chain complexes, from the corresponding model structure on the category $s\mathbf{Pre}_R$ of simplicial modules given by Theorem 8.6.

A morphism $f : C \to D$ of $\mathrm{Ch}_+(\mathbf{Pre}_R)$ is said to be a *local weak equivalence* (respectively *cofibration*, *injective fibration*) if the induced map $f_* : \Gamma C \to \Gamma D$ is a local weak equivalence (respectively cofibration, injective fibration) of simplicial R-modules.

Similar definitions are made for chain complexes in sheaves of R-modules, with respect to the injective model structure on sheaves of simplicial R-modules.

We then have the following corollary of Theorem 8.6:

Corollary 8.7

1) *With the above definitions, the category $\mathrm{Ch}_+(\mathbf{Pre}_R)$ of ordinary chain complexes in R-modules satisfies the axioms for a proper closed simplicial model category. This model structure is cofibrantly generated. All cofibrations are monomorphisms.*
2) *With the above definitions, the category $\mathrm{Ch}_+(\mathbf{Shv}_R)$ of ordinary chain complexes in sheaves in R-modules satisfies the axioms for a proper closed simplicial model category. This model structure is cofibrantly generated. All cofibrations are monomorphisms.*
3) *The inclusion and associated sheaf functors define a Quillen equivalence*

$$L^2 : \mathrm{Ch}_+(\mathbf{Pre}_R) \leftrightarrows \mathrm{Ch}_+(\mathbf{Shv}_R) : i$$

between the (injective) model structures of parts 1) and 2).

Remark 8.8 Every injective fibration $p : C \to D$ of $\mathrm{Ch}_+(\mathbf{Pre}_R)$ corresponds to an injective fibration $p_* : \Gamma C \to \Gamma D$ of simplicial R-modules. The map p_* is a Kan fibration in each section (Lemma 5.12), so that the maps $p : C_n \to D_n$ are surjective in all sections for $n \geq 1$ [32, III.2.11]. The traditional identification of fibrations of simplicial abelian groups with chain complex morphisms that are surjective in non-zero degrees fails for the injective model structures of Theorem 8.6 and Corollary 8.7. Chain complex morphisms $C \to D$ which are local epimorphisms in nonzero degrees correspond to local fibrations under the Dold–Kan correspondence.

The identification of cofibrant chain complexes with complexes of projective modules also fails for the injective model structures.

The simplicial model structure of Theorem 8.6 for the category of simplicial R-modules can be enriched follows:

8.1 Chain Complexes

Lemma 8.9 *Suppose that the maps $i : C \to D$ and $j : E \to F$ are cofibrations of simplicial R-modules. Then the induced map*

$$(i, j) : (D \otimes E) \cup (C \otimes F) \to D \otimes F$$

is a cofibration, which is a local weak equivalence if either i or j is a local weak equivalence.

Proof Fix a choice of cofibration i. The class of cofibrations j for which the map (i, j) is a cofibration (respectively trivial cofibration) is closed under pushout, composition and retract. It follows that it suffices to check that the map

$$(Ri, Rj) : (R(B) \otimes R(U)) \cup (R(A) \otimes R(V)) \to R(B) \otimes R(V)$$

which is induced by cofibrations $i : A \to B$ and $j : U \to V$ of simplicial presheaves is a cofibration, which is trivial if either i or j is a trivial cofibration of simplicial presheaves. The map (Ri, Rj) is isomorphic to the map $R(i, j)$ which one obtains by applying the functor R to the simplicial presheaf map

$$(i, j) : (B \times U) \cup (A \times V) \to B \times V,$$

which map is a cofibration which is trivial if either i or j is trivial, by Corollary 4.41.

Corollary 8.10 *The bifunctor*

$$(A, K) \mapsto A \otimes K$$

preserves local weak equivalences in simplicial R-modules A and simplicial presheaves K.

Proof The functor $(A, K) \mapsto A \otimes K$ preserves sectionwise weak equivalences in A. Thus, for example, it suffices to show that the map $dg \otimes 1 : A \otimes K \to B \otimes K$ is a local weak equivalence if $g : A \to B$ is a local weak equivalence of cofibrant objects A. But this follows from Lemma 8.9.

The map $1 \otimes g : A \otimes K \to A \otimes L$ is a local weak equivalence if $g : K \to L$ is a local weak equivalence of simplicial presheaves, since $g_* : R(K) \to R(L)$ is a local weak equivalence of cofibrant simplicial R-modules by Lemma 8.2, and we can assume that the simplicial R-module A is cofibrant.

We close this section with a bounded subobject statement for simplicial R-modules that will be used later. Observe that there is no requirement for any of the monomorphisms in the statement of the following Lemma to be cofibrations.

Suppose that α is a regular cardinal such that $\alpha > |\operatorname{Mor}(\mathcal{C})|$ and $\alpha > |R|$.

Lemma 8.11 *Suppose that A and X are subobjects of a simplicial R-module Y, such that the inclusion $X \to Y$ is a local weak equivalence and such that A is α-bounded. Then there is an α-bounded subobject B of Y such that $A \subset B$ and the map $B \cap X \to B$ is a local weak equivalence.*

Proof The homology sheaves $\tilde{H}_*(Y/X)$ are trivial, and there is a presheaf isomorphism

$$\varinjlim_{B \subset Y} H_*(B/(B \cap X)) \xrightarrow{\cong} H_*(Y/X)$$

where B varies over the filtered category of α-bounded subobjects of Y.

All classes $\gamma \in H_*(Y/X)(U)$ are locally trivial. It follows that there is an α-bounded subobject $B_1 \subset Y$ with $A \subset B$ such that all elements of $H_*(A/(A \cap X))(U)$ map to locally trivial classes of $H_*(B_1/(B_1 \cap X))$.

Continue inductively, to produce a countable ascending sequence of subobjects

$$A \subset B_1 \subset B_2 \subset \dots$$

such that all elements of $H_*(B_i/(B_i \cap X)$ map to locally trivial elements of

$$H_*(B_{i+1}/(B_{i+1} \cap X)).$$

Set $B = \cup_i B_i$. By construction, all elements of the presheaves $H_*(B/(B \cap X))$ are locally trivial, so that the maps

$$H_*(B \cap X) \to H_*(X)$$

induce sheaf isomorphisms.

8.2 The Derived Category

Every ordinary chain complex C can be identified with an unbounded chain complex $\tau^*(C)$ by putting 0 in negative degrees. The right adjoint of the resulting functor τ^* is the *good truncation* $D \mapsto \tau(D)$ at level 0, where

$$\tau(D)_n = \begin{cases} \ker(\partial : D_0 \to D_{-1}) & \text{if } n = 0, \text{ and} \\ D_n & \text{if } n > 0. \end{cases}$$

The functor τ^* is fully faithful.

If D is an unbounded complex and $n \in \mathbb{Z}$, then the shifted complex $D[n]$ is defined by

$$D[n]_p = D_{p+n}.$$

If C is an ordinary chain complex and $n \in \mathbb{Z}$, define the *shifted complex* $C[n]$ by

$$C[n] = \tau(\tau^*(C)[n]).$$

8.2 The Derived Category

If $n > 0$, then $C[-n]$ is the complex with $C[-n]_p = C_{p-n}$ for $p \geq n$ and $C[-n]_p = 0$ for $p < n$, while $C[n]$ is the complex with $C[n]_p = C_{p+n}$ for $p > 0$ and

$$C[n]_0 = \ker(\partial : C_n \to C_{n-1}).$$

There is an adjunction isomorphism

$$\hom(C[-n], D) \cong \hom(C, D[n])$$

for ordinary chain complexes C and D, and all $n > 0$.

The functor $C \mapsto C[-1]$ is a *suspension functor* for ordinary chain complexes C, while $C \mapsto C[1]$ is a *loop functor*.

A *spectrum* D in chain complexes consists of ordinary chain complexes D^n, $n \geq 0$, together with chain complex maps

$$\sigma : D^n[-1] \to D^{n+1}$$

called *bonding maps*. A *map of spectra* $f : D \to E$ in chain complexes consists of chain complex maps $f : D^n \to E^n$ which respect structure in the sense that the diagrams

$$\begin{array}{ccc} D^n[-1] & \xrightarrow{\sigma} & D^{n+1} \\ {\scriptstyle f[-1]} \downarrow & & \downarrow {\scriptstyle f} \\ E^n[-1] & \xrightarrow{\sigma} & E^{n+1} \end{array}$$

commute. We shall write $\mathbf{Spt}(\mathbf{Ch}_+(\ ))$ to denote the corresponding category of spectra, wherever it occurs. For example, $\mathbf{Spt}(\mathbf{Ch}_+(\mathbf{Pre}_R))$ is the category of spectra in chain complexes of R-modules.

Example 8.12 Suppose that E is an unbounded chain complex. There is a canonical map

$$\sigma : \tau(E)[-1] \to \tau(E[-1])$$

which is defined by the diagram

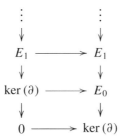

Replacing E by $E[-n]$ gives maps

$$\sigma : \tau(E[-n]))[-1] \to \tau(E[-n-1]).$$

These are the bonding maps for a spectrum object $\tau_*(E)$ with

$$\tau_*(E)^n = \tau(E[-n]).$$

Thus, every unbounded chain complex E defines a spectrum object $\tau_*(E)$ in chain complexes. The assignment $E \mapsto \tau_*(E)$ is functorial.

Example 8.13 If D is a spectrum object in chain complexes, the maps

$$\tau^*(D^n)[-1] = \tau^*(D^n[-1]) \to \tau^*(D^{n+1})$$

have adjoints $\tau^*(D^n) \to \tau^*(D^{n+1})[1]$ in the category of unbounded complexes. Write $\tau^*(D)$ for the colimit of the diagram

$$\tau^*(D^0) \to \tau^*(D^1)[1] \to \tau^*(D^2)[2] \to \cdots$$

in the unbounded chain complex category. Then it is not hard to see that, for the spectrum object $\tau_*(\tau^*(D))$, the complex $\tau_*(\tau^*(D))^n$ is naturally isomorphic to the colimit of the diagram of chain complexes

$$D^n \to D^{n+1}[1] \to D^{n+2}[2] \to \cdots$$

and that the "adjoint bonding maps"

$$\tau_*(\tau^*(D))^n \to \tau_*(\tau^*(D))^{n+1}[1]$$

are the isomorphisms which are determined by the diagrams

$$\begin{array}{ccccccc} D^n & \to & D^{n+1}[1] & \to & D^{n+2}[2] & \to & \cdots \\ \downarrow & & \downarrow & & \downarrow & & \\ D^{n+1}[1] & \to & D^{n+2}[2] & \to & D^{n+3}[3] & \to & \cdots \end{array}$$

There is a canonical map

$$\eta : D \to \tau_*(\tau^*(D))$$

that is defined by maps to colimits. One usually writes

$$Q(D) = \tau_*(\tau^*(D)).$$

Lemma 8.14 *The suspension functor $C \mapsto C[-1]$ preserves cofibrations of presheaves of chain complexes.*

8.2 The Derived Category

Proof It is enough to show that the functor $X \mapsto NR(X)[-1]$ takes cofibrations of simplicial presheaves X to cofibrations of $\text{Ch}_+(\textbf{Pre}_R)$.

There is an identification
$$R(X) = R_\bullet(X_+),$$
where $R_\bullet(X_+)$ is the reduced part of the complex $R(X_+)$ associated to $X_+ = X \sqcup \{*\}$, pointed by $*$. The functor $Y \mapsto R_\bullet(Y)$ is left adjoint to the forgetful functor from $s\textbf{Pre}_R$ to pointed simplicial presheaves, and therefore preserves cofibrations.

There is a natural isomorphism
$$\overline{W}(R_\bullet(Y)) \cong R_\bullet(\Sigma Y),$$
where ΣY is the Kan suspension of the pointed simplicial presheaf Y, and the Kan suspension preserves cofibrations of pointed simplicial sets (or presheaves) [32, III.5]. The isomorphism
$$N(\overline{W}(R_\bullet(Y))) \cong NR_\bullet(Y)[-1]$$
defines the simplicial R-module $\overline{W}(R_\bullet(Y))$.

Say that a map $f : E \to F$ of spectra in chain complexes is a *strict weak equivalence* (respectively *strict fibration*) if all maps $f : E^n \to F^n$ are local weak equivalences (respectively injective fibrations).

A *cofibration* is a map $i : A \to B$ of spectrum objects such that

1) the map $A^0 \to B^0$ is a cofibration of chain complexes, and
2) all induced maps
$$B^n[-1] \cup_{A^n[-1]} A^{n+1} \to B^{n+1}$$
are cofibrations.

It follows from Lemma 8.14 that if $i : A \to B$ is a cofibration of spectrum objects then all component maps $i : A^n \to B^n$ are cofibrations of chain complexes.

Lemma 8.15 *With the definitions of strict equivalence, strict fibration and cofibration given above, the category* $\textbf{Spt}(\text{Ch}_+(\textbf{Pre}_R))$ *satisfies the axioms for a proper closed simplicial model category.*

The proof of this result is a formality. The model structure of Lemma 8.15 is the *strict model structure* for the category of spectra in chain complexes.

Say that a map $f : A \to B$ of spectrum objects in chain complexes is a *stable equivalence* if the induced map $f_* : Q(A) \to Q(B)$ is a strict equivalence.

A map $f : A \to B$ is a stable equivalence of spectrum objects if and only if the induced map $f_* : \tau^*(A) \to \tau^*(B)$ of unbounded complexes is a homology sheaf isomorphism. A map $g : E \to F$ of unbounded complexes induces a stable equivalence $g_* : \tau_*(E) \to \tau_*(F)$ if and only if g is a homology sheaf isomorphism.

A map $p : C \to D$ of spectrum objects is said to be a *stable fibration* if it has the right lifting property with respect to all maps which are cofibrations and stable equivalences.

Proposition 8.16 *The classes of cofibrations, stable equivalences and stable fibrations give the category* $\mathbf{Spt}(\mathrm{Ch}_+(\mathbf{Pre}_R))$ *the structure of a proper closed simplicial model category.*

Proof The proof uses a method of Bousfield and Friedlander [13], [32, X.4]. The result is a formal consequence of the following assertions:

A4 The functor Q preserves strict weak equivalences.
A5 The maps η_{QC} and $Q(\eta_C)$ are strict weak equivalences for all spectrum objects C.
A6 The class of stable equivalences is closed under pullback along all stable fibrations, and is closed under pushout along all cofibrations.

Only the last of these statements requires proof, but it is a consequence of long exact sequence arguments in homology in the unbounded chain complex category. One uses Lemma 8.14 to verify the cofibration statement. The fibration statement is proved by showing that every stable fibration $p : C \to D$ is a strict fibration, and so the induced map $\tau^*(C) \to \tau^*(D)$ of unbounded complexes is a local epimorphism in all degrees.

The model structure of Proposition 8.16 is the *stable model structure* for the category of spectrum objects in chain complexes of R-modules. The associated homotopy category

$$\mathrm{Ho}(\mathbf{Spt}(\mathrm{Ch}_+(\mathbf{Pre}_R)))$$

is the *derived category* for the category of presheaves (or sheaves) of R-modules.

There are multiple models for the stable category, or equivalently for the derived category of R-modules. Here are two examples:

1) A *Kan spectrum* object E in simplicial R-modules consists of simplicial R-modules $A^n, n \geq 0$, together with bonding maps

$$\overline{W}(A^n) \to A^{n+1}.$$

A morphism $E \to F$ of Kan spectrum objects consists of morphisms $E^n \to F^n$ of simplicial R-modules, which respect bonding maps in the obvious sense. The category of Kan spectrum objects is equivalent to the category of spectrum objects in chain complexes of R-modules via the Dold–Kan correspondence, and therefore inherits all homotopical structure from that category.

2) A *spectrum* object F in simplicial R-modules consists of simplicial R-modules $F^n, n \geq 0$, together with bonding maps

$$S^1 \otimes A^n \to A^{n+1}.$$

Here, $S^1 = \Delta^1/\partial \Delta^1$ is the simplicial circle. A morphism $E \to F$ of spectrum objects consists of morphisms $E^n \to F^n$ of simplicial R-modules which respect bonding maps. Write $\mathbf{Spt}(s\mathbf{Pre}_R)$ for the corresponding category.

There is a strict model structure on $\mathbf{Spt}(s\mathbf{Pre}_R)$, for which the weak equivalences and fibrations are defined levelwise. A map $E \to F$ of spectrum objects is a stable equivalence if and only if the underlying map $u(E) \to u(F)$ of presheaves of spectra is a stable equivalence—see Sect. 10.1. The (strict) cofibrations and the stable equivalences together determine a stable model structure, by standard methods as in [61], or by using the localization methods of Chap. 10 — see Example 10.88.

The existence of the natural homotopy equivalence

$$\overline{W}(A) \simeq S^1 \otimes A$$

for simplicial R-modules A [61, 4.1] leads to an equivalence of the strict structure for spectrum objects in simplicial R-modules with spectrum objects in chain complexes of R-modules, and to a Quillen equivalence of the associated stable model structures. This is proved in Theorem 4.6 of [61]. It follows that the stable category of spectrum objects in chain complexes of R-modules and the stable categories for the two categories of spectrum objects in simplicial R-modules all give Quillen equivalent models for the derived category.

There is also a category of symmetric spectrum objects in simplicial R-modules, with a stable model structure which gives another Quillen equivalent model for the derived category. See Sect. 11.7 as well as [61]. This category of symmetric spectrum objects is a good setting for constructing tensor products, just as the category of symmetric spectra is a good setting for constructing smash products.

8.3 Abelian Sheaf Cohomology

Suppose that C is a chain complex, with associated simplicial abelian object $\Gamma(C)$, and that X is a simplicial presheaf. Recall that the cocycle category $h(X, \Gamma(C))$ has for objects all pairs of maps

$$X \xleftarrow[\simeq]{f} U \xrightarrow{g} \Gamma(C).$$

The morphisms $(f, g) \to (f', g')$ of $h(X, \Gamma(C))$ are the commutative diagrams of simplicial set maps

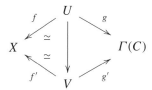

The category $h(X, \Gamma(C))$ is isomorphic, via adjunctions, to two other categories:

1) the category whose objects are all pairs

$$X \xleftarrow[\simeq]{f} U, \ \mathbb{Z}(U) \xrightarrow{g} \Gamma(C),$$

where $\mathbb{Z}(U)$ is the free simplicial abelian presheaf on U and g is a morphism of simplicial abelian presheaves, and

2) the category whose objects are all pairs

$$X \xleftarrow[\simeq]{f} U, \ N\mathbb{Z}(U) \xrightarrow{g} C,$$

where N is the normalized chains functor and g is a morphism of chain complexes.

Write $\pi(C, D)$ for the abelian group of chain homotopy classes of maps $C \to D$ between chain complexes C and D, and write $[\alpha]$ for the chain homotopy class of a morphism $\alpha : C \to D$.

There is a category $h_M(X, C)$ whose objects are all pairs

$$X \xleftarrow[\simeq]{f} U, \ \mathbb{Z}(U) \xrightarrow{[g]} C,$$

where $\mathbb{Z}(U)$ is the Moore complex associated to the simplicial abelian object $\mathbb{Z}(U)$ (having the same name), and $[g]$ is a chain homotopy class of morphisms of chain complexes. A morphism $\theta : (f, [g]) \to (f', [g'])$ is a simplicial presheaf map θ which makes the diagrams

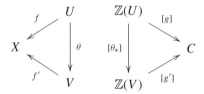

commute.

Recall that there are natural chain maps $i : N\mathbb{Z}(U) \to \mathbb{Z}(U)$ and $p : \mathbb{Z}(U) \to N\mathbb{Z}(U)$ such that $p \cdot i$ is the identity on $N\mathbb{Z}(U)$ and that $i \cdot p$ is naturally chain homotopic to the identity on the Moore complex $\mathbb{Z}(U)$ [32, III.2.4].

The category $h_M(X, C)$ can then be identified up to isomorphism, via precomposition with the natural map i, with the category whose objects are the pairs

$$X \xleftarrow[\simeq]{f} U, \ N\mathbb{Z}(U) \xrightarrow{[g]} C,$$

and whose morphisms $\theta : (f, [g]) \to (f', [g'])$ are maps θ of simplicial presheaves such that the diagrams

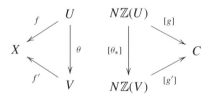

commute.

8.3 Abelian Sheaf Cohomology

Lemma 8.17 *Suppose given chain maps $\alpha, \beta : N\mathbb{Z}(U) \to C$ which are chain homotopic, and suppose that $f : U \to X$ is a local weak equivalence of simplicial presheaves. Then the cocycles (f, α) and (f, β) represent the same element of $\pi_0 h(X, \Gamma(C))$.*

Proof Chain homotopies are defined by path objects for the projective model structure on the category of simplicial presheaves (with sectionwise weak equivalences). Choose a projective cofibrant model $\pi : W \to U$ for this model structure. If there is a chain homotopy $\alpha \simeq \beta : N\mathbb{Z}(U) \to C$, then the composite maps $\alpha_* \eta \pi$ and $\beta_* \eta \pi$ are left homotopic for some choice of cylinder $W \otimes I$ for W in the projective model structure.

This means that there is a diagram

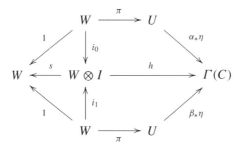

where the maps s, i_0, i_1 are all part of the cylinder object structure for $W \otimes I$, and are sectionwise weak equivalences. It follows that

$$(f, \alpha_* \eta) \sim (f\pi, \alpha_* \eta\pi) \sim (f\pi s, h) \sim (f\pi, \beta_* \eta\pi) \sim (f, \beta_* \eta)$$

in $\pi_0 h(X, \Gamma(C))$.

As noted previously, we can identify $h(X, \Gamma(C))$ with the category of cocycles

$$X \xleftarrow{f} U, \ N\mathbb{Z}(U) \xrightarrow{\alpha} C,$$

where f is a local weak equivalence of simplicial presheaves and α is a chain map. Every such cocycle determines an object

$$X \xleftarrow{f} U, \ N\mathbb{Z}(U) \xrightarrow{[\alpha]} C,$$

of $h_M(X, C)$. This assignment defines a functor

$$\psi : h(X, \Gamma(C)) \to h_M(X, C).$$

Lemma 8.18 *The functor ψ induces an isomorphism*

$$[X, \Gamma(C)] \cong \pi_0 h(X, \Gamma(C)) \xrightarrow[\cong]{\psi_*} \pi_0 h_M(X, C).$$

Proof If the chain maps $\alpha, \beta : N\mathbb{Z}(U) \to C$ are chain homotopic and $f : U \to X$ is a local weak equivalence, then the cocycles (f, α) and (f, β) are in the same path component of $h(X, \Gamma(C))$, by Lemma 8.17. The assignment

$$(f, [\alpha]) \mapsto [(f, \alpha)]$$

therefore defines a function

$$\gamma : \pi_0 h_M(X, C) \to \pi_0 h(X, \Gamma(C)),$$

and one checks that γ is the inverse of ψ_*.

The isomorphism

$$\pi_0 h_M(X, C) \cong [X, \Gamma(C)]$$

which results from Lemma 8.18 is a chain complex variant of the Verdier hypercovering theorem (Theorem 6.12). This result allows one to represent morphisms in the homotopy category taking values in simplicial abelian presheaves by chain homotopy classes of maps.

The simplicial presheaf $\Gamma(C)$ is locally fibrant. As in the proof of Theorem 6.12, there is a category $H_h(X, C)$ whose objects are pairs $([q], [\alpha])$ where $[q]$ is a simplicial homotopy class of a hypercover $q : U \to X$ and $[\alpha]$ is a chain homotopy class of a map $\alpha : N\mathbb{Z}(U) \to C$. A morphism $([q], [\alpha]) \to ([q'], [\beta])$ in $H_h(X, C)$ is a simplicial homotopy class of maps $[\theta] : U \to V$ such that the diagrams

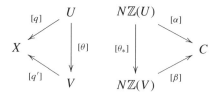

commute.

Recall that $h_{hyp}(X, \Gamma(C))$ is the full subcategory of $h(X, \Gamma(C))$ on those cocycles

$$X \xleftarrow{q} U \xrightarrow{\alpha} \Gamma(C)$$

such that the weak equivalence q is a hypercover, or equivalently those pairs of maps

$$X \xleftarrow{q} U, \; N\mathbb{Z}(U) \xrightarrow{\beta} C,$$

such that β is a chain map and q is a hypercover. There is a functor

$$\gamma : h_{hyp}(X, \Gamma(C)) \to H_h(X, C)$$

8.3 Abelian Sheaf Cohomology

which is defined by $(q, \beta) \mapsto ([q], [\beta])$, and there is a commutative diagram

$$
\begin{array}{ccc}
\pi_0 h_{hyp}(X, \Gamma(C)) & \xrightarrow{\gamma_*} & \pi_0 H_h(X, C) \\
\cong \downarrow & & \downarrow \omega \\
\pi_0 h(X, \Gamma(C)) & \xrightarrow{\cong} & [X, \Gamma(C)]
\end{array}
\qquad (8.2)
$$

The isomorphisms of the diagram come from Theorem 6.5 and Lemma 6.14.

The displayed function ω is defined by sending a class $[([q], [\alpha])]$ to the composite map $(\Gamma(\alpha) \cdot \eta) \cdot q^{-1}$ in the homotopy category. Here, $\Gamma(\alpha)$ denotes the composite map

$$\mathbb{Z}(U) \cong \Gamma(N\mathbb{Z}(U)) \xrightarrow{\Gamma(\alpha)} \Gamma(C)$$

of simplicial abelian presheaves, and h is the canonical simplicial presheaf map $\eta : U \to \mathbb{Z}(U)$, otherwise known as the Hurewicz homomorphism. This function ω is well defined by Lemma 8.17.

The function γ_* in (8.2) is plainly surjective, but it is also injective by the commutativity of the diagram. It follows that all functions in the diagram are bijections.

The resulting bijection

$$\pi_0 H_h(X, C) \cong [X, \Gamma(C)],$$

and the observation that the natural chain homotopy equivalence $N\mathbb{Z}(U) \to \mathbb{Z}(U)$ induces an isomorphism

$$\pi(N\mathbb{Z}(U), C) \cong \pi(\mathbb{Z}(U), C)$$

in chain homotopy classes of maps, together give the following result:

Proposition 8.19 *Suppose that X is a simplicial presheaf and that C is a presheaf of chain complexes. Then there are isomorphisms*

$$[X, \Gamma(C)] \cong \pi_0 H_h(X, C) = \varinjlim_{[p]:U \to X} \pi(\mathbb{Z}(U), C).$$

Remark 8.20 The identification

$$[X, \Gamma(C)] \cong \varinjlim_{[p]:U \to X} \pi(\mathbb{Z}(U), C)$$

of Proposition 8.19 is an older form of Lemma 8.18, which appeared as Theorem 2.1 in [49] with an argument that used the Verdier hypercovering theorem. The displayed colimit happens to be filtered by a calculus of fractions argument [16], but that observation is irrelevant for the proof which is given here.

It is well known that the abelian sheaf category on a small site has enough injectives, as does the abelian presheaf category. For a simple proof, observe that an abelian presheaf I is injective if and only if it has the right lifting property with respect to all inclusions of subobjects $B \subset \mathbb{Z}(U)$, $U \in \mathcal{C}$, so that one can show that there is is an inclusion $A \subset I$ with I injective by using a small object argument.

We shall identify (ordinary) cochain complexes with unbounded chain complexes which are concentrated in degrees $n \leq 0$.

Suppose that A is a sheaf of abelian groups, and let $A \to J$ be an injective resolution of A, where of course J is a cochain complex. As usual, if $n \geq 0$, then $A[-n]$ is the chain complex consisting of A concentrated in degree n. We consider the shifted chain map $A[-n] \to J[-n]$.

The simplicial abelian sheaf

$$K(A, n) = \Gamma(A[-n])$$

which corresponds to the chain complex $A[-n]$ under the Dold–Kan correspondence is the *Eilenberg–Mac Lane object* associated to A and n.

It is an abuse, but write

$$K(D, n) = \Gamma(\tau(D[-n]))$$

for all chain complexes D, where $\tau(D[-n])$ is the good truncation of $D[-n]$ in nonnegative degrees. The simplicial abelian object $K(D, n)$ is not an Eilenberg–Mac Lane object in general, because it potentially has more than one nontrivial sheaf of homotopy groups.

There are isomorphisms

$$\pi(\tau^*(C), D[-n]) \cong \pi(C, \tau(D[-n])), \tag{8.3}$$

which are natural in ordinary chain complexes C and unbounded complexes D, where $\tau^*(C)$ is the unbounded complex which is constructed from C by putting 0 in all negative degrees.

Suppose that C is an ordinary chain complex and that K is a cochain complex. Form the bicomplex

$$\hom(C, K)_{p,q} = \hom(C_{-p}, K_q)$$

with the obvious induced differentials:

$$\partial' = \partial_C^* : \hom(C_{-p}, K_q) \to \hom(C_{-p-1}, K_q)$$
$$\partial'' = (-1)^p \partial_{K*} : \hom(C_{-p}, K_q) \to \hom(C_{-p}, K_{q-1}).$$

8.3 Abelian Sheaf Cohomology

Then $\hom(C, K)$ is a third quadrant bicomplex with total complex $\mathrm{Tot}_\bullet \hom(C, K)$ defined by

$$\mathrm{Tot}_{-n} \hom(C, K) = \bigoplus_{p+q=-n} \hom(C_{-p}, K_q)$$
$$= \bigoplus_{0 \le p \le n} \hom(C_p, K_{-n+p}),$$

for $n \ge 0$. The complex $\mathrm{Tot}_\bullet \hom(C, K)$ is concentrated in negative degrees.

Lemma 8.21 *There are natural isomorphisms*

$$H_{-n}(\mathrm{Tot}_\bullet \hom(C, K)) \cong \pi(\tau^*(C), K[-n]).$$

Proof Write $(f_0, f_1, \ldots, f_n)$ for a typical element of

$$\mathrm{Tot}_{-n} \hom(C, K) = \bigoplus_{0 \le p \le n} \hom(C_p, K_{-n+p}).$$

Then

$$\partial(f_0, \ldots, f_n) = (g_0, \ldots, g_{n+1}),$$

where

$$g_k = \begin{cases} \partial f_0 & \text{if } k = 0, \\ f_{k-1}\partial + (-1)^k \partial f_k & \text{if } 0 < k < n+1, \text{ and} \\ f_n \partial & \text{if } k = n+1. \end{cases}$$

Set

$$\alpha(k) = \begin{cases} 1 & \text{if } k = 0, \text{ and} \\ \sum_{j=1}^{k-1} j & \text{if } k \ge 1. \end{cases}$$

Then the maps $(-1)^{\alpha(k)} f_k$ define a chain map $\tau^*(C) \to K[-n]$.

Suppose that

$$\partial(s_0, \ldots s_{n-1}) = (f_0, \ldots, f_n).$$

Then the maps $(-1)^{\alpha(k)} s_k$ define a chain homotopy from the chain map $(-1)^{\alpha(k)} f_k$ to the 0 map.

Lemma 8.22 *Suppose that J is a cochain complex of injective sheaves, and that $f : C \to D$ is a homology isomorphism of ordinary chain complexes objects. Then the induced morphism of cochain complexes*

$$\mathrm{Tot}_\bullet \hom(D, J) \xrightarrow{f^*} \mathrm{Tot}_\bullet \hom(C, J)$$

is a homology isomorphism.

Proof The functors hom $(\,,J_{-q})$ are exact, and there are isomorphisms

$$H_{-p}\hom(C,J_{-q}) \cong \hom(\tilde{H}_p(C),J_{-q}),$$

which are natural in chain complexes C. It follows, using Lemma 8.21, that there is a spectral sequence with

$$E_1^{p,q} \cong \hom(\tilde{H}_p(C),J_{-q}) \Rightarrow \pi(\tau^*(C),J[-p-q]) \cong H_{-p-q}\operatorname{Tot}_\bullet \hom(C,J).$$

which is natural in C. The claim follows from a comparison of such spectral sequences.

Corollary 8.23 *Suppose that J is a cochain complex of injective sheaves. Then every local weak equivalence $f : X \to Y$ of simplicial presheaves induces an isomorphism*

$$\pi(N\mathbb{Z}Y,\tau(J[-n])) \xrightarrow{\cong} \pi(N\mathbb{Z}X,\tau(J[-n]))$$

in chain homotopy classes for all $n \geq 0$.

Proof This result follows from Lemma 8.22 and the existence of the natural isomorphisms (8.3).

Again, let J be a cochain complex of injective sheaves. As in the proof of Lemma 8.18, there is a well defined abelian group homomorphism

$$\gamma : \pi(N\mathbb{Z}X,\tau(J[-n])) \to \pi_0 h(X,K(J,n))$$

which takes a chain homotopy class $[\alpha]$ to the element $[(1, \Gamma(\alpha) \cdot \eta]$, where the map $\Gamma(\alpha) : \mathbb{Z}(X) \to K(J,n)$ is induced by α under the Dold–Kan correspondence, and $\eta : X \to \mathbb{Z}(X)$ is the Hurewicz map. This morphism is natural in simplicial presheaves X.

Lemma 8.24 *Suppose that J is a cochain complex of injective sheaves. Then we have the following:*

1) The map

$$\gamma : \pi(N\mathbb{Z}X,\tau(J[-n])) \to \pi_0 h(X,K(J,n)).$$

is an isomorphism.
2) The canonical map

$$c : \pi(N\mathbb{Z}X,\tau(J[-n])) \to [N\mathbb{Z}X,\tau(J[-n])]$$

is an isomorphism.
3) The simplicial abelian sheaf $K(J,n) = \Gamma(\tau(J[-n]))$ satisfies descent.

Recall that a simplicial presheaf X on a site $\mathcal{C}$ satisfies descent if some (hence any) injective fibrant model $j : X \to Z$ is a sectionwise weak equivalence in the

8.3 Abelian Sheaf Cohomology

sense that the simplicial set maps $j : X(U) \to Z(U)$ are weak equivalences for all objects U of $\mathcal{C}$.

A simplicial R-module A satisfies descent (even as a simplicial presheaf) if some injective fibrant model $A \to Z$ in simplicial R-modules is a sectionwise weak equivalence.

Proof For statement 1), suppose that the pair of morphisms

$$X \xleftarrow{f}_{\simeq} U, \ N\mathbb{Z}(U) \xrightarrow{g} \tau(J[-n])$$

defines an object of the cocycle category $h(X, K(J, n))$. Then there is a unique chain homotopy class $[v] : N\mathbb{Z}X \to J[-n]$ such that $[v \cdot f_*] = [g]$ since f is a local weak equivalence, by Corollary 8.23. This chain homotopy class $[v]$ is also independent of the choice of representative for the path component of (f, g) in the cocycle category. We therefore have a well-defined function

$$\omega : \pi_0 h(X, K(J, n)) \to \pi(N\mathbb{Z}X, \tau(J[-n])).$$

The composites $\omega \cdot \gamma$ and $\gamma \cdot \omega$ are identity morphisms.

For statement 2), observe that there is a commutative diagram

$$\begin{array}{ccc} \pi(N\mathbb{Z}X, \tau(J[-n])) & \xrightarrow{\gamma}_{\cong} & \pi_0 h(X, K(J, n)) \\ {\scriptstyle c}\downarrow & & \downarrow{\scriptstyle \phi}\cong \\ [N\mathbb{Z}X, \tau(J[-n])] & \xrightarrow{\cong} & [X, K(J, n)] \end{array}$$

where ϕ is the isomorphism of Theorem 6.5, and the bottom isomorphism is induced by the Dold–Kan correspondence.

Suppose that $j : \tau(J[-n]) \to E$ is an injective fibrant model for $K(J, n)$ in sheaves of chain complexes. Statement 3) says that the induced maps $j : \tau(J[-n])(U) \to E(U)$ of chain complexes are homology isomorphisms, for all $U \in \mathcal{C}$.

To prove this, first observe that there is a commutative diagram

$$\begin{array}{ccc} \pi(N\mathbb{Z}X, \tau(J[-n])) & \xrightarrow{c}_{\cong} & [N\mathbb{Z}X, \tau(J[-n])] \\ {\scriptstyle j_*}\downarrow & & \downarrow{\scriptstyle j_*}\cong \\ \pi(N\mathbb{Z}X, E) & \xrightarrow{\cong}_{c} & [N\mathbb{Z}X, E] \end{array}$$

for each simplicial presheaf X, in which the top occurrence of the canonical map c is an isomorphism by statement 2), and the bottom occurrence is an isomorphism since the chain complex object $N\mathbb{Z}X$ is cofibrant and E is injective fibrant. It follows that the map

$$j_* : \pi(N\mathbb{Z}X, \tau(J[-n])) \to \pi(N\mathbb{Z}X, E)$$

of chain homotopy classes is an isomorphism for all simplicial presheaves X. There is a split short exact sequence

$$0 \to N\mathbb{Z}* \to N\mathbb{Z}(\Delta^m/\partial\Delta^m) \to \mathbb{Z}[-m] \to 0$$

of chain complexes for all $m \geq 0$, and it follows that the maps

$$j_* : \pi(L_U(\mathbb{Z}[-m]), \tau(J[-n])) \to \pi(L_U(\mathbb{Z}[-m]), E)$$

are isomorphisms for all $m, n \geq 0$ and $U \in \mathcal{C}$. Recall that

$$L_U(A) = A \otimes \hom(\ , U)$$

is the left adjoint of the U-sections functor $A \mapsto A(U)$.

The chain complex maps

$$\tau(J[-n])(U) \to E(U)$$

are therefore homology isomorphisms for all $U \in \mathcal{C}$, and statement 3) is proved.

The following result is now a corollary of Lemma 8.24:

Theorem 8.25 *Suppose that A is a sheaf of abelian groups on $\mathcal{C}$, and let $A \to J$ be an injective resolution of A in the category of abelian sheaves. Let X be a simplicial presheaf on $\mathcal{C}$. Then there is an isomorphism*

$$\pi(N\mathbb{Z}X, \tau(J[-n])) \cong [X, K(A, n)].$$

This isomorphism is natural in X.

The *sheaf cohomology group* $H^n(\mathcal{C}, A)$ for an abelian sheaf A on a site $\mathcal{C}$ is traditionally defined by

$$H^n(\mathcal{C}, A) = H_{-n}(\Gamma_* J)$$

where $A \to J$ is an injective resolution of A concentrated in negative degrees and Γ_* is the global sections functor (i.e. inverse limit). But $\Gamma_* Y = \hom(*, Y)$ for any Y, where $*$ is the one-point simplicial presheaf, and there are isomorphisms

$$H^n(\mathcal{C}, A) \cong \pi(\mathbb{Z}*, \tau(J[-n])) \cong [*, K(A, n)].$$

The second displayed isomorphism is a consequence of Theorem 8.25. We have proved.

Theorem 8.26 *Suppose that A is an abelian sheaf on a site $\mathcal{C}$. Then there is an isomorphism*

$$H^n(\mathcal{C}, A) \cong [*, K(A, n)]$$

which is natural in abelian sheaves A.

8.3 Abelian Sheaf Cohomology

Suppose that A is an abelian *presheaf* on $\mathcal{C}$ and that X is a simplicial presheaf. Write
$$H^n(X, A) = [X, K(A, n)] \cong [\mathbb{Z}(X), K(A, n)],$$
and say that this group is the *n*th *cohomology group* of X with coeffients in A.

The associated sheaf map
$$\eta : K(A, n) \to L^2 K(A, n) \cong K(\tilde{A}, n)$$
is a local weak equivalence, so there is a canonical isomorphism
$$H^n(X, A) \cong H^n(X, \tilde{A}).$$

Write
$$\tilde{H}_n(X, \mathbb{Z}) = \tilde{H}_n(N\mathbb{Z}X) \cong \tilde{H}_n(\mathbb{Z}X)$$
and call this object the *n*th *integral homology sheaf* of the simplicial presheaf X. It is also common to write
$$\tilde{H}_n(X) = \tilde{H}_n(X, \mathbb{Z})$$
for the integral homology sheaves of X.

If A is an abelian presheaf, write
$$\tilde{H}_n(X, A) = \tilde{H}_n(N\mathbb{Z}(X) \otimes A) \cong \tilde{H}_n(\mathbb{Z}(X) \otimes A)$$
for the *n*th *homology sheaf of X with coefficients in A*.

We have the following trivial consequence of the definitions, which is listed for emphasis.

Lemma 8.27 *Suppose that the simplicial presheaf map $f : X \to Y$ induces a homology sheaf isomorphism*
$$\tilde{H}_*(X, \mathbb{Z}) \cong \tilde{H}_*(Y, \mathbb{Z}).$$

Then f induces an isomorphism
$$H^*(Y, A) \to H^*(X, A)$$
for all presheaves abelian groups A.

The following result is a consequence of Lemma 8.21 and Theorem 8.25:

Corollary 8.28 *Suppose that X is a simplicial presheaf and that A is a presheaf of abelian groups. Then there is a spectral sequence, with*
$$E_2^{p,q} = \mathrm{Ext}^q(\tilde{H}_p(X, \mathbb{Z}), \tilde{A}) \Rightarrow H^{p+q}(X, A). \qquad (8.4)$$

The spectral sequence (8.4) is the *universal coefficients spectral sequence* for the cohomology groups $H^*(X, A)$.

Example 8.29 Suppose that X is a simplicial set and that A is an abelian sheaf on a small site $\mathcal{C}$. The cohomology $H^*(\Gamma^*X, A)$ of the constant simplicial presheaf Γ^*X with coefficients in A is what Grothendieck would call a *mixed cohomology theory* [35]. In this case, the universal coefficients spectral sequence has a particularly simple form, in that there is a short exact sequence

$$0 \to \bigoplus_{p+q=n} \mathrm{Ext}^1\left(H_{p-1}(X, \mathbb{Z}), H^q(\mathcal{C}, A)\right) \to H^{p+q}(\Gamma^*X, A)$$

$$\to \bigoplus_{p+q=n} \hom\left(H_p(X, \mathbb{Z}), H^q(\mathcal{C}, A)\right) \to 0.$$

The existence of this sequence is best proved with the standard argument that leads to the classical universal coefficients theorem: apply the functor $\hom(\ ,\Gamma_*I)$ to the short split exact sequence of chain complexes

$$0 \to Z(\mathbb{Z}X) \to \mathbb{Z}X \to B(\mathbb{Z}X)[-1] \to 0.$$

where the complexes $Z(\mathbb{Z}X)$ and $B(\mathbb{Z}X)$ consist of cycles and boundaries, respectively, with 0 differentials.

Suppose that R is a presheaf of commutative rings with unit. There are R-linear versions of all results so far encountered in this section. In particular, there is an R-linear universal coefficients spectral sequence:

Lemma 8.30 *Suppose that X is a simplicial presheaf and that A is a presheaf of R-modules. Then there is a spectral sequence, with*

$$E_2^{p,q} = \mathrm{Ext}_R^q\left(\tilde{H}_p(X, \tilde{R}), \tilde{A}\right) \Rightarrow H^{p+q}(X, A).$$

We also have the following R-linear analogue of Lemma 8.27.

Lemma 8.31 *Suppose that the simplicial presheaf map $f : X \to Y$ induces a homology sheaf isomorphism*

$$\tilde{H}_*(X, R) \cong \tilde{H}_*(Y, R).$$

Then f induces an isomorphism

$$H^*(Y, A) \to H^*(X, A)$$

for all presheaves of R-modules A.

Lemma 8.31 is again a trivial consequence of the definitions. This was not so initially [49]—the sheaf theoretic version of this result was originally derived as a consequence of a statement analogous to Lemma 8.30.

8.3 Abelian Sheaf Cohomology

Recall that, if X and Y are simplicial presheaves, then the internal function complex $\mathbf{Hom}(X, Y)$ is the simplicial presheaf with sections defined by

$$\mathbf{Hom}(X, Y)(U) = \mathbf{hom}(X|_U, Y|_U) \cong \mathbf{hom}(X \times U, Y).$$

We then have the following enhancement of Theorem 8.25:

Proposition 8.32 *Suppose that A is a presheaf of abelian groups, and that X is a simplicial presheaf. Suppose that the map*

$$j : K(A, n) \to FK(A, n)$$

is an injective fibrant model of $K(A, n)$. Then there are isomorphisms

$$\pi_j \mathbf{Hom}(X, FK(A, n))(U) \cong \begin{cases} H^{n-j}(X|_U, A|_U) & 0 \leq j \leq n \\ 0 & j > n. \end{cases}$$

for all $U \in \mathcal{C}$.

Proof There are isomorphisms

$$\pi_0 \mathbf{Hom}(X, FK(A, n))(U) \cong [X|_U, FK(A|_U, n)] \cong H^n(X|_U, A|_U),$$

since $FK(A, n)|_U$ is an injective fibrant model of $K(A|_U, n)$ by Corollary 5.26 and Theorem 8.26.

The associated sheaf map

$$\eta : K(A, 0) \to K(\tilde{A}, 0)$$

is an injective fibrant model for the constant simplicial presheaf $K(A, 0)$ by Lemma 5.11, and there is an isomorphism

$$\mathbf{Hom}(X, K(\tilde{A}, 0)) \cong \mathbf{Hom}(\tilde{\pi}_0(X), \tilde{A}),$$

where the latter is identified with a constant simplicial sheaf. It follows that the sheaves $\tilde{\pi}_j \mathbf{Hom}(X, K(\tilde{A}, 0))$ vanish for $j > 0$.

There is a sectionwise fibre sequence

$$K(A, n-1) \to WK(A, n-1) \xrightarrow{p} K(A, n)$$

where the simplicial abelian presheaf $WK(A, n-1)$ is sectionwise contractible.

Take an injective fibrant model

$$\begin{array}{ccc} WK(A, n-1) & \xrightarrow{j} & FWK(A, n-1) \\ p \downarrow & & \downarrow q \\ K(A, n) & \xrightarrow{j} & FK(A, n) \end{array}$$

for the map p. This means that the maps labelled j are local weak equivalences, $FK(A,n)$ is injective fibrant and q is an injective fibration. Let $F = q^{-1}(0)$. Then the simplicial presheaf F is injective fibrant and the induced map

$$K(A, n-1) \to F$$

is a local weak equivalence, by Lemma 4.37. Write $FK(A, n-1)$ for F.

We have injective (hence sectionwise) fibre sequences

$$\mathbf{Hom}(X, FK(A, n-1)) \to \mathbf{Hom}(X, FWK(A, n-1)) \to \mathbf{Hom}(X, FK(A, n))$$

by Lemma 5.12 and the enriched simplicial model structure of Corollary 5.19. The map

$$\mathbf{Hom}(X, FWK(A, n-1)) \to \mathbf{Hom}(X, *) \cong *$$

is a trivial injective fibration, and is therefore a sectionwise trivial fibration. It follows that there are isomorphisms

$$\pi_j \mathbf{Hom}(X, FK(A,n))(U) \cong \pi_{j-1} \mathbf{Hom}(X, FK(A, n-1))(U)$$

for $j \geq 1$ and all $U \in \mathcal{C}$, so that

$$\pi_j \mathbf{Hom}(X, FK(A,n))(U) \cong H^{n-j}(X|_U, \tilde{A}|_U)$$

for $1 \leq j \leq n$ and $\pi_j \mathbf{Hom}(X, FK(A,n))(U) = 0$ for $j > n$, by induction on n.

Corollary 8.33 *Suppose that A is a presheaf of abelian groups, and that*

$$j : K(A,n) \to FK(A,n)$$

is an injective fibrant model of $K(A,n)$. Then there are isomorphisms

$$\pi_j FK(A,n)(U) \cong \begin{cases} H^{n-j}(\mathcal{C}/U, \tilde{A}|_U) & 0 \leq j \leq n \\ 0 & j > n. \end{cases}$$

for all $U \in \mathcal{C}$.

We have a Quillen adjunction

$$\tau^* : \mathrm{Ch}_+(R) \leftrightarrows \mathrm{Ch}(R) : \tau$$

defined by the good truncation functor τ and its left adjoint, in which both functors preserve weak equivalences. It follows that there is a natural isomorphism

$$[\tau^*(C), D[-n]] \cong [C, \tau(D[-n])].$$

for all ordinary complexes C and unbounded complexes D. We also know that there is a natural isomorphism

$$[\mathbb{Z}(X), B] \cong [X, u(B)]$$

8.3 Abelian Sheaf Cohomology

for all simplicial presheaves X and simplicial abelian group objects B. It follows that there is an isomorphism

$$[X, K(A, n)] \cong [\tau^*(\mathbb{Z}(X)), H(A)[-n]]$$

for all abelian presheaves A, relating morphisms in the simplicial presheaf homotopy category with morphisms in the full derived category, so that there is a natural isomorphism

$$H^n(X, A) \cong [\tau^*(\mathbb{Z}(X)), H(A)[-n]].$$

More generally, if D is an unbounded complex and X is a simplicial presheaf, we define

$$\mathbb{H}^n(X, D) = [\tau^*(\mathbb{Z}(X)), D[-n]] \cong [\mathbb{Z}(X), \tau(D[-n])] \cong [X, \Gamma(\tau(D[-n]))],$$

and say that this invariant is the *nth hypercohomology group* of X with coefficients in the complex D. In principle, $\mathbb{H}^n(X, D)$ is morphisms in a stable category from a suspension spectrum object $\tau^*(\mathbb{Z}(X))$ to a shifted spectrum $D[-n]$.

If D is an ordinary chain complex and $n \geq 0$, then there is an isomorphism

$$\mathbb{H}^n(X, \tau^*(D)) \cong [\mathbb{Z}(X), D[-n]]$$

where the displayed morphisms are in the derived category of ordinary chain complexes. It follows that there is a natural isomorphism

$$\mathbb{H}^n(X, H(A)) \cong [\mathbb{Z}(X), H(A)[-n]] = H^n(X, A)$$

for all presheaves of abelian groups A.

There is a weak equivalence

$$\overline{W}A = \Gamma(NA[-1]) \simeq S^1 \otimes A = d(BA)$$

(see Corollary 9.39 below) which is natural in simplicial abelian presheaves A. It follows that there are natural weak equivalences

$$\Gamma(NA[-n]) \simeq S^n \otimes A$$

and

$$\Gamma(NA[n]) \simeq \Omega^n A$$

for all $n \geq 0$.

Write

$$\Omega^n A = A \otimes S^{-n}$$

for $n < 0$. Then there are natural isomorphisms

$$\mathbb{H}^n(X, \tau^*(NA)) \cong [\mathbb{Z}(X), \Gamma(NA[-n])] \cong [\mathbb{Z}(X), \Omega^{-n}A]$$

for simplicial presheaves X and simplicial abelian presheaves A.

We also write

$$\mathbb{H}^n(X, A) = [X, \Omega^{-n}A] \tag{8.5}$$

for simplicial presheaves X and simplicial abelian presheaves A. This group is the *nth hypercohomology group* of X with coefficients in the simplicial abelian presheaf A.

Suppose that $A \to K$ is a fibrant model in the category of simplicial abelian presheaves. There are isomorphisms

$$\pi_p \mathbf{hom}(X, K) \cong [X, \Omega^p K] \cong [\mathbb{Z}(X), K[p]],$$

and it follows that there is an isomorphism

$$\pi_p \mathbf{hom}(X, K) \cong \mathbb{H}^{-p}(X, A).$$

The final result of this section gives a large class of calculational examples—it says that our definition of the cohomology of a simplicial presheaf agrees with the classical description of the cohomology of a simplicial scheme.

Lemma 8.34 *Suppose that S is a simplicial object in $\mathcal{C}$ and that A is an abelian sheaf on $\mathcal{C}$, and let $\mathcal{C}/S$ be the site fibred over the simplicial object S. Then there are isomorphisms*

$$H^n(S, A) \cong H^n(\mathcal{C}/S, A|_S).$$

These isomorphisms are natural in abelian sheaves A.

Proof This result is a consequence of Proposition 5.29.

Suppose that $j : K(A, n) \to FK(A, n)$ is an injective fibrant model on $\mathcal{C}$, and choose an injective fibrant model $FK(A, n)|_S \to W$ on the site $\mathcal{C}/S$. Then Proposition 5.29 says that there is a weak equivalence

$$\mathbf{hom}(S, FK(A, n)) \simeq \mathbf{hom}(*, W).$$

The simplicial presheaf W is an injective fibrant model for the restricted simplicial presheaf $K(A|_S, n)$, and so there are isomorphisms

$$H^n(S, A) \cong \pi_0 \mathbf{hom}(S, FK(A, n)) \cong \pi_0 \mathbf{hom}(*, W) \cong H^n(\mathcal{C}/S, A|_S).$$

Remark 8.35 Both the statement of Lemma 8.34 and its proof are prototypical.

A similar argument shows that the étale cohomology group $H_{et}^n(S, A)$ of a simplicial T-scheme S, which is traditionally defined to be $H^n(et|_S, A|_S)$ for an abelian sheaf A on the big étale site [26], can be defined by

$$H_{et}^n(S, A) = [S, K(A, n)]$$

as morphisms in the injective homotopy category of simplicial presheaves or sheaves on the big site $(Sch|_T)_{et}$.

Here, $et|_S$ is the fibred étale site whose objects are the étale morphisms $\phi : U \to S_n$, and whose morphisms are diagrams of scheme homomorphisms of the form of (5.3), where the vertical maps are étale—this is usually what is meant by the étale site of a simplicial scheme S.

One uses the ideas of Example 5.28 to show that the restriction $Z|_S$ of an injective fibrant object Z to the site $et|_S$ satisfies descent. The remaining part of the argument for the weak equivalence

$$\mathbf{hom}(S, Z) \simeq \mathbf{hom}(*, W),$$

where $Z|_S \to W$ is an injective fibrant model on $et|_S$, is formal.

A different argument is available for the étale cohomological analogue of Corollary 8.34 if one's sole interest is a cohomology isomorphism: see [49].

Analogous techniques and results hold for other standard algebraic geometric topologies, such as the flat or Nisnevich topologies.

8.4 Products and Pairings

The category of *pointed simplicial presheaves* on a site $\mathcal{C}$ is the slice category $*/s\mathbf{Pre}(\mathcal{C})$. The objects $* \to X$ alternatively be viewed as pairs (X, x), where X is a simplicial presheaf and x is a choice of vertex in the global sections simplicial set

$$\Gamma_* X = \varprojlim_{U \in \mathcal{C}} X(U).$$

A *pointed map* $f : (X, x) \to (Y, y)$ is a simplicial presheaf map $f : X \to Y$ such that $f_*(x) = y$ in global sections, or equivalently such that the diagram

commutes. One also writes $s\mathbf{Pre}_*(\mathcal{C})$ to denote this category.

All slice categories for $s\mathbf{Pre}(\mathcal{C})$ inherit injective model structures from the injective model structure for simplicial presheaves—see Remark 6.16. In the case at hand, a pointed map $(X, x) \to (Y, y)$ is a local weak equivalence (respectively cofibration, injective fibration) if the underlying map $f : X \to Y$ is a local weak equivalence (respectively cofibration, injective fibration) of simplicial presheaves. The notation $[X, Y]_*$ denotes morphisms in the pointed homotopy category

$$\mathrm{Ho}\,(s\mathbf{Pre}_*(\mathcal{C})).$$

The functor $q : s\mathbf{Pre}_*(\mathcal{C}) \to s\mathbf{Pre}(\mathcal{C})$ forgets the base point. One usually writes $Y = q(Y)$ for the underlying simplicial presheaf of an object Y. The left adjoint $X \mapsto X_+$ of this functor is defined by adding a disjoint base point: $X_+ = X \sqcup \{*\}$. The functor q and its left adjoint form a Quillen adjunction, and there is a bijection

$$[X_+, Y]_* \cong [X, Y].$$

Every simplicial abelian presheaf B is canonically pointed by 0, so there is an isomorphism

$$[X_+, B]_* \cong [X, B].$$

In particular, cohomology groups can be computed in the pointed homotopy category via the natural isomorphism

$$H^n(X, A) = [X, K(A, n)] \cong [X_+, K(A, n)]_*,$$

where the simplicial abelian presheaf $K(A, n)$ is pointed by 0.

The *smash product* $X \wedge Y$ of two pointed simplicial presheaves is formed as in pointed simplicial sets:

$$X \wedge Y := (X \times Y)/(X \vee Y)$$

where the wedge $X \vee Y$ is the coproduct of X and Y in the pointed category.

If X is a pointed simplicial presheaf, $\mathbb{Z}_\bullet(X)$ is the cokernel of the map

$$\mathbb{Z}(*) \to \mathbb{Z}(X)$$

which is defined by the base point of X, as in Remark 8.1. The object $\mathbb{Z}_\bullet(X)$ is the *reduced free simplicial abelian group* associated to X, and the homology sheaves

$$\tilde{H}_*(\mathbb{Z}_\bullet(X) \otimes A)$$

are the *reduced homology sheaves* of X with coefficients in the abelian presheaf A.

The isomorphism

$$\mathbb{Z}(X) \otimes \mathbb{Z}(Y) \xrightarrow{\cong} \mathbb{Z}(X \times Y)$$

8.4 Products and Pairings

of simplicial abelian presheaves induces an isomorphism

$$\mathbb{Z}_\bullet(X) \otimes \mathbb{Z}_\bullet(Y) \xrightarrow{\cong} \mathbb{Z}_\bullet(X \wedge Y) \tag{8.6}$$

which is natural in pointed simplicial presheaves X and Y.

Suppose that A is a presheaf of abelian groups, and let $S^n \otimes A$ denote the simplicial abelian presheaf $\mathbb{Z}_\bullet(S^n) \otimes A$. Here, S^n is the n-fold smash power

$$S^n = S^1 \wedge \cdots \wedge S^1$$

of the simplicial circle $S^1 = \Delta^1/\partial\Delta^1$.

The simplicial abelian presheaf $S^n \otimes A$ has a unique nontrivial homology presheaf, namely

$$H_n(S^n \otimes A) \cong A,$$

and the good truncation functor τ_n at level n in chain complexes defines homology presheaf isomorphisms

$$S^n \otimes A \xleftarrow{\cong} \tau_n(S^n \otimes A) \xrightarrow{\cong} \Gamma(A[-n])$$

of chain complex objects. It follows that the simplicial abelian presheaf $S^n \otimes A$ is naturally locally equivalent to the Eilenberg–Mac Lane object $K(A, n)$.

The natural isomorphism (8.6) induces a natural isomorphism

$$(S^n \otimes A) \otimes (S^m \otimes B) \xrightarrow{\cong} S^{n+m} \otimes (A \otimes B), \tag{8.7}$$

and a pairing

$$(S^n \otimes A) \wedge (S^m \otimes B) \to (S^n \otimes A) \otimes (S^m \otimes B) \xrightarrow{\cong} S^{n+m} \otimes (A \otimes B)$$

of pointed simplicial presheaves. This pairing can be rewritten as a map

$$\cup : K(A, n) \wedge K(B, m) \to K(A \otimes B, n + m) \tag{8.8}$$

in the pointed homotopy category.

The pairing (8.8), in any of its equivalent forms, is the cup product pairing. It induces the *external cup product*

$$\cup : H^n(X, A) \times H^m(Y, B) \to H^{n+m}(X \times Y, A \otimes B), \tag{8.9}$$

of cohomology groups of simplicial presheaves, which we now describe in terms of cocycles.

Suppose that E and F are presheaves of simplicial abelian groups. There is a natural map

$$\cup : E \wedge F \to E \otimes F$$

of pointed simplicial presheaves which takes values in the degreewise tensor product. Given cocycles

$$X \xleftarrow{u}_{\simeq} U \xrightarrow{f} E, \quad Y \xleftarrow{v}_{\simeq} V \xrightarrow{g} F,$$

there is a cocycle

$$X \times Y \xleftarrow{u \times v} U \times V \xrightarrow{(f \wedge g)_*} E \otimes F,$$

where $(f \wedge g)_*$ is the composite

$$U \times V \to (U \times V)_+ \cong U_+ \wedge V_+ \xrightarrow{f \wedge g} E \wedge F \xrightarrow{\cup} E \otimes F.$$

The assignment

$$((f, u), (g, v)) \mapsto ((f \wedge g)_*, u \times v)$$

is functorial in the cocycles (f, u) and (g, v), and defines a functor

$$h(X, E) \times h(Y, F) \to h(X \times Y, E \otimes F).$$

The induced map in path components gives the cup product pairing

$$\cup : [X, E] \times [Y, F] \to [X \times Y, E \otimes F].$$

If R is a presheaf of commutative rings with unit, then precomposition with the diagonal $\Delta : X \to X \times X$ and composition with the multiplication $R \otimes R \to R$, applied to the pairing (8.9), together define the pairing

$$H^n(X, R) \times H^m(X, R) \to H^{n+m}(X \times X, R \otimes R) \to H^{n+m}(X, R),$$

which is the *cup product* for the cohomology of the simplicial presheaf X with coefficients in the presheaf of rings R.

The cup product ring structure on $H^*(X, R)$ is associative, and has a two-sided multiplicative identity which is defined by the composite

$$X \to * \xrightarrow{1} R,$$

where the global section 1 is the multiplicative identity of the presheaf of rings R. The resulting ring structure on the cohomology $H^*(X, R)$ is graded commutative, since R is commutative and the twist isomorphism

$$S^p \wedge S^q \xrightarrow{\tau}_{\cong} S^q \wedge S^p$$

is multiplication by $(-1)^{pq}$ in the homotopy category.

In particular, the cohomology $H^*(X, R)$ of a simplicial presheaf X with coefficients in a commutative unitary ring R has the structure of a graded commutative ring.

Remark 8.36 Note the level of generality. Cup products are defined for cohomology of simplicial presheaves having all abelian presheaf coefficients, on all Grothendieck sites.

It is an exercise to show that cup products are preserved by inverse image functors associated with geometric morphisms.

We can go further. Suppose that E is a presheaf of simplicial abelian groups. It is an exercise to show that there is a natural pairing

$$\Omega^{-p}(E) \otimes \Omega^{-q}(F) \to \Omega^{-p-q}(E \otimes F)$$

for all integers p and q and all presheaves of simplicial abelian groups E and F, which generalizes the pairings of (8.7). This pairing induces a cup product pairing

$$\mathbb{H}^p(X, E) \otimes \mathbb{H}^q(X, F) \to \mathbb{H}^{p+q}(X, E \otimes F) \tag{8.10}$$

in hypercohomology for all simplicial presheaves X, simplicial abelian presheaves E and F, and integers p and q.

8.5 Localized Chain Complexes

As in Sect. 7.2, suppose that $\mathcal{F}$ is a set of cofibrations of simplicial presheaves, such that the following conditions hold:

C1: the set $\mathcal{F}$ contains all members of the generating set J of trivial cofibrations for the injective model structure,

C2: if the map $i : A \to B$ is a member of $\mathcal{F}$, and $j : C \to D$ is an α-bounded cofibration, then the cofibration

$$(B \times C) \cup (A \times D) \to B \times D$$

is a member of $\mathcal{F}$.

In this section, we describe the $\mathcal{F}$-local homotopy theory for simplicial R-modules. The method of construction is parallel to that for the $\mathcal{F}$-local homotopy theory of Theorem 7.18, but occurs within the model structure for the category $s\mathbf{Pre}_R$ of simplicial R-modules which is given by Theorem 8.6.

Of course, model structures on the category of simplicial R-modules give model structures on the category of presheaves of chain complexes of R-modules, via the Dold–Kan correspondence.

The injective model structure of Theorem 8.6 is a proper closed simplicial model structure for the category $s\mathbf{Pre}_R$ of simplicial R-modules, which is cofibrantly generated. The generating sets I and J for the classes of cofibrations and trivial cofibrations are given by the maps $R(A) \to R(B)$ which are induced by the α-bounded cofibrations, respectively trivial α-bounded cofibrations $A \to B$ of simplicial presheaves.

Every cofibration of $s\mathbf{Pre}_R$ is a monomorphism, and the category has epi-monic factorizations.

Write $R(\mathcal{F})$ for the set of cofibrations $R(C) \to R(D)$ of $s\mathbf{Pre}_R$ which are associated to the cofibrations $C \to D$ of $\mathcal{F}$. This is the set of cofibrations which is used to construct a functor

$$L_{\mathcal{F}} := L_{R(\mathcal{F})} : s\mathbf{Pre}_R \to s\mathbf{Pre}_R,$$

by using the methods of Sect. 7.1.

Recall the notational convention that

$$E \otimes K := E \otimes R(K)$$

for all simplicial R-modules E and simplicial sets K. The maps $A \otimes \Delta^n \to B$ are the n-simplices of the function complex $\mathbf{hom}(A, B)$.

It is an exercise to show that the map

$$(F \otimes \partial \Delta^n) \cup (E \otimes \Delta^n) \to F \otimes \Delta^n$$

is in $R(\mathcal{F})$ if the map $E \to F$ is in $R(\mathcal{F})$.

A map $p : A \to B$ of simplicial R-modules is said to be $\mathcal{F}$-*injective* if it has the right lifting property with respect to all members of the set $R(\mathcal{F})$, or equivalently if the underlying simplicial presheaf map $u(A) \to u(B)$ is $\mathcal{F}$-injective in the sense of Sect. 7.2.

A simplicial R-module Z is $\mathcal{F}$-injective if and only if the underlying simplicial presheaf $u(Z)$ is $\mathcal{F}$-injective, or equivalently fibrant (ie. $\mathcal{F}$-fibrant) for the $\mathcal{F}$-local model structure on the simplicial presheaf category of Theorem 7.18. In particular, Z is an injective fibrant simplicial R-module, and its underlying simplicial presheaf $u(Z)$ is injective fibrant.

Following Sect. 7.1, a map $g : A \to B$ of simplicial R-modules is an $\mathcal{F}$-*equivalence* if the induced map $g_c : A_c \to B_c$ of cofibrant models induces a weak equivalence

$$g^* : \mathbf{hom}(B_c, Z) \to \mathbf{hom}(A_c, Z)$$

of function complex objects for all $\mathcal{F}$-injective simplicial R-modules Z.

Choose a regular cardinal β such that $\beta > |R|$, $\beta > |\mathcal{F}|$, $\beta > |B|$ for all morphisms $A \to B$ in $\mathcal{F}$, and that $\beta > \alpha > |\mathrm{Mor}(\mathcal{C})|$.

The choice of β is similar to that of Sect. 7.2, except one also insists that it is an upper bound for the cardinality $|R|$ of the presheaf of rings R.

Suppose that λ is a cardinal such that $\lambda > 2^\beta$.

As in Sect. 7.2, every map $f : X \to Y$ of simplicial R-modules has a functorial system of factorizations

8.5 Localized Chain Complexes

for $s < \lambda$ which is defined by partial solutions of the lifting property for maps in $\mathcal{F}$. These factorizations are defined by successive pushouts, by analogy with the system of factorizations of the same name of Sect. 7.1. The colimit of these factorizations has the form

where the map f_λ has the right lifting property with respect to all $R(C) \to R(D)$ in $\mathcal{F}$, and i_λ is in the saturation of $\mathcal{F}$.

Define a functor $L_\mathcal{F}$ from the category of simplicial R-modules to itself by setting

$$L_\mathcal{F}(X) = E_\lambda(X \to 0).$$

The following result is a version of Lemma 7.16 for simplicial R-modules:

Lemma 8.37

1) Suppose that the assignment $t \mapsto X_t$ defines a diagram of simplicial R-modules, indexed by $\gamma > 2^\beta$. Then the map

$$\varinjlim_{t < \gamma} L_\mathcal{F}(X_t) \to L_\mathcal{F}(\varinjlim_{t < \gamma} X_t)$$

is an isomorphism.

2) Suppose that ζ is a cardinal with $\zeta > \beta$, and let $B_\zeta(X)$ be the family of subobjects of X having cardinality less than ζ. Then the map

$$\varinjlim_{Y \in B_\zeta(X)} L_\mathcal{F}(Y) \to L_\mathcal{F}(X)$$

is an isomorphism.

3) The functor $X \mapsto L_\mathcal{F}(X)$ preserves monomorphisms.

4) Suppose that A and B are subobjects of X. Then the natural map

$$L_\mathcal{F}(A \cap B) \to L_\mathcal{F}(A) \cap L_\mathcal{F}(B)$$

is an isomorphism.

5) If $|X| \leq 2^\mu$ where $\mu \geq \lambda$ then $|L_\mathcal{F}(X)| \leq 2^\mu$.

Proof It suffices to prove statements 1)–4) with $L_\mathcal{F}(X)$ replaced by $E_1(X)$, which is the first stage in the construction. There is a pushout diagram

$$\begin{array}{ccc}
\bigoplus_\mathcal{F} (R(C) \otimes \hom(C, X)) & \longrightarrow & X \\
\downarrow & & \downarrow \\
\bigoplus_\mathcal{F} (R(D) \otimes \hom(C, X)) & \longrightarrow & E_1 X
\end{array}$$

To prove statement 1), one shows that the map

$$\varinjlim_{t<\gamma} hom(C, X_t) \to hom\,(C, \varinjlim_{t<\gamma} X_t)$$

is an isomorphism, since C is α-bounded.

The proof of statement 2) is similar. One shows that the map

$$\varinjlim_{Y \in B_\zeta(X)} hom\,(C, Y) \to hom\,(C, X)$$

is an isomorphism by using the fact that the image of a map $C \to X$ is contained in a subobject $Y \subset X$ with $|Y| < \beta < \zeta$. The object X is also a colimit of subobjects Y with $|Y| < \zeta$.

Statement 3) is a consequence of the fact a monomorphism $X \to Y$ induces injective functions $hom\,(C, X) \to hom\,(C, Y)$. It follows that, for the diagram of monomorphisms

$$\begin{array}{ccc} \oplus_\mathcal{F} R(C) \otimes hom\,(C, X) & \longrightarrow & \oplus_\mathcal{F} R(D) \otimes hom\,(C, X) \\ \downarrow & & \downarrow \\ \oplus_\mathcal{F} R(C) \otimes hom\,(C, Y) & \longrightarrow & \oplus_\mathcal{F} R(D) \otimes hom\,(C, Y) \end{array}$$

the induced map on cokernels of the vertical maps is a monomorphism, and so the induced map $E_1 X \to E_1 Y$ is a monomorphism.

The cokernel $C(X)$ of the monomorphism $X \to E_1 X$ has the form

$$C_1(X) = \oplus_\mathcal{F} (R(D)/R(C)) \otimes hom\,(C, X),$$

and it follows that the map

$$C(A \cap B) \to C(A) \cap C(B)$$

is an isomorphism. One then uses an element chase to show that the monomorphism

$$E_1(A \cap B) \to E_1(A) \cap E_1(B)$$

is surjective in all sections, and we have proved statement 4).

The proof of statement 5) is the same as for the corresponding statement of Lemma 7.16: the key point is that the cardinality of the set $hom\,(C, X)$ is bounded by

$$(2^\mu)^\beta = 2^{\mu \cdot \beta} = 2^\mu.$$

Say that a map $X \to Y$ of simplicial R-modules is an $L_\mathcal{F}$-*equivalence* if the induced map of simplicial R-modules $L_\mathcal{F}(X) \to L_\mathcal{F}(Y)$ is a local weak equivalence.

8.5 Localized Chain Complexes

Let κ be the successor cardinal for 2^μ, where μ is cardinal of statement 5) of Lemma 8.37. Then κ is a regular cardinal, and Lemma 8.37 implies that if a simplicial R-module X is κ-bounded then $L_{\mathcal{F}}(X)$ is κ-bounded.

The following result is the bounded monomorphism property for $L_{\mathcal{F}}$-equivalences of simplicial R-modules.

Lemma 8.38 *Suppose given a diagram of monomorphisms*

of $s\mathbf{Pre}_R$ such that i is an $L_{\mathcal{F}}$-equivalence and the object A is κ-bounded. Then there is a κ-bounded subobject B of Y such that $A \subset B$, factorization $A \subset B \subset Y$ of j by monomorphisms such that B is κ-bounded and the map $B \cap X \to B$ is an $L_{\mathcal{F}}$-equivalence.

Proof The proof proceeds by analogy with the proof of Lemma 7.17, using Lemma 8.11 and various features of Lemma 8.37.

There is an induced diagram of monomorphisms

$$\begin{array}{ccc} & & L_{\mathcal{F}}X \\ & & \downarrow {\scriptstyle i_*} \\ L_{\mathcal{F}}A & \xrightarrow{j_*} & L_{\mathcal{F}}Y \end{array}$$

in which the map i_* is a local weak equivalence of R-modules and $L_{\mathcal{F}}A$ is κ-bounded. Then, by Lemma 8.11, there is a κ-bounded subobject $A_0 \subset L_{\mathcal{F}}X$ such that the map $A_0 \cap L_{\mathcal{F}}X \to A_0$ is a local weak equivalence of simplicial R-modules.

There is a κ-bounded subobject $B_0 \subset Y$ such that $A \subset B_0$ and $A_0 \subset L_{\mathcal{F}}B_0$. In effect, $L_{\mathcal{F}}Y$ is a sum of subobjects $L_{\mathcal{F}}F$ associated to the κ-bounded subobjects $F \subset Y$, and every section of the κ-bounded object B_0 is in some such $L_{\mathcal{F}}F$.

Continue inductively. Find a κ-bounded object A_1 such that $L_{\mathcal{F}}B_0 \subset A_1$ and the map $A_1 \cap L_{\mathcal{F}}X \to L_{\mathcal{F}}Y$ is a local weak equivalence, and find a κ-bounded subobject B_1 of Y with $B_0 \subset B_1$ and $A_1 \subset L_{\mathcal{F}}B_1$. Repeat the construction λ times, and set $B = \varinjlim_{s<\lambda} B_s$.
Then the map

$$L_{\mathcal{F}}(B \cap X) \cong L_{\mathcal{F}}B \cap L_{\mathcal{F}}X \to L_{\mathcal{F}}B$$

coincides the map

$$\varinjlim_s (A_s \cap L_{\mathcal{F}}X) \to \varinjlim_s A_s,$$

which is a local weak equivalence.

The natural map $\eta : X \to L_{\mathcal{F}}X$ is a cofibration and an $\mathcal{F}$-equivalence by Lemma 7.8, and the object $L_{\mathcal{F}}(X)$ is $\mathcal{F}$-injective. It follows from Lemma 7.6 that a map $X \to Y$ of simplicial R-modules is an $\mathcal{F}$-equivalence if and only if the induced map $L_{\mathcal{F}}(X) \to L_{\mathcal{F}}(Y)$ is a local weak equivalence.

Following Sect. 7.1, a map $p : X \to Y$ of simplicial R-modules is an $\mathcal{F}$-fibration if it has the right lifting property with respect to all maps which are cofibrations and $\mathcal{F}$-equivalences.

We then have a proof of the following result, which gives the $\mathcal{F}$-local model structure for the category of simplicial R-modules. Theorem 8.39 is a special case of Theorem 7.10, via Lemma 8.38. The left properness for the $\mathcal{F}$-local structure is a consequence of Lemma 7.8.

Theorem 8.39 *Suppose that R is a presheaf of commutative unitary rings on a small Grothendieck site $\mathcal{C}$. Let $\mathcal{F}$ be a set of cofibrations of simplicial presheaves which satisfies the conditions* **C1** *and* **C2**. *Then the category $s\mathbf{Pre}_R$ of simplicial R-modules, together with the classes of cofibrations, $\mathcal{F}$-equivalences and $\mathcal{F}$-fibrations, satisfies the axioms for a cofibrantly generated closed simplicial model category. This model structure is left proper.*

Remark 8.40 The forgetful and free R-module functors

$$R : s\mathbf{Pre}(\mathcal{C}) \leftrightarrows s\mathbf{Pre}_R : u$$

define a Quillen adjunction for the $\mathcal{F}$-local model structures on the respective categories.

This is a consequence of the observation that a simplicial R-module Z is $\mathcal{F}$-injective if and only if the underlying simplicial presheaf $u(Z)$ is $\mathcal{F}$-injective. It follows that the free R-module functor R preserves $\mathcal{F}$-equivalences, as well as cofibrations. It also follows that a simplicial R-module Z is $\mathcal{F}$-fibrant if and only if its underlying simplicial presheaf $u(Z)$ is $\mathcal{F}$-fibrant.

There is no claim that the forgetful functor u either preserves or reflects $\mathcal{F}$-weak equivalences in the unstable setting. Something can be said, however, in stable homotopy theory. See Lemma 10.93 in Sect. 10.7.

The $\mathcal{F}$-local model structure on simplicial R-modules is an enriched structure, in the sense that we have the following:

Lemma 8.41 *Suppose that the maps $i : C \to D$ and $j : E \to F$ are cofibrations of simplicial R-modules. Then the induced map*

$$(i, j) : (D \otimes E) \cup (C \otimes F) \to D \otimes F$$

is a cofibration which is an $\mathcal{F}$-equivalence if either i of j is an $\mathcal{F}$-equivalence.

Proof The proof of this result follows that of Lemma 7.23, which is the corresponding result for simplicial presheaves. We must account for the fact that not every simplicial R-module is cofibrant. It suffices, however, to assume that the objects C, D, E and F are cofibrant, by a (sectionwise) left properness argument.

In effect, the cofibrant model $\pi : A_c \to A$ of a simplicial R-module A can be defined by a trivial injective fibration π, which is a sectionwise weak equivalence. The map $\pi \otimes 1 : A_c \otimes B \to A \otimes B$ is then a sectionwise equivalence for all simplicial R-modules B.

The map (i, j) is a cofibration by Lemma 8.9. For a fixed cofibration i, one uses Lemma 7.8 to show that the class of all cofibrations j such that (i, j) is an $\mathcal{F}$-equivalence is closed under composition, pushout and retracts.

Suppose that the simplicial presheaf map $i : A \to B$ is in $\mathcal{F}$ and that $j : C \to D$ is an α-bounded cofibration of simplicial presheaves. Then the simplicial presheaf map (i, j) is in $\mathcal{F}$, so that the induced map of simplicial R-modules

$$(i_*, j_*) : (R(B) \otimes R(C)) \cup (R(A) \otimes R(S)) \to R(B) \otimes R(D)$$

is an $\mathcal{F}$-equivalence. It follows that the map (i_*, j) is an $\mathcal{F}$ for all cofibrations of simplicial R-modules j, if i is a member of $\mathcal{F}$. It also follows that the map (i, j) is an $\mathcal{F}$-equivalence if i is in the saturation of $R(\mathcal{F})$ and j is a cofibration of simplicial R-modules.

In particular, the canonical $\mathcal{F}$-fibrant model $j : A \to L_{\mathcal{F}}A$ is in the saturation of $R(\mathcal{F})$ so that the map

$$j \otimes 1 : A \otimes C \to L_{\mathcal{F}}(A) \otimes C$$

is a cofibration and an $\mathcal{F}$-equivalence for all cofibrant simplicial R-modules C.

Suppose that $g : C \to D$ is an $\mathcal{F}$-equivalence of cofibrant simplicial R-modules, and that A is a cofibrant simplicial R-module. Form the diagram

$$\begin{array}{ccc} A \otimes C & \xrightarrow{1 \otimes g} & A \otimes D \\ {\scriptstyle 1 \otimes j} \downarrow & & \downarrow {\scriptstyle 1 \otimes j} \\ A \otimes L_{\mathcal{F}}C & \xrightarrow{1 \otimes g_*} & A \otimes L_{\mathcal{F}}D \end{array}$$

Then the maps $1 \otimes j$ are $\mathcal{F}$-equivalences by the previous paragraph, and the map $g_* : L_{\mathcal{F}}C \to L_{\mathcal{F}}D$ is a sectionwise equivalence since g is an $\mathcal{F}$-equivalence. It follows that the map $1 \otimes g$ is an $\mathcal{F}$-equivalence.

Finally, for cofibrations $i : C \to D$ and $j : E \to F$ (between cofibrant objects), suppose that i is an $\mathcal{F}$-equivalence and form the diagram

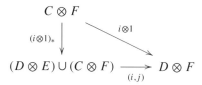

Then the maps $(i \otimes 1)_*$ and $i \otimes 1$ are trivial cofibrations for the $\mathcal{F}$-local structure, so that the map (i, j) is an $\mathcal{F}$-equivalence.

Example 8.42 Following Example 7.19, suppose that $f : A \to B$ is a cofibration of simplicial presheaves, and let $\mathcal{F} = \langle f \rangle$ be the smallest set of cofibrations of simplicial presheaves which contains f and the generating set J of trivial cofibrations, and satisfies the closure property **C2**. See also Example 7.15.

The $\langle f \rangle$-equivalences of simplicial R-modules are called f-*equivalences*, and the $\langle f \rangle$-fibrations are called f-*fibrations*. The $\langle f \rangle$-model structure of Theorem 8.39 is the f-*local model structure* for the category of simplicial R-modules. We shall also write

$$L_f(X) := L_{\langle f \rangle}(X)$$

for simplicial R-modules X.

We shall focus on the f-local theories for simplicial R-modules henceforth, for the reason discussed in Example 7.19: each $\mathcal{F}$-local theory is an f-local theory for some cofibration f.

Example 8.43 Suppose that S is a scheme which is Noetherian and of finite dimension, and let $(Sm|_S)_{Nis}$ be the category of smooth schemes of finite type over S, equipped with the Nisnevich topology, as in Example 7.20. Let $f : * \to \mathbb{A}^1$ be the 0-section of the affine line over S.

The f-local model structure on the category $s\mathbf{Pre}_R(Sm|_S)_{Nis}$ of simplicial R-modules which is given by Theorem 8.39 is the *motivic model structure* for the category of simplicial R-modules, or for the equivalent category of presheaves of chain complexes of R-modules. The weak equivalences and fibrations for this model structure are, respectively, the *motivic weak equivalences* and the *motivic fibrations* of simplicial R-modules.

The relationship between motivic weak equivalences of simplicial R-modules and motivic weak equivalences of simplicial presheaves can be a bit subtle. The free R-module and forgetful functors determine a Quillen adjunction

$$R : s\mathbf{Pre}((Sm|_S)_{Nis}) \rightleftarrows s\mathbf{Pre}_R((Sm|_S)_{Nis}) : u,$$

by construction, and the free R-module functor preserves motivic weak equivalences, but it is far from clear that the forgetful functor u either preserves or creates motivic weak equivalences in general.

That said, suppose that Y is a simplicial R-module whose underlying simplicial presheaf satisfies motivic descent, and let $j : Y \to Z$ be an injective fibrant model in simplicial R-modules for the Nisnevich topology. Then j is an injective fibrant model in the simplicial presheaf category, and is therefore a sectionwise equivalence by the Nisnevich descent theorem. The simplicial presheaf underlying Z is motivic fibrant, and so Y is sectionwise weakly equivalent to a motivic fibrant model in the simplicial R-module category.

In other words, there are both Nisnevich and motivic descent theorems for the simplicial R-module category.

We will need to know later that the map $A \to A \otimes R(\mathbb{A}^1)$ which is induced by the 0-section $* \to \mathbb{A}^1$ is a motivic weak equivalence for all simplicial R-modules A on $(Sm|_S)_{Nis}$.

To see this, observe that there is a projective model structure on $s\mathbf{Mod}_R$ for which the weak equivalences and fibrations are defined sectionwise (prove this directly or use Corollary 8.45 below). We can therefore assume that the simplicial R-module A is projective cofibrant.

We then show that the map $A \to A \otimes R(\mathbb{A}^1)$ has the left lifting property with respect to all motivic fibrations of simplicial R-modules $p : X \to Y$. Equivalently, the map

$$\mathbf{Hom}(\mathbb{A}^1, X) \to X \times_Y \mathbf{Hom}(\mathbb{A}^1, Y)$$

has the right lifting property with respect to the projective cofibration $0 \to A$. But this is so, since the maps

$$X(\mathbb{A}^1 \times U) \to X(U) \times_{Y(U)} Y(\mathbb{A}^1 \times U)$$

are trivial fibrations by construction of the motivic model structure: all maps

$$(\partial \Delta^n \times \mathbb{A}^1 \times U) \cup (\Delta^n \times U) \to \Delta^n \times \mathbb{A}^1 \times U$$

are trivial cofibrations in the motivic model structure.

Observe that a simplicial R-module X is motivic fibrant if

1) X is injective fibrant for the Nisnevich topology, and
2) all induced maps $X(U \times \mathbb{A}^1) \to X(U)$ are weak equivalences of simplicial abelian groups.

The second condition is equivalent (in the presence of condition 1)) to the assertion that the injective fibration

$$\mathbf{Hom}(U \times \mathbb{A}^1, X) \to \mathbf{Hom}(U, X)$$

is a local weak equivalence for all smooth k-schemes U.

8.6 Linear Simplicial Presheaves

Suppose that R is an ordinary commutative ring with identity. Let $\mathbf{Mod}_R$ denote the category of R-modules, and write $s\mathbf{Mod}_R$ for the category of simplicial R-modules.

Suppose that $\mathcal{C}$ is a small category, and that $\mathcal{A}$ is a small category which is enriched in R-modules. Suppose that there is a functor $\phi : \mathcal{C} \to \mathcal{A}$ which is the identity on objects.

The assumption that $\mathcal{A}$ is enriched in R-modules means that all hom objects $\mathcal{A}(U, V)$ are R-modules and that the composition law is defined by a bilinear pairing

$$\mathcal{A}(U, V) \otimes_R \mathcal{A}(V, W) \to \mathcal{A}(U, W).$$

An $\mathcal{A}$-*linear simplicial presheaf* $X : \mathcal{A}^{op} \to s\mathbf{Mod}_R$ consists of simplicial R-modules $X(U)$, $U \in \mathcal{A}$, together with R-module homomorphisms

$$X(V) \otimes_R \mathcal{A}(U, V) \to X(U)$$

which satisfy the usual properties: the law of composition in $\mathcal{A}$ is respected, as are all identities.

Write $s\mathbf{Mod}_R^{\mathcal{A}}$ for the category of all $\mathcal{A}$-linear simplicial presheaves $\mathcal{A}^{op} \to s\mathbf{Mod}_R$ which take values in simplicial R-modules.

The morphisms $\phi : X \to Y$ of the category $s\mathbf{Mod}_R^{\mathcal{A}}$ are the $\mathcal{A}$-linear natural transformations. Such a morphism ϕ consists of homomorphisms of simplicial R-modules $\phi : X(U) \to Y(U), U \in \mathrm{Ob}(\mathcal{A})$, which respect the action of $\mathcal{A}$.

Say that a map $f : A \to B$ of $\mathcal{A}$-linear simplicial presheaves is a *sectionwise equivalence* (respectively *sectionwise fibration*) if the maps $f : A(U) \to B(U)$ are weak equivalences (respectively fibrations) of simplicial R-modules for all objects $U \in \mathcal{A}$. A *projective cofibration* of $\mathcal{A}$-linear simplicial presheaves is a map which has the left lifting property with respect to all maps which are sectionwise fibrations and sectionwise equivalences.

Suppose that A is a simplicial R-module and that $U \in \mathrm{Ob}(\mathcal{A})$. Then the assignment

$$L_U(A) = A \otimes \hom_{\mathcal{A}}(\ , U)$$

defines the left adjoint of the U-sections functor $s\mathbf{Mod}_R^{\mathcal{A}} \to s\mathbf{Ab}$, which is defined by $X \mapsto X(U)$.

The functor L_U takes weak equivalences of cofibrant simplicial R-modules (which are homotopy equivalences) to sectionwise weak equivalences of $s\mathbf{Mod}_R^{\mathcal{A}}$, and takes cofibrations of simplicial R-modules to projective cofibrations of $s\mathbf{Mod}_R^{\mathcal{A}}$.

It follows that a map $p : A \to B$ of $s\mathbf{Mod}_R^{\mathcal{A}}$ is a sectionwise fibration, respectively trivial sectionwise fibration, if and only if it has the right lifting property with respect to all maps

$$L_U(R(\Lambda_k^n)) \to L_U(R(\Delta^n)),$$

respectively with respect to all maps

$$L_U(R(\partial \Delta^n)) \to L_U(R(\Delta^n)).$$

It also follows that every cofibration (respectively trivial cofibration) $K \subset L$ of simplicial sets induces projective cofibrations (respectively trivial projective cofibrations)

$$L_U(R(K)) \to L_U(R(L))$$

of $s\mathbf{Mod}_R^{\mathcal{A}}$.

8.6 Linear Simplicial Presheaves

Suppose that A is an $\mathcal{A}$-linear simplicial presheaf and that K is a simplicial set. As before, write

$$A \otimes K := A \otimes R(K).$$

The following result gives the *projective model structure* for the category of $\mathcal{A}$-linear simplicial presheaves.

Lemma 8.44 *The category $s\mathbf{Mod}_R^{\mathcal{A}}$ of $\mathcal{A}$-linear simplicial presheaves, together with the classes of sectionwise weak equivalences, sectionwise fibrations and projective cofibrations, has the structure of a proper closed simplicial model category. This model structure is cofibrantly generated. Every projective cofibration is a monomorphism.*

Proof Limits and colimits for the category $s\mathbf{Mod}_R^{\mathcal{A}}$ of $\mathcal{A}$-linear simplicial presheaves are formed sectionwise, as in the presheaf category $s\mathbf{Ab}^{\mathcal{C}}$. The category $s\mathbf{Mod}_R^{\mathcal{A}}$ is complete and cocomplete.

The weak equivalence axiom **CM2** and the retract axiom **CM3** are easily verified. Every map $f : A \to B$ has factorizations

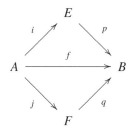

where p is a fibration and the cofibration i is in the saturation of the maps

$$L_U(R(\Lambda_k^n)) \to L_U(R(\Delta^n)), \tag{8.11}$$

and q is trivial fibration and j is a cofibration, by a standard small object argument. Each of the maps (8.11) is a strong deformation retraction in sections, and strong deformation retractions are closed under pushout in the simplicial R-module category. It follows that the map i is a sectionwise weak equivalence as well as a cofibration. We have therefore verified the factorization axiom **CM5**.

The argument for the lifting axiom **CM4** is standard: every trivial cofibration is a retract of a morphism in the saturation of the maps $L_U(R(\Lambda_k^n)) \to L_U(R(\Delta^n))$, and therefore has the left lifting property with respect to all fibrations.

By the same argument, every cofibration $A \to B$ consists of maps $A(U) \to B(U)$ which are cofibrations of simplicial R-modules. In particular, every cofibration is a monomorphism.

The function complex $\mathbf{hom}(A, B)$ has n-simplices consisting of the homomorphisms $A \otimes \Delta^n \to B$. Given $\mathcal{A}$-linear simplicial presheaf A and a simplicial set K, the $\mathcal{A}$-linear simplicial presheaf A^K is defined in sections by

$$A^K(U) = \mathbf{hom}(K, A(U)).$$

If $p : A \to B$ is a fibration of $s\mathbf{Mod}_R^{\mathcal{A}}$ and $i : K \to L$ is a cofibration of simplicial sets, then all maps

$$(i^*, p_*) : A^L(U) \to A^K(U) \times_{B^K(U)} B^L(U)$$

are fibrations of simplicial R-modules for all objects U of $\mathcal{C}$, and these fibrations are trivial if either i or p is trivial. The simplicial model axiom **SM7** follows.

Properness follows from properness for the model structure on the category of simplicial R-modules. It is immediate from the proof of the factorization axiom **CM5** that this model structure on $s\mathbf{Mod}_R^{\mathcal{A}}$ is cofibrantly generated.

There is an analogous projective model structure for the category $s\mathbf{Pre}_R$ of presheaves simplicial R-modules on $\mathcal{C}$, which is defined sectionwise. A map $A \to B$ of presheaves of simplicial R-modules is a *sectionwise weak equivalence* (respectively *sectionwise fibration*) if all maps $A(U) \to B(U)$ are weak equivalences (respectively fibrations) of simplicial R-modules, and a *projective cofibration* is a map which has the left lifting property with respect to all trivial fibrations. Then we have the following:

Corollary 8.45 *The category $s\mathbf{Pre}_R$ of presheaves of simplicial R-modules on $\mathcal{C}$, together with the classes of sectionwise weak equivalences, sectionwise fibrations and projective cofibrations, satisfies the axioms for a proper closed simplicial model category. This model structure is cofibrantly generated. Every cofibration is a monomorphism.*

Proof The category $\mathcal{C}$ determines a category $R(\mathcal{C})$, having the same objects, and with

$$R(\mathcal{C})(U, V) = R(\mathcal{C}(U, V)),$$

the free R-module on $\mathcal{C}(U, V)$, and there is an isomorphism of categories

$$s\mathbf{Pre}_R \cong s\,\mathrm{Mod}_R^{R(\mathcal{C})}.$$

Now use Lemma 8.44.

Alternatively, one can give an argument for Corollary 8.45 which is completely analogous to the proof of Lemma 8.44.

The model structure of Corollary 8.45 is the *projective model structure* for presheaves of simplicial R-modules. That structure is defined in such a way that the free R-module functor $X \mapsto R(X)$ and the forgetful functor u form a Quillen adjunction

$$R : s\mathbf{Pre}(\mathcal{C}) \leftrightarrows s\mathbf{Pre}_R : u$$

between the projective model structures for simplicial presheaves and simplicial R-modules.

Precomposition with the functor $\phi : \mathcal{C} \to \mathcal{A}$ induces a functor

$$\phi_* : s\mathbf{Mod}_R^{\mathcal{A}} \to s\mathbf{Pre}_R$$

8.6 Linear Simplicial Presheaves

This functor has a left adjoint

$$\phi^* : s\mathbf{Pre}_R \to s\mathbf{Mod}_R^{\mathcal{A}}$$

which is defined by

$$\phi^*(X) = \varinjlim L_U(R(\Delta^n)),$$

where the colimit varies over the simplices

$$R(\Delta^n) \otimes R(\hom(\ ,U)) \to X$$

of X.

Every projective cofibration $A \to B$ of simplicial presheaves induces a projective cofibration

$$\phi^*R(A) \to \phi^*R(B)$$

of $\mathcal{A}$-linear simplicial presheaves, by adjointness.

Suppose that α is a regular cardinal such that $\alpha > |R|$, $\alpha > |\text{Mor}(\mathcal{C})|$ and $\alpha > |\text{Mor}(\mathcal{A})|$.

If the simplicial presheaf B is α-bounded, then the presheaf of simplicial R-modules $R(B)$ is α-bounded, and therefore has an α-bounded collection of simplices

$$R(\Delta^n) \otimes R(\hom(\ ,U)) \to R(B).$$

The $\mathcal{A}$-linear simplicial presheaves $R(\Delta^n) \otimes \hom_{\mathcal{A}}(\ ,U)$ are also α-bounded. The $\mathcal{A}$-linear simplicial presheaf $\phi^*R(B)$ is therefore α-bounded, since it is a colimit of α-bounded objects on an α-bounded diagram. All cofibrations of $s\mathbf{Mod}_R^{\mathcal{A}}$ are monomorphisms, so that the cofibration $\phi^*R(A) \to \phi^*R(B)$ is α-bounded if the projective cofibration $A \to B$ is α-bounded.

The localization techniques of Chap. 7 apply, without modification, to the projective model structure on the category $s\mathbf{Mod}_R^{\mathcal{A}}$ of $\mathcal{A}$-linear simplicial presheaves, in the presence of the functor $\phi : \mathcal{C} \to \mathcal{A}$, where $\mathcal{C}$ is a Grothendieck site.

Suppose henceforth (by making the cardinal α sufficiently large) that $\mathcal{F}$ is an α-bounded set of α-bounded projective cofibrations in $s\mathbf{Pre}(\mathcal{C})$.

We shall also suppose that the set of cofibrations $\mathcal{F}$ satisfies the following conditions:

1) Every member $A \to B$ of the set $\mathcal{F}$ has a projective cofibrant source object.
2) the set $\mathcal{F}$ contains the set of all maps

$$\Lambda_k^n \times \hom(\ ,U) \to \Delta^n \times \hom(\ ,U),$$

which set generates the class of trivial cofibrations for the projective model structure on $s\mathbf{Pre}(\mathcal{C})$,

3) if $C \to D$ is a member of $\mathcal{F}$, then so are all induced maps

$$(D \times \partial \Delta^n) \cup (C \times \Delta^n) \to D \times \Delta^n.$$

Write $\phi^* R(\mathcal{F})$ for the set of morphisms

$$\phi^* R(C) \to \phi^* R(D),$$

which are induced by morphisms $C \to D$ of $\mathcal{F}$. The members of $\phi^* R(\mathcal{F})$ are α-bounded projective cofibrations of $s\mathbf{Mod}_R^{\mathcal{A}}$ with projective cofibrant source objects.

Following Sect. 7.1, say that a map $p : X \to Y$ of $s\mathbf{Mod}_R^{\mathcal{A}}$ is $\mathcal{F}$-*injective* if it has the right lifting property with respect to all members of the set $\phi^* R(\mathcal{F})$. An object Z is $\mathcal{F}$-injective if the map $Z \to 0$ is $\mathcal{F}$-injective.

All $\mathcal{F}$-injective maps are sectionwise fibrations of $s\mathbf{Mod}_R^{\mathcal{A}}$, by construction.

A map $E \to F$ of $s\mathbf{Mod}_R^{\mathcal{A}}$ is an $\mathcal{F}$-*equivalence* if some cofibrant replacement $E_c \to F_c$ induces a weak equivalence

$$\mathbf{hom}(F_c, Z) \to (E_c, Z)$$

for all $\mathcal{F}$-injective objects Z.

Start with a cardinal $\lambda > 2^\alpha$. We repeat the general construction of the $\mathcal{F}$-injective model $X \to L_\mathcal{F}(X)$ for $\mathcal{A}$-linear simplicial presheaves X, seen most recently in Sect. 8.5.

Every map $g : X \to Y$ of $\mathcal{A}$-linear simplicial presheaves has a functorial factorization

such that the map p is $\mathcal{F}$-injective and the map i is a cofibration in the saturation of the set $\phi^* R(\mathcal{F})$, by the usual small object argument. This small object construction terminates after λ steps.

Write

$$L_\mathcal{F}(X) = E_\lambda(X \to 0).$$

Say that a map $X \to Y$ of $\mathcal{A}$-linear simplicial presheaves is an $L_\mathcal{F}$-*equivalence* if it induces a sectionwise weak equivalence $L_\mathcal{F}(X) \to L_\mathcal{F}(Y)$.

The following result is the analogue, for $\mathcal{A}$-linear simplicial presheaves, of Lemma 8.38, and has the same proof.

Lemma 8.46

1) Suppose that the assignment $t \mapsto X_t$ defines a diagram of $\mathcal{A}$-linear simplicial presheaves, indexed by $\gamma > 2^\alpha$. Then the map

$$\varinjlim_{t < \gamma} L_\mathcal{F}(X_t) \to L_\mathcal{F}(\varinjlim_{t < \gamma} X_t)$$

8.6 Linear Simplicial Presheaves

is an isomorphism.

2) *Suppose that ζ is a cardinal with $\zeta > \alpha$, and let $B_\zeta(X)$ be the family of subobjects of X having cardinality less than ζ. Then the map*

$$\varinjlim_{Y \in B_\zeta(X)} L_\mathcal{F}(Y) \to L_\mathcal{F}(X)$$

is an isomorphism.

3) *The functor $X \mapsto L_\mathcal{F}(X)$ preserves monomorphisms.*

4) *Suppose that E and F are subobjects of X. Then the natural map*

$$L_\mathcal{F}(E \cap F) \to L_\mathcal{F}(E) \cap L_\mathcal{F}(F)$$

is an isomorphism.

5) *If $|X| \leq 2^\mu$ where $\mu \geq \lambda$ then $|L_\mathcal{F}(X)| \leq 2^\mu$.*

By construction, a map $X \to Y$ of $s\mathbf{Mod}_R^\mathcal{A}$ is an $\mathcal{F}$-equivalence if and only if the induced map $L_\mathcal{F}(X) \to L_\mathcal{F}(Y)$ is a sectionwise weak equivalence—see Sect. 7.1.

Let κ be the successor cardinal for 2^μ, where μ is cardinal of statement 5) of Lemma 8.46. Then κ is a regular cardinal, and Lemma 8.46 implies that if a simplicial R-module X is κ-bounded then $L_\mathcal{F}(X)$ is κ-bounded. The following result is a consequence of Lemma 8.46, in the same way that Lemma 8.38 is a consequence of Lemma 8.37.

Lemma 8.47 *Suppose given a diagram of monomorphisms*

of $s\mathbf{Mod}_R^\mathcal{A}$ such that i is an $\mathcal{F}$-equivalence and A is κ-bounded. Then there is a factorization $A \subset B \subset Y$ of j by monomorphisms such that B is κ-bounded and the map $B \cap X \to B$ is an $\mathcal{F}$-equivalence.

The projective model structure for the category $s\mathbf{Mod}_R^\mathcal{A}$ of $\mathcal{A}$-linear simplicial presheaves of Lemma 8.44, with sectionwise weak equivalences and sectionwise fibrations, satisfies conditions **M1**—**M6** of Sect. 7.1. In particular, the condition **M6** says that if $A \to B$ is a sectionwise weak equivalence and K is a simplicial set, then the map

$$A \otimes K \to B \otimes K$$

is a sectionwise weak equivalence. This is a consequence of the corresponding condition for the standard model structure on the category of simplicial R-modules.

The functor

$$L_\mathcal{F} : s\mathbf{Mod}_R^\mathcal{A} \to s\mathbf{Mod}_R^\mathcal{A}$$

satisfies condition **L1** and **L2** of Sect. 7.1 by construction (Lemma 7.6 and Lemma 7.8, respectively). The functor $L_{\mathcal{F}}$ satisfies condition **L3** by Lemma 8.47.

Recall that a map $A \to B$ of $\mathcal{A}$-linear simplicial presheaves is an $\mathcal{F}$-equivalence if and only if it induces a sectionwise weak equivalence $L_{\mathcal{F}} A \to L_{\mathcal{F}} B$. An $\mathcal{F}$-fibration of $\mathcal{A}$-linear simplicial presheaves is a map which has the right lifting property with respect to all projective cofibrations which are $\mathcal{F}$-equivalences.

Theorem 7.10 therefore applies to the projective model structure on the category of $\mathcal{A}$-linear simplicial presheaves and the functor $L_{\mathcal{F}}$, giving the following:

Theorem 8.48 *Suppose that $\phi : \mathcal{C} \to \mathcal{A}$ is a functor, where $\mathcal{C}$ is a Grothendieck site and $\mathcal{A}$ is a small R-linear category, and such that the functor ϕ is the identity on objects. Suppose that $\mathcal{F}$ is a set of projective cofibrations of simplicial presheaves on $\mathcal{C}$ which satisfies conditions 1)–3) above.*

Then the category $s\mathbf{Mod}_R^{\mathcal{A}}$ of $\mathcal{A}$-linear simplicial presheaves, together with the classes of projective cofibrations, $\mathcal{F}$-equivalences and $\mathcal{F}$-fibrations, satisfies the axioms for a left proper closed simplicial model category. This model category is cofibrantly generated.

The model structure of Theorem 8.48 is the $\mathcal{F}$-*local model structure* for the category of $\mathcal{A}$-linear simplicial presheaves.

Example 8.49 Suppose that k is a perfect field. Following [78], the category Cor_k is the additive category of finite correspondences over k.

The objects of this category are the k-schemes which are smooth and separated over k. The group of morphisms $Cor_k(X, Y)$ is the group of finite correspondences: if X is connected, $Cor_k(X, Y)$ is freely generated as an abelian group by the elementary correspondences, which are the irreducible closed subsets $W \subset X \times Y$ whose irreducible integral subschemes are finite and surjective over X. If $f : X \to Y$ is a morphism of smooth k-schemes (and X is connected), then the graph $\Gamma(f) \subset X \times Y$ is an elementary correspondence.

The composition law

$$Cor_k(X, Y) \otimes Cor_k(Y, Z) \to Cor_k(X, Z)$$

is defined by intersection pairing. The assignment $f \mapsto \Gamma(f)$ defines a functor

$$\gamma : Sm|_k \to Cor_k,$$

here called the *graph functor*, which is the identity on objects.

A *simplicial presheaf with transfers* is a member of the category

$$s\mathbf{PST}(k) := s\mathbf{Mod}_{\mathbb{Z}}^{Cor_k}$$

of Cor_k-linear simplicial presheaves. These are simplicial objects in the category $\mathbf{PST}(k)$ of *presheaves with transfers*.

As in Example 7.22, suppose that $\mathcal{F}$ is a set of projective cofibrations of simplicial presheaves on $(Sm|_S)_{Nis}$ which contains the following:

8.6 Linear Simplicial Presheaves

1) the members of the generating set

$$\Lambda_k^n \times \hom(\ ,U) \to \Delta^n \times \hom(\ ,U)$$

for the trivial cofibrations of the projective model structure,

2) projective cofibrant replacements of the maps

$$U \cup_{\phi^{-1}(U)} V \to T$$

which are associated to all elementary distinguished squares (7.4), and

3) the map

$$\emptyset \to \hom(\ ,\emptyset).$$

from the empty simplicial presheaf to the presheaf which is represented by the empty scheme.

We further require that the set $\mathcal{F}$ is generated by this list of cofibrations, subject to the closure property that if the map $A \to B$ is in $\mathcal{F}$, then all maps

$$(B \times \partial \Delta^n) \cup (A \times \Delta^n) \to B \times \Delta^n$$

are in $\mathcal{F}$.

We use Theorem 8.48 to form the $\mathcal{F}$-local model structure on the category $s\mathbf{PST}(k)$ of simplicial presheaves with transfers.

By construction, a presheaf with transfers Z is $\mathcal{F}$-fibrant if and only if the underlying simplicial abelian presheaf $\gamma_*(Z)$ is $\mathcal{F}$-injective, and hence satisfies Nisnevich descent (Theorem 5.39).

Every map $A \to B$ of the set $\mathcal{F}$ restricts to a map $\gamma^*\mathbb{Z}(A) \to \gamma^*\mathbb{Z}(B)$ which is a local weak equivalence for the Nisnevich topology.

This claim is proved by verifying that the maps which generate $\mathcal{F}$ have this property. For this, one uses the observation that the class of cofibrations $A \to B$ of simplicial presheaves with transfers which induce local weak equivalences $\gamma_* A \to \gamma_* B$ of simplicial abelian presheaves is closed under pushout, filtered colimits and the formation of the maps

$$(B \otimes \partial \Delta^n) \cup (A \otimes \Delta^n) \to B \otimes \Delta^n.$$

The interesting part of the argument is verifying that the map

$$Cor_k(\ ,U) \cup_{Cor_k(\ ,\phi^{-1}(U))} Cor_k(\ ,V) \to Cor_k(\ ,T)$$

induces a Nisnevich local equivalence for any elementary distinguished square

$$\begin{array}{ccc} \phi^{-1}(U) & \longrightarrow & V \\ \downarrow & & \downarrow \\ U & \longrightarrow & T \end{array}$$

as in (5.8). This follows from a result of Suslin and Voevodsky [97, Prop. 4.3.9], which implies that the sequence

$$0 \to Cor_k(,\phi^{-1}(U)) \to Cor_k(,U) \oplus Cor_k(,V) \to Cor_k(,T) \to 0$$

induces a short exact sequence of Nisnevich sheaves.

It follows that a map $X \to Y$ of simplicial presheaves with transfers is an $\mathcal{F}$-equivalence if and only if the underlying simplicial abelian presheaf map $\gamma_*(X) \to \gamma_*(Y)$ is a Nisnevich weak equivalence.

Add the set of 0-section maps $U \to U \times \mathbb{A}^1$ to the generating set for $\mathcal{F}$ to form a new set of cofibrations $\mathcal{F}_{\mathbb{A}^1}$. Then all maps

$$Cor_k(,U) \to Cor_k(,U \times \mathbb{A}^1)$$

are formally inverted, and the corresponding $\mathcal{F}_{\mathbb{A}^1}$-local structure is the *motivic* or $\mathbb{A}^1$-*local model structure* on the category $s\mathbf{PST}(k)$ of simplicial presheaves with transfers. The associated homotopy category for this model structure is Voevodsky's category $\mathbf{DM}_{Nis}^{eff,-}(k)$ of *effective motives* over the field k. [78, Def. 14.1]

Taking products of elementary correspondences defines a pairing

$$Cor_k(U,X) \otimes Cor_k(V,Y) \to Cor_k(U \times V, X \times Y), \qquad (8.12)$$

and hence a pairing

$$Cor_k(U,X) \otimes Cor_k(U,Y) \to Cor_k(U \times U, X \times Y) \xrightarrow{\Delta^*} Cor_k(U, X \times Y),$$

where $\Delta : U \to U \times U$ is the diagonal map. It follows that there is an algebraic homotopy

$$\gamma_* Cor_k(,U \times \mathbb{A}^1) \otimes \mathbb{Z}(\mathbb{A}^1) \to \gamma_* Cor_k(,U \times \mathbb{A}^1 \times \mathbb{A}^1) \to \gamma_* Cor_k(,U \times \mathbb{A}^1)$$

from the identity on $\gamma_* Cor_k(,U \times \mathbb{A}^1)$ to the self map which is induced by the composite

$$U \times \mathbb{A}^1 \xrightarrow{pr} U \xrightarrow{f} U \times \mathbb{A}^1.$$

The maps

$$\gamma_* Cor_k(,U \times \mathbb{A}^1) \to \gamma_* Cor_k(,U \times \mathbb{A}^1) \otimes \mathbb{Z}(\mathbb{A}^1)$$

are motivic weak equivalences of simplicial abelian presheaves (see Example 8.43), and so all maps

$$\gamma_*(Cor_k(,U)) \to \gamma_*(Cor_k(,U \times \mathbb{A}^1))$$

are motivic weak equivalences of simplicial abelian presheaves.

8.6 Linear Simplicial Presheaves

Again, the class of all cofibrations $A \to B$ of simplicial presheaves with transfers which induce motivic weak equivalences $\gamma_* A \to \gamma_* B$ of simplicial abelian presheaves is closed under pushout, filtered colimits, and the formation of the maps

$$(B \otimes \partial \Delta^n) \cup (A \otimes \Delta^n) \to B \otimes \Delta^n.$$

It follows that the canonical map $A \to L_{\mathcal{F}_{\mathbb{A}^1}}(A)$ of simplicial presheaves with transfers induces a motivic weak equivalence $\gamma_*(A) \to \gamma_*(L_{\mathcal{F}_{\mathbb{A}^1}}(A))$ of the underlying simplicial abelian presheaves.

The object $\gamma_*(L_{\mathcal{F}_{\mathbb{A}^1}}(A))$ satisfies motivic descent by construction, and is therefore sectionwise equivalent to any motivic fibrant model of $\gamma_*(A)$. It follows that a map $E \to F$ of simplicial presheaves with transfers is an $\mathcal{F}_{\mathbb{A}^1}$-equivalence if and only if the underlying map $\gamma_*(E) \to \gamma_*(F)$ is a motivic weak equivalence of simplicial abelian presheaves in the sense of Example 8.43.

The motivic model structure for the category $s\mathbf{PST}(k)$ of simplicial presheaves with transfers first appeared in [89].

Say that a presheaf of simplicial abelian groups X is $\mathbb{A}^1$-*local* if some (hence any) Nisnevich fibrant model Z of X is motivic fibrant. A comparison of hypercohomology spectral sequences shows that X is $\mathbb{A}^1$-local if the maps

$$H^p_{Nis}(U, \tilde{H}_q(X)) \to H^p_{Nis}(U \times \mathbb{A}^1, \tilde{H}_q(X))$$

in Nisnevich cohomology are isomorphisms for all p, q and all S-schemes U. In this case, one says that the homology sheaves of X are *homotopy invariant*.

Suppose that K is a simplicial presheaf with transfers. Then K has an associated *singular complex*

$$C_*(K) = d(\mathbf{Hom}(\mathbb{A}^\bullet, K))$$

in $s\mathbf{PST}(k)$, where $\mathbb{A}^\bullet$ is the cosimplicial scheme made up of the affine spaces $\mathbb{A}^n$, $n \geq 0$, in the usual way, and d is the diagonal functor from bisimplicial abelian groups to simplicial abelian groups.

The singular complex $C_*(K)$ is $\mathbb{A}^1$-local, in the sense that the underlying simplicial presheaf $\gamma_*(C_*(K))$ is $\mathbb{A}^1$-local [78, 14.9]. This is a consequence of a nontrivial result [78, 13.8], which asserts that if a presheaf F with transfers is homotopy invariant, then all associated Nisnevich cohomology presheaves $H^q(\ , \tilde{F})$ are homotopy invariant.

Suppose that X is a simplicial presheaf on the smooth Nisnevich site for k, and let $x : * \to X$ be a global choice of base point for X.

Define the *free simplicial presheaf with transfers* $\mathbb{Z}_{tr}(X)$ by setting

$$\mathbb{Z}_{tr}(X) := \varinjlim_{\Delta^n \times \hom(\ ,U) \to X} \mathbb{Z}(\Delta^n) \otimes Cor_k(\ , U)$$

where the colimit is defined on the category of simplices of X. The object $\mathbb{Z}_{tr}(X)$ is the free simplicial presheaf with transfers on the simplicial presheaf X. The base point x determines a splitting of the map $\mathbb{Z}_{tr}(X) \to \mathbb{Z}_{tr}(*)$, and we define $\mathbb{Z}_{tr}(X, x)$

to be the cokernel of the map $\mathbb{Z}_{tr}(*) \to \mathbb{Z}_{tr}(X)$ which is induced by the base point x.

The *motive* $\mathbb{Z}(n)$ is defined as a chain complex object in presheaves with transfers by the shifted complex

$$\mathbb{Z}(n) = N(C_*\mathbb{Z}_{tr}((\mathbb{G}_m)^{\wedge n}, e))[-n],$$

where N is the normalized chains functor and the multiplicative group $\mathbb{G}_m$ is pointed by the identity e. The motive $\mathbb{Z}(n)$ can be expressed as a simplicial presheaf with transfers by

$$\mathbb{Z}(n) = C_*\mathbb{Z}_{tr}((\mathbb{G}_m)^{\wedge n}, e) \otimes (S^1)^{\otimes n},$$

where we write $S^1 = \mathbb{Z}_\bullet(S^1)$ for the reduced free simplicial abelian group associated to the pointed simplicial set S^1.

The *motivic cohomology groups* $H^p(X, \mathbb{Z}(q))$ are hypercohomology groups, which are defined by

$$H^p(X, \mathbb{Z}(q)) = \mathbb{H}^p(X, \gamma_*(\mathbb{Z}(q))),$$

with respect to the Nisnevich topology [78, Ex. 13.11].

The pairings (8.12) induce natural pairings

$$\mathbb{Z}_{tr}(X, x) \otimes \mathbb{Z}_{tr}(Y, y) \to \mathbb{Z}_{tr}(X \wedge Y, *)$$

for pointed simplicial presheaves X and Y. These pairings specialize to pairings of motives

$$\mathbb{Z}(q) \otimes \mathbb{Z}(q') \to \mathbb{Z}(q + q'),$$

which induce the cup product

$$H^p(X, \mathbb{Z}(q)) \otimes H^r(X, \mathbb{Z}(s)) \to H^{p+r}(X, \mathbb{Z}(q + s))$$

in motivic cohomology. This cup product is a special case of the hypercohomology pairing (8.10).

Chapter 9
Non-abelian Cohomology

The homotopy theoretic approach to non-abelian cohomology had its origins in [53], in a study of characteristic classes in Galois cohomology for quadratic forms and orthogonal representations of Galois groups.

Suppose that k is a field such that $char(k) \neq 2$. As of the late 1980s, it was well known that the set of isomorphism classes of non-degenerate bilinear forms A over k of rank n could be identified with the set $H^1_{et}(k, O_n)$ of étale torsors for the k-group-scheme O_n of automorphisms of the trivial form. The point of departure of [53] was the discovery of an identification

$$H^1_{et}(k, O_n) \cong [*, BO_n],$$

of the set of isomorphism classes O_n-torsors with morphisms $[*, BO_n]$ in the homotopy category of simplicial sheaves on the étale site for the field k, so that quadratic forms become homotopy classes of maps of simplicial sheaves. This isomorphism was derived as a special case of a general identification

$$H^1(\mathcal{C}, G) \cong [*, BG] \tag{9.1}$$

of isomorphism classes of G-torsors and sheaves of groups G with the set of morphisms $[*, BG]$ in the homotopy category of simplicial sheaves (or presheaves) on an arbitrary site $\mathcal{C}$.

To put it in a different way, classical non-abelian H^1 has a homotopy classification in the simplicial sheaf category. The original proof of this result used the Verdier hypercovering theorem in an essential way, and was built on ideas from étale homotopy theory of Friedlander [27] and Dwyer-Friedlander [24].

We verify the identification (9.1) in the first section of this chapter, with a very different approach that starts with a homotopy theoretic method of defining G-torsors for a sheaf of groups G. Explicitly, a G-torsor is a sheaf F with G-action such that the corresponding Borel construction $EG \times_G F$ is acyclic in the sense that the map $EG \times_G F \to *$ is a local weak equivalence. It is an exercise to show that this definition is equivalent to the classical requirement that the G-sheaf F is a G-torsor if and only if the G-action is principal and transitive.

Every G-torsor F defines a "canonical cocycle"

$$* \xleftarrow{\simeq} EG \times_G F \to BG,$$

in simplicial sheaves, while every cocycle

$$* \xleftarrow{\simeq} U \xrightarrow{f} BG$$

defines a G-torsor F, which is the sheaf of path components of the homotopy fibre of the map f. One shows that these two constructions are inverse to each other up to equivalence, so that there is a bijection

$$H^1(\mathcal{C}, G) = \pi_0(G - \textbf{tors}) \cong \pi_0 h(*, BG) \cong [*, BG],$$

which gives the identification (9.1). This result is Theorem 9.8 below. The well-known fact that the category $G - \textbf{tors}$ of G-equivariant maps between G-torsors is a groupoid appears as Lemma 9.4, with a simple homotopy theoretic proof.

One can, more generally, say that an internally defined sheaf-valued functor F on a sheaf of groupoids H is an H-torsor if the induced map

$$\underrightarrow{\text{holim}}_H F \to *$$

is a local weak equivalence, and then the method of proof of Theorem 9.8 generalizes, in Corollary 9.15, to give an identification

$$\pi_0(H - \textbf{tors}) \cong [*, BH],$$

of the path components of the groupoid of H-torsors with morphisms in the homotopy category. This result is a consequence of Theorem 9.14, which asserts that the canonical cocycle construction defines a weak equivalence

$$\phi : B(H - \textbf{tors}) \xrightarrow{\simeq} Bh(*, BH)$$

of simplicial sets between the nerve of the H-torsor groupoid and the nerve of the cocycle category $h(*, BH)$. This map ϕ is global sections of a weak equivalence of simplicial presheaves

$$B(H - \textbf{Tors}) \to B\mathbb{H}(*, BH),$$

which is later parlayed, in Corollary 9.27, into a cocycle description of the stack associated to the sheaf of groupoids H.

Stacks have become homotopy theoretic objects, with the introduction of the injective model structures for both presheaves and sheaves of groupoids on a site $\mathcal{C}$. The model structure for presheaves of groupoids first appeared in Hollander's thesis [43], and is straightforward to define and derive in the presence of the injective model structures for simplicial presheaves. The corresponding structure for sheaves of groupoids was introduced by Joyal and Tierney [72].

Say that a map $G \to H$ of presheaves of groupoids is a local weak equivalence (respectively injective fibration) if the induced map $BG \to BH$ is a local weak equivalence (respectively injective fibration) of simplicial presheaves, while the cofibrations are defined by a lifting property as they must be. The model structures for presheaves of groupoids and sheaves of groupoids appear in Propositions 9.19 and 9.20 in Sect. 9.2. These model structures are Quillen equivalent.

We show in Proposition 9.28 that a sheaf of groupoids H is a stack in the classical sense (i.e. satisfies effective descent) if and only if it satisfies descent for the injective model structure. In the present context, a presheaf of groupoids H satisfies descent if and only if every injective fibrant model $H \to H'$ consists of equivalences of groupoids $H(U) \to H'(U)$ in all sections. Given this observation, we can identify stacks with injective fibrant presheaves of groupoids, or more broadly with homotopy types of presheaves or sheaves of groupoids.

For example (see Lemma 9.24), the quotient stack for a group action $G \times X \to X$ on a sheaf X can be identified with the translation groupoid $E_G X$, and hence with the Borel construction

$$EG \times_G X = B(E_G X)$$

in simplicial sheaves. We also show, in Lemma 9.25 that sheaves of groupoids G and H are locally weakly equivalent if and only if they are Morita equivalent.

Opinions vary on what the theory of higher stacks should be. The original circle of ideas arose in an effort to classify higher order cohomological phenomena in algebraic geometry. Early formulations by Giraud and others in the 1970s created a geometric interpretation of the classification of stacks in terms of symmetries in 2-groupoids. Some modern approaches to the subject start with a definition that is much less geometric: a higher stack from this point of view is a simplicial presheaf, so that higher stacks become essentially everything—see [42] and [93], for example. The theory that is described here is a departure from this point of view, in that symmetries are again the focus.

The base category for higher stack theory that we use is the category $s_0\mathbf{Gpd}$ of groupoids enriched in simplicial sets (or simplicial groupoids with discrete objects). Categories enriched in simplicial sets have been extensively used in homotopy coherence theory, and homotopy theoretic structures have existed for them for some time. The homotopy theory of groupoids enriched in simplicial sets is the subject of Sect. 9.3. Local versions of it appear in subsequent sections.

In many instances, the results of Sect. 9.3 come directly from the literature, for example from [32]. Generally, weak equivalences $G \to H$ of groupoids enriched in simplicial sets can be measured by the method of Dwyer and Kan [22] by comparing sets of path components and by showing that all simplicial set maps $G(x,y) \to H(f(x), f(y))$ are weak equivalences, or by using either the diagonal of the classifying space $H \mapsto dB(H)$ or the Eilenberg–Mac Lane $\overline{W}$ functor $H \mapsto \overline{W}H$.

The Eilenberg–Mac Lane functor $\overline{W}$ is the easiest thing to use for generating model structures, because it has a well-behaved left adjoint given by the loop groupoid functor $X \mapsto G(X)$. That said, the Eilenberg–Mac Lane functor has a long history of being fussy to define and manipulate. One can use the cocycle methods of [32], or say that the n-simplices of $\overline{W}H$ are the maps $G(\Delta^n) \to H$ of $s_0\mathbf{Gpd}$ subject to having a definition of the loop groupoid functor G, but the properties of the functor $\overline{W}$ that one needs are rather tricky to derive in both of these approaches.

There is a fix for this problem that has evolved in the category theory literature in recent years, most recently in the work of Stevenson [94], and which goes back to ideas that were introduced by Artin-Mazur [3] and Illusie [48]. Specifically, if X is a simplicial set, there is a bisimplicial set $\text{Dec}(X)$ with (p,q)-bisimplices defined by

$$\text{Dec}(X)_{p,q} = \hom(B(\mathbf{p} * \mathbf{q}), X)$$

where $\mathbf{p} * \mathbf{q}$ is the poset join of the ordinal numbers $\mathbf{p}$ and $\mathbf{q}$. The resulting functor $X \mapsto \text{Dec}(X)$ is Illusie's total décalage functor. Its right adjoint is the Artin-Mazur total simplicial set functor T: if Y is a bisimplicial set then the n-simplices of $T(Y)$ are the bisimplicial set maps $\text{Dec}(\Delta^n) \to Y$. The key point, for us, is that there is a natural isomorphism

$$T(BH) \cong \overline{W}H$$

for groupoids H enriched in simplicial sets. It is easy to see, with this definition, that there is a natural map

$$\Phi : d(BH) \to T(BH) = \overline{W}H.$$

We show in Proposition 9.38 that this map is a weak equivalence. This result is a consequence of a more general result for bisimplicial sets (Proposition 9.38) that is proved in [94], and we reproduce that proof here in somewhat different language.

With these definitions and equivalences in hand, it is relatively painless to show (Theorem 9.43) that there is a $\overline{W}$-model structure on the category of groupoids enriched in simplicial sets, in which a map $G \to H$ is a weak equivalence (respectively fibration) if the map $\overline{W}G \to \overline{W}H$ is a weak equivalence (respectively fibration) of simplicial sets.

This model structure for groupoids enriched in simplicial sets is promoted in Theorem 9.50 of Sect. 9.4 to a local $\overline{W}$-model structure for presheaves of groupoids enriched in simplicial sets, with analogous definitions of weak equivalence and fibration: a map $G \to H$ is a local weak equivalence (respectively fibration) if the induced map $\overline{W}G \to \overline{W}H$ is a local weak equivalence (respectively injective fibration) of simplicial presheaves. Furthermore, the functor $\overline{W}$ and the loop groupoid functor G determine a Quillen equivalence between the injective model structure on simplicial presheaves and the $\overline{W}$-model structure on presheaves of groupoids enriched in simplicial sets. The homotopy types of presheaves of groupoids enriched in simplicial presheaves, for this model structure, are the ∞-stacks.

The $\overline{W}$-model structure of Theorem 9.50 specializes to an n-equivalence model structure for presheaves of groupoids enriched in simplicial sets (Theorem 9.56) and to an injective model structure for presheaves of 2-groupoids (Theorem 9.57).

We construct a Postnikov section P_n for a groupoid H enriched in simplicial sets: the simplicial groupoid $P_n(H)$ has the same set of objects as does H, and its morphism simplicial set is the n_{th} Moore–Postnikov section $P_n \operatorname{Mor}(H)$ of the morphism simplicial set Mor (H) of H. This construction is functorial, and applies to presheaves of groupoids enriched in simplicial sets. A map $G \to H$ of such objects is a local n-equivalence if the map $P_n G \to P_n H$ is a local weak equivalence of presheaves of groupoids enriched in simplicial sets. The cofibrations for the n-equivalence structure are the cofibrations for the $\overline{W}$-structure, and the fibrations are defined by a lifting property. The homotopy types of presheaves of groupoids enriched in simplicial sets within the n-equivalence model structure are the $(n + 1)$-stacks.

In particular, the 1-stacks, or homotopy types within the 0-equivalence model structure, are equivalent to homotopy types of presheaves of groupoids, or stacks, by Lemma 9.58. The 2-stacks are equivalent to homotopy types of presheaves of 2-groupoids by Proposition 9.59. See also Remark 9.60.

In the present language, a gerbe is a locally connected homotopy type. The model structure for presheaves of 2-groupoids of Theorem 9.57 is used in Sect. 9.5 to give a homotopy classification within presheaves of 2-groupoids for gerbes with automorphism sheaves in a fixed family of groups $\mathcal{F}$ in Corollary 9.68, and a homotopy classification of extensions of sheaves of groups in Corollary 9.72.

These results are elements of Giraud's theory [31] of non-abelian H^2. They are consequences of the main result of the section, Theorem 9.66, which is a homotopy classification within 2-groupoids of extensions of presheaves of groupoids with kernels in a family of sheaves of groups $\mathcal{F}$, suitably defined. The classifying object in all cases is a 2-groupoid object **Iso**($\mathcal{F}$), which consists of the objects of the family $\mathcal{F}$, the isomorphisms of these objects, and the homotopies of the isomorphisms.

9.1 Torsors

Suppose that G is a sheaf of groups. A G-*torsor* is traditionally defined to be a sheaf F with a free G-action such that the canonical map $F/G \to *$ is an isomorphism in the sheaf category, where $*$ is the terminal sheaf.

The Borel construction $EG \times_G F$ is the nerve of a sheaf of groupoids, which is given in each section by the translation category for the action of $G(U)$ on $F(U)$—see Example 2.7. It follows that all sheaves of higher homotopy groups for $EG \times_G F$ vanish. The requirement that the action $G \times F \to F$ is free means that the isotropy (or stabilizer) subgroups of G for the action are trivial in all sections, which is equivalent to requiring that all sheaves of fundamental groups for the object $EG \times_G F$ are trivial. Finally, there is an isomorphism of sheaves

$$\tilde{\pi}_0(EG \times_G F) \cong F/G.$$

These observations together imply the following:

Lemma 9.1 *A sheaf F with G-action is a G-torsor if and only if the simplicial sheaf map*

$$EG \times_G F \to *$$

is a local weak equivalence.

Example 9.2 If G is a sheaf of groups, then $EG = EG \times_G G$ is contractible in each section, so that the map

$$EG \times_G G \to *$$

is a local weak equivalence, and so G is a G-torsor. This object is often called the *trivial G-torsor*.

Example 9.3 Suppose that L/k is a finite Galois extension of fields with Galois group G. Let $C(L)$ be the Čech resolution for the étale covering $\operatorname{Sp}(L) \to \operatorname{Sp}(k)$, as in Example 4.17. There is an isomorphism of simplicial schemes

$$C(L) \cong EG \times_G \operatorname{Sp}(L),$$

while the simplicial presheaf map $C(L) \to *$ on $Sch|_k$ is a local weak equivalence for the étale topology. The k-scheme $\operatorname{Sp}(L)$ represents a G-torsor for all of the standard étale sites for the field k.

The category $G-\mathbf{tors}$ of G-torsors is the category whose objects are all G-torsors and whose maps are all G-equivariant maps between them.

Lemma 9.4 *Suppose that G is a sheaf of groups. Then the category $G-\mathbf{tors}$ of G-torsors is a groupoid.*

Proof If $f : F \to F'$ is a map of G-torsors, then f is induced as a map of fibres by the comparison of local fibrations

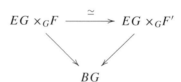

The map $f : F \to F'$ of fibres is a weak equivalence of constant simplicial sheaves by Lemma 5.20 and properness of the injective model structure for simplicial sheaves, and is therefore an isomorphism of sheaves.

Remark 9.5 Suppose that F is a G-torsor, and that the map $F \to *$ has a (global) section $\sigma : * \to X$. Then σ extends, by multiplication, (also uniquely) to a G-equivariant map

$$\sigma_* : G \to F,$$

with $\sigma_*(g) = g \cdot \sigma(*)$ for $g \in G(U)$. This map is an isomorphism of torsors, so that F is trivial with trivializing isomorphism σ_*. Conversely, if $\tau : G \to F$ is a map of torsors, then F has a global section $\tau(e)$. Thus a G-torsor F is trivial in the sense that it is G-equivariantly isomorphic to G if and only if it has a global section.

9.1 Torsors

Example 9.6 Suppose that X is a topological space. The category of sheaves on $op|_X$ can be identified up to equivalence with the category of local homeomorphisms $Y \to X$ over X.

If G is a discrete group, then G represents the sheaf $G \times X \to X$ given by projection. A sheaf with G-action consists of a map $Y \to X$ together with a G-action $G \times Y \to Y$ such that the map $Y \to X$ is G-equivariant for the trivial G-action on X. Such a thing is a G-torsor if the action $G \times Y \to Y$ is free and the map $Y/G \to X$ is an isomorphism. The latter implies that X has an open covering $i: U \subset X$ such that there are liftings

Torsors are stable under pullback along continuous maps, and the map $U \times_X Y \to U$ is a G-torsor over U. The map σ induces a global section σ_* of this map, so that the pulled back torsor is trivial, and there is a commutative diagram

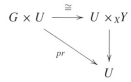

where the displayed isomorphism is G-equivariant. It follows that a G-torsor over X is a principal G-bundle over X, and conversely.

Example 9.7 Suppose that U is an object of a small site $\mathcal{C}$. The restriction functor

$$\mathbf{Shv}(\mathcal{C}) \to \mathbf{Shv}(\mathcal{C}/U),$$

written $F \mapsto F|_U$, is exact, and therefore takes G-torsors to $G|_U$-torsors. The global sections of $F|_U$ coincide with the elements of the set $F(U)$, so that a G-torsor F trivializes over U if and only if $F(U) \ne \emptyset$, or if and only if there is a diagram

The map $F \to *$ is a local epimorphism, so there is a covering family $U_\alpha \to *$ (such that $\bigsqcup U_\alpha \to *$ is a local epimorphism) with $F(U_\alpha) \ne \emptyset$. In other words, every torsor F trivializes over some covering family of the point $*$.

Suppose that the picture

$$* \xleftarrow{\simeq} Y \xrightarrow{\alpha} BG$$

is an object of the cocycle category $h(*, BG)$ in simplicial presheaves, and forms the pullback

$$\begin{array}{ccc} \mathrm{pb}(Y) & \longrightarrow & Y \\ \downarrow & & \downarrow \alpha \\ EG & \xrightarrow{\pi} & BG \end{array}$$

where $EG = B(G/*) = EG \times_G G$ and $\pi : EG \to BG$ is the canonical map. Then $\mathrm{pb}(Y)$ inherits a G-action from the G-action on EG, and the map

$$EG \times_G \mathrm{pb}(Y) \to Y$$

is a sectionwise weak equivalence (this is a consequence of Lemma 9.9 below). Also, the square is homotopy cartesian in sections, so there is a local weak equivalence

$$G|_U \to \mathrm{pb}(Y)|_U$$

where $Y(U) \ne \emptyset$. It follows that the natural map $\mathrm{pb}(Y) \to \tilde{\pi}_0 \mathrm{pb}(Y)$ is a G-equivariant local weak equivalence, and hence that the maps

$$EG \times_G \tilde{\pi}_0 \mathrm{pb}(Y) \leftarrow EG \times_G \mathrm{pb}(Y) \to Y \simeq *$$

are natural local weak equivalences. In particular, the G-sheaf $\tilde{\pi}_0 \mathrm{pb}(Y)$ is a G-torsor.

We therefore have a functor

$$h(*, BG) \to G - \mathbf{tors},$$

which is defined by sending $* \xleftarrow{\simeq} Y \to BG$ to the object $\tilde{\pi}_0 \mathrm{pb}(Y)$. The Borel construction defines a functor

$$G - \mathbf{tors} \to h(*, BG)$$

in which the G-torsor X is sent to the (canonical) cocycle

$$* \xleftarrow{\simeq} EG \times_G X \to BG.$$

It is elementary to check (see also the proof of Theorem 9.14 below) that these two functors induce a bijection

$$\pi_0 h(*, BG) \cong \pi_0(G - \mathbf{tors}).$$

The set $\pi_0(G - \mathbf{tors})$ is the set of isomorphism classes of G-torsors, while we know from Theorem 6.5 that there is an isomorphism

$$\pi_0 h(*, BG) \cong [*, BG].$$

9.1 Torsors   255

The non-abelian invariant $H^1(\mathcal{C}, G)$ is traditionally defined to be the collection of isomorphism classes of G-torsors. We have therefore proved the following:

Theorem 9.8 *Suppose that G is a sheaf of groups on a small Grothendieck site $\mathcal{C}$. Then there is a bijection*

$$[*, BG] \cong H^1(\mathcal{C}, G).$$

Theorem 9.8 was first proved, by a different method, in [53].

The following result completes the proof of Theorem 9.8:

Lemma 9.9 *Suppose that I is a small category and that $p : X \to BI$ is a simplicial set map. Let the pullback diagrams*

$$\begin{array}{ccc} \mathrm{pb}\,(X)(i) & \longrightarrow & X \\ \downarrow & & \downarrow p \\ B(I/i) & \longrightarrow & BI \end{array}$$

define the I-diagram $i \mapsto \mathrm{pb}\,(X)(i)$. Then the resulting map

$$\omega : \varinjlim\mathrm{holim}_{i \in I}\, \mathrm{pb}\,(X)(i) \to X$$

is a weak equivalence.

Proof The simplicial set

$$\varinjlim\mathrm{holim}_{i \in I}\, \mathrm{pb}\,(X)(i)$$

is the diagonal of a bisimplicial set whose (n, m)-bisimplices are pairs

$$(x, i_0 \to \cdots \to i_n \to j_0 \to \cdots \to j_m)$$

where $x \in X_n$, the morphisms are in I, and $p(x)$ is the string

$$i_0 \to \cdots \to i_n.$$

The map

$$\omega : \varinjlim\mathrm{holim}_{i \in I}\, \mathrm{pb}\,(X)(i) \to X$$

takes such an (n, m)-bisimplex to $x \in X_n$. The fibre of ω over x in vertical degree n can be identified with the simplicial set $B(i_n/I)$, which is contractible.

Example 9.10 Suppose that k is a field such that $char(k) \neq 2$. Identify the orthogonal group $O_n = O_{n,k}$ with a sheaf of groups on the big étale site $(Sch|_k)_{et}$ for k. The non-abelian cohomology object $H^1_{et}(k, O_n)$ coincides with the set of isomorphism classes of non-degenerate symmetric bilinear forms over k of rank n. Thus, every such form q determines a morphism $* \to BO_n$ in the simplicial (pre)sheaf homotopy category on the site $(Sch|_k)_{et}$, and this morphism determines the form q up to isomorphism.

There is a ring isomorphism

$$H^*_{et}(BO_{n,k}, \mathbb{Z}/2) \cong H^*_{et}(k, \mathbb{Z}/2)[HW_1, \ldots, HW_n]$$

where the polynomial generator HW_i has degree i (see [53]). The generator HW_i is characterized by mapping to the ith elementary symmetric polynomial $\sigma_i(x_1, \ldots, x_n)$ under the map

$$H^*(BO_{n,k}, \mathbb{Z}/2) \to H^*(\Gamma^* B\mathbb{Z}/2^{\times n}, \mathbb{Z}/2) \cong H^*_{et}(k, \mathbb{Z}/2)[x_1, \ldots, x_n],$$

which is induced by the inclusion $\mathbb{Z}/2^{\times n} \subset O_n(k)$ of the diagonal subgroup.

Every symmetric bilinear form α determines a map $\alpha : * \to BO_n$ in the simplicial presheaf homotopy category, and therefore induces a map

$$\alpha^* : H^*_{et}(BO_n, \mathbb{Z}/2) \to H^*_{et}(k, \mathbb{Z}/2),$$

and $HW_i(\alpha) = \alpha^*(HW_i)$ is the ith *Hasse–Witt class* of α.

The ring $H^*_{et}(k, \mathbb{Z}/2)$ can otherwise be characterized as the mod-2 cohomology of the absolute Galois group of k.

One can show that $HW_1(\alpha)$ is the pullback of the determinant $BO_n \to B\mathbb{Z}/2$, and $HW_2(\alpha)$ is the classical Hasse–Witt invariant of α.

The Steenrod algebra is used to calculate the relation between Hasse–Witt and Stiefel–Whitney classes for orthogonal Galois representations. This calculation uses the Wu formulas for the action of the Steenrod algebra on elementary symmetric polynomials—see [53, 54].

Example 9.11 Suppose that S is a scheme. The general linear group Gl_n represents a sheaf of groups on the étale site $(Sch|_S)_{et}$ and the sheaf of groups $\mathbb{G}_m$ can be identified with the centre of Gl_n via the diagonal imbedding $\mathbb{G}_m \to GL_n$. There is a short exact sequence

$$e \to \mathbb{G}_m \to Gl_n \xrightarrow{p} PGl_n \to e$$

of sheaves of groups on $(Sch|_S)_{et}$. The projective general linear group PGl_n can be identified with the group scheme of automorphisms $Aut(M_n)$ of the scheme of $(n \times n)$-matrices M_n, and the homomorphism p takes an invertible matrix A to the automorphism defined by conjugation by A.

Since $\mathbb{G}_m$ is a central subgroup of Gl_n, there is an induced action

$$B\mathbb{G}_m \times BGl_n \to BGl_n$$

of the simplicial abelian group $B\mathbb{G}_m$ on the simplicial sheaf BGl_n, and there is an induced sectionwise (hence local) fibre sequence associated to the sequence of bisimplicial objects

$$BGl_n \to EB\mathbb{G}_m \times_{B\mathbb{G}_m} BGl_n \xrightarrow{\pi} BB\mathbb{G}_m \simeq K(\mathbb{G}_m, 2)$$

9.1 Torsors

after taking diagonals.

In effect, if $A \times X \to X$ is an action of a connected simplicial abelian group A on a connected simplicial set X, then all sequences

$$X \to A^{\times n} \times X \to A^{\times n}$$

are fibre sequences of connected simplicial sets, so that the sequence

$$X \to EA \times_A X \to BA$$

of bisimplicial set maps induces a fibre sequence of simplicial sets after taking diagonals, by a theorem of Bousfield and Friedlander [32, IV.4.9].

The $B\mathbb{G}_m$ action on BGl_n is free, so there is a local weak equivalence

$$EB\mathbb{G}_m \times_{B\mathbb{G}_m} BGl_n \xrightarrow{\simeq} BPGl_n.$$

It follows that the map π induces a function

$$H^1_{et}(S, PGl_n) = [*, BPGl_n] \xrightarrow{d:=\pi_*} [*, K(\mathbb{G}_m, 2)] = H^2_{et}(S, \mathbb{G}_m).$$

The object $H^1_{et}(S, Gl_n)$ is the set of isomorphism classes of vector bundles over S of rank n, and the set $H^1_{et}(S, PGl_n)$ is the set of isomorphisms classes of rank n^2 Azumaya algebras. The map

$$p_* : [*, BGl_n] \to [*, BPGl_n]$$

takes a vector bundle E to the Azumaya algebra **End**(E), which is defined by the sheaf of endomorphisms of the S-module E.

Recall that the *Brauer group* $Br(S)$ is the abelian group of similarity classes of Azumaya algebras over S: the Azumaya algebras A and B are similar if there are vector bundles E and F such that there are isomorphisms

$$A \otimes \mathbf{End}(E) \cong B \otimes \mathbf{End}(F).$$

The group structure on $Br(S)$ is induced by tensor product of Azumaya algebras.

In more detail, tensor product of modules induces a comparison of exact sequences

$$\begin{array}{ccccccccc}
e & \to & \mathbb{G}_m \times \mathbb{G}_m & \to & Gl_n \times Gl_m & \xrightarrow{p \times p} & PGl_n \times PGl_m & \to & e \\
& & {\scriptstyle +}\downarrow & & {\scriptstyle \otimes}\downarrow & & {\scriptstyle \otimes}\downarrow & & \\
e & \to & \mathbb{G}_m & \to & Gl_{nm} & \xrightarrow{p} & PGl_{nm} & \to & e
\end{array}$$

where $+$ is the group structure on $\mathbb{G}_m$, and the induced map

$$\otimes : [*, BPGl_n] \times [*, BPGl_m] \to [*, BPGl_{nm}]$$

defines the tensor product of Azumaya algebras. We also have induced commutative diagrams

$$
\begin{CD}
[*, BPGl_n] \times [*, BPGl_m] @>{d \times d}>> H^2_{et}(S, \mathbb{G}_m) \times H^2_{et}(S, \mathbb{G}_m) \\
@V{\otimes}VV @VV{+}V \\
[*, BPGl_{nm}] @>>{d}> H^2_{et}(S, \mathbb{G}_m)
\end{CD}
$$

in which the displayed pairing on $H^2(S, \mathbb{G}_m)$ is the abelian group addition. It follows that the collection of morphisms

$$d : [*, BPGl_n] \to H^2_{et}(S, \mathbb{G}_m)$$

defines a group homomorphism

$$d : Br(S) \to H^2_{et}(S, \mathbb{G}_m).$$

This homomorphism d is a monomorphism: if $d(A) = 0$ for some Azumaya algebra A, then there is an isomorphism $A \cong \mathbf{End}(E)$ for some vector bundle E by the exactness of the sequence

$$[*, BGl_n] \xrightarrow{p_*} [*, BPGl_n] \xrightarrow{d} H^2_{et}(S, \mathbb{G}_m),$$

so that A represents 0 in the Brauer group.

Finally, if S is connected, then the Brauer group $Br(S)$ consists of torsion elements. As in [79, IV.2.7], this follows from the existence of the diagram of short exact sequences of sheaves of groups

$$
\begin{CD}
@. e @. e @. @. \\
@. @VVV @VVV @. @. \\
e @>>> \mu_n @>>> \mathbb{G}_m @>{\times n}>> \mathbb{G}_m @>>> e \\
@. @VVV @VVV @VV{1}V @. \\
e @>>> Sl_n @>>> Gl_n @>>{det}> \mathbb{G}_m @>>> e \\
@. @VVV @VVV @. @. \\
@. PGl_n @>>{1}> PGl_n @. @. \\
@. @VVV @VVV @. @. \\
@. e @. e @. @.
\end{CD}
$$

on the étale site $(Sch|_S)_{et}$, where μ_n is the subgroup of n-torsion elements in $\mathbb{G}_m$. The vertical sequence on the left is a central extension, so that there is a map d :

9.1 Torsors

$[*, BPGl_n] \to H^*_{et}(S, \mu_n)$ that fits into a commutative diagram

$$\begin{array}{ccc} [*, BPGl_n] & \xrightarrow{d} & H^2_{et}(S, \mu_n) \\ {\scriptstyle 1}\downarrow & & \downarrow \\ [*, BPGl_n] & \xrightarrow{d} & H^2_{et}(S, \mathbb{G}_m) \end{array}$$

and $H^2_{et}(S, \mu_n)$ is an n-torsion abelian group. It follows that the Brauer group $Br(S)$ consists of torsion elements if the scheme S has finitely many components.

The assertion that there is monomorphism

$$d : Br(S) \to H^2_{et}(S, \mathbb{G}_m)_{tors}$$

into the torsion part of $H^2_{et}(S, \mathbb{G}_m)$ is a well-known theorem of étale cohomology theory [79, IV.2.5]. The distinctive feature of the present discussion is the use of easily defined fibre sequences of simplicial sheaves to produce the map d in place of an appeal to non-abelian H^2 invariants.

Suppose that I is a small category. A functor $X : I \to \mathbf{Set}$ consists of sets $X(i)$, $i \in \mathrm{Ob}(I)$ and functions $\alpha_* : X(i) \to X(j)$ for $\alpha : i \to j$ in $\mathrm{Mor}(I)$ such that $\alpha_* \beta_* = (\alpha \cdot \beta)_*$ for all composable pairs of morphisms in I and $(1_i)_* = 1_{X(i)}$ for all objects i of I.

The sets $X(i)$ can be collected together to give a set

$$\pi : X = \bigsqcup_{i \in \mathrm{Ob}(I)} X(i) \to \bigsqcup_{i \in \mathrm{Ob}(I)} = \mathrm{Ob}(I)$$

and the assignments $\alpha \mapsto \alpha_*$ can be collectively rewritten as a commutative diagram

$$\begin{array}{ccc} X \times_{\pi,s} \mathrm{Mor}(I) & \xrightarrow{m} & X \\ {\scriptstyle pr}\downarrow & & \downarrow{\scriptstyle \pi} \\ \mathrm{Mor}(I) & \xrightarrow{t} & \mathrm{Ob}(I) \end{array} \quad (9.2)$$

where $s, t : \mathrm{Mor}(I) \to \mathrm{Ob}(I)$ are the source and target maps, respectively, and the diagram

$$\begin{array}{ccc} X \times_{\pi,s} \mathrm{Mor}(I) & \xrightarrow{pr} & \mathrm{Mor}(I) \\ \downarrow & & \downarrow{\scriptstyle s} \\ X & \xrightarrow{\pi} & \mathrm{Ob}(I) \end{array}$$

is a pullback.

The notation is awkward, but the composition laws for the functor X translate into the commutativity of the diagrams

$$\begin{CD} X \times_{\pi,s} \text{Mor}(I) \times_{t,s} \text{Mor}(I) @>{1 \times m_I}>> X \times_{\pi,s} \text{Mor}(I) \\ @V{m \times 1}VV @VV{m}V \\ X \times_{\pi,s} \text{Mor}(I) @>>{m}> X \end{CD} \qquad (9.3)$$

and

$$\begin{CD} X @>{e_*}>> X \times_{\pi,s} \text{Mor}(I) \\ @. @VV{m}V \\ @. X \end{CD} \qquad (9.4)$$

with a diagonal map 1 from X to X.

Here, m_I is the composition law of the category I, and the map e_* is uniquely determined by the commutative diagram

$$\begin{CD} X @>{\pi}>> \text{Ob}(I) @>{e}>> \text{Mor}(I) \\ @V{1}VV @. @VV{s}V \\ X @>>{\pi}> @. \text{Ob}(I) \end{CD}$$

where the map e picks out the identity morphisms of I.

Thus, a functor $X : I \to \textbf{Set}$ consists of a function $\pi : X \to \text{Ob}(I)$ together with an action $m : X \times_{\pi,s} \text{Mor}(I) \to X$ making the diagrams (9.2), (9.3) and (9.4) commute. This is the *internal description* of a functor, which can be used to define functors on category objects.

Specifically, suppose that H is a sheaf of groupoids on a site $\mathcal{C}$. Then a *sheaf-valued functor* X on H, or more commonly an *H-functor*, consists of a sheaf map $\pi : X \to \text{Ob}(H)$, together with an action morphism $m : X \times_{\pi,s} \text{Mor}(H) \to X$ in sheaves such that the diagrams corresponding to (9.2), (9.3) and (9.4) commute in the sheaf category.

Alternatively, X consists of set-valued functors

$$X(U) : H(U) \to \textbf{Sets}$$

with $x \mapsto X(U)_x$ for $x \in \text{Ob}(H(U))$, together with functions

$$\phi^* : X(U)_x \to X(V)_{\phi^*(x)}$$

for each $\phi : V \to U$ in $\mathcal{C}$, such that the assignment

$$U \mapsto X(U) = \bigsqcup_{x \in \text{Ob}(H(U))} X(U)_x, \quad U \in \mathcal{C},$$

9.1 Torsors

defines a sheaf and the diagrams

$$
\begin{array}{ccc}
X(U)_x & \xrightarrow{\alpha_*} & X(U)_y \\
\phi^* \downarrow & & \downarrow \phi^* \\
X(V)_{\phi^*(x)} & \xrightarrow{(\phi^*(\alpha))_*} & X(V)_{\phi^*(y)}
\end{array}
$$

commute for each $\alpha : x \to y$ of $\mathrm{Mor}\,(H(U))$ and all $\phi : V \to U$ of $\mathcal{C}$.

From this alternative point of view, it is easy to see that an H-functor X defines a natural simplicial (pre)sheaf homomorphism

$$p : \varinjlim_H X \to BH.$$

One makes the construction sectionwise.

Remark 9.12 The homotopy colimit construction for H-functors is a direct generalization of the Borel construction for sheaves Y with actions by sheaves of groups G: the simplicial sheaf $EG \times_G Y$ is the homotopy colimit $\varinjlim_G Y$.

Say that an H-functor X is an *H-torsor* if the canonical map

$$\varinjlim_H X \to *$$

is a local weak equivalence.

A *morphism* $f : X \to Y$ of H-torsors is a natural transformation of H-functors, namely a sheaf morphism

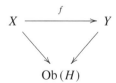

fibred over $\mathrm{Ob}\,(H)$, which respects the multiplication maps. Write $H - \mathbf{tors}$ for the category of H-torsors and the natural transformations between them.

The diagram

$$
\begin{array}{ccc}
X & \longrightarrow & \varinjlim_H X \\
\pi \downarrow & & \downarrow p \\
\mathrm{Ob}\,(H) & \longrightarrow & BH
\end{array}
$$

is homotopy cartesian in each section by Quillen's Theorem B [32, IV.5.2] (more specifically, Lemma 9.46) since H is a (pre)sheaf of groupoids, and is therefore homotopy cartesian in simplicial sheaves. It follows that a morphism $f : X \to Y$ of

H-torsors specializes to a weak equivalence $X \to Y$ of constant simplicial sheaves, which is therefore an isomorphism.

We therefore have the following generalization of Lemma 9.4:

Lemma 9.13 *Suppose that H is a sheaf of groupoids. Then the category $H-$**tors** of H-torsors is a groupoid.*

Every H-torsor X has an associated cocycle

$$ * \xleftarrow{\simeq} \underrightarrow{\holim}_H X \xrightarrow{p} BH, $$

called the *canonical cocycle*, and this association defines a functor

$$ \phi : H - \mathbf{tors} \to h(*, BH) $$

taking values in the simplicial sheaf cocycle category.

Now suppose given a cocycle

$$ * \xleftarrow{\simeq} Y \xrightarrow{g} BH $$

in simplicial sheaves, and form the pullback diagrams

$$ \begin{array}{ccc} \mathrm{pb}(Y)(U)_x & \longrightarrow & Y(U) \\ \downarrow & & \downarrow g \\ B(H(U)/x) & \longrightarrow & BH(U) \end{array} $$

of simplicial sets for each $x \in \mathrm{Ob}(H(U))$ and $U \in \mathcal{C}$. Set

$$ \mathrm{pb}(Y)(U) = \bigsqcup_{x \in \mathrm{Ob}(H(U))} \mathrm{pb}(Y)(U)_x. $$

Then the resulting simplicial presheaf map $\mathrm{pb}(Y) \to \mathrm{Ob}(H)$ specifies an internally defined functor on H that takes values in simplicial presheaves. There is a sectionwise weak equivalence

$$ \underrightarrow{\holim}_H \mathrm{pb}(Y) \to Y \simeq * $$

by Lemma 9.9, and the diagram

$$ \begin{array}{ccc} \mathrm{pb}(Y) & \longrightarrow & \underrightarrow{\holim}_H \mathrm{pb}(Y) \\ \downarrow & & \downarrow \\ \mathrm{Ob}(H) & \longrightarrow & BH \end{array} $$

is a sectionwise homotopy cartesian since H is a sheaf of groupoids. It follows that the natural transformation

$$ \mathrm{pb}(Y) \to \tilde{\pi}_0(Y) $$

9.1 Torsors

of simplicial presheaf-valued functors on H is a local weak equivalence. In summary, we have local weak equivalences

$$\varinjlim_H \tilde{\pi}_0 \operatorname{pb}(Y) \simeq \varinjlim_H \operatorname{pb}(Y) \simeq Y \simeq *,$$

so that the sheaf-valued functor $\tilde{\pi}_0 \operatorname{pb}(Y)$ on H is an H-torsor. These constructions are natural on $h(*, BH)$ and there is a functor

$$\psi : h(*, BH) \to H - \textbf{tors}.$$

Theorem 9.14 *Suppose that H is a sheaf of groupoids. Then the functors ϕ and ψ induce a homotopy equivalence*

$$B(H - \textbf{tors}) \simeq Bh(*, BH).$$

Corollary 9.15 *The functors ϕ and ψ induce a bijection*

$$\pi_0(H - \textbf{tors}) \cong [*, BH].$$

For the proof of Theorem 9.14, it is convenient to use a trick for diagrams of simplicial sets that are indexed by groupoids.

Suppose that Γ is a small groupoid, and let $s\textbf{Set}^\Gamma$ be the category of Γ-diagrams in simplicial sets. Let $s\textbf{Set}/B\Gamma$ be the category of simplicial set morphisms $Y \to B\Gamma$. The homotopy colimit defines a functor

$$\varinjlim_\Gamma : s\textbf{Set}^\Gamma \to s\textbf{Set}/B\Gamma.$$

This functor sends a diagram $X : \Gamma \to s\textbf{Set}$ to the canonical map $\varinjlim_\Gamma X \to B\Gamma$. On the other hand, given a simplicial set map $Y \to B\Gamma$, the collection of pullback diagrams

$$\begin{array}{ccc} \operatorname{pb}(Y)_x & \longrightarrow & Y \\ \downarrow & & \downarrow \\ B(\Gamma/x) & \longrightarrow & B\Gamma \end{array}$$

defines an Γ-diagram $\operatorname{pb}(Y) : \Gamma \to s\textbf{Set}$ that is functorial in $Y \to B\Gamma$.

Lemma 9.16 *Suppose that Γ is a groupoid. Then the functors*

$$\operatorname{pb} : s\textbf{Set}/B\Gamma \leftrightarrows s\textbf{Set}^\Gamma : \varinjlim_\Gamma$$

form an adjoint pair: pb *is left adjoint to* $\varinjlim_\Gamma$.

Proof Suppose that X is a Γ-diagram and that $p : Y \to B\Gamma$ is a simplicial set over $B\Gamma$. Suppose given a natural transformation

$$f : \operatorname{pb}(Y)_n \to X_n.$$

and let x be an object of Γ. An element of $(\mathrm{pb}\,(Y)_x)_n$ can be identified with a pair

$$(y, a_0 \to \cdots \to a_n \xrightarrow{\alpha} x)$$

where the string of arrows is in Γ and $p(y)$ is the string $a_0 \to \cdots \to a_n$. The map f is uniquely determined by the images of the elements

$$f(y, a_0 \to \cdots \to a_n \xrightarrow{1} a_n)$$

in $X_n(a_n)$. Since Γ is a groupoid, an element $z \in X(a_n)$ uniquely determines an element

$$(z_0, a_0) \to (z_1, a_1) \to \cdots \to (z_n, a_n)$$

with $z_n = z$. It follows that there is a natural bijection

$$\hom_\Gamma(\mathrm{pb}\,(Y)_n, X_n) \cong \hom_{B\Gamma_n}(Y_n, (\varinjlim{}_\Gamma X)_n).$$

Extend simplicially to get the adjunction isomorphism

$$\hom_\Gamma(\mathrm{pb}\,(Y), X) \cong \hom_{B\Gamma}(Y, \varinjlim{}_\Gamma X).$$

Proof (*Proof of Theorem 9.14*) It follows from Lemma 9.16 that the functor ψ is left adjoint to the functor ϕ.

Suppose that K is an ordinary groupoid, and that $x \in \mathrm{Ob}\,(K)$. The groupoid K/x has a terminal object and hence determines a cocycle

$$* \xleftarrow{\simeq} B(K/x) \to BK$$

in simplicial sets.

If $a \in \mathrm{Ob}\,(K)$, then in the pullback diagram

$$\begin{array}{ccc} \mathrm{pb}(B(K/x))(a) & \longrightarrow & B(K/x) \\ \downarrow & & \downarrow \\ B(K/a) & \longrightarrow & BK \end{array}$$

the object $\mathrm{pb}\,(B(K/x))(a)$ is the nerve of a groupoid whose objects are the diagrams

$$a \xleftarrow{\alpha} b \xrightarrow{\beta} x$$

in K, and whose morphisms are the diagrams

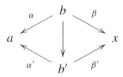

9.1 Torsors

In the presence of such a picture, $\beta \cdot \alpha^{-1} = \beta' \cdot (\alpha')^{-1}$. There are uniquely determined diagrams

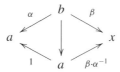

for each object $a \xleftarrow{\alpha} b \xrightarrow{\beta} x$. It follows that there is a natural bijection

$$\pi_0 \operatorname{pb}(B(K/x)(a) \cong \hom_K(a, x)$$

and that

$$\operatorname{pb}(B(K/x))(a) \to \pi_0 \operatorname{pb}(B(K/x))(a)$$

is a natural weak equivalence.

It also follows that there are weak equivalences

$$\begin{array}{c} \varinjlim_{a \in K} \operatorname{pb}(B(K/x))(a) \xrightarrow{\simeq} B(K/x) \simeq * \\ \simeq \downarrow \\ \varinjlim_{a \in K} \hom_K(a, x) \end{array}$$

so that the functor $a \mapsto \hom_K(a, x)$ defines a K-torsor. Here, the function

$$\beta_* : \hom_K(a, x) \to \hom_K(b, x)$$

induced by $\beta : a \to b$ is precomposition with β^{-1}.

To put it in a different way, each $x \in K$ determines a K-torsor $a \mapsto \hom_K(a, x)$, which we will call $\hom_K(\ , x)$ and there is a functor

$$K \to K - \mathbf{tors},$$

which is defined by $x \mapsto \hom_K(\ , x)$.

Observe that the maps $\hom_K(\ , x) \to Y$ classify elements of $Y(x)$ for all functors $Y : K \to \mathbf{Set}$.

In general, every global section x of a sheaf of groupoids H determines an H-torsor $\hom_H(\ , x)$ which is constructed sectionwise according to the recipe above. In particular, this is the torsor associated by the pullback construction to the cocycle

$$* \xleftarrow{\simeq} B(H/x) \to BH.$$

The torsors $\hom_H(\ , x)$ are the *trivial torsors* for the sheaf of groupoids H.

There is a functor

$$j : \Gamma_* H \to H - \mathbf{tors}$$

that is defined by $j(x) = \hom_H(\ , x)$. Recall that Γ_* is the global sections functor, so that $\Gamma_* H$ is the groupoid of global sections of the sheaf of groupoids H.

The torsor isomorphisms

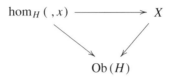

(or *trivializations*) are in bijective correspondence with global sections of X that map to $x \in \mathrm{Ob}(H)$ under the structure map $X \to \mathrm{Ob}(H)$.

Torsors and cocycle categories behave well with respect to restriction. If $\phi : V \to U$ is a morphism of the underlying site $\mathcal{C}$, then composition with ϕ defines a functor

$$\phi_* : \mathcal{C}/V \to \mathcal{C}/U,$$

and composition with ϕ_* determines a restriction functor

$$\phi^* : \mathbf{Pre}(\mathcal{C}/U) \to \mathbf{Pre}(\mathcal{C}/V),$$

which takes $F|_U$ to $F|_V$ for any presheaf F on $\mathcal{C}$. All restriction functors take sheaves to sheaves and are exact. Thus, ϕ^* takes an $H|_U$-torsor to an $H|_V$ torsor. In particular, there is an identification

$$\phi^* \hom_{H|_U}(\ , x) = \hom_{H|_V}(\ , x_V)$$

for all $x \in H(U)$. The functor ϕ^* also preserves cocycles.

Every H-torsor X has sections along some cover, since the map $\varinjlim_H X \to *$ is a local weak equivalence. It follows that every H-torsor is locally trivial.

Remark 9.17 We have not discussed the size of the objects that are involved in Theorem 9.14. The statement of that result makes no sense unless the cocycle and torsor categories are small in some sense.

Suppose that α is a regular cardinal such that $\alpha > |\mathrm{Mor}(\mathcal{C})|$ and $\alpha > |\mathrm{Mor}(H)|$. If the H-torsor F has $F(U) \neq \emptyset$, then the restriction $F|_U$ has global sections and hence is isomorphic to some $\hom_{H|_U}(\ , x)$, so the set $F(U)$ is α-bounded. It follows that F is α-bounded as a sheaf. We can therefore assume that the category $H - \mathbf{tors}$ is small.

The canonical cocycle functor

$$\phi : H - \mathbf{tors} \to h(*, BH)$$

takes values in the full subcategory $h(*, BH)_\alpha$ of cocycles

$$* \xleftarrow{\simeq} Y \to BH$$

of cocycles that are bounded by α in the sense that $|Y| < \alpha$.

Thus, it is more correct to assert in Theorem 9.14 that the groupoid $H - \mathbf{tors}$ is small and that there are homotopy equivalences

$$B(H - \mathbf{tors}) \simeq Bh(*, BH)_\alpha.$$

if the cardinal α is larger than $|\operatorname{Mor}(H)|$.

It follows that the map

$$Bh(*, BH)_\alpha \to Bh(*, BH)_\gamma$$

is a weak equivalence for all $\gamma \geq \alpha$.

See also Propositions 6.7 and 6.10.

There is a presheaf of groupoids $H - \mathbf{Tors}$ on the site $\mathcal{C}$ with

$$H - \mathbf{Tors}(U) = H|_U - \mathbf{tors}$$

and a presheaf of categories $\mathbb{H}(*, BH)$ with

$$\mathbb{H}(*, BH)(U) = h(*, BH|_U).$$

There are functors

$$H \xrightarrow{j} H - \mathbf{Tors} \xrightarrow{\phi} \mathbb{H}(*, BH)$$

where ϕ induces a sectionwise weak equivalence

$$B(H - \mathbf{Tors}) \xrightarrow{\simeq} B\mathbb{H}(*, BH)$$

by Theorem 9.14, and the composite map $\phi \cdot j$ is defined by sending an object $x \in H(U)$ to the cocycle

$$* \xleftarrow{\simeq} B(H|_U/x) \to BH|_U.$$

9.2 Stacks and Homotopy Theory

Write

$$\mathbf{Pre(Gpd)} = \mathbf{Pre(Gpd}(\mathcal{C}))$$

for the category of presheaves of groupoids on a small site $\mathcal{C}$.

Say that a morphism $f : G \to H$ of presheaves of groupoids is a *local weak equivalence* (respectively *injective fibration*) if and only if the induced map $f_* : BG \to BH$ is a local weak equivalence (respectively injective fibration) of simplicial presheaves. A morphism $i : A \to B$ of presheaves of groupoids is a *cofibration* if it has the left lifting property with respect to all trivial fibrations.

The fundamental groupoid functor $X \mapsto \pi(X)$ is left adjoint to the nerve functor. It follows that every cofibration $A \to B$ of simplicial presheaves induces a cofibration $\pi(A) \to \pi(B)$ of presheaves of groupoids. The class of cofibrations $A \to B$ is closed under pushout along arbitrary morphisms $A \to G$, because cofibrations are defined by a left lifting property.

There is a function complex construction for presheaves of groupoids: the simplicial set $\mathbf{hom}(G, H)$ has for n-simplices, all morphisms

$$\phi : G \times \pi(\Delta^n) \to H.$$

There is a natural isomorphism

$$\hom(G, H) \cong \hom(BG, BH)$$

that sends the simplex ϕ to the composite

$$BG \times \Delta^n \xrightarrow{1 \times \eta} BG \times B\pi(\Delta^n) \cong B(G \times \pi(\Delta^n)) \xrightarrow{\phi_*} BH.$$

Note that the fundamental groupoid $\pi(\Delta^n)$ is the trivial groupoid on the set $\{0, 1, \ldots, n\}$.

Preservation of local weak equivalences is a basic property of the fundamental groupoid functor.

Lemma 9.18 *The functor $X \mapsto B\pi(X)$ preserves local weak equivalences of simplicial presheaves.*

Proof The nerve functor B takes sheaves of groupoids to simplicial sheaves, and commutes with the formation of associated sheaves. It follows that there is a natural isomorphism

$$L^2(B\pi(X)) \cong B(L^2\pi(X)).$$

It therefore suffices to show that the functor $Y \mapsto B\tilde{\pi}Y$ preserves local weak equivalences of simplicial sheaves, where $\tilde{\pi}Y = L^2(\pi(Y))$ is the sheaf theoretic fundamental groupoid.

The natural composite

$$Y \to \mathrm{Ex}^\infty Y \to L^2 \mathrm{Ex}^\infty Y$$

induces a local weak equivalence $B\tilde{\pi}Y \to B\tilde{\pi}(L^2 \mathrm{Ex}^\infty Y)$, so it is enough to show that a local weak equivalence $f : X \to Y$ of locally fibrant simplicial sheaves induces a local weak equivalence $B\tilde{\pi}(X) \to B\tilde{\pi}(Y)$.

Suppose that $p : \mathbf{Shv}(\mathcal{B}) \to \mathbf{Shv}(\mathcal{C})$ is a Boolean localization.

9.2 Stacks and Homotopy Theory

The nerve functor B commutes with the direct image p_* and the inverse image p^* functors up to natural isomorphism, for sheaves of groupoids on the sites $\mathcal{B}$ and $\mathcal{C}$, respectively. It follows that there is a natural isomorphism

$$\tilde{\pi}(p^*Y) \cong p^*\tilde{\pi}(Y),$$

and so we can assume that the map $f : X \to Y$ of locally fibrant simplicial sheaves is defined on $\mathcal{B}$. But f is a sectionwise weak equivalence of sheaves of Kan complexes, so the simplicial presheaf map $B\pi(X) \to B\pi(Y)$ is a sectionwise weak equivalence. The map $B\tilde{\pi}(X) \to B\tilde{\pi}(Y)$ of associated simplicial sheaves is therefore a local weak equivalence.

The following result first appeared in [43]:

Proposition 9.19 *With the definitions of local weak equivalence, injective fibration and cofibration given above, the category* **Pre(Gpd)** *satisfies the axioms for a proper closed simplicial model category.*

Proof The injective model structure for the category $s\mathbf{Pre}(\mathcal{C})$ is cofibrantly generated. It follows that every morphism $f : G \to H$ has a factorization

such that j is a cofibration and p is a trivial fibration.

The other factorization statement for **CM5** can be proved the same way, provided one knows that if $i : A \to B$ is a trivial cofibration of simplicial presheaves and the diagram

$$\begin{array}{ccc} \pi(A) & \longrightarrow & G \\ {\scriptstyle i_*}\downarrow & & \downarrow{\scriptstyle i'} \\ \pi(B) & \longrightarrow & H \end{array}$$

is a pushout, then the map i' is a local weak equivalence. But one can prove the corresponding statement for ordinary groupoids, and the general case follows by a Boolean localization argument.

The claim is proved for ordinary groupoids by observing that in all pushout diagrams

$$\begin{array}{ccc} \pi(\Lambda_k^n) & \longrightarrow & G \\ {\scriptstyle i_*}\downarrow & & \downarrow{\scriptstyle i'} \\ \pi(\Delta^n) & \longrightarrow & H \end{array}$$

the map i_* is an isomorphism for $n \geq 2$ and is the inclusion of a strong deformation retraction if $n = 1$. The classes of isomorphisms and strong deformation retractions are both closed under pushout in the category of groupoids.

All other closed model axioms are trivial to verify, as is right properness. Left properness follows from the fact that all presheaves of groupoids are cofibrant, since the adjunction map

$$\epsilon : \pi(BG) \to G$$

is an isomorphism.

One verifies the simplicial model axiom **SM7** by using the natural isomorphism

$$\pi(BG) \cong G,$$

with Lemma 9.18, and the fact that the fundamental groupoid functor preserves products.

One can make the same definitions for sheaves of groupoids: say that a map $f : G \to H$ of sheaves of groupoids is a *local weak equivalence* (respectively *injective fibration*) if the associated simplicial sheaf map $f_* : BG \to BH$ is a local weak equivalence (respectively injective fibration) of simplicial sheaves. *Cofibrations* are defined by a left lifting property, as before.

Write **Shv(Gpd)** for the category of sheaves of groupoids on the site $\mathcal{C}$. The forgetful functor i and associated sheaf functor L^2 induce an adjoint pair

$$L^2 : \mathbf{Pre}(\mathbf{Gpd}) \leftrightarrows \mathbf{Shv}(\mathbf{Gpd}(\mathcal{C})) : i.$$

According to the definitions, the forgetful functor i preserves fibrations and trivial fibrations. The canonical map $\eta : BG \to iL^2 BG$ is always a local weak equivalence. The method of proof of Proposition 9.19 and formal nonsense now combine to prove the following:

Proposition 9.20

(1) With these definitions, the category **Shv(Gpd)** *of sheaves of groupoids satisfies the axioms for a proper closed simplicial model category.*

(2) The adjoint pair

$$L^2 : \mathbf{Pre}(\mathbf{Gpd}) \leftrightarrows \mathbf{Shv}(\mathbf{Gpd}) : i$$

forms a Quillen equivalence.

The model structures of Propositions 9.19 and 9.20 are the *injective model structures* for presheaves and sheaves of groupoids on a site $\mathcal{C}$, respectively—they are Quillen equivalent.

Here is a characterization of local weak equivalence of sheaves of groupoids that is frequently useful:

Lemma 9.21 *A map $f : G \to H$ of sheaves of groupoids is a local weak equivalence if and only if the maps*

$$\mathrm{Mor}\,(G) \to (\mathrm{Ob}\,(G) \times \mathrm{Ob}\,(G)) \times_{(\mathrm{Ob}\,(H) \times \mathrm{Ob}\,(H))} \mathrm{Mor}\,(H) \qquad (9.5)$$

9.2 Stacks and Homotopy Theory

and

$$\tilde{\pi}_0(G) \to \tilde{\pi}_0(H) \qquad (9.6)$$

are isomorphisms.

Proof A map $G \to H$ of ordinary groupoids induces a weak equivalence $BG \to BH$ if and only if the induced functions

$$\mathrm{Mor}\,(G) \to (\mathrm{Ob}\,(G) \times \mathrm{Ob}\,(G)) \times_{(\mathrm{Ob}\,(H) \times \mathrm{Ob}\,(H))} \mathrm{Mor}\,(H)$$

and

$$\pi_0(G) \to \pi_0(H)$$

are isomorphisms.

Suppose that $p : \mathbf{Shv}(\mathcal{B}) \to \mathbf{Shv}(\mathcal{C})$ is a Boolean localization. Then the simplicial sheaf map $BG \to BH$ is a local weak equivalence if and only if the map $Bp^*G \to Bp^*H$ is a local weak equivalence. All nerves BG of sheaves of groupoids G are sheaves of Kan complexes, and are therefore locally fibrant. Thus, the map $Bp^*G \to Bp^*H$ is a local weak equivalence if and only if it is a sectionwise weak equivalence, and this is true if and only if the inverse image functor p^* takes the maps (9.5) and (9.6) to isomorphisms (Proposition 4.28).

Corollary 9.22 *A map $f : G \to H$ of presheaves of groupoids is a local weak equivalence if and only if the map*

$$\mathrm{Mor}\,(G) \to (\mathrm{Ob}(G) \times \mathrm{Ob}(G)) \times_{(\mathrm{Ob}\,(H) \times \mathrm{Ob}\,(H))} \mathrm{Mor}\,(H) \qquad (9.7)$$

is a local isomorphism and

$$\pi_0(G) \to \pi_0(H)$$

is a local epimorphism.

Proof Suppose that the conditions hold for a map $f : G \to H$ of presheaves of groupoids. Then the induced map $f_* : \tilde{G} \to \tilde{H}$ of sheaves of groupoids is fully faithful in each section since the sheaf map induced by (9.7) is an isomorphism. But then it follows that the presheaf map $\pi_0(\tilde{G}) \to \pi_0(\tilde{H})$ is a monomorphism, and so its associated sheaf map $\tilde{\pi}_0(\tilde{G}) \to \tilde{\pi}_0(\tilde{H})$ is an isomorphism. The map $f_* : \tilde{G} \to \tilde{H}$ is therefore a local weak equivalence by Lemma 9.21, so that the map $f : G \to H$ is a local weak equivalence.

The converse is again a consequence of Lemma 9.21.

Part (1) of Proposition 9.20 was first proved in [71]. This was a breakthrough result, in that it enabled the definition of stacks as homotopy theoretic objects. Specifically, a sheaf of groupoids H is said to be a *stack* if it satisfies descent for the injective model structure on $\mathbf{Shv}(\mathbf{Gpd}(\mathcal{C}))$.

In other words, H is a stack if and only if every injective fibrant model $j : H \to H'$ is a sectionwise weak equivalence.

Remark 9.23 Classically, stacks are defined to be sheaves of groupoids that satisfy the effective descent condition. The effective descent condition, which is described below, is equivalent to the homotopical descent condition for sheaves of groupoids—this is proved in Proposition 9.28.

Observe that if $j : H \to H'$ is an injective fibrant model in sheaves (or presheaves) of groupoids, then the induced map $j_* : BH \to BH'$ is an injective fibrant model in simplicial presheaves. Thus, H is a stack if and only if the simplicial presheaf BH satisfies descent.

Every injective fibrant object is a stack, because injective fibrant objects satisfy descent. This means that every injective fibrant model $j : G \to H$ of a sheaf of groupoids G is a *stack completion*. This model j can be constructed functorially, since the injective model structure on **Shv**(**Gpd**($\mathcal{C}$)) is cofibrantly generated. We can therefore speak unambiguously about the stack completion of a sheaf of groupoids G—the stack completion is also called the *associated stack*.

Similar definitions can also be made for presheaves of groupoids. In particular, we say that a presheaf of groupoids G is a stack if it satisfies descent, or equivalently if the simplicial presheaf BG satisfies descent. This means, effectively, that stacks are identified with homotopy types of presheaves or sheaves of groupoids, within the respective injective model structures.

Some of the most common examples of stacks come from group actions. Suppose that $G \times X \to X$ is an action of a sheaf of groups G on a sheaf X. Then the Borel construction $EG \times_G X$ is the nerve of a sheaf of groupoids $E_G X$. The stack completion

$$j : E_G X \to X/G$$

is called the *quotient stack*.

A *G-torsor over X* is a G-equivariant map $P \to X$ where P is a G-torsor. A morphism of G-torsors over X is a commutative diagram

of G-equivariant morphisms, where P and P' are G-torsors. Write $G - \mathbf{tors}/X$ for the corresponding groupoid.

If $P \to X$ is a G-torsor over X, then the induced map of Borel constructions

$$* \xleftarrow{\simeq} EG \times_G P \to EG \times_G X$$

9.2 Stacks and Homotopy Theory

is an object of the cocycle category

$$h(*, EG \times_G X),$$

and the assignment is functorial. Conversely, if the diagram

$$* \xleftarrow{\simeq} U \to EG \times_G X$$

is a cocycle, then the induced map

$$\tilde{\pi}_0 \operatorname{pb}(U) \to \tilde{\pi}_0 \operatorname{pb}(EG \times_G X) \xrightarrow[\cong]{\epsilon} X$$

is a G-torsor over X. Here, as above, the pullback functor pb is defined by pulling back over the canonical map $EG \to BG$.

The functors

$$\tilde{\pi}_0 \operatorname{pb} : h(*, EG \times_G X) \leftrightarrows G-\mathbf{tors}/X : EG \times_G ?$$

are adjoint, and we have proved the following:

Lemma 9.24 *There is a weak equivalence*

$$B(G-\mathbf{tors}/X) \simeq Bh(*, EG \times_G X).$$

In particular, there is an induced bijection

$$\pi_0(G-\mathbf{tors}/X) \cong [*, EG \times_G X].$$

Lemma 9.24 was proved by a different method in [59]. There is a generalization of this result, having essentially the same proof, for the homotopy colimit $\underrightarrow{\operatorname{holim}}_H X$ of a diagram X on a sheaf of groupoids H—see [65].

The sheaves of groupoids G and H are said to be *Morita equivalent* if there is a diagram

$$G \xleftarrow{p} K \xrightarrow{q} H$$

of morphisms of sheaves of groupoids such that the induced maps p_* and q_* in the diagram

$$BG \xleftarrow{p_*} BK \xrightarrow{q_*} BH$$

are local trivial fibrations of simplicial sheaves.

Clearly, if G and H are Morita equivalent, then they are weakly equivalent for the injective model structure.

Conversely, if $f : G \to H$ is a local weak equivalence, take the cocycle

$$G \xrightarrow{(1,f)} G \times H$$

and find a factorization

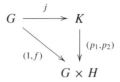

such that j is a local weak equivalence and (p_1, p_2) is an injective fibration. Then the induced map

$$BK \xrightarrow{(p_{1*}, p_{2*})} BG \times BH$$

is an injective hence local fibration, and the projection maps $BG \times BH \to BG$ and $BG \times BH \to BH$ are local fibrations since BG and BH are locally fibrant. It follows that the maps

$$G \xleftarrow{p_1} K \xrightarrow{p_2} H$$

define a Morita equivalence.

We have shown the following:

Lemma 9.25 *Suppose that G and H are sheaves of groupoids. Then G and H are locally weakly equivalent if and only if they are Morita equivalent.*

Categories of cocycles and torsors can both be used to construct models for the associated stack. The precise statement appears in Corollary 9.27, which is a corollary of the proof of the following:

Proposition 9.26 *Suppose that H is a sheaf of groupoids on a small site $\mathcal{C}$. Then the induced maps*

$$BH \xrightarrow{j_*} B(H - \mathbf{Tors}) \quad (9.8)$$
$$\downarrow \phi_*$$
$$B\mathbb{H}(*, BH)$$

are local weak equivalences of simplicial presheaves.

Proof The map ϕ_* is a sectionwise equivalence for all sheaves of groupoids H by Theorem 9.14, and the morphism j is fully faithful in all sections.

Suppose that H is an injective fibrant sheaf of groupoids. We show that the morphisms in the diagram (9.8) are sectionwise weak equivalences of simplicial presheaves.

For this, it suffices to show that the maps

$$j_* : \pi_0 BH(U) \to \pi_0 B(H - \mathbf{Tors})(U)$$

9.2 Stacks and Homotopy Theory

are surjective under the assumption that H is injective fibrant. By a restriction argument (which uses Corollary 5.26), we can assume that the site $\mathcal{C}$ has a terminal object t and show that the function

$$\pi_0 BH(t) \to \pi_0 B\mathbb{H}(*, BH)(t) = \pi_0 Bh(*, BH)$$

is surjective. Recall that the image of an element $x : * \to BH$ under the map $\phi_* \cdot j_* : BH(t) \to h(*, BH)$ is the cocycle

$$* \leftarrow B(H/x) \to BH.$$

In every cocycle

$$* \xleftarrow{s} U \xrightarrow{f} BH,$$

the map s is a local weak equivalence, so there is a homotopy commutative diagram

since BH is injective fibrant. This means that the cocycles (s, f), (s, xs) and $(1, x)$ are all in the same path component of the cocycle category $h(*, BH)$. Finally, there is a morphism

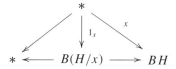

in $h(*, BH)$.

For the general case, suppose that $i : H \to H'$ is an injective fibrant model for H. In the diagram

$$\begin{array}{ccc}
BH & \xrightarrow{i_*} & BH' \\
{\scriptstyle j} \downarrow & & \downarrow {\scriptstyle j} \simeq \\
B(H - \mathbf{Tors}) & \longrightarrow & B(H' - \mathbf{Tors}) \\
{\scriptstyle \phi_*} \downarrow \simeq & & \simeq \downarrow {\scriptstyle \phi_*} \\
B\mathbb{H}(*, BH) & \xrightarrow[i_*]{\simeq} & B\mathbb{H}(*, BH')
\end{array}$$

the indicated maps are sectionwise weak equivalences: use the paragraphs above for the vertical maps on the right, Corollary 6.11 for the bottom i_* (see also Remark 9.17), and Theorem 9.14 for ϕ_*. The map $i_* : BH \to BH'$ is a local weak equivalence since the map i is an injective fibrant model. It follows that the map $j_* : BH \to B(H - \mathbf{Tors})$ is a local weak equivalence.

Corollary 9.27 *Suppose that H is a sheaf of groupoids. Then the maps*

$$j : H \to H - \mathbf{Tors}$$

and

$$\phi \cdot j : H \to \mathbb{H}(*, BH)$$

are models for the stack completion of H.

Suppose that $R \subset \hom(\ , U)$ is a covering sieve, and also write R for the full subcategory on $\mathcal{C}/U$ whose objects are the members $\phi : V \to U$ of the sieve. Following Giraud [31], an *effective descent datum* $x : R \to H$ on the sieve R for the sheaf of groupoids H consists of

(1) objects $x_\phi \in H(V)$, one for each object $\phi : V \to U$ of R, and
(2) morphisms $x_\phi \xrightarrow{\alpha_*} \alpha^*(x_\psi)$ in $H(V)$, one for each morphism

of R,

such that the diagram

$$\begin{array}{ccc} x_\phi & \xrightarrow{\alpha_*} & \alpha^*(x_\psi) \\ (\beta\alpha)_* \downarrow & & \downarrow \alpha^*(\beta_*) \\ (\beta\alpha)^*(x_\zeta) & = & \alpha^*\beta^*(x_\zeta) \end{array}$$

commutes for each composable pair of morphisms

$$V \xrightarrow{\alpha} W \xrightarrow{\beta} W'$$

9.2 Stacks and Homotopy Theory

in R.

There is a functor $R \to \mathbf{Pre}(\mathcal{C})$ that takes an object $\phi : V \to U$ to the representable functor $\hom(\ ,V)$. The corresponding translation object E_R is a presheaf of categories on $\mathcal{C}$: the presheaf of objects for E_R is the disjoint union

$$\bigsqcup_{\phi:V\to U} \hom(\ ,V),$$

and the presheaf of morphisms for E_R is the disjoint union

$$\bigsqcup_{\substack{V \xrightarrow{\alpha} W \\ \phi \searrow \swarrow \psi \\ U}} \hom(\ ,V),$$

which is indexed over the morphisms of R.

An effective descent datum $x : R \to H$ can be identified with a functor $x : E_R \to H$ of presheaves of categories. A *morphism* of effective descent data is a natural transformation of such functors. Write $\hom(E_R, H)$ for the corresponding groupoid of effective descent data on the sieve R in H.

Any refinement $S \subset R$ of covering sieves induces a restriction functor

$$\hom(E_R, H) \to \hom(E_S, H)$$

and, in particular, the inclusion $R \subset \hom(\ ,U)$ induces a functor

$$H(U) \to \hom(E_R, H). \tag{9.9}$$

One says that the sheaf of groupoids H satisfies the *effective descent* condition if an only if the map (9.9) is an equivalence of groupoids for all covering sieves $R \subset \hom(\ ,U)$ and all objects U of the site $\mathcal{C}$.

The effective descent condition is the classical criterion for a sheaf of groupoids to be a stack, and we have the following:

Proposition 9.28 *A sheaf of groupoids H is a stack if and only if it satisfies the effective descent condition.*

Proof Suppose that H is a stack. The effective descent condition is an invariant of sectionwise equivalence of groupoids, so it suffices to assume that H is injective fibrant. The nerve of the groupoid $\hom(E_R, H)$ may be identified up to isomorphism with the function complex $\mathbf{hom}(BE_R, BH)$. There is a canonical local weak equivalence $BE_R \to U$ (see Lemma 9.29 below), and so the induced map

$$BH(U) \to \mathbf{hom}(BE_R, BH)$$

is a weak equivalence of simplicial sets. This means, in particular, that the homomorphism

$$H(U) \to \hom(E_R, H)$$

is an equivalence of groupoids. It follows that H satisfies the effective descent condition.

Suppose, conversely, that the sheaf of groupoids H satisfies the effective descent condition, and let $j : H \to H'$ be an injective fibrant model in sheaves of groupoids. We must show that the map j is a sectionwise equivalence of groupoids.

Write $\tilde{U}$ for the sheaf that is associated to the representable presheaf $\hom(\ , U)$. There is a natural function

$$\psi : \pi_0 H(U) \to \pi_0 h(\tilde{U}, BH),$$

which is defined by sending the homotopy class of a map $\alpha : \hom(\ , U) \to BH$ to the path component of the cocycle

$$\tilde{U} \xleftarrow{1} \tilde{U} \xrightarrow{\alpha_*} BH,$$

where α_* is the map of simplicial sheaves associated to the presheaf map α. This function ψ is a bijection if the sheaf of groupoids H is injective fibrant.

The map j is fully faithful in all sections since it is a local equivalence between sheaves of groupoids. It therefore suffices to show that the induced map $\pi_0 H(U) \to \pi_0 H'(U)$ is surjective for each $U \in \mathcal{C}$. In view of the commutativity of the diagram

$$\begin{array}{ccc} \pi_0 H(U) & \xrightarrow{\psi} & \pi_0 h(\tilde{U}, BH) \\ \downarrow & & \downarrow \cong \\ \pi_0 H'(U) & \xrightarrow[\cong]{\psi} & \pi_0 h(\tilde{U}, BH') \end{array}$$

it further suffices to show that each function

$$\psi : \pi_0 H(U) \to \pi_0 h(\tilde{U}, BH)$$

is surjective.

Suppose that

$$\tilde{U} \xleftarrow[\simeq]{f} V \xrightarrow{g} BH$$

is a cocycle in simplicial sheaves.

The fundamental groupoid sheaf $\tilde{\pi}(V)$ can be identified up to isomorphism with the Čech groupoid $C(p)$ for the local epimorphism $p : V_0 \to \tilde{U}$ (Example 4.17). In effect, the canonical map $\tilde{\pi}(V) \to C(p)$ is fully faithful and is an isomorphism on objects in all sections. It follows that there is a map of cocycles

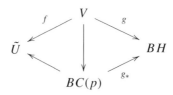

9.2 Stacks and Homotopy Theory

Let $R \subset \hom(\ ,U)$ be the covering sieve of all maps $\phi : V \to U$ that lift to V_0, and pick a lifting $\sigma_\phi : V \to V_0$ for each such ϕ. The morphisms σ_ϕ define a morphism

$$\sigma : \bigsqcup_{V \xrightarrow{\phi} U \in R} V \to V_0.$$

If $\alpha : \phi \to \psi$ is a morphism of R, then there is a diagram

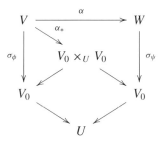

The collection of these maps α_* defines a morphism

$$\bigsqcup_{\phi \xrightarrow{\alpha} \psi} V \to V_0 \times_U V_0,$$

and we have defined a functor $\sigma : E_R \to C(p)$. There is a corresponding diagram

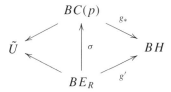

Finally, the assumption that H satisfies effective descent means that there is a homotopy commutative diagram

$$\begin{array}{ccc} BE_R & \xrightarrow{g'} & BG \\ \downarrow & \nearrow & \\ \tilde{U} & & \end{array}$$

and it follows that the original cocycle (f, g) is in the path component of a cocycle of the form

$$\tilde{U} \xleftarrow{1} \tilde{U} \to BH.$$

Lemma 9.29 *Suppose that $R \subset \hom(\ ,U)$ is a covering sieve. Then the canonical map*

$$BE_R \to U$$

of simplicial presheaves is a local weak equivalence.

Proof Suppose that $W \in \mathcal{C}$, and consider the induced map of W-sections

$$\bigsqcup_{\phi_0 \to \cdots \to \phi_n} \hom(W, V_0) \to \hom(W, U).$$

The fibre F_ϕ over a fixed morphism $\phi : W \to U$ is the nerve of the category of factorizations

$$\begin{array}{ccc} & & V \\ & \nearrow & \downarrow \psi \\ W & \xrightarrow{\phi} & U \end{array}$$

of ϕ with $\psi \in R$. If $\phi : W \to U$ is a member of R, then this category is non-empty and has an initial object, namely the picture

$$\begin{array}{ccc} & & W \\ & \nearrow^{1} & \downarrow \phi \\ W & \xrightarrow{\phi} & U \end{array}$$

The fibre F_ϕ is empty if ϕ is not a member of R.

In all cases, there is a covering sieve $S \subset \hom(\ , W)$ such that $\phi \cdot \psi$ is in R for all $\psi \in S$.

9.3 Groupoids Enriched in Simplicial Sets

This section presents a collection of basic results from the homotopy theory of groupoids enriched in simplicial sets, which will be needed later to develop the local homotopy theory of presheaves of groupoids enriched in simplicial sets.

Write $s_0\mathbf{Gpd}$ for the category of groupoids enriched in simplicial sets.

Groupoids enriched in simplicial sets can be viewed as simplicial groupoids G with simplicially discrete objects. A map $f : G \to H$ of $s_0\mathbf{Gpd}$ is the obvious thing: it is a map of simplicial groupoids. The earliest model structure for this category is due to Dwyer and Kan [23] who state that

(1) the map f is a *weak equivalence* if the map

$$\text{Mor}(G) \to (\text{Ob}(G) \times \text{Ob}(G)) \times_{\text{Ob}(H) \times \text{Ob}(H)} \text{Mor}(H) \qquad (9.10)$$

9.3 Groupoids Enriched in Simplicial Sets

is a weak equivalence of simplicial sets and the function $f_* : \pi_0 G \to \pi_0 H$ is an isomorphism,

(2) the map f is a *fibration* if the map (9.10) is a Kan fibration and the map $f : G_0 \to H_0$ is a fibration of groupoids,

(3) the *cofibrations* are those maps that have the left lifting property with respect to all trivial fibrations.

Here, for a groupoid G enriched in simplicial sets, we define $\pi_0 G$ to be the set of path components $\pi_0(G_0)$ of the groupoid G_0, and call it the *path component set* of G. Observe that vertices x, y of G represent the same element of $\pi_0(G_0)$ if and only if the simplicial set $G(x, y)$ is non-empty. It follows that there are isomorphisms

$$\pi_0 G = \pi_0(G_0) \cong \pi_0(G_n)$$

for $n \geq 0$.

Recall that a weak equivalence (respectively fibration) $G \to H$ of ordinary groupoids is a map such that the induced simplicial set map $BG \to BH$ is a weak equivalence (respectively Kan fibration). Most of the lifting properties that are ostensibly necessary for a fibration of groupoids are automatic: an argument on coskeleta shows that a map $G \to H$ is a fibration if and only if the map $BG \to BH$ has the path lifting property.

According to the present definition of fibration, all groupoids enriched in simplicial sets are fibrant, since all simplicial groups are Kan complexes and all ordinary groupoids are fibrant.

Theorem 9.30 (Dwyer–Kan) *With the definitions given above, the category s_0**Gpd** of groupoids enriched in simplicial sets has the structure of a cofibrantly generated, right proper closed model category.*

The model structure of Theorem 9.30 will be called the *Dwyer–Kan model structure* for groupoids enriched in simplicial sets. A proof of this result appears in [32, V.7.6].

In summary, every simplicial set K has an associated free simplicial groupoid $F'(K)$ on two objects, and one characterizes the simplicial set map (9.10) associated to a map $p : G \to H$ of s_0**Gpd** as a fibration or trivial cofibration by the map $p : G \to H$ having a right lifting property with respect to maps $F'(K) \to F'(L)$ that are induced by the usual sets of cofibrations $K \to L$ of simplicial sets. The question of whether or not the map $G_0 \to H_0$ is a fibration of groupoids is detected by a separate lifting property. The cofibrant generation is implicit in this approach. All objects of s_0**Gpd** are fibrant, since all simplicial groups and the nerves BH of groupoids are Kan complexes, and right properness follows by a standard result that is due to [32, II.8.5].

There are various equivalent ways to specify weak equivalences of groupoids enriched in simplicial sets.

First of all, suppose that $f : G \to H$ is a morphism such that the map (9.10) is a weak equivalence of simplicial sets and the map $\pi_0 G \to \pi_0 H$ is surjective. Suppose that x, y are objects of G such that $[f(x)] = [f(y)]$ in $\pi_0(H)$. Then the simplicial set $H(f(x), f(y))$ is non-empty so that $G(x, y)$ is non-empty since the map $G(x, y) \to H(f(x), f(y))$ is a weak equivalence, and it follows that $[x] = [y]$ in $\pi_0(G)$.

Thus a map $f : G \to H$ is a weak equivalence of $s_0\mathbf{Gpd}$, which is a weak equivalence if and only if the following conditions are satisfied:

(a) the map
$$\mathrm{Mor}\,(G) \to (\mathrm{Ob}(G) \times \mathrm{Ob}(G)) \times_{\mathrm{Ob}(H) \times \mathrm{Ob}(H)} \mathrm{Mor}\,(H)$$
is a weak equivalence of simplicial sets, and

(b) the function
$$f_* : \pi_0 G \to \pi_0 H$$
is surjective.

One says that a morphism $f : G \to H$ that satisfies condition (1) is *homotopically full and faithful*. We say that f is *surjective in path components* if it satisfies condition (2).

The object f^*H of $s_0\mathbf{Gpd}$ has $\mathrm{Ob}(f^*H) = \mathrm{Ob}(G)$, while the pullback diagram

$$\begin{array}{ccc} \mathrm{Mor}\,(f^*H) & \longrightarrow & \mathrm{Mor}\,(H) \\ \downarrow & & \downarrow {\scriptstyle (s,t)} \\ \mathrm{Ob}(G) \times \mathrm{Ob}(G) & \xrightarrow{f \times f} & \mathrm{Ob}(H) \times \mathrm{Ob}(H) \end{array}$$

defines the simplicial set $\mathrm{Mor}\,(f^*H)$. There is a natural commutative diagram

of groupoids enriched in simplicial sets. The map α_f is the identity on objects, while the map β_f is the identity on individual morphism simplicial sets.

It follows that the map $f : G \to H$ is a weak equivalence if and only if α_f is a weak equivalence (equivalently, f is homotopically full and faithful) and β_f is a weak equivalence (equivalently, f is surjective in path components).

Every object G of $s_0\mathbf{Gpd}$ is a simplicial groupoid, and therefore has a functorially associated bisimplicial set BG.

Recall [32, IV.3.3] that a map $X \to Y$ of bisimplicial sets is said to be a *diagonal weak equivalence* if the induced map $d(X) \to d(Y)$ of diagonal simplicial sets is a weak equivalence. We use the diagonal simplicial set $d(BG)$ to measure the homotopy type of the bisimplicial set BG.

9.3 Groupoids Enriched in Simplicial Sets

In all that follows, we say that a commutative square

of bisimplicial set maps is homotopy cartesian if the associated diagram

$$\begin{array}{ccc} d(X) & \longrightarrow & d(Z) \\ \downarrow & & \downarrow \\ d(Y) & \longrightarrow & d(W) \end{array}$$

of diagonal simplicial sets is homotopy cartesian in the usual sense.

Suppose again that H is a groupoid enriched in simplicial sets. Write $\operatorname{Mor}_l(H)$ for the simplicial groupoid (which is *not* a groupoid enriched in simplicial sets) whose objects are the morphisms $a \to b$ of H and whose morphisms are the commutative diagrams

Then $\operatorname{Mor}_l(H) = \bigsqcup_{a \in \operatorname{Ob}(H)} a/H$, and there is a functor $\operatorname{Mor}_l(H) \to H$ that is defined by taking the diagram above to the morphism $b \to b'$.

There is a similarly defined simplicial groupoid $\operatorname{Mor}_r(H)$ whose objects are the morphisms of H again, and whose morphisms are the diagrams

There is again a canonical functor $\operatorname{Mor}_r(H) \to H$, and there is a decomposition

$$\operatorname{Mor}_r(H) = \bigsqcup_{c \in \operatorname{Ob}(H)} H/c.$$

Form the pullback diagram

$$
\begin{array}{ccc}
P(H) & \longrightarrow & \mathrm{Mor}_r(H) \\
\downarrow & & \downarrow \\
\mathrm{Mor}_l(H) & \longrightarrow & H
\end{array}
$$

in simplicial groupoids. It follows from Corollary 9.48 below that the induced diagram

$$
\begin{array}{ccc}
BP(H) & \longrightarrow & B\,\mathrm{Mor}_r(H) \\
\downarrow & & \downarrow \\
B\,\mathrm{Mor}_l(H) & \longrightarrow & BH
\end{array}
$$

is homotopy cartesian.

The simplicial groupoid $P(H)$ has objects (in the various simplicial degrees) consisting of strings $a \to b \to c$; its morphisms are the commutative diagrams

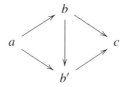

The components of P are contractible in each simplicial degree n, and there is an isomorphism

$$\pi_0 P(H)_n \xrightarrow{\cong} \mathrm{Mor}(H)_n$$

that respects the simplicial structure. There are also canonical equivalences

$$B\,\mathrm{Mor}_r(H) \xrightarrow{\simeq} \mathrm{Ob}(H) \text{ and } B\,\mathrm{Mor}_r(H) \xrightarrow{\simeq} \mathrm{Ob}(H).$$

We have proved the following:

Lemma 9.31 *There is a natural homotopy cartesian diagram*

$$
\begin{array}{ccc}
BP(H) & \longrightarrow & B\,\mathrm{Mor}_r(H) \\
\downarrow & & \downarrow \\
B\,\mathrm{Mor}_l(H) & \longrightarrow & BH
\end{array}
$$

9.3 Groupoids Enriched in Simplicial Sets

and natural diagonal equivalences $B\,\mathrm{Mor}_l\,(H) \xrightarrow{\simeq} \mathrm{Ob}\,(H)$, $B\,\mathrm{Mor}_r\,(H) \xrightarrow{\simeq} \mathrm{Ob}\,(H)$, and $BP(H) \xrightarrow{\simeq} \mathrm{Mor}\,(H)$ for groupoids H enriched in simplicial sets.

Lemma 9.31 has the following basic consequence:

Proposition 9.32 *A map $f : G \to H$ is a weak equivalence of $s_0\mathbf{Gpd}$ if and only if it induces a diagonal equivalence $BG \to BH$ of bisimplicial sets.*

Proof Suppose that the bisimplicial set map $BG \to BH$ is a diagonal equivalence. The map $\pi_0 G \to \pi_0 H$ is isomorphic to the map $\pi_0 BG \to \pi_0 BH$ and is therefore a bijection. Lemma 9.31 implies that the map $\mathrm{Mor}\,(G) \to \mathrm{Mor}\,(f^*H)$ is a weak equivalence, so that f is homotopically full and faithful. It follows that f is a weak equivalence of $s_0\mathbf{Gpd}$.

Conversely, suppose that $f : G \to H$ is a weak equivalence of groupoids enriched in simplicial sets. Then f is homotopically full and faithful, so that map $\mathrm{Mor}\,(G) \to \mathrm{Mor}\,(f^*H)$ is a weak equivalence of simplicial sets. The induced maps

$$\mathrm{Mor}\,(G) \times_{\mathrm{Ob}(G)} \mathrm{Mor}\,(G) \times_{\mathrm{Ob}(G)} \cdots \times_{\mathrm{Ob}(G)} \mathrm{Mor}\,(G)$$
$$\downarrow$$
$$\mathrm{Mor}\,(f^*H) \times_{\mathrm{Ob}(G)} \mathrm{Mor}\,(f^*H) \times_{\mathrm{Ob}(G)} \cdots \times_{\mathrm{Ob}(G)} \mathrm{Mor}\,(f^*H)$$

of iterated fibre products $\mathrm{Ob}\,(G)$ are therefore weak equivalences, since $\mathrm{Ob}\,(G)$ is a discrete simplicial set, so that the map $BG \to Bf^*H$ is a diagonal equivalence.

All groupoid maps $\beta_f : f^*H_n \to H_n$ are fully faithful, and are surjective in path components since f is surjective in path components. These groupoid maps are therefore weak equivalences. It follows that the map $Bf^*H \to BH$ is a diagonal weak equivalence, and so the map $BG \to BH$ is also a diagonal weak equivalence.

We shall also use the Eilenberg–Mac Lane $\overline{W}$ construction for simplicial groupoids. This functor is defined and its basic properties are displayed in [32], but the purely cocycle-theoretic definition of $\overline{W}$ that is given there can be rather awkward to manipulate. We begin here by presenting an alternative description of this functor that has evolved in recent years in the category theory literature [18, 94], and starts with a general construction for bisimplicial sets.

To make the notation of the following easier to deal with, observe that for every finite totally ordered poset P, there is a unique order-preserving isomorphism $\mathbf{n} \cong P$. We shall deemphasize the cardinality of such posets P notationally, by writing

$$\Delta^P = BP \cong B\mathbf{n} = \Delta^n,$$

and say that Δ^P is the *standard P-simplex*. We shall also say that a simplicial set map $\Delta^P \to X$ is a P-simplex of X.

This notation is particularly useful when discussing poset joins: if P and Q are finitely totally ordered posets, then the join $P * Q$ is the unique finite totally ordered poset such that every element of P is less than every element of Q. Thus, for example, a functor $\mathbf{p} * \mathbf{q} \to C$, taking values in a small category C, can be identified with a string of morphisms

$$a_0 \to a_1 \to \cdots \to a_p \to b_0 \to b_1 \to \cdots \to b_q$$

in the category C, and can be identified with a $\mathbf{p}*\mathbf{q}$-simplex $\Delta^{\mathbf{p}*\mathbf{q}} \to BC$ of the nerve BC. There is an isomorphism of totally ordered posets

$$\mathbf{p} * \mathbf{q} \cong \mathbf{p}+\mathbf{q}+\mathbf{1},$$

which we do not use explicitly.

We shall also steal some notation from the combinatorics literature, and write $[i, j] \subset P$ for the *interval* of elements s of a poset P such that $i \leq s \leq j$. This only makes sense, of course, when $i \leq j$ in P.

There is a functor

$$\text{Dec} : s\mathbf{Set} \to s^2\mathbf{Set}$$

that is defined for a simplicial set X by

$$\text{Dec}(X)_{p,q} = \hom(\Delta^{\mathbf{p}*\mathbf{q}}, X).$$

The functor Dec is the *total décalage functor* of Illusie [48].

Example 9.33 Suppose that C is a small category. Then $\text{Dec}(BC)$ is the bisimplicial set whose (p,q)-bisimplices are strings of arrows

$$a_0 \to a_1 \to \cdots \to a_p \to b_0 \to b_1 \to \cdots \to b_q.$$

This gadget has been used systematically for a long time since Quillen's paper [87]. It can be identified with the bisimplicial set underlying the homotopy colimit

$$\varinjlim_{j \in C} B(C/j),$$

which is weakly equivalent to BC.

Example 9.34 The (p,q)-bisimplices of $\text{Dec}(\Delta^n) = \text{Dec}(B\mathbf{n})$ are strings of relations

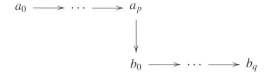

in $\mathbf{n}$ with $a_p \leq b_0$. Suppose that $r \leq s \leq n$, and write $F_{r,s} \subset \text{Dec}(\Delta^n)$ for the subcomplex of all such strings with $a_p \leq r$ and $s \leq b_0$. Then $F_{r,s}$ is generated by the string

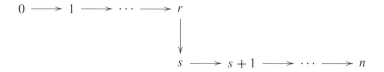

9.3 Groupoids Enriched in Simplicial Sets

that defines a bisimplex $\tau_{r,s} : \Delta^{[0,r],[s,n]} \to \mathrm{Dec}(\Delta^n)$, given by the map $[0,r] * [s,n] \to \mathbf{n}$ that is defined by the inclusion of the two intervals. The induced map

$$\tau_{r,s} : \Delta^{[0,r],[s,n]} \to F_{r,s}$$

is an isomorphism of bisimplicial sets.

Suppose that $r' \leq r \leq s \leq s'$. Then there is an identification

$$F_{r',s'} = F_{r,s'} \cap F_{r',s}.$$

It follows that for $r \leq s$, there is an identification

$$F_{r,r} \cap F_{s,s} = F_{r,s}.$$

Thus, there is a coequalizer

$$\bigsqcup_{r \leq s \leq n} \Delta^{[0,r],[s,n]} \rightrightarrows \bigsqcup_{r \leq n} \Delta^{[0,r],[r,n]} \xrightarrow{\tau} \mathrm{Dec}(\Delta^n) \qquad (9.11)$$

in bisimplicial sets.

If a system of bisimplicial set maps $f_r : F_{r,r} \to X$ coincides on all intersections

$$F_{r,r+1} = F_{r,r} \cap F_{r+1,r+1},$$

then these maps coincide on all intersections

$$F_{r,s} = F_{r,r} \cap F_{r+1,r+1} \cap \cdots \cap F_{s,s}.$$

It follows that (9.11) can be rewritten as the coequalizer

$$\bigsqcup_{r \leq n-1} \Delta^{[0,r],[r+1,n]} \rightrightarrows \bigsqcup_{r \leq n} \Delta^{[0,r],[r,n]} \xrightarrow{\tau} \mathrm{Dec}(\Delta^n).$$

A bisimplicial set map $f : \mathrm{Dec}(\Delta^n) \to X$ can therefore be identified with a system of bisimplices $x_{[0,r],[r,n]} \in X_{[0,r],[r,n]}$ such that

$$d_0^v(x_{[0,r],[r,n]}) = d_{r+1}^h(x_{[0,r+1],[r+1,n]})$$

for $r \leq n - 1$. Observe that the vertical face d_0^v is induced by the inclusion of intervals $[r+1, n] \subset [r, n]$ and the horizontal face d_{r+1}^h is induced by the inclusion of $[0, r] \subset [0, r+1]$.

The *total simplicial set* $T(X)$ that is associated to a bisimplicial set X is defined to have n-simplices given by the maps

$$\mathrm{Dec}(\Delta^n) \to X.$$

The resulting total simplicial set functor

$$T : s^2\mathbf{Set} \to s\mathbf{Set}$$

is right adjoint to the total décalage functor Dec. The total simplicial set construction was introduced by Artin and Mazur [3]—see also [18] and [94].

Example 9.35 Suppose that C is a simplicial category, and let BC be the bisimplicial set consisting of the nerves BC_q of the various categories C_q. We identify $BC_{p,q} = (BC_q)_p$ with strings of arrows of length p in C_q.

A bisimplicial set map $\alpha : \mathrm{Dec}(\Delta^n) \to BC$ consists of strings

$$\alpha_{[0,r],[r,n]} : a_{0,[r,n]} \to \cdots \to a_{r,[r,n]}$$

in $C_{[r,n]}$ such that

$$d_0^v a_{0,[r,n]} \to \cdots \to d_0^v a_{r,[r,n]} = a_{0,[r+1,n]} \to \cdots \to a_{r,[r+1,n]}.$$

It follows that α consists of morphisms

$$d_0^v a_{r,[r,n]} \to a_{r+1,[r+1,n]} \text{ in } C_{[r+1,n]}, \ 0 \le r \le n-1.$$

This set of morphisms forms a cocycle in the sense of [32, Sect. V.7], albeit written backward.

Equivalently, from (9.11), a map $\sigma : \mathrm{Dec}(\Delta^n) \to BC$ consists of a collection of functors $\sigma_r : [0,r] \to C_{[r,n]}, 0 \le r \le n$ such that if $s \le r \le n$, then the diagram of functors

$$\begin{array}{ccc} [0,r] & \xrightarrow{\sigma_r} & C_{[r,n]} \\ {\scriptstyle i}\uparrow & & \uparrow{\scriptstyle j^*} \\ [0,s] & \xrightarrow{\sigma_s} & C_{[s,n]} \end{array} \qquad (9.12)$$

commutes, where $i : [0,s] \subset [0,r]$ and $j : [r,n] \subset [s,n]$ are the respective segment inclusions.

If $\sigma : \mathrm{Dec}(\Delta^n) \to BC$ is a bisimplical set map and $\theta : \mathbf{m} \to \mathbf{n}$ is an ordinal number map, then $\theta^*(\sigma)$ is defined by the set of composite functors

$$[0,s] \xrightarrow{\theta_*} [0,\theta(s)] \xrightarrow{\sigma_{\theta(s)}} C_{[\theta(s),n]} \xrightarrow{\theta^*} C_{[s,m]},$$

where θ induces ordinal number maps $\theta : [0,s] \to [0,\theta(s)]$ and $\theta : [s,m] \to [\theta(s),n]$.

Example 9.36 Suppose that A is a small category, and identify A with a simplicial category that is simplicially discrete. Then $A_r = A$ for all r, and a functor $\sigma : [0,n] \to A$ restricts to functors $\sigma_r : [0,r] \subset A$ that are defined by the composites

$$[0,r] \subset [0,n] \xrightarrow{\sigma} A,$$

and the assignment $\sigma \mapsto (\sigma_r)$ defines a natural map

$$\psi : BA \to \overline{W}A.$$

9.3 Groupoids Enriched in Simplicial Sets

Every such family (σ_r) is completely determined by the functor $\sigma_n : [0, n] \to A$, and it follows that the natural map ψ is an isomorphism.

Each $(m + 1)$-simplex

$$\sigma : a_0 \to a_1 \to a_2 \to \cdots \to a_{m+1}$$

of Δ^n determines an ordinal number map

$$\sigma_* : [0, 1] * [1, m + 1] \to \mathbf{n},$$

which is defined by the string of morphisms

$$\begin{array}{ccc} a_0 & \longrightarrow & a_1 \\ & & \downarrow \\ a_1 & \longrightarrow \cdots \longrightarrow & a_{m+1} \end{array}$$

in $\mathbf{n}$. The map σ_* is the map $a_0 \to a_1$ on $[0, 1]$ and is $d_0\sigma$ on $[1, m + 1]$, and defines an element of $\mathrm{Dec}\,(\Delta^n)_{1,m}$.

Suppose that H is a groupoid enriched in simplicial sets. A bisimplicial set map $f : \mathrm{Dec}\,(\Delta^n) \to BH$ is specified by the images $f(\sigma_*)$ of the bisimplices σ_*. If $\sigma = s_0(\tau)$ for an m-simplex τ, then the image $f(\sigma_*)$ is an identity of the groupoid $H_m = H_{[1,m+1]}$.

Suppose that $\theta : [1, k + 1] \to [1, m + 1]$ is an ordinal number map. Then the $(2, k)$-bisimplex

$$\begin{array}{ccccc} (\sigma, \theta) : a_0 & \longrightarrow & a_1 & \longrightarrow & a_{\theta(1)} \\ & & & & \downarrow \\ & & a_{\theta(1)} & \longrightarrow \cdots \longrightarrow & a_{\theta(k+1)} \end{array}$$

defines a 2-simplex of $BH_{[1,k+1]}$. It follows that there is a commutative diagram

$$\begin{array}{ccc} & f(d_1(\sigma,\theta)_*) & \\ f(a_0) & \longrightarrow & f(a_{\theta(1)}) \\ \theta^*(f(\sigma_*)) \downarrow & \nearrow f(d_0(\sigma,\theta)_*) & \\ f(a_1) & & \end{array}$$

in the groupoid $H_{[1,k+1]}$, and hence a relation

$$\theta^*(f(\sigma_*)) = f(d_0(\sigma, \theta)_*)^{-1} f(d_1(\sigma, \theta)_*)$$

in $H_{[1,k+1]} = H_k$.

Write $G(\Delta^n)_m$ for the groupoid enriched in simplicial sets which is the free groupoid on the graph of $(m+1)$-simplices $\sigma : \sigma(0) \to \sigma(1)$ of Δ^n, modulo the relation that $s_0\tau$ is the identity on $\tau(0)$. If $\theta : [1, k+1] \to [1, m+1]$ is an ordinal number map, we specify

$$\theta^*(\sigma) = d_0(\sigma, \theta)^{-1} d_1(\sigma, \theta),$$

where (σ, θ) is the simplex

$$a_0 \to a_1 \to a_{\theta(1)} \to \cdots \to a_{\theta(k+1)}.$$

Then $G(\Delta^n)$ is a groupoid enriched in simplicial sets, and there is an isomorphism

$$\hom(G(\Delta^n), H) \cong \hom(\Delta^n, \overline{W}(H))$$

for all objects H of $s_0\mathbf{Gpd}$.

The definition of the simplicial groupoids $G(\Delta^n)$ extends to a functor

$$G : s\mathbf{Set} \to s_0\mathbf{Gpd}.$$

Following [32], we say that $G(X)$ is the *loop groupoid* of the simplicial set X. This functor is left adjoint to the *Eilenberg–Mac Lane functor*

$$\overline{W} : s_0\mathbf{Gpd} \to s\mathbf{Set}$$

—see [32, V.7.7].

In view of the adjointness, we can define the Eilenberg–Mac Lane functor for a groupoid H enriched in simplicial sets by specifying

$$\overline{W}(H)_n = \hom(G(\Delta^n), H).$$

By the discussion above, there is an isomorphism

$$\hom(G(\Delta^n), H) \cong \hom(\operatorname{Dec}(\Delta^n), BH),$$

and one can show that this map respects the simplicial structure. We therefore have the following:

Proposition 9.37 *There is an isomorphism of simplicial sets*

$$\overline{W}(H) \cong T(BH),$$

which is natural in groupoids H enriched in simplicial sets.

There is a bisimplicial set map

$$\phi : \operatorname{Dec}(\Delta^n) \to \Delta^{n,n}$$

that takes a morphism $(\theta, \tau) : \mathbf{r} * \mathbf{s} \to \mathbf{n}$ to the pair of maps $(\theta, \tau) : (\mathbf{r}, \mathbf{s}) \to (\mathbf{n}, \mathbf{n})$. This map is natural in ordinal numbers $\mathbf{n}$, and hence defines a morphism of

9.3 Groupoids Enriched in Simplicial Sets

cosimplicial bisimplicial sets. Precomposing with the map ϕ determines a natural simplicial set map

$$\Phi = \phi^* : d(X) \to T(X),$$

which is natural in bisimplicial sets X.

We shall prove the following:

Proposition 9.38 *The natural map*

$$\Phi : d(X) \to T(X)$$

is a weak equivalence, for all bisimplicial sets X.

Corollary 9.39

(1) The natural map

$$\Phi : d(BH) \to T(BH) \cong \overline{W}(H)$$

is a weak equivalence for all groupoids H enriched in simplicial sets.

(2) The functor $H \mapsto \overline{W}H$ preserves and reflects weak equivalences of groupoids H enriched in simplicial sets.

Remark 9.40 The definition

$$\overline{W}(C) := T(BC)$$

extends the definition of the Eilenberg–Mac Lane functor $\overline{W}$ to simplicial categories C. Statement (1) of Corollary 9.39 is a special case of the assertion that the map

$$\Phi : d(BC) \to \overline{W}(C) = T(BC)$$

is a weak equivalence for all simplicial categories C.

Proposition 9.38 is a main result of [94], and is called the "generalized Eilenberg–Zilber theorem" in that paper. We give the same proof here, in somewhat different language.

Proof Proof of Proposition 9.38. The diagonal functor has a right adjoint

$$d_* : s\mathbf{Set} \to s^2\mathbf{Set},$$

where $d_*(X)$ is the bisimplicial set with

$$d_*(X)_{p,q} = \hom(\Delta^p \times \Delta^q, X).$$

The adjoint $\Phi_* : X \to d_*T(X)$ of the map $\Phi : d(X) \to T(X)$ takes a bisimplex $\Delta^{p,q} \to X$ to a simplicial set map $\Delta^p \times \Delta^q \to T(X)$, or equivalently a map

$$\mathrm{Dec}(\Delta^p) \times \mathrm{Dec}(\Delta^q) \cong \mathrm{Dec}(\Delta^p \times \Delta^q) \to X.$$

This assignment is defined by precomposition with the map

$$\zeta : \mathrm{Dec}\,(\Delta^p) \times \mathrm{Dec}\,(\Delta^q) \to \Delta^{p,q},$$

which takes a pair $(\alpha_1, \alpha_2) : \mathbf{r} * \mathbf{s} \to \mathbf{p}$, $(\beta_1, \beta_2) : \mathbf{r} * \mathbf{s} \to \mathbf{q}$ of (r,s)-bisimplices to the bisimplex $\alpha_1 \times \beta_2 : \mathbf{r} \times \mathbf{s} \to \mathbf{p} \times \mathbf{q}$. The bisimplex $\Delta^{p,q}$ has the form

$$\Delta^{p,q} = p_1^* \Delta^p \times p_2^* \Delta^q,$$

where $p_1^* \Delta^p_{r,s} = \Delta^p_r$ and $p_2^* \Delta^q_{r,s} = \Delta^q_s$. Furthermore, the map ζ has the form

$$\zeta = \zeta_1 \times \zeta_2 : \mathrm{Dec}\,(\Delta^p) \times \mathrm{Dec}\,(\Delta^q) \to p_1^* \Delta^p \times p_2^* \Delta^q,$$

where ζ_1 takes the bisimplex $(\alpha, \beta) : \mathbf{r} * \mathbf{s} \to \mathbf{p}$ to the simplex $\alpha : \mathbf{r} \to \mathbf{p}$ and ζ_2 takes the bisimplex $(\omega, \gamma) : \mathbf{r} * \mathbf{s} \to \mathbf{q}$ to the simplex $\gamma : \mathbf{s} \to \mathbf{q}$.

We shall prove that the map $\Phi_* : X \to d_* T(X)$ induces a weak equivalence $d(\Phi_*) : d(X) \to d(d_* T X)$ of simplicial sets. This suffices to prove the proposition since there is a commutative diagram

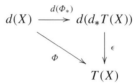

and the map ϵ is a weak equivalence by Lemma 9.49 below.

For each bisimplicial set U, there is a simplicial set $\mathbf{hom}_{Dec}(U, X)$ whose n-simplices are bisimplicial set maps $U \times \mathrm{Dec}\,(\Delta^n) \to X$. The functor Dec has a right adjoint, namely T, and it follows that there is a natural bijection

$$\mathrm{hom}\,(K, \mathbf{hom}_{Dec}(U, X)) \cong \mathrm{hom}\,(U \times \mathrm{Dec}\,(K), X).$$

In particular, every map $K \times \Delta^1 \to \mathbf{hom}_{Dec}(U, X)$ can be identified with a map

$$U \times \mathrm{Dec}\,(K) \times \mathrm{Dec}\,(\Delta^1) \to X.$$

The bisimplicial set map $\zeta_1 : \mathrm{Dec}\,(\Delta^p) \to p_1^* \Delta^p$ sends the bisimplex $(\phi, \gamma) : \mathbf{r} * \mathbf{s} \to \mathbf{p}$ to the map $\phi : \mathbf{r} \to \mathbf{p}$. This map has a section $\sigma_1 : p_1^*(\Delta^p) \to \mathrm{Dec}\,(\Delta^p)$ which sends the map $\phi : \mathbf{r} \to \mathbf{p}$ to the map $(\phi, p) : \mathbf{r} * \mathbf{s} \to \mathbf{p}$ which is constant at the vertex p on the join factor $\mathbf{s}$. The diagrams

$$\begin{array}{ccccccccc}
\phi(0) & \to & \cdots & \to & \phi(r) & \to & \psi(0) & \to & \cdots & \to & \psi(s) \\
\downarrow & & & & \downarrow & & \downarrow & & & & \downarrow \\
\phi(0) & \to & \cdots & \to & \phi(r) & \to & p & \to & \cdots & \to & p
\end{array}$$

9.3 Groupoids Enriched in Simplicial Sets

define a map $\mathrm{Dec}\,(\Delta^p) \to \mathrm{Dec}\,(\mathbf{hom}(\Delta^1, \Delta^p))$, which induces a map

$$h : \mathrm{Dec}\,(\Delta^p) \times \mathrm{Dec}\,(\Delta^1) \to \mathrm{Dec}\,(\Delta^p).$$

The map h determines a homotopy

$$\mathbf{hom}_{Dec}(\,\mathrm{Dec}\,(\Delta^p), X) \times \Delta^1 \to \mathbf{hom}_{Dec}(\,\mathrm{Dec}\,(\Delta^p), X)$$

from the identity to the map defined by composition with $\sigma_1 \cdot \zeta_1$. It follows that the map

$$\zeta_1^* : \mathbf{hom}_{Dec}(\,\mathrm{Dec}\,(\Delta^p), X) \to \mathbf{hom}_{Dec}(p_1^*\Delta^p, X)$$

is a weak equivalence.

This is true for all p, and it follows that the bisimplicial set map

$$(\zeta_1 \times 1)^* : \mathrm{hom}\,(p_1^*(\Delta) \times \mathrm{Dec}\,(\Delta), X) \to \mathrm{hom}\,(\,\mathrm{Dec}\,(\Delta) \times \mathrm{Dec}\,(\Delta), X)$$

is a weak equivalence. The assertion that the map ζ_2 induces a diagonal weak equivalence

$$(1 \times \zeta_2)^* : \mathrm{hom}\,(p_1^*(\Delta) \times p_2^*(\Delta), X) \to \mathrm{hom}\,(p_1^*(\Delta) \times \mathrm{Dec}\,(\Delta), X)$$

has an essentially analogous proof.

If X is a *reduced simplicial set* in the sense that it has only one vertex, then the object $G(X)$ is necessarily a simplicial group.

The Dwyer–Kan model structure on the category $s_0\mathbf{Gpd}$ of groupoids enriched in simplicial sets is an extension of a model structure on the category $s\mathbf{Gr}$ of simplicial groups [32, V.5.2], for which a map $f : G \to H$ of simplicial groups is a *weak equivalence* (respectively *fibration*) if the underlying map of simplicial sets is a weak equivalence (respectively Kan fibration). All maps $G(A) \to G(B)$ that are induced by cofibrations $A \to B$ of reduced simplicial sets are cofibrations for this model structure on simplicial groups [32, V.6.1].

Suppose that X is a reduced simplicial set. It is a fundamental classical result [32, V.5.10, 94, Theorem 21] that the space X_η in the pullback diagram

$$\begin{array}{ccc} X_\eta & \longrightarrow & WG(X) \\ \downarrow & & \downarrow \pi \\ X & \xrightarrow{\eta} & \overline{W}G(X) \end{array}$$

is contractible, where π is the universal principal $G(X)$-bundle over $\overline{W}G(X)$ in simplicial sets. The simplicial set X is reduced, and the space $\overline{W}G(X)$ is connected. It follows that the adjunction map $\eta : X \to \overline{W}G(X)$ is a weak equivalence. This assertion is in fact equivalent to the claim that the space X_η is contractible.

The space $WG(X)$ is contractible, and it follows that there is a natural weak equivalence $G(X) \simeq \Omega(X)$, where $\Omega(X)$ is the derived loop space of X. It also follows that the loop group functor $X \mapsto G(X)$ preserves weak equivalences of reduced simplicial sets.

The *universal principal G-bundle* $\pi : WG \to \overline{W}G$ for a simplicial group G can be constructed in a modern way, by setting

$$W(G) = \overline{W}(G/*).$$

The simplicial groupoid $G/*$ is formed by taking the slice groupoid $G_n/*$ in each simplicial degree n. It is a consequence of Proposition 9.38 that the space WG is contractible. The map π is a Kan fibration since it is a quotient of principal G-space by the G-action [32, V.2.7].

We can now prove the following:

Lemma 9.41 *Suppose that the simplicial group homomorphism $p : G \to H$ is a fibration (respectively trivial fibration) for the Dwyer–Kan model structure on $s_0\mathbf{Gpd}$. Then the induced map $p_* : \overline{W}G \to \overline{W}H$ is a Kan fibration (respectively trivial Kan fibration) of simplicial sets.*

Proof The loop group functor preserves cofibrations and weak equivalences of reduced simplicial sets. The map $p_* : \overline{W}G \to \overline{W}H$ is therefore a fibration of reduced simplicial sets. We can conclude that p_* is a Kan fibration if we can show that the group homomorphism $p : G_0 \to H_0$ is surjective, by [32, V.6.9], but this is so since $p : G_0 \to H_0$ is a fibration of groupoids and therefore has the path lifting property.

If p is a trivial fibration, then $p_* : \overline{W}G \to \overline{W}H$ is a trivial fibration of reduced simplicial sets, and is therefore weak equivalence.

All groupoids enriched in simplicial sets are fibrant for the Dwyer–Kan model structure. It therefore follows from Lemma 9.41 that the simplicial set $\overline{W}G$ is a Kan complex for each simplicial group G.

Lemma 9.42 *The functor $\overline{W} : s_0\mathbf{Gpd} \to s\mathbf{Set}$ preserves fibrations and weak equivalences.*

Proof This result appears in [32, V.7.8]. We give a somewhat different proof of the fibration statement here.

Suppose that $f : G \to H$ is a weak equivalence of $s_0\mathbf{Gpd}$. Choose a representative x for each element $[x] \in \pi_0(G) \cong \pi_0 H$. Choose paths $a \to x$ in G for each $a \in [x]$, and then choose paths $b \to f(x)$ in H for each object $b \in H$ that is outside the image of f. Do this for all path components $[x]$. Then these paths, taken together, define a simplicial homotopy equivalence between the map f and a map

$$\bigsqcup_{[x] \in \pi_0 G} G(x,x) \to \bigsqcup_{[x] \in \pi_0 G} H(f(x), f(x)).$$

The functor $\overline{W}$ takes homotopy equivalences defined by paths to homotopy equivalences, and preserves disjoint unions. All maps $\overline{W}G(x,x) \to \overline{W}H(f(x), f(x))$

9.3 Groupoids Enriched in Simplicial Sets

are weak equivalences since $\overline{W}$ preserves weak equivalences of simplicial groups. It follows that $\overline{W}$ preserves weak equivalences of groupoids enriched in simplicial sets.

Suppose that $\sigma : \text{Dec}(\Delta^n) \to BH$ is a simplex of $\overline{W}(H) = T(BH)$ in the sense of (9.12) with image vertices x_i, and suppose that there is a path $\omega_i : x_i \to y_i$ in H_0 for each i. The paths ω_i define homotopies

$$h_r : [0, r] \times \mathbf{1} \to G_{[r,n]}$$

that start at various $\sigma_r : [0, r] \to G_{[r,n]}$, and such that the diagrams

$$\begin{array}{ccc} [0, r] \times \mathbf{1} & \xrightarrow{h_r} & G_{[r,n]} \\ {\scriptstyle i \times 1} \uparrow & & \uparrow {\scriptstyle j^*} \\ [0, s] \times \mathbf{1} & \xrightarrow{h_s} & G_{[s,n]} \end{array}$$

commute. It follows that the composites

$$[0, r] \xrightarrow{1 \times 1} [0, r] \times \mathbf{1} \xrightarrow{h_r} G_{[r,n]}$$

form a simplex $\text{Dec}(\Delta^n) \to BG$, which (following [32]) is said to be *cocycle conjugate* to σ.

Suppose that the map $p : G \to H$ is a fibration for the Dwyer–Kan model structure on the category $s_0\mathbf{Gpd}$, and consider the lifting problem

$$\begin{array}{ccc} \Lambda^n_k & \longrightarrow & \overline{W}G \\ \downarrow & \nearrow & \downarrow {\scriptstyle p_*} \\ \Delta^n & \longrightarrow & \overline{W}H \end{array} \qquad (9.13)$$

We can assume that the objects G and H are path connected. Pick an object $x \in G$. Then the map $G(x, x) \to H(p(x), p(x))$ is a strong deformation retract of p, and is therefore a fibration of $s_0\mathbf{Gpd}$. The lifting problem (9.13) is cocycle conjugate to a lifting problem

$$\begin{array}{ccc} \Lambda^n_k & \longrightarrow & \overline{W}G(x, x) \\ \downarrow & \nearrow & \downarrow \\ \Delta^n & \longrightarrow & \overline{W}H(p(x), p(x)) \end{array} \qquad (9.14)$$

which lifting problem has a solution by Lemma 9.41. The lifting problems (9.13) and (9.14) are equivalent, so that the lifting problem (9.13) has a solution.

It follows that the simplicial set map $p_* : \overline{W}G \to \overline{W}H$ is a Kan fibration.

Lemma 9.42 and Corollary 9.39 lead to a model structure on the category $s_0\mathbf{Gpd}$ of groupoids enriched in simplicial sets, called the $\overline{W}$-*model structure*, for which the weak equivalences $f : G \to H$ are those maps that induce weak equivalences $\overline{W}G \to \overline{W}H$ (or are weak equivalences for the Dwyer–Kan structure, or induce weak equivalences $d(BG) \to d(BH)$), and a $\overline{W}$-*fibration* is a map $p : G \to H$ such that the induced map $\overline{W}G \to \overline{W}H$ is a Kan fibration.

Theorem 9.43

(1) With the definitions of weak equivalence and $\overline{W}$-fibration given above, the category $s_0\mathbf{Gpd}$ of groupoids enriched in simplicial sets has the structure of a right proper, cofibrantly generated, closed model category.

(2) The adjunction

$$G : s\mathbf{Set} \leftrightarrows s_0\mathbf{Gpd} : \overline{W}$$

defines a Quillen equivalence between the $\overline{W}$-model structure on $s_0\mathbf{Gpd}$ and the standard model structure on the category of simplicial sets.

Proof A map $p : G \to H$ of $s_0\mathbf{Gpd}$ is a $\overline{W}$-fibration (respectively trivial $\overline{W}$-fibration) if and only if it has the right lifting property with respect to all maps $G(\Lambda_k^n) \to G(\Delta^n)$ (respectively all maps $G(\partial \Delta^n) \to G(\Delta^n)$).

If $i : A \to B$ is a cofibration of simplicial sets, then $i_* : G(A) \to G(B)$ is a cofibration for the Dwyer–Kan structure, on account of Lemma 9.42. If the map i is a trivial cofibration and the diagram

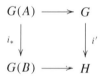

is a pushout in $s_0\mathbf{Gpd}$, then i' has the left lifting property with respect to all Dwyer–Kan fibrations, and is therefore a weak equivalence. One finishes the proof of statement (1) with standard small object arguments. The right properness is a consequence of the fact that the functor $\overline{W}$ preserves pullbacks, along with right properness for the standard model structure on simplicial sets.

The functor G preserves weak equivalences since it preserves trivial cofibrations. The functor $\overline{W}$ preserves weak equivalences by definition.

As noted above, it is a classical fact that the adjunction map $\eta : Y \to \overline{W}G(Y)$ is a weak equivalence for reduced simplicial sets Y [32, V.5.10]. The functor $X \mapsto \overline{W}G(X)$ preserves weak equivalences of simplicial sets Y, by Lemma 9.42. Every simplicial set X is weakly equivalent to a disjoint union of reduced simplicial sets, so it follows that the adjunction map $\eta : X \to \overline{W}G(X)$ is a weak equivalence for

9.3 Groupoids Enriched in Simplicial Sets

all simplicial sets X. A triangle identity argument then shows that the natural map $\epsilon : G\overline{W}(H) \to H$ is a weak equivalence for all H in $s_0\mathbf{Gpd}$.

Suppose that H is a groupoid enriched in simplicial sets. The *path component groupoid* $\pi_0(H)$ has the same objects as H, and has morphism sets specified by

$$\pi_0(H)(x, y) = \pi_0(H(x, y))$$

for all objects x, y of H. There is a canonical map $H \to \pi_0(H)$, and this map is initial among all morphisms $H \to K$ of $s_0\mathbf{Gpd}$ such that K is a groupoid.

Lemma 9.44 *The functor $H \mapsto \pi_0(H)$ preserves weak equivalences of groupoids enriched in simplicial sets.*

Proof Suppose that $f : G \to H$ is a weak equivalence of $s_0\mathbf{Gpd}$. Then all simplicial set maps $G(x, y) \to H(f(x), f(y))$ are weak equivalences, so the induced functions

$$\pi_0 G(x, y) \to \pi_0 H(f(x), f(y))$$

are bijections. The natural groupoid map $G_0 \to \pi_0(G)$ induces an isomorphism $\pi_0(G_0) \cong \pi_0(\pi_0(G))$. It follows that the induced function $\pi_0(\pi_0(G)) \to \pi_0(\pi_0(H))$ of sets of path components of path component groupoids is a bijection.

Corollary 9.45 *There is a natural weak equivalence*

$$\pi(\overline{W}H)) \simeq \pi_0(H)$$

relating the fundamental groupoid $\pi(\overline{W}H)$ of the space $\overline{W}H$ to the path component groupoid $\pi_0(H)$ of H, for all groupoids H enriched in simplicial sets.

Proof The natural map $\epsilon : G(\overline{W}H) \to H$ is a weak equivalence of $s_0\mathbf{Gpd}$, by the proof of Theorem 9.43. The map ϵ therefore induces weak equivalence

$$\epsilon_* : \pi_0 G(\overline{W}H) \to \pi_0 H$$

by Lemma 9.44.

There are natural isomorphisms

$$\hom(\pi_0 GX, L) \cong \hom(GX, L) \cong \hom(X, \overline{W}L) \cong \hom(X, BL) \cong \hom(\pi X, L)$$

by adjointness (see also Example 9.36), for all spaces X and groupoids L. It follows that there is a natural isomorphism

$$\pi_0 GX \cong \pi X,$$

which is natural in simplicial sets X. The composite map

$$\pi(\overline{W}H)) \cong \pi_0 G(\overline{W}H) \xrightarrow{\epsilon_*} \pi_0(H)$$

is the desired equivalence.

We close this section by proving a few technical results. The first of these is the enriched version of the central technical lemma behind Quillen's Theorem B. The standard form of this result appears, for example, as Lemma IV.5.7 of [32]. The following result is due to Moerdijk, and first appeared in [80].

Lemma 9.46 *Suppose that C is a small category enriched in simplicial sets and that $X : C \to s\mathbf{Set}$ is an enriched functor such that all morphisms $a \to b$ of C_0 induce weak equivalences $X(a) \to X(b)$. Then the pullback diagram*

$$\begin{array}{ccc} X(a) & \longrightarrow & \underrightarrow{\operatorname{holim}}_C X \\ \downarrow & & \downarrow \\ * & \xrightarrow{a} & BC \end{array} \qquad (9.15)$$

is homotopy cartesian.

Proof We show that the diagram (9.15) induces a homotopy cartesian diagram of simplicial sets on application of the diagonal functor.

The bisimplicial set BC is a homotopy colimit of its bisimplices, and it suffices to show [32, IV.5] that every map

$$\begin{array}{ccc} \Delta^{r,s} & \xrightarrow{\sigma} & BC \\ \theta \downarrow & \nearrow \tau & \\ \Delta^{k,l} & & \end{array}$$

induces a (diagonal) weak equivalence

$$\sigma^{-1} \underrightarrow{\operatorname{holim}} X \xrightarrow{\theta_*} \tau^{-1} \underrightarrow{\operatorname{holim}} X$$

of pullbacks over the respective bisimplices.

Every bisimplex $\sigma : \Delta^{k,l} \to BC$ is determined by a string of arrows

$$\sigma : a_0 \xrightarrow{\alpha_1} a_1 \xrightarrow{\alpha_2} a_2 \cdots \xrightarrow{\alpha_k} a_k$$

of length k in C_l, where C_l is the category in simplicial degree l in the simplicial category C. In *horizontal degree n*, this bisimplex determines a simplicial set map

$$\bigsqcup_{\gamma:\mathbf{n}\to\mathbf{k}} \Delta^l \to BC_n = \bigsqcup_{c_0,c_1,\ldots,c_n} C(c_0,c_1) \times \cdots \times C(c_{n-1},c_n).$$

On the summand corresponding to $\gamma : \mathbf{n} \to \mathbf{k}$, this map resticts to the composite

$$\gamma^*(\sigma) : \Delta^l \to C(a_{\gamma(0)}, a_{\gamma(1)}) \times \cdots \times C(a_{\gamma(n-1)}, a_{\gamma(n)}) \to BC_n.$$

9.3 Groupoids Enriched in Simplicial Sets

The simplicial set ($\varinjlim \operatorname{holim} X)_n$ in horizontal degree n has the form

$$(\varinjlim \operatorname{holim} X)_n = \bigsqcup_{c_0, c_1, \ldots, c_n} X(c_0) \times C(c_0, c_1) \times \cdots \times C(c_{n-1}, c_n).$$

It follows that (in horizontal degree n) there is an identification

$$(\sigma^{-1} \varinjlim \operatorname{holim} X)_n = \bigsqcup_{\gamma: \mathbf{n} \to \mathbf{k}} X(a_{\gamma(0)}) \times \Delta^l.$$

The map $(1, \theta) : \Delta^{k,r} \to \Delta^{k,l}$ induces the simplicial set map

$$\bigsqcup_{\gamma: \mathbf{n} \to \mathbf{k}} X(a_{\gamma(0)}) \times \Delta^r \to \bigsqcup_{\gamma: \mathbf{n} \to \mathbf{k}} X(a_{\gamma(0)}) \times \Delta^l$$

in horizontal degree n which is specified on summands by

$$1 \times \theta : X(a_{\gamma(0)}) \times \Delta^r \to X(a_{\gamma(0)}) \times \Delta^l.$$

This map is a weak equivalence, and so there is a diagonal weak equivalence

$$(\sigma(1 \times \theta))^{-1} \varinjlim \operatorname{holim} X \to \sigma^{-1} \varinjlim \operatorname{holim} X.$$

In particular, any vertex $\Delta^0 \to \Delta^l$ determines a weak equivalence

$$\bigsqcup_{\gamma: \mathbf{n} \to \mathbf{k}} X(a_{\gamma(0)}) \to \bigsqcup_{\gamma: \mathbf{n} \to \mathbf{k}} X(a_{\gamma(0)}) \times \Delta^l.$$

A bisimplicial set map $(\theta, \gamma) : \Delta^{r,s} \to \Delta^{k,l}$ and any choice of vertex $v : \Delta^0 \to \Delta^s$ together induce a commutative diagram of bisimplicial set maps

$$\begin{array}{ccc} \Delta^{r,0} & \xrightarrow{(1,v)} & \Delta^{r,s} \\ {\scriptstyle (\theta,1)} \downarrow & & \downarrow {\scriptstyle (\theta,\gamma)} \\ \Delta^{k,0} & \xrightarrow{(1,\gamma(v))} & \Delta^{k,l} \end{array}$$

It therefore suffices to assume that all diagrams of bisimplices

$$\begin{array}{ccc} \Delta^{r,0} & \xrightarrow{\sigma} & BC \\ {\scriptstyle (\theta,1)} \downarrow & \nearrow {\scriptstyle \tau} & \\ \Delta^{k,0} & & \end{array}$$

induce diagonal weak equivalences

$$\sigma^{-1} \varinjlim \operatorname{holim} X \xrightarrow{\theta_*} \tau^{-1} \varinjlim \operatorname{holim} X.$$

300   9 Non-abelian Cohomology

The simplicial functor $X : C \to \mathbf{S}$ restricts to an ordinary functor $X_0 : C_0 \to \mathbf{S}$ via the identification of the category C_0 with a discrete simplicial subobject of the simplicial category C. There is a pullback diagram

$$\begin{array}{ccc} \underrightarrow{\mathrm{holim}}\, X_0 & \longrightarrow & \underrightarrow{\mathrm{holim}}\, X \\ \downarrow & & \downarrow \\ BC_0 & \longrightarrow & BC \end{array}$$

and all bisimplices $\Delta^{k,0} \to BC$ factor through the inclusion $BC_0 \to BC$. Each morphism $a \to b$ of C_0 induces a weak equivalence

$$X_0(a) = X(a) \xrightarrow{\simeq} X(b) = X_0(b)$$

by assumption. It therefore follows from the standard argument for ordinary functors taking values in simplicial sets that all induced maps

$$\sigma^{-1}\, \underrightarrow{\mathrm{holim}}\, X \xrightarrow{\theta_*} \tau^{-1}\, \underrightarrow{\mathrm{holim}}\, X$$

are weak equivalences of simplicial sets.

In the setting of Lemma 9.46, write

$$X = \bigsqcup_{a \in \mathrm{Ob}(C)} X(a)$$

and let $\pi : X \to \mathrm{Ob}(C)$ be the canonical map. Then Lemma 9.46 implies the following:

Corollary 9.47 *Suppose that C is a category enriched in simplicial sets and that $X : C \to s\mathbf{Set}$ is a functor such that all arrows $a \to b$ of C_0 induce weak equivalences $X(a) \to X(b)$. Then the pullback diagram*

$$\begin{array}{ccc} X & \longrightarrow & \underrightarrow{\mathrm{holim}}_C\, X \\ \downarrow & & \downarrow \\ \mathrm{Ob}(C) & \longrightarrow & BC \end{array}$$

is homotopy cartesian.

Corollary 9.48 *Suppose that $f : G \to H$ is a morphism of groupoids enriched in simplicial sets. Then the pullback diagram*

$$\begin{array}{ccc} B(f/x) & \longrightarrow & BG \\ \downarrow & & \downarrow f_* \\ B(H/x) & \longrightarrow & BH \end{array}$$

9.3 Groupoids Enriched in Simplicial Sets

is homotopy cartesian for each objects x of H.

Proof This result is a consequence of Quillen's diagram [87, p.98, 32, (5.12)]GJ

$$
\begin{array}{ccccc}
B(f/x) & \longrightarrow & \underrightarrow{\text{holim}}_H\, B(f/x) & \xrightarrow{\simeq} & BG \\
\downarrow & & \downarrow & & \downarrow f_* \\
B(H/x) & \longrightarrow & \underrightarrow{\text{holim}}_H\, B(H/x) & \xrightarrow{\simeq} & BH \\
\simeq \downarrow & & \downarrow \simeq & & \\
* & \xrightarrow{\ x\ } & BH & &
\end{array}
$$

together with Lemma 9.46. The conditions of Lemma 9.46 for the enriched diagram $x \mapsto B(f/x)$ are automatically satisfied, since H is a groupoid enriched in simplicial sets.

Write $(\Delta \times \Delta)/X$ for the category of maps $\Delta^p \times \Delta^q \to X$, for a simplicial set X. The category Δ/X is the simplex category for X.

We use Quillen's Theorem A [87] in the proof of the following result. Theorem A says that if $f : C \to D$ is a functor between small categories such that either all spaces $B(f/a)$ or all spaces $B(a/f)$ are contractible, then the induced map $f_* : BC \to BD$ of nerves is a weak equivalence. Theorem A is an easy consequence of the fact that BC is a homotopy colimit of the spaces $B(f/a)$, and of the spaces $B(a/f)$.

Lemma 9.49 *The natural map*

$$\epsilon : d(d_*(X)) = \varinjlim_{\Delta^p \times \Delta^q \to X} \Delta^p \times \Delta^q \to X$$

is a weak equivalence of simplicial sets, for all simplicial sets X.

This result is proved in [80], under the additional assumption that X is a Kan complex, by using a different technique.

Proof Suppose that $L(X)$ is the category whose objects are the strings of maps

$$\Delta^n \to \Delta^r \times \Delta^s \to X, \qquad (9.16)$$

and whose morphisms are the commutative diagrams

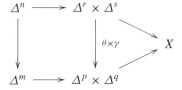

There are functors $p : L(X) \to \Delta/X$ and $q : L(X) \to (\Delta \times \Delta)/X$ that take the object (9.16) to the composite

$$\Delta^n \to X$$

and the map

$$\Delta^r \times \Delta^s \to X,$$

respectively.

If $\sigma : \Delta^m \to X$ is an object of the simplex category Δ/X, then the category σ/p has an initial object of the form

$$\begin{array}{c} \Delta^m \\ {\scriptstyle 1}\downarrow \\ \Delta^m \xrightarrow{\Delta} \Delta^m \times \Delta^m \longrightarrow X \end{array}$$

and is therefore contractible. It follows from Quillen's Theorem A that the map $p_* : B(L(X)) \to B(\Delta/X)$ is a weak equivalence.

If $\sigma : \Delta^p \times \Delta^q \to X$ is an object of $(\Delta \times \Delta)/X$, then the category q/σ has objects consisting of commutative diagrams

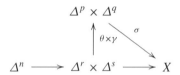

This category is homotopy equivalent to the simplex category $\Delta/(\Delta^p \times \Delta^q)$, with the homotopy provided by the family of maps $\theta \times \gamma$, and is therefore contractible. It follows, again from Quillen's Theorem A, that the map $q_* : B(L(X)) \to B((\Delta \times \Delta)/X)$ is a weak equivalence.

There is a commutative diagram

$$\begin{array}{ccc} \underset{\Delta^n \to \Delta^r \times \Delta^s \to X}{\mathrm{holim}} \Delta^n & \xrightarrow{p_*} & \underset{\Delta^n \to X}{\mathrm{holim}} \Delta^n \\ {\scriptstyle \simeq}\downarrow & & \downarrow{\scriptstyle \simeq} \\ \underset{\Delta^n \to \Delta^r \times \Delta^s \to X}{\mathrm{holim}} \Delta^r \times \Delta^s & \longrightarrow & X \\ {\scriptstyle q_*}\downarrow & \nearrow & \\ \underset{\Delta^r \times \Delta^s \to X}{\mathrm{holim}} \Delta^r \times \Delta^s & & \end{array}$$

9.3 Groupoids Enriched in Simplicial Sets

The maps p_* and q_* are weakly equivalent to the weak equivalences $p_* : B(L(X)) \to B(\Delta/X)$ and $q_* : B(L(X)) \to B((\Delta \times \Delta)/X)$, respectively. It follows that the map

$$\varinjlim\nolimits_{\Delta^r \times \Delta^s \to X} \Delta^r \times \Delta^s \to X$$

is a weak equivalence.

The map

$$\varinjlim\nolimits_{\Delta^p \times \Delta^q \to X} \Delta^p \times \Delta^q \to \varinjlim\nolimits_{\Delta^p \times \Delta^q \to X} \Delta^p \times \Delta^q$$

is a weak equivalence.

To see this, suppose that

$$\Delta^n \xrightarrow{(\theta,\gamma)} \Delta^p \times \Delta^q \xrightarrow{\sigma} X \qquad (9.17)$$

is an object of the translation category of n-simplices. Then the diagram

$$\begin{array}{ccc}
\Delta^n & \xrightarrow{\Delta} & \Delta^n \times \Delta^n \\
& {\scriptstyle (\theta,\gamma)} \searrow & \downarrow {\scriptstyle \theta \times \gamma} \\
& & \Delta^p \times \Delta^q \\
& & \downarrow {\scriptstyle \sigma} \\
& & X
\end{array}$$

commutes, and the composites

$$\Delta^n \times \Delta^n \xrightarrow{\zeta \times \omega} \Delta^p \times \Delta^q \xrightarrow{\sigma} X$$

coincide for all objects

$$\Delta^n \xrightarrow{(\zeta,\omega)} \Delta^p \times \Delta^q \xrightarrow{\sigma} X$$

in the component of the object (9.17). It follows that the corresponding object

$$\Delta^n \xrightarrow{\Delta} \Delta^n \times \Delta^n \to X$$

is terminal in the component of the object (9.17).

All components of the translation category of n-simplices are therefore contractible, and so the map

$$\varinjlim\nolimits_{\Delta^p \times \Delta^q \to X} (\Delta^p \times \Delta^q)_n \to \varinjlim\nolimits_{\Delta^p \times \Delta^q \to X} (\Delta^p \times \Delta^q)_n$$

is a weak equivalence for all n.

9.4 Presheaves of Groupoids Enriched in Simplicial Sets

Write $\mathbf{Pre}(s_0\mathbf{Gpd})$ for the category of presheaves of groupoids enriched in simplicial sets on a site $\mathcal{C}$.

The functors $\overline{W}$ and G relating simplicial sets and groupoids enriched in simplicial sets of the previous section define an adjoint pair of functors

$$G : s\mathbf{Pre}(\mathcal{C}) \leftrightarrows \mathbf{Pre}(s_0\mathbf{Gpd}) : \overline{W}$$

in the obvious way: if X is a simplicial presheaf, then $G(X)$ is the presheaf of simplicial groupoids, which is defined in sections by

$$G(X)(U) = G(X(U)), \ U \in \mathcal{C}.$$

The presheaf-level functor $\overline{W}$ has a similar, sectionwise definition.

In what follows, $\tilde{G}(X)$ will denote the sheaf of groupoids enriched in simplicial sets, which is associated to the presheaf object $G(X)$, for a simplicial presheaf X.

Say that a map $G \to H$ of such presheaves is a *local weak equivalence* if the map holds the following equivalent conditions:

(1) the map $\overline{W}G \to \overline{W}H$ is a local weak equivalence of simplicial presheaves,
(2) the map $BG \to BH$ is a diagonal local weak equivalence of bisimplicial presheaves.

The equivalence of these two conditions is a consequence of Corollary 9.39.

A map $p : G \to H$ of $\mathbf{Pre}(s_0\mathbf{Gpd})$ is said to be a $\overline{W}$-*fibration* if the map $\overline{W}G \to \overline{W}H$ is an injective fibration of simplicial presheaves. *Cofibrations* for this category are defined by the left lifting property with respect to trivial fibrations.

If G is a simplicial object in sheaves of groupoids, then the object $\overline{W}G$ is a simplicial sheaf. This follows from the description of n-simplices of $\overline{W}G$ (e.g. as in (9.12)) as strings of arrows in the sheaves of groupoids G_q which satisfy matching conditions.

Theorem 9.50

(1) With the definitions of local weak equivalence, $\overline{W}$-fibration and cofibration given above, the category $\mathbf{Pre}(s_0\mathbf{Gpd})$ of presheaves of groupoids enriched in simplicial sets has the structure of a right proper, cofibrantly generated, closed model category.
(2) The adjoint functors

$$G : s\mathbf{Pre}(\mathcal{C}) \leftrightarrows \mathbf{Pre}(s_0\mathbf{Gpd}) : \overline{W}$$

define a Quillen equivalence of this model structure on $\mathbf{Pre}(s_0\mathbf{Gpd})$ with the injective model structure on the category of simplicial presheaves.

The model structure of Theorem 9.50 is the $\overline{W}$-*model structure* for the category of presheaves of groupoids enriched in simplicial sets. The following lemma is the key step in its proof:

9.4 Presheaves of Groupoids Enriched in Simplicial Sets

Lemma 9.51 *Suppose given a pushout diagram*

$$\begin{array}{ccc} \tilde{G}(A) & \longrightarrow & G \\ {\scriptstyle i_*}\downarrow & & \downarrow{\scriptstyle i'} \\ \tilde{G}(B) & \longrightarrow & H \end{array} \qquad (9.18)$$

in sheaves of groupoids enriched in simplicial sets, where $i : A \to B$ is a local trivial cofibration of simplicial sheaves. Then the map i' is a local weak equivalence.

Lemma 9.51 implies that if we have a pushout diagram

$$\begin{array}{ccc} G(A) & \longrightarrow & G \\ {\scriptstyle i_*}\downarrow & & \downarrow{\scriptstyle i'} \\ G(B) & \longrightarrow & H \end{array}$$

in $\mathbf{Pre}(s_0\mathbf{Gpd})$ associated to a local trivial cofibration of simplicial presheaves $i : A \to B$, then the map i' is a local weak equivalence.

In effect, the induced map $i : \tilde{A} \to \tilde{B}$ is a local trivial cofibration of simplicial sheaves, while $\tilde{G}(\tilde{X})$ is naturally isomorphic to the sheaf associated to the presheaf $G(X)$. It then follows from Lemma 9.51 that the map $i'_* : \tilde{G} \to \tilde{H}$ is a local weak equivalence, and so the original map $i' : G \to H$ is a local weak equivalence.

The proof of Theorem 9.50 follows, according to the method displayed for the proof of Theorem 9.43. Observe that a map $p : G \to H$ is a $\overline{W}$-fibration (respectively trivial $\overline{W}$-fibration) if and only if it has the right lifting property with respect to all maps $G(A) \to G(B)$ induced by the set of generators $A \to B$ for the class of trivial cofibrations (respectively cofibrations) of the simplicial presheaf category $s\mathbf{Pre}(\mathcal{C})$.

Proof (*Proof of Lemma 9.51*) The functor $G \mapsto d(BG)$ commutes with the formation of associated sheaves. It follows that the map $dB(\tilde{G}A) \to dB(\tilde{G}B)$ is locally weakly equivalent to the map $dB(GA) \to dB(GB)$, and is therefore locally weakly equivalent to the map $\overline{W}(GA) \to \overline{W}(GB)$, and hence to the map $A \to B$. The map $dB(\tilde{G}A) \to dB(\tilde{G}B)$ is locally weakly equivalent $\overline{W}(\tilde{G}A) \to \overline{W}(\tilde{G}B)$. It follows that the map $\overline{W}(\tilde{G}A) \to \overline{W}(\tilde{G}B)$ is a local weak equivalence.

Suppose that $p : \mathbf{Shv}(\mathcal{B}) \to \mathbf{Shv}(\mathcal{C})$ is a Boolean localization (or a suitable collection of stalks). Recall from Lemma 4.27 that the associated inverse image functor preserves and reflects local weak equivalences of simplicial sheaves.

There is a natural isomorphism

$$p^*(dB(G)) \cong dB(p^*(G))$$

for all sheaf objects G, so that in the pushout

$$\begin{array}{ccc} p^*\tilde{G}(A) & \longrightarrow & p^*G \\ {\scriptstyle i_*}\downarrow & & \downarrow{\scriptstyle i''} \\ p^*\tilde{G}(B) & \longrightarrow & p^*H \end{array}$$

the map i_* is a local weak equivalence. It is enough to show that i'' is a local weak equivalence over $\mathcal{B}$, because the simplicial sheaf map $dB(G) \to dB(H)$ is a local weak equivalence if and only if the map $p^*dB(G) \to p^*dB(H)$ is a local weak equivalence. It therefore suffices to assume that our diagram (9.18) lives in sheaves on the complete Boolean algebra $\mathcal{B}$.

We can also assume that the simplicial sheaves A and B are locally fibrant. To see this, use the diagram

$$\begin{array}{ccccc} \tilde{G}(A) & \xrightarrow{\tilde{G}\eta} & \tilde{G}\overline{W}\tilde{G}(A) & \xrightarrow{\tilde{\epsilon}} & \tilde{G}(A) \\ \downarrow & & \downarrow & & \downarrow \\ \tilde{G}(B) & \xrightarrow[\tilde{G}\eta]{} & \tilde{G}\overline{W}\tilde{G}B & \xrightarrow[\tilde{\epsilon}]{} & \tilde{G}(B) \end{array}$$

in which the horizontal composites are identities. The map $\overline{W}\tilde{G}(A) \to \overline{W}\tilde{G}(B)$ has a factorization

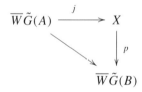

in simplicial sheaves, where p is a trivial injective fibration and j is a cofibration. Then the map $B \to \overline{W}\tilde{G}(B)$ lifts to X, and it follows that the map $i_* : \tilde{G}(A) \to \tilde{G}(B)$ is a retract of the map $j_* : \tilde{G}\overline{W}\tilde{G}(A) \to \tilde{G}(X)$, which is induced by the trivial cofibration $j : \overline{W}\tilde{G}(A) \to X$ of locally fibrant simplicial sheaves. The fact that the simplicial sheaf $\overline{W}\tilde{G}(A)$ is locally fibrant is a consequence of Lemma 9.41.

Finally, in the pushout diagram

$$\begin{array}{ccc} G(A) & \longrightarrow & G \\ {\scriptstyle i_*}\downarrow & & \downarrow \\ G(B) & \longrightarrow & H' \end{array}$$

9.4 Presheaves of Groupoids Enriched in Simplicial Sets

of presheaves of groupoids enriched in simplicial sets on $\mathcal{B}$, the map $A \to B$ is a sectionwise trivial cofibration of simplicial sheaves because A and B are locally fibrant (Lemma 4.24), so that $G(A) \to G(B)$ is a sectionwise trivial cofibration of presheaves of groupoids enriched in simplicial sets. It follows that the map $G \to H'$ is also a sectionwise trivial cofibration, and so the map $i' : G \to H$ of associated sheaves is a local weak equivalence.

We shall also need the local version of Proposition 9.32:

Proposition 9.52 *A map $f : G \to H$ of* **Pre**$(s_0\mathbf{Gpd})$ *is a local weak equivalence if and only if the following conditions hold:*

(1) the induced diagram of simplicial presheaf maps

$$\begin{array}{ccc} \mathrm{Mor}(G) & \longrightarrow & \mathrm{Mor}(H) \\ \downarrow & & \downarrow \\ \mathrm{Ob}(G) \times \mathrm{Ob}(G) & \longrightarrow & \mathrm{Ob}(H) \times \mathrm{Ob}(H) \end{array} \qquad (9.19)$$

is homotopy cartesian for the injective model structure,
(2) the sheaf map

$$\tilde{\pi}_0 G \to \tilde{\pi}_0 H$$

is an isomorphism.

Proof Form the object f^*H of **Pre**$(s_0\mathbf{Gpd})$ via the pullback

$$\begin{array}{ccc} \mathrm{Mor}(f^*H) & \longrightarrow & \mathrm{Mor}(H) \\ \downarrow & & \downarrow {\scriptstyle (s,t)} \\ \mathrm{Ob}(G) \times \mathrm{Ob}(G) & \xrightarrow{f \times f} & \mathrm{Ob}(H) \times \mathrm{Ob}(H) \end{array}$$

as in the proof of Proposition 9.32. There is a commutative diagram

in presheaves of groupoids enriched in simplicial sets.

The map $G \to f^*H$ is a local weak equivalence if and only if the map $\mathrm{Mor}(G) \to \mathrm{Mor}(f^*H)$ is a local weak equivalence of simplicial presheaves.

To verify this claim, suppose that the map $BG \to B(f^*H)$ is a local weak equivalence. According to Lemma 9.31, the natural diagram

$$\begin{array}{ccc} BP(H) & \longrightarrow & B\operatorname{Mor}_r(H) \\ \downarrow & & \downarrow \\ B\operatorname{Mor}_l(H) & \longrightarrow & BH \end{array}$$

is sectionwise homotopy cartesian, and is therefore homotopy cartesian for the injective model structure by Lemma 5.20, for all objects H of $\mathbf{Pre}(s_0\mathbf{Gpd})$. There are also natural sectionwise (hence local) weak equivalences

$$BP(H) \xrightarrow{\simeq} \operatorname{Mor}(H),\ B\operatorname{Mor}_r(H) \xrightarrow{\simeq} \operatorname{Ob}(H),\ \text{and}\ B\operatorname{Mor}_l(H) \xrightarrow{\simeq} \operatorname{Ob}(H).$$

It follows that the simplicial presheaf map $\operatorname{Mor}(G) \to \operatorname{Mor}(f^*H)$ is a local weak equivalence.

Conversely, suppose that the map $\operatorname{Mor}(G) \to \operatorname{Mor}(f^*H)$ is a local weak equivalence. Then the simplicial presheaf maps

$$BG_n \to B(f^*H)_n$$

are finite iterated products of the map $\operatorname{Mor}(G) \to \operatorname{Mor}(f^*H)$ over the discrete object $\operatorname{Ob}(H)$, and are therefore local weak equivalences by Lemma 4.39. It follows that the map $BG \to Bf^*H$ is a local weak equivalence.

The map $f^*H \to H$ is fully faithful in all simplicial degrees, and there are natural isomorphisms $\tilde{\pi}_0(H) \cong \tilde{\pi}_0(H_n)$ for all n. It follows that the map $B(f^*H_n) \to BH_n$ is a local weak equivalence for each n if and only if the map $\tilde{\pi}_0(f^*H) \to \tilde{\pi}_0(H)$ is an epimorphism of sheaves (since it is already monic). It follows that the map $f^*H \to H$ is a local weak equivalence if and only if the sheaf map $\tilde{\pi}_0(G) \to \tilde{\pi}_0(H)$ is an isomorphism.

It follows from Lemma 5.20 that the map $G \to f^*(H)$ is a local weak equivalence if and only if the diagram (9.19) is homotopy cartesian for the injective model structure, since the map

$$\operatorname{Ob}(G) \times \operatorname{Ob}(G) \to \operatorname{Ob}(H) \times \operatorname{Ob}(H)$$

of simplicially discrete objects is a sectionwise Kan fibration.

If the map $dBG \to dBH$ is a local weak equivalence, then the map $\tilde{\pi}_0(G) \to \tilde{\pi}_0(H)$ is a sheaf isomorphism, so that the map $f^*H \to H$ is a local weak equivalence. It follows that the map $G \to f^*H$ is a local weak equivalence so that the map $\operatorname{Mor}(G) \to \operatorname{Mor}(f^*H)$ is a local weak equivalence, and the diagram (9.19) is homotopy cartesian.

Conversely, suppose that the sheaf map $\tilde{\pi}_0 G \to \tilde{\pi}_0 H$ is an isomorphism and the diagram (9.19) is homotopy cartesian. Then the simplicial presheaf map $\operatorname{Mor}(G) \to \operatorname{Mor}(f^*H)$ is a local weak equivalence, and so the map $G \to f^*H$ is a local

9.4 Presheaves of Groupoids Enriched in Simplicial Sets

weak equivalence. The assumption that the map $\tilde{\pi}_0 G \to \tilde{\pi}_0 H$ is an isomorphism of sheaves implies that the map $f^*H \to H$ is a local weak equivalence, and it follows that the map $f : G \to H$ is a local weak equivalence.

We have the following corollary of the proof of Proposition 9.52:

Corollary 9.53 *A map $f : G \to H$ of presheaves of groupoids enriched in simplicial sets is a local weak equivalence if and only if the induced maps $G \to f^*H$ and $f^*H \to H$ are local weak equivalences.*

Recall from Sect. 5.6 that the derived Postnikov section $\mathbf{P}_n X$ of a simplicial presheaf X is defined by

$$\mathbf{P}_n X = P_n(\mathrm{Ex}^\infty X),$$

where $Y \mapsto P_n Y$ is the classical natural Postnikov section construction for simplicial sets Y. Lemma 5.46 says, in part, that the functor $X \mapsto \mathbf{P}_n X$ preserves local weak equivalences of simplicial presheaves X.

The canonical map

$$P_n X \to P_n(\mathrm{Ex}^\infty X) = \mathbf{P}_n X$$

is a sectionwise equivalence if X is a presheaf of Kan complexes. In particular, if H is a presheaf of groupoids enriched in simplicial sets, then the morphism object $\mathrm{Mor}\,(H)$ is a presheaf of Kan complexes, so that the natural simplicial presheaf map

$$P_n \,\mathrm{Mor}\,(H) \to \mathbf{P}_n \,\mathrm{Mor}\,(H)$$

is a sectionwise, hence local weak equivalence.

Suppose that G is a groupoid enriched in simplicial sets. The object $P_n G$ is the object of $s_0\mathbf{Gpd}$ with $\mathrm{Ob}\,(P_n G) = \mathrm{Ob}\,(G)$, and with

$$\mathrm{Mor}\,(P_n G) = P_n \,\mathrm{Mor}\,(G).$$

In other words, $\mathrm{Mor}\,(P_n G)$ is the disjoint union

$$\bigsqcup_{x,y \in \mathrm{Ob}\,(G)} P_n G(x, y),$$

and the composition laws $G(x, y) \times G(y, z) \to G(x, z)$ induce composition laws $P_n G(x, y) \times P_n G(y, z) \to P_n G(x, z)$ (since the functor P_n preserves finite products), such that the obvious diagram commutes. This construction is functorial in G, and the map $\pi : \mathrm{Mor}\,(G) \to P_n \,\mathrm{Mor}\,(G)$ defines a natural map

$$\eta : G \to P_n G$$

of groupoids enriched in simplicial sets, which is the identity on objects.

This construction extends to presheaves of groupoids enriched in simplicial sets: there is a functor

$$P_n : \mathbf{Pre}(s_0\mathbf{Gpd}) \to \mathbf{Pre}(s_0\mathbf{Gpd}),$$

with $P_n H$ defined in sections by

$$(P_n H)(U) = P_n(H(U)).$$

There is a natural map

$$\eta : H \to P_n H$$

that is defined in sections by the corresponding map for groupoids enriched in simplicial sets.

Say that a map $G \to H$ of $\mathbf{Pre}(s_0\mathbf{Gpd})$ is a *local n-equivalence* if the induced map $P_n G \to P_n H$ is a local weak equivalence of $\mathbf{Pre}(s_0\mathbf{Gpd})$, and that G is an *n-type* if the map $\eta : G \to P_n G$ is a local weak equivalence.

An *n-fibration* is a map that has the right lifting property with respect to all maps that are cofibrations for the $\overline{W}$-model structure and local n-equivalences.

Lemma 9.54

(1) An object H of $\mathbf{Pre}(s_0\mathbf{Gpd})$ is an n-type if and only if $\overline{W}H$ is an $(n+1)$-type in simplicial presheaves.

(2) A map $G \to H$ is a local n-equivalence if and only if the map $\overline{W}G \to \overline{W}H$ is a local $(n+1)$-equivalence of simplicial presheaves.

Proof The natural maps

$$dB(P_n G) \xrightarrow{p} P_{n+1}(dB(P_n G)) \xleftarrow{P_{n+1}(p)} P_{n+1}(dBG)$$

are weak equivalences for all groupoids G enriched in simplicial sets, on account of the homotopy cartesian diagrams

$$\begin{array}{ccc} dBP(G) & \longrightarrow & dB\,\mathrm{Mor}_r(G) \\ \downarrow & & \downarrow \\ dB\,\mathrm{Mor}_l(G) & \longrightarrow & dBG \end{array}$$

and the natural weak equivalences

$$dBP(G) \xrightarrow{\simeq} \mathrm{Mor}(G),\ dB\,\mathrm{Mor}_r(G) \xrightarrow{\simeq} \mathrm{Ob}(G),\ \text{and } dB\,\mathrm{Mor}_l(G) \xrightarrow{\simeq} \mathrm{Ob}(G)$$

of Lemma 9.31.

9.4 Presheaves of Groupoids Enriched in Simplicial Sets 311

The maps
$$dB(P_nH) \xrightarrow{p} P_{n+1}(dB(P_nH)) \xleftarrow{P_{n+1}(p)} P_{n+1}(dBH)$$

are therefore local weak equivalences for all presheaves H of groupoids enriched in simplicial sets. It follows from Corollary 9.39 that the natural maps

$$\overline{W}(P_nH) \xrightarrow{p} P_{n+1}(\overline{W}(P_nH)) \xleftarrow{\overline{W}(p)} P_{n+1}(\overline{W}H) \qquad (9.20)$$

are also local weak equivalences for all H in $\mathbf{Pre}(s_0\mathbf{Gpd})$.

There is a commutative diagram

so that the simplicial presheaf $\overline{W}H$ is a local $(n+1)$-type if and only if H is a local n-type.

It follows from the naturality of the local equivalences in (9.20) that the map $P_nG \to P_nH$ is a local weak equivalence of $\mathbf{Pre}(s_0\mathbf{Gpd})$ if and only if the map $P_{n+1}(\overline{W}G) \to P_{n+1}(\overline{W}H)$ is a local weak equivalence of simplicial presheaves. This proves statement 2).

Lemma 9.55 *The following statements hold within the category* $\mathbf{Pre}(s_0\mathbf{Gpd})$ *of presheaves of groupoids enriched in simplicial sets:*

A4 *The functor P_n preserves local weak equivalences.*
A5 *The maps $\eta, P_n(\eta) : P_nG \to P_nP_nG$ are local weak equivalences.*
A6 *Suppose given a pullback diagram*

$$\begin{array}{ccc} A & \xrightarrow{\alpha} & G \\ \downarrow & & \downarrow p \\ B & \xrightarrow{\beta} & H \end{array}$$

in $\mathbf{Pre}(s_0\mathbf{Gpd})$ *such that p is a $\overline{W}$-fibration. Suppose that the maps $\eta : H \to P_nH$ and $\beta_* : P_nB \to P_nH$ are local weak equivalences. Then the map $\alpha_* : P_nA \to P_nG$ is a local weak equivalence.*

Proof If $f : G \to H$ is a local weak equivalence of presheaves of groupoids enriched in simplicial sets, then the map $\tilde{\pi}_0G \to \tilde{\pi}_0H$ is an isomorphism of sheaves, and the simplicial presheaf map $\operatorname{Mor}(G) \to f^* \operatorname{Mor}(H)$ is a local weak equivalence. The map

$$P_n \operatorname{Mor}(G) \to P_n f^* \operatorname{Mor}(H) \cong f^* P_n \operatorname{Mor}(H)$$

is a local weak equivalence of simplicial presheaves by Lemma 5.47. It follows from Proposition 9.52 that the map $P_n G \to P_n H$ is a local weak equivalence, and we have proved statement **A4**.

Statement **A5** follows from basic properties of the Postnikov section functor $X \mapsto P_n X$ for Kan complexes X: the maps $p, P_n(p) : P_n G \to P_n P_n G$ are sectionwise equivalences since $\operatorname{Mor}(G)$ is a presheaf of Kan complexes, and both maps are the identity on objects.

Statement **A6** follows from Lemma 9.54 which implies that the induced diagram

$$\begin{array}{ccc} \overline{W}\Lambda & \longrightarrow & \overline{W}G \\ \downarrow & & \downarrow \\ \overline{W}B & \longrightarrow & \overline{W}H \end{array}$$

satisfies the conditions of Lemma 5.47.

Theorem 9.56

(1) *The category* **Pre**(s_0**Gpd**), *together with the classes of cofibrations, local n-equivalences and n-fibrations, satisfies the axioms for a right proper, cofibrantly generated, closed model category.*

(2) *A map* $p : G \to H$ *is an n-fibration if and only if it is a $\overline{W}$-fibration and the diagram*

$$\begin{array}{ccc} G & \xrightarrow{\eta} & P_n G \\ p \downarrow & & \downarrow p_* \\ H & \xrightarrow{\eta} & P_n H \end{array}$$

is homotopy cartesian for the $\overline{W}$-structure on **Pre**(s_0**Gpd**) *of Theorem 9.50.*

Proof The proof uses the $\overline{W}$-model structure of Theorem 9.50 along with Lemma 9.55, and has the same formal structure as the proof of Theorem 5.49.

The generating set for the trivial cofibrations of this model structure are the maps $G(A) \to G(B)$ that are induced by α-bounded cofibrations $A \to B$ of simplicial presheaves that are local $(n + 1)$-equivalences, by Lemmas 9.54 and 5.50. As usual, α is a regular cardinal which is larger than the cardinality of the set of morphisms of the underlying site.

9.4 Presheaves of Groupoids Enriched in Simplicial Sets

Say that the model structure of Theorem 9.56 on $s_0\mathbf{Gpd}$ is the *n-equivalence structure*.

A *2-groupoid* H is a groupoid object in groupoids, such that the groupoid $\mathrm{Ob}\,(H)$ is discrete. Write $2-\mathbf{Gpd}$ for the resulting category of 2-groupoids.

The fundamental groupoid and classifying space (nerve) functors define an adjoint pair

$$\pi : s\mathbf{Set} \leftrightarrows \mathbf{Gpd} : B.$$

Both functors preserve finite products, and hence induce an adjoint pair of functors

$$\pi : s_0\mathbf{Gpd} \leftrightarrows 2-\mathbf{Gpd} : B.$$

Explicitly, if G is a groupoid enriched in simplicial sets, then $\pi(G)$ is the 2-groupoid with

$$\mathrm{Mor}\,(\pi(G)) = \pi(\mathrm{Mor}\,(G)) \text{ and } \mathrm{Ob}(\pi(G)) = \mathrm{Ob}(G).$$

If H is a 2-groupoid, then BH is the groupoid enriched in simplicial sets with

$$\mathrm{Mor}\,(BH) = B(\mathrm{Mor}\,(H)) \text{ and } \mathrm{Ob}(BH) = \mathrm{Ob}(H).$$

Write $\mathbf{Pre}(2-\mathbf{Gpd})$ for the category of presheaves of 2-groupoids. Then the fundamental groupoid and classifying space functors define an adjoint pair

$$\pi : \mathbf{Pre}(s_0\mathbf{Gpd}) \leftrightarrows \mathbf{Pre}(2-\mathbf{Gpd}) : B \qquad (9.21)$$

relating presheaves of groupoids enriched in simplicial sets with presheaves of 2-groupoids. The fundamental groupoid functor π is left adjoint to the classifying space functor B in all contexts.

We will show in Theorem 9.57 that the adjunction (9.21) induces a model structure on the category $\mathbf{Pre}(2-\mathbf{Gpd})$ of presheaves of 2-groupoids.

Say that a map $f : G \to H$ of $\mathbf{Pre}(2-\mathbf{Gpd})$ is a *local weak equivalence* (respectively *fibration*) if the induced map $BG \to BH$ is a local weak equivalence (respectively $\overline{W}$-fibration) of $\mathbf{Pre}(s_0\mathbf{Gpd})$. Cofibrations of $\mathbf{Pre}(2-\mathbf{Gpd})$ are defined by a left lifting property. In particular, every cofibration $A \to B$ of $\mathbf{Pre}(s_0\mathbf{Gpd})$ induces a cofibration $\pi(A) \to \pi(B)$ of presheaves of 2-groupoids.

The fundamental groupoid functor $\pi : s\mathbf{Pre}(\mathcal{C}) \to \mathbf{Pre}(\mathbf{Gpd})$ preserves local weak equivalences of simplicial presheaves, by Lemma 9.18, and it follows from Proposition 9.52 that the functor $\pi : \mathbf{Pre}(s_0\mathbf{Gpd}) \to \mathbf{Pre}(2-\mathbf{Gpd})$ preserves local weak equivalences.

One forms colimits in $2-\mathbf{Gpd}$ as in $s_0\mathbf{Gpd}$. In particular, suppose given a diagram of 2-groupoids where $i : A \to B$ is a morphism of $s_0\mathbf{Gpd}$. Consider the pushout

$$\begin{array}{ccc} A & \longrightarrow & BG \\ {\scriptstyle i}\downarrow & & \downarrow{\scriptstyle j} \\ B & \longrightarrow & E \end{array}$$

in $s_0\mathbf{Gpd}$. Then the diagram

$$\begin{array}{ccc} \pi A & \longrightarrow & G \\ {\scriptstyle i_*}\downarrow & & \downarrow{\scriptstyle j_*} \\ \pi B & \longrightarrow & \pi E \end{array} \quad (9.22)$$

is a pushout in 2-groupoids.

The same construction obtains in presheaves of 2-groupoids. It follows that if the diagram (9.22) is a pushout in $\mathbf{Pre}(2 - \mathbf{Gpd})$ and i is a trivial cofibration of $\mathbf{Pre}(s_0\mathbf{Gpd})$, then j_* is a local weak equivalence of presheaves of 2-groupoids, since the functor π preserves local weak equivalences.

Theorem 9.57

(1) With the definitions of local weak equivalence, fibration and cofibration given above, the category $\mathbf{Pre}(2 - \mathbf{Gpd})$ of presheaves of 2-groupoids has the structure of a right proper closed model category.

(2) The adjoint pair of functors

$$\pi : \mathbf{Pre}(s_0\mathbf{Gpd}) \leftrightarrows \mathbf{Pre}(2 - \mathbf{Gpd}) : B$$

defines a Quillen adjunction.

Theorems 9.50 and 9.57 first appeared in the Ph.D. thesis of Z. Luo [73].

Proof Statement (2) is obvious from the definitions, once statement (1) is proved.

A map $p : G \to H$ of $\mathbf{Pre}(2-\mathbf{Gpd})$ is a fibration (respectively trivial fibration) if and only if it has the right lifting property with respect to the maps $\pi A \to \pi B$ that are induced by trivial cofibrations (respectively cofibrations) $A \to B$ for the $\overline{W}$-model structure on $\mathbf{Pre}(s_0\mathbf{Gpd})$ of Theorem 9.50. Pushouts of maps $\pi A \to \pi B$ induced by trivial cofibrations $A \to B$ of $\mathbf{Pre}(s_0\mathbf{Gpd})$ along arbitrary maps $\pi A \to G$ are trivial cofibrations of presheaves of 2-groupoids, and it follows that every map $f : G \to H$

9.4 Presheaves of Groupoids Enriched in Simplicial Sets

of presheaves of 2-groupoids has factorizations

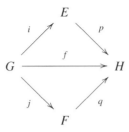

such that i is a trivial cofibration that has the left lifting property with respect to all fibrations and p is a fibration, and such that j is a cofibration and p is a trivial fibration. This gives the factorization axiom **CM5**, and the lifting axiom **CM4** follows in the usual way.

The model structure on presheaves of 2-groupoids is right proper, since the $\overline{W}$-model structure on $\mathbf{Pre}(s_0\mathbf{Gpd})$ is right proper.

Various definitions can be made. In particular, the 2-groupoid object $\pi G(X)$ is the *fundamental 2-groupoid* of a space (or simplicial presheaf) X.

The path component groupoid functor induces a functor

$$\pi_0 : \mathbf{Pre}(s_0\mathbf{Gpd}) \to \mathbf{Pre}(\mathbf{Gpd}),$$

which takes values in presheaves of groupoids. There is a natural weak equivalence

$$\pi(\overline{W}H) \simeq \pi_0 H$$

by Lemma 9.44, so it follows from Lemma 9.18 that the functor $H \mapsto \pi_0 H$ preserves local weak equivalences of presheaves of groupoids enriched in simplicial sets.

The path component functor π_0 has a right adjoint

$$R : \mathbf{Pre}(\mathbf{Gpd}) \to \mathbf{Pre}(s_0\mathbf{Gpd})$$

which takes a groupoid to the associated constant simplicial groupoid. There is a natural isomorphism

$$\pi_0 H \cong \pi_0 P_0 H$$

for groupoids H enriched in simplicial sets, and the functor π_0 preserves local weak equivalences of presheaves of groupoids enriched in simplicial sets. It follows that the path component functor π_0 preserves 0-equivalences. There is a natural isomorphism

$$\pi_0 G(X) \cong \pi(X)$$

from the proof of Corollary 9.45, and it follows that the functor π_0 preserves cofibrations. The adjunction

$$\pi_0 : \mathbf{Pre}(s_0\mathbf{Gpd}) \leftrightarrows \mathbf{Pre}(\mathbf{Gpd}) : R$$

therefore defines a Quillen adjunction between the 0-equivalence model structure on **Pre**(s_0**Gpd**) and the injective model structure on the category **Pre**(**Gpd**) of presheaves of groupoids of Proposition 9.19.

The canonical map $\eta : H \to R(\pi_0 H)$ induces an isomorphism of groupoids $\pi_0 H \cong \pi_0 R(\pi_0 H)$, so that the map $\eta_* : P_0 H \to P_0 R(\pi_0 H)$ is a local weak equivalence. The map η is therefore a 0-equivalence. The canonical map $\epsilon : \pi_0 R(G) \to G$ is an isomorphism for all groupoids G. It follows that the adjunction defined by the functors π_0 and R defines a Quillen equivalence between the 0-equivalence model structure on s_0**Gpd** and the ordinary model structure on **Gpd**.

We have proved the following:

Lemma 9.58 *The adjoint pair of functors*

$$\pi_0 : \mathbf{Pre}(s_0\mathbf{Gpd}) \leftrightarrows \mathbf{Pre}(\mathbf{Gpd}) : R$$

defines a Quillen equivalence between the 0-equivalence model structure on the category **Pre**(s_0**Gpd**) *of presheaves of groupoids enriched in simplicial sets and the injective model structure on the category* **Pre**(**Gpd**) *of presheaves of groupoids.*

The natural map $\eta : BH \to P_1 BH$ is a sectionwise equivalence, for all presheaves of 2-groupoids H. It follows in particular that BH is a local 1-type for all objects H of **Pre**(2 − **Gpd**). This observation extends to the following result:

Proposition 9.59 *The adjunction*

$$\pi : \mathbf{Pre}(s_0\mathbf{Gpd}) \leftrightarrows \mathbf{Pre}(2 - \mathbf{Gpd}) : B$$

defines a Quillen equivalence between the 1-equivalence structure on **Pre**(s_0**Gpd**) *and the model structure on* **Pre**(2 − **Gpd**) *of Theorem 9.57.*

Proof There is a natural commutative diagram

$$\begin{array}{ccc} G & \xrightarrow{\eta} & P_1 G \\ {\scriptstyle \eta}\downarrow & \simeq & \downarrow {\scriptstyle P_1 \eta} \\ B\pi G & \xrightarrow[\eta]{\simeq} & P_1 B\pi G \end{array}$$

for all objects G of **Pre**(s_0**Gpd**), in which the indicated maps are sectionwise weak equivalences. It follows that the functor π takes local 1-equivalences to local weak equivalences of presheaves of 2-groupoids. The functor π also preserves cofibrations, so that the adjunction is a Quillen adjunction.

The map $\eta : G \to B\pi G$ is a 1-equivalence for all objects G of **Pre**(s_0**Gpd**), and the map $\epsilon : \pi BH \to H$ is an isomorphism for all presheaves of 2-groupoids H.

Remark 9.60 An $(n + 1)$-stack G is a presheaf of groupoids enriched in simplicial sets, such that some (hence any) fibrant model $G \to H$ in the n-equivalence model

9.4 Presheaves of Groupoids Enriched in Simplicial Sets

structure of Theorem 9.56 is a sectionwise equivalence. In other words, G should satisfy descent with respect to the n-equivalence model structure.

(1) Suppose that G is a fibrant object for the 0-equivalence model structure on presheaves of groupoids enriched in simplicial sets. Suppose that $j : \pi_0 G \to H$ is a fibrant model in presheaves of groupoids. Then all maps in the picture

$$G \to R\pi_0 G \xrightarrow{j_*} RH$$

are 0-equivalences. The composite map $G \to RH$ is therefore a 0-equivalence of fibrant models for G in 0-equivalence model structure. It follows from a standard argument (i.e. every trivial fibration has the right lifting property with respect to all cofibrations) that the map $G \to RH$ is a sectionwise equivalence of presheaves of groupoids enriched in simplicial sets. All 1-stacks are represented by stacks up to sectionwise equivalence in this sense.

(2) Similarly, suppose that K is a fibrant object in the 1-equivalence model structure. Let $j : \pi(K) \to L$ be a fibrant model in presheaves of 2-groupoids. Then the composite

$$K \to B(\pi(K)) \xrightarrow{j_*} BL$$

consists of 1-equivalences, and the objects BL is fibrant for the 1-equivalence model structure on presheaves of groupoids enriched in simplicial sets. The composite map $K \to BL$ is therefore a sectionwise weak equivalence. All 2-stacks are therefore represented by fibrant presheaves of 2-groupoids up to sectionwise equivalence.

We also have the following:

Proposition 9.61 *The adjunction*

$$G : s\mathbf{Pre}(\mathcal{C}) \leftrightarrows \mathbf{Pre}(s_0\mathbf{Gpd}) : \overline{W}$$

induces a Quillen equivalence between the n-equivalence structure on $\mathbf{Pre}(s_0\mathbf{Gpd})$ *of Theorem 9.56 and the $(n+1)$-equivalence structure of Theorem 5.49 on the category $s\mathbf{Pre}(\mathcal{C})$ of simplicial presheaves.*

Proof We have the natural weak equivalences

$$\overline{W}(P_n G) \xrightarrow{\eta} P_{n+1}(\overline{W}(P_n G)) \xleftarrow{\overline{W}(\eta)} P_{n+1}(\overline{W} G)$$

from (9.20).

Suppose that $X \to Y$ is a map of presheaves of Kan complexes such that the map $P_{n+1} X \to P_{n+1} Y$ is a local weak equivalence. Then there is a commutative diagram

$$\begin{array}{ccc} P_{n+1} X & \longrightarrow & P_{n+1} Y \\ \simeq \downarrow & & \downarrow \simeq \\ P_{n+1} \overline{W} G X & \longrightarrow & P_{n+1} \overline{W} G Y \end{array}$$

where the vertical maps are induced by the sectionwise weak equivalences $X \to \overline{W}GX$ and $Y \to \overline{W}GY$ (see the proof of Theorem 9.43). The map along the bottom of the diagram is weakly equivalent to the map $\overline{W}P_n GX \to \overline{W}P_n GY$, and so the map $GX \to GY$ is an n-equivalence.

The functor G therefore takes $(n+1)$-equivalences of simplicial presheaves to n-equivalences of $\mathbf{Pre}(s_0\mathbf{Gpd})$. This functor also preserves cofibrations, so that the adjoint functors G and $\overline{W}$ form a Quillen equivalence between the n-equivalence structure on $\mathbf{Pre}(s_0\mathbf{Gpd})$ and the $(n+1)$-equivalence structure on the simplicial presheaf category $s\mathbf{Pre}(\mathcal{C})$. The canonical maps $X \to \overline{W}G(X)$ and $G\overline{W}H \to H$ are sectionwise equivalences, and are therefore $(n+1)$-equivalences and n-equivalences, respectively.

9.5 Extensions and Gerbes

This section is a description of the relation between presheaves of groupoids and presheaves of 2-groupoids on a small Grothendieck site $\mathcal{C}$. The general idea, which is central to all discussions of non-abelian H^2 invariants dating from the time of Giraud's book [31] in the early 1970s, is that 2-groupoids can be used to classify various classes of groupoids up to weak equivalence.

Traditionally, although it is not quite said this way in the literature, a gerbe G is a locally connected stack. This means that G is a sheaf of groupoids which satisfies descent (a stack), and has a path component sheaf $\tilde{\pi}_0 G$ that is trivial in the sense that the canonical map $\tilde{\pi}_0 G \to *$ to the terminal sheaf is an isomorphism. From the point of view developed in this chapter, stacks are local homotopy types of presheaves of groupoids, and we broaden the classical definition a bit to say that a *gerbe* is a locally connected presheaf of groupoids.

In standard homotopy theory, a connected groupoid G is equivalent as a category with some group: the group in question is the group $\pi_1(BG, x) = G(x, x)$ of automorphisms of some object x—any object x will do, because G is connected, so that all automorphism groups $G(x, x)$ are isomorphic.

More generally, if H is a presheaf of groupoids, then the automorphism sheaves of H (the sheaves of fundamental groups of BH) are fibred over the presheaf $\mathrm{Ob}(H)$ of objects of H, and could more generally belong to some family of groups, which we specify up to isomorphism as a group object $\mathcal{F} \to S$ in sheaves fibred over a sheaf S. Thus we can speak of gerbes (locally connected presheaves of groupoids) with automorphism sheaves in a family of sheaves of groups $\mathcal{F}$.

One of the main results of this section gives a homotopy classification of the local weak equivalence classes of gerbes with automorphism sheaves in a family $\mathcal{F}$ as the set of morphisms

$$[*, \mathbf{Iso}(\mathcal{F})]$$

in the homotopy category of presheaves of 2-groupoids, where $\mathbf{Iso}(\mathcal{F})$ is the 2-groupoid object consisting of S, the isomorphisms $\mathcal{F}_x \xrightarrow{\cong} \mathcal{F}_y$, and their homotopies.

9.5 Extensions and Gerbes

This homotopy classification result is Corollary 9.68—it gives a description of the non-abelian cohomology object $H^2(\mathcal{C}, \mathcal{F})$ for the underlying site $\mathcal{C}$.

Corollary 9.68 is a special case of a more general result that classifies essentially surjective groupoid homomorphisms $q: G \to H$ with kernels K in a family $\mathcal{F}$. We say that the map q is essentially surjective if the map $\mathrm{im}(q) \to H$ defined on the image of q is a local weak equivalence of presheaves of groupoids.

We denote the category of essentially surjective morphisms $q: G \to H$ with kernels in $\mathcal{F}$ by $\mathbf{Ext}(H, \mathcal{F})$, and then Theorem 9.66 says (this is formally stated in Corollary 9.67) that there is a natural bijection

$$[H, \mathbf{Iso}(\mathcal{F})] \cong \pi_0 \mathbf{Ext}(H, \mathcal{F}),$$

where the morphisms are again in the homotopy category of presheaves of 2-groupoids. Corollary 9.68 is the case of Theorem 9.66 which corresponds to H being the trivial groupoid $*$.

The method of proof (and statement) of Theorem 9.66 is cocycle theoretic. Every object $q: G \to H$ in $\mathbf{Ext}(H, \mathcal{F})$ determines resolution 2-groupoid $R(q) \xrightarrow{\simeq} H$ and a cocycle

$$H \xleftarrow{\simeq} R(q) \to \mathbf{Iso}(\mathcal{F})$$

in presheaves of 2-groupoids, and every cocycle

$$H \xleftarrow{\simeq} A \xrightarrow{F} \mathbf{Iso}(\mathcal{F})$$

defines an essentially surjective groupoid homomorphisms $E_A F \to H$ with kernels in the family $\mathcal{F}$, where $E_A F$ is a Grothendieck construction which is associated to the map F. Theorem 9.66 is proved by showing that these two constructions are inverse to each other.

Suppose now that H is a presheaf of groups and that L is a sheaf of groups. Theorem 9.66 has a pointed version, which specializes in Corollary 9.72 to an identification of isomorphism classes of extensions of sheaves of groups

$$e \to L \to G \to \tilde{H} \to e$$

with the set of morphisms

$$[BH, \overline{W}\mathbf{Iso}(L)]_*$$

in the homotopy category of pointed simplicial presheaves, where we have identified the sheaf of groups L with the family $L \to *$ over a point.

Corollaries 9.68 and 9.72 have been known in some form for quite a while. Corollary 9.68 is the homotopy classification of gerbes—it generalizes a result of Breen for singleton families [15], while the cocycle theoretic technique for its proof was introduced in [66]. Corollary 9.72 generalizes the cocycle classification of extensions of groups that appears in [65]; the classification of extensions of groups is a classical result.

The novel feature of the present exposition is the demonstration that these two streams of results have a common source.

A *family of presheaves of groups* $\mathcal{F} \to S$ over a presheaf S is a group object in the category of presheaves over S. We shall often write $\mathcal{F}$ to denote the family $\mathcal{F} \to S$.

The fibre $\mathcal{F}_x$ of the family $\mathcal{F}$ over $x \in S(U)$ is defined by the pullback diagram

$$\begin{array}{ccc} \mathcal{F}_x & \longrightarrow & \mathcal{F}|_U \\ \downarrow & & \downarrow \\ * & \xrightarrow{x} & S|_U \end{array}$$

in the category of presheaves on the site $\mathcal{C}/U$. Recall that the restriction $F|_U$ of a presheaf F on $\mathcal{C}$ is defined by the composite

$$(\mathcal{C}/U)^{op} \to \mathcal{C}^{op} \xrightarrow{F} \mathbf{Set}.$$

The fibre $\mathcal{F}_x$ inherits the structure of a presheaf of groups on $\mathcal{C}/U$ from the family $\mathcal{F} \to S$.

Example 9.62 The family of presheaves of groups $\mathrm{Aut}(G) \to \mathrm{Ob}(G)$ is defined, for a presheaf of groupoids G, by the pullback diagram

$$\begin{array}{ccc} \mathrm{Aut}(G) & \longrightarrow & \mathrm{Mor}(G) \\ \downarrow & & \downarrow {\scriptstyle (s,t)} \\ \mathrm{Ob}(G) & \xrightarrow{\Delta} & \mathrm{Ob}(G) \times \mathrm{Ob}(G) \end{array}$$

where s and t are source and target, respectively, and Δ is the diagonal map.

A *morphism* $\mathcal{F} \to \mathcal{F}'$ of families of presheaves of groups is a commutative diagram of presheaf maps

$$\begin{array}{ccc} \mathcal{F} & \xrightarrow{f} & \mathcal{F}' \\ \downarrow & & \downarrow \\ S & \xrightarrow{f} & S' \end{array} \qquad (9.23)$$

such that the induced sheaf map

$$\mathcal{F} \to S \times_{S'} \mathcal{F}' \qquad (9.24)$$

defines a morphism of group objects over S.

9.5 Extensions and Gerbes

If the map (9.24) induces an isomorphism of associated sheaves, or equivalently if the diagram (9.23) induces a pullback diagram of associated sheaves, then we say that the morphism of families $\mathcal{F} \to \tilde{\mathcal{F}}$ is a *pseudo-isomorphism*.

Example 9.63 If the morphism $f : G \to H$ of presheaves of groupoids is a local weak equivalence, then the induced diagram

$$\begin{array}{ccc} \mathrm{Aut}(G) & \longrightarrow & \mathrm{Aut}(H) \\ \downarrow & & \downarrow \\ \mathrm{Ob}(G) & \longrightarrow & \mathrm{Ob}(H) \end{array}$$

defines a pseudo-isomorphism of families.

If the family $\mathcal{F} \to S$ is a morphism of sheaves, we say that it is a *family of sheaves of groups*.

Example 9.64 Every family of presheaves of groups has an associated family of sheaves of groups, and the associated sheaf map

$$\begin{array}{ccc} \mathcal{F} & \xrightarrow{\eta} & \tilde{\mathcal{F}} \\ \downarrow & & \downarrow \\ S & \xrightarrow{\eta} & \tilde{S} \end{array}$$

is a pseudo-isomorphism of families.

Every family $\mathcal{F} \to S$ of presheaves of groups has an associated presheaf of 2-groupoids **Iso**($\mathcal{F}$). In sections, the 0-cells are the elements of $S(U)$, the 1-cells $\alpha : x \to y$ are the isomorphisms $\alpha : \mathcal{F}_x \xrightarrow{\cong} \mathcal{F}_y$ of presheaves of groups on $\mathcal{C}/U$, and the 2-cells are the homotopies $h : \alpha \to \beta$. Such homotopies are defined by conjugation with elements $h \in \mathcal{F}_y(U)$.

Suppose given a pseudo-isomorphism

$$\begin{array}{ccc} \mathcal{F} & \xrightarrow{f} & \mathcal{G} \\ \downarrow & & \downarrow \\ S & \xrightarrow{f} & T \end{array}$$

where $\mathcal{G} \to T$ is a family of sheaves of groups. Then this map induces a morphism

$$f_* : \mathbf{Iso}(\mathcal{F}) \to \mathbf{Iso}(\mathcal{G}).$$

This map f_* is the presheaf map $f : S \to T$ on 0-cells, and is defined on 1-cells $\alpha : x \to y$ by the conjugation diagrams

$$\begin{array}{ccc} \tilde{\mathcal{F}}_x & \xrightarrow{\alpha} & \tilde{\mathcal{F}}_y \\ f \downarrow \cong & & \cong \downarrow f \\ \mathcal{G}_{f(x)} & \xdashrightarrow{f(\alpha)} & \mathcal{G}_{f(y)} \end{array}$$

in the category of sheaves of groups. If $h \in \mathcal{F}_y(U)$ defines a homotopy of 1-cells $\alpha, \beta : x \to y$, then $f(h) \in \mathcal{G}_{f(y)}(U)$ defines a homotopy of the 1-cells $f(\alpha), f(\beta)$.

The *resolution 2-groupoid* $R(q)$ for a groupoid morphism $q : E \to F$ is a 2-groupoid which has the same objects and 1-cells as E, and there is a 2-cell $g \to h$ between morphisms $g, h : x \to y$ in $R(q)$ if and only if $q(g) - q(h)$ in F. There is a canonical morphism

$$q_* : R(q) \to F$$

of 2-groupoids which is defined by the morphism q on 0-cells and 1-cells. Here, the groupoid F is identified with a 2-groupoid that has only identity 2-cells.

The path component groupoid $\pi_0 R(q)$ of $R(q)$ is the *image* of q: its objects are the same as those of E, and the set of morphisms $\pi_0 R(q)(x, y)$ can be identified with the image of the function

$$E(x, y) \to F(q(x), q(y)).$$

Write

$$\operatorname{im}(q) = \pi_0 R(q),$$

and observe that the 2-groupoid morphism

$$R(q) \to \pi_0 R(q) = \operatorname{im}(q)$$

is a weak equivalence.

There is a canonical morphism $E \to R(q)$, and the composite

$$E \to R(q) \xrightarrow{q_*} F$$

is the morphism q. The composite

$$E \to R(q) \to \operatorname{im}(q)$$

induces an isomorphism $\pi_0 E \cong \pi_0(\operatorname{im}(q))$.

9.5 Extensions and Gerbes

The resolution 2-groupoid construction is natural in morphisms of groupoids, so that there is a presheaf $R(p)$ of 2-groupoids for any morphism $p : G \to H$ of presheaves of groupoids, along with a presheaf of groupoids im (p) and natural maps

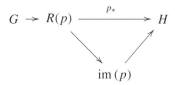

such that the top composite $G \to R(p) \to H$ is the map p. The presheaf of groupoids im (p) is the image of p in the category of presheaves of groupoids. The map $R(p) \to \text{im}(p)$ is a weak equivalence of 2-groupoids in each section, and the induced map $\pi_0 G \to \pi_0(\text{im}(p))$ is an isomorphism of presheaves.

Say that a morphism $p : G \to H$ of sheaves of groupoids is *essentially surjective* if the induced map im $(p) \to H$ is a local weak equivalence of presheaves of groupoids, or equivalently if the induced map $p_* : R(p) \to H$ is a local weak equivalence of presheaves of 2-groupoids.

It is a consequence of Lemma 9.21 that the morphism im $(p) \to H$ of presheaves of groupoids is a local weak equivalence if and only if it is fully faithful in the sense that the map

$$\text{Mor}(\text{im}(p)) \to (\text{Ob}(G) \times \text{Ob}(G)) \times_{\text{Ob}(H) \times \text{Ob}(H)} \text{Mor}(H) \qquad (9.25)$$

induces an isomorphism of associated sheaves, and the map $\tilde{\pi}_0(\text{im}(p)) \to \tilde{\pi}_0(H)$ is a sheaf epimorphism. The map (9.25) is a monomorphism by construction, and so the morphism $p : G \to H$ is essentially surjective if and only if the presheaf morphism

$$\text{Mor}(G) \to (\text{Ob}(G) \times \text{Ob}(G)) \times_{\text{Ob}(H) \times \text{Ob}(H)} \text{Mor}(H)$$

is a local epimorphism and the map

$$\tilde{\pi}_0(G) \to \tilde{\pi}_0(H)$$

is a sheaf epimorphism.

It follows in particular that if the morphism $p : G \to H$ is essentially surjective, then it induces an isomorphism $\tilde{\pi}_0(G) \xrightarrow{\cong} \tilde{\pi}_0(H)$.

A *kernel* of a morphism $p : G \to H$ of presheaves of groupoids is a morphism of families

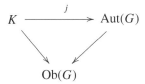

such that the diagram

$$
\begin{array}{ccc}
K & \xrightarrow{j} & \mathrm{Aut}(G) \\
\downarrow & & \downarrow p_* \\
\mathrm{Ob}(G) & \xrightarrow[p_* \cdot e]{} & \mathrm{Aut}(H)
\end{array}
$$

is a pullback. There is a canonically defined section $e : \mathrm{Ob}(G) \to \mathrm{Aut}(G)$ of the map $\mathrm{Aut}(G) \to \mathrm{Ob}(G)$ which picks out identities in all sections.

Suppose henceforth that $\mathcal{F} \to S$ is a family of sheaves of groups.

Given a morphism $p : G \to H$, a *kernel for p in the family $\mathcal{F}$* is a kernel $j : K \to \mathrm{Aut}(G)$ together with a pseudo-isomorphism of families $\alpha : K \to \mathcal{F}$.

An essentially surjective morphism $p : G \to H$ of presheaves of groupoids, together with a choice of kernel

in the family $\mathcal{F}$, is a member of the category $\mathbf{Ext}(H, \mathcal{F})$. A morphism $f : (p, j, \alpha) \to (p', j', \alpha')$ of this category is a local weak equivalence $f : G \to G'$ such that the diagram of morphisms of sheaves of groupoids

$$
\begin{array}{ccc}
G & \xrightarrow{f} & G' \\
& \searrow p \quad \swarrow p' & \\
& H &
\end{array}
$$

commutes and the diagram of family morphisms

$$
\begin{array}{ccc}
K & \xrightarrow{f_*} & K' \\
& \searrow \alpha \quad \swarrow \alpha' & \\
& \mathcal{F} &
\end{array}
$$

commutes. The induced map $f_* : K \to K'$ of families is necessarily a pseudo-isomorphism.

Suppose that (p, j, α) is an object of $\mathbf{Ext}(H, \mathcal{F})$. There is a morphism

$$F'(p) : R(p) \to \mathbf{Iso}(K)$$

9.5 Extensions and Gerbes

of presheaves of 2-groupoids which is the identity on 0-cells, takes the 1-cell $\alpha : x \to y$ to the isomorphism $K_x \to K_y$ of presheaves of groups which is defined by conjugation by α, and takes the 2-cell $\alpha \to \beta$ to the element $\beta\alpha^{-1}$ of $K_y(U)$. Write $F(p)$ for the composite

$$R(p) \xrightarrow{F'(p)} \mathbf{Iso}(K) \xrightarrow{\alpha_*} \mathbf{Iso}(\mathcal{F}).$$

If $f : (p, j, \alpha) \to (p', j', \alpha')$ is a morphism of $\mathbf{Ext}(H, \mathcal{F})$, then there is a commutative diagram of morphisms of presheaves of 2-groupoids

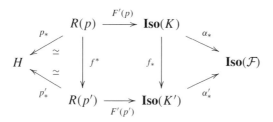

It follows that the assignment $(p, j, \alpha) \mapsto (p_*, F(p))$ defines a functor

$$\phi : \mathbf{Ext}(H, \mathcal{F}) \to h(H, \mathbf{Iso}(\mathcal{F})),$$

where $h(H, \mathbf{Iso}(\mathcal{F}))$ is the category of cocycles from H to $\mathbf{Iso}(\mathcal{F})$ in the category of presheaves of 2-groupoids, for the model structure of Theorem 9.57.

Suppose given a cocycle

$$H \xleftarrow[\simeq]{g} A \xrightarrow{F} \mathbf{Iso}(\mathcal{F})$$

in presheaves of 2-groupoids, where H is a presheaf of groupoids. There is a presheaf of 2-groupoids $\mathbf{E}_A F$, which is associated to the map

$$F : A \to \mathbf{Iso}(\mathcal{F}),$$

whose 0-cells are the 0-cells of A. The 1-cells $x \to y$ of $\mathbf{E}_A F(U)$ are pairs (α, f) consisting of a 1-cell $\alpha : x \to y$ of $A(U)$ and an element f of the group $\mathcal{F}_{F(y)}(U)$. A 2-cell $(\alpha, f) \to (\beta, g)$ of $\mathbf{E}_A F(U)$ is a 2-cell $h : \alpha \to \beta$ of $A(U)$ such that the diagram

commutes in the group $\mathcal{F}_{F(y)}(U)$.

Given 1-cells

$$x \xrightarrow{(\alpha,f)} y \xrightarrow{(\beta,g)} z,$$

their composite is the 1-cell $(\beta\alpha, g\beta_*(f)) : x \to z$.

Given 2-cells $h : (\alpha, f) \to (\alpha', f')$ from x to y and $k : (\beta, g) \to (\beta', g')$ from y to z, there is a commutative diagram

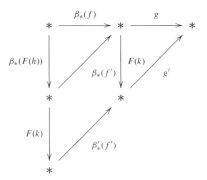

in the group $\mathcal{F}_{F(z)}(U)$. The element $F(k)\beta_*(F(h)) \in \mathcal{F}_{F(z)}(U)$ is the image of the horizontal composite $k*h : \beta\cdot\alpha \to \beta'\cdot\alpha'$ of $A(U)$ under the 2-groupoid homomorphism F, and so we have a 2-cell

$$k * h : (\beta\alpha, g\beta_*(f)) \to (\beta'\alpha', g'\beta'_*(f')).$$

This 2-cell is the horizontal composite of the 2-cells k and h of $\mathbf{E}_A F(U)$. Vertical composition of 2-cells $x \to y$ in $\mathbf{E}_A F(U)$ is easily defined.

Write

$$E_A F = \pi_0(\mathbf{E}_A(F)).$$

In other words, $E_A(F)$ is the path component groupoid of the 2-groupoid object $\mathbf{E}_A(F)$. The presheaf of groupoids $E_A F$ is the *Grothendieck construction* for the map F.

There is a morphism $\pi : \mathbf{E}_A F \to A$ of presheaves of 2-groupoids which is the identity on objects, is defined on 1-cells by $(\alpha, f) \mapsto \alpha$, and takes a 2-cell $h : (\alpha, f) \to (\beta, g)$ to the underlying 2-cell $h : \alpha \to \beta$ of A. The induced morphism

$$\pi_* : E_A F \to \pi_0 A$$

of path component groupoids is the identity on objects and is full in all sections. Write g_* for the composite

$$E_A F \to \pi_0 A \xrightarrow{\simeq} H.$$

This morphism g_* of presheaves of groupoids is essentially surjective.

9.5 Extensions and Gerbes

There is a homomorphism

$$j : \mathcal{F}_{F(x)} \to E_A F_x$$

of presheaves of groups, which is defined in sections by the assignment $g \mapsto [(1_x, g)]$. Set $K(F)_x = \mathcal{F}_{F(x)}$ for all 0-cells x of A. We therefore have a pseudo-isomorphism of families $\alpha_F : K(F) \to \mathcal{F}$ which is defined by pullback along the presheaf morphism $F : \mathrm{Ob}(A) \to S$. We also have a map of families $j_F : K(F) \to \mathrm{Aut}(G)$.

Lemma 9.65 *The sequence of homomorphisms of presheaves of groups*

$$e \to \mathcal{F}_{F(x)} \xrightarrow{i} E_A F_x \xrightarrow{\pi_*} \pi_0 A_x$$

is exact.

Proof We show that the composite $\pi_* \cdot i$ is trivial and that i is a monomorphism.

An element of the kernel of the map

$$\pi_* : E_A F_x \to \pi_0 A_x$$

is represented by a 1-cell $(\alpha, f) : x \to x$ such that there is a 2-cell $h : \alpha \to 1_x$ in A. Form the picture

of group elements in $\mathcal{F}_{F(x)}$. It follows that $[(\alpha, f)] = [(1_x, \theta)]$, which is in the image of i.

If $[(1_x, g)] = [(1_x, g')]$ for $g, g' \in \mathcal{F}_{F(x)}$, there is a 2-cell $h : 1_x \to 1_x$ such that the diagram of group elements

commutes in $\mathcal{F}_{F(x)}$. But there is only one 2-cell $1_x \to 1_x$ in A, namely the identity, so that $F(h) = e$ and $g = g'$.

It follows from Lemma 9.65 that the family maps

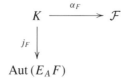

define a kernel in the family $\mathcal{F}$ for the essentially surjective groupoid homomorphism $g_* : E_A F \to H$.

The assignment $(g, F) \mapsto (g_*, j_F, \alpha_F)$ is functorial in cocycles, and hence defines a functor
$$\psi : h(H, \mathbf{Iso}(\mathcal{F})) \to \mathbf{Ext}(H, \mathcal{F}).$$

The functors ϕ and ψ are natural with respect to pseudo-isomorphisms of families $\mathcal{F} \to \mathcal{F}'$ in sheaves of groups.

Start with the essentially surjective groupoid morphism $p : G \to H$ with kernel

in $\mathcal{F}$, and form the corresponding cocycle
$$H \xleftarrow{\simeq} R(p) \xrightarrow{F'(p)} \mathbf{Iso}(K) \xrightarrow{\alpha_*} \mathbf{Iso}(\mathcal{F}).$$

The 2-groupoid $\mathbf{E}_{R(p)} F'(p)$ has the 0-cells of G as 0-cells, 1-cells $(\alpha, f) : x \to y$ consisting of 1-cells $\alpha : x \to y$ of G, and $f \in K_y$ and 2-cells $(\alpha, f) \to (\beta, g)$ consisting of those 2-cells $h : \alpha \to \beta$ ($p(\alpha) = p(\beta)$) such that the diagram

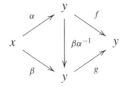

commutes in K_y. The corresponding diagram

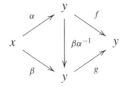

commutes in G, and it follows that the assignment $(\alpha, f) \mapsto f \cdot \alpha$ defines a 2-groupoid morphism $\mathbf{E}_{R(p)} F'(p) \to G$. The induced groupoid morphism
$$\epsilon : E_{R(p)} F'(p) \to G$$

9.5 Extensions and Gerbes

is an isomorphism, and there is a commutative diagram

$$
\begin{array}{ccc}
E_{R(p)}F'(p) & \xrightarrow{\epsilon}_{\cong} & G \\
& \searrow^{\pi_*} \quad \downarrow^{p} & \\
& H &
\end{array}
\tag{9.26}
$$

of groupioid homomorphisms. The kernel of the map $E_{R(p)}F'(p) \to H$ is the map $j' : K \to \text{Aut}(E_{R(p)}F'(p))$ which is specified by the inclusions $f \mapsto [(e, f)]$ for $f \in K_x$. There is a commutative diagram of family morphisms

$$
\begin{array}{ccc}
& K & \\
{}^{j'}\swarrow & & \searrow^{j} \\
\text{Aut}(E_{R(p)}F'(p)) & \xrightarrow[\epsilon_{ast}]{\cong} & \text{Aut}(G)
\end{array}
$$

The groupoid homomorphism ϵ defines a morphism $\psi\phi(p, j, 1) \to (p, j, 1)$ of $\mathbf{Ext}(H, K)$. It follows that

$$\psi_*\phi_*[(p, j, \alpha)] = [(p, j, \alpha)]$$

in $\pi_0\mathbf{Ext}(H, \mathcal{F})$ for any object (p, j, α) of $\mathbf{Ext}(H, \mathcal{F})$.

Suppose now that

$$H \xleftarrow[\cong]{g} A \xrightarrow{F} \mathbf{Iso}(\mathcal{F})$$

is a cocycle, and form the diagram

$$
\begin{array}{ccc}
E_A F & \longleftarrow & \mathbf{E}_A F \\
{}^{\pi_*}\downarrow & & \downarrow^{\pi} \\
H & \xleftarrow{g} & A
\end{array}
$$

The map π_* is defined by the assignment $[(\alpha, f)] \mapsto g(\alpha)$. There is a commutative diagram of 2-groupoid morphisms

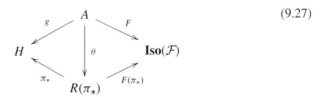

(9.27)

where θ is the identity on 0-cells, and takes a 1-cell $\alpha : x \to y$ to the 1-cell $[(\alpha, e)] : x \to y$.

For the effect of θ on 2-cells, suppose given 1-cells $\alpha, \beta : x \to y$ of A and a 2-cell $h : \alpha \to \beta$. Then α and β have the same image in the path component groupoid $\pi_0 A$, so that $g(\alpha) = g(\beta)$. It follows that there is a unique 2-cell $h_* : [(\alpha, e)] \to [(\beta, e)]$ in $R(\pi_*)$, and $\theta(h) = h_*$.

There is an identity

$$(\alpha, e)(1_y, f)(\alpha^{-1}, e) = (1_y, \alpha_*(f))$$

in 1-cells of $\mathbf{E}_A F$, so that F and $F(\pi_*)\theta$ coincide on 1-cells. Also, there is an identity

$$[(\alpha, F(h))] = [(\beta, e)]$$

in $E_A F$ for each 2-cell $h : \alpha \to \beta$ of A. It follows that

$$[(\beta\alpha^{-1}, e)] = [(1_y, F(h))]$$

so that the 2-cell h has the same image under F and the composite $F(\pi_*)\theta$.

It follows that

$$\phi_* \psi_*([(g, F)]) = [(g, F)]$$

for all objects (g, F) of the cocycle category $h(H, \mathbf{Iso}(\mathcal{F}))$.

We have therefore proved the following:

Theorem 9.66 *Suppose that H is a presheaf of groupoids and that $\mathcal{F}$ is a family of sheaves of groups, on a fixed Grothendieck site $\mathcal{C}$. Then the functions ϕ and ψ define a natural bijection*

$$\pi_0 \mathbf{Ext}(H, \mathcal{F}) \cong \pi_0 h(H, \mathbf{Iso}(\mathcal{F})).$$

Corollary 9.67 *Suppose that H is a presheaf of groupoids and that $\mathcal{F}$ is a family of sheaves of groups. Then there are natural bijections*

$$\pi_0 \mathbf{Ext}(H, \mathcal{F}) \cong [H, \mathbf{Iso}(\mathcal{F})] \cong [BH, \overline{W}\mathbf{Iso}(\mathcal{F})],$$

where $[H, \mathbf{Iso}(\mathcal{F})]$ is morphisms in the homotopy category of presheaves of 2-groupoids, and $[BH, \overline{W}\mathbf{Iso}(\mathcal{F})]$ is morphisms in the homotopy category of simplicial presheaves.

Proof There is a natural bijection

$$\pi_0 h(H, \mathbf{Iso}(\mathcal{F})) \cong [H, \mathbf{Iso}(\mathcal{F})]$$

by Theorem 6.5. The bijection

$$[H, \mathbf{Iso}(\mathcal{F})] \cong [BH, \overline{W}\mathbf{Iso}(\mathcal{F})]$$

9.5 Extensions and Gerbes

is induced by the functor $\overline{W}$—see Theorem 9.43. There is a natural weak equivalence
$$BH \cong dBH \xrightarrow{\simeq} \overline{W}H$$
for all presheaves of groupoids H, by Corollary 9.39.

Let $*$ be the trivial presheaf of groupoids with one object—this object is terminal in the category of presheaves of groupoids.

Suppose that G is a presheaf of groupoids. Then the canonical map $G \to *$ is essentially surjective if and only if G is locally connected in the sense that the sheaf $\tilde{\pi}_0(G)$ is terminal. In other words, the map $G \to *$ is essentially surjective if and only if its associated stack is a gerbe in the traditional sense. This characterization is invariant of the homotopy type of G in the injective model structure for presheaves of groupoids, and we shall say, more generally, that a *gerbe* is a presheaf of groupoids, which is locally connected.

Suppose that G is a gerbe, and let $\mathcal{F}$ be a family of sheaves of groups. A kernel for the map $G \to *$ in the family $\mathcal{F}$ is a a pseudo-isomorphism $\alpha : \text{Aut}(G) \to \mathcal{F}$. Recall that the map α determines a cocycle
$$* \xleftarrow{\simeq} R(G) \xrightarrow{\alpha_*} \mathbf{Iso}(\mathcal{F}).$$

Here, $R(G)$ has the same 0-cells and 1-cells as G, and there is a unique 2-cell between any two 1-cells.

From a cocycle-theoretic point of view, the choice of the pseudo-isomorphism α does not matter, because any two choices α, α' determine maps of cocycles

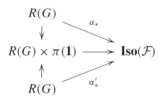

over $*$ relating the cocycles associated to α and α'.

It follows that the set $\pi_0\mathbf{Ext}(*, \mathcal{F})$ can be identified with weak equivalence classes of locally connected (pre)sheaves of groupoids G admitting a pseudo-isomorphism $\text{Aut}(G) \to \mathcal{F}$. These objects are otherwise known as $\mathcal{F}$-*gerbes*, and the set of weak equivalence classes of $\mathcal{F}$-gerbes is the non-abelian cohomology object $H^2(\mathcal{C}, \mathcal{F})$.

We have proved the following:

Corollary 9.68 *There are natural isomorphisms*
$$H^2(\mathcal{C}, \mathcal{F}) \cong \pi_0 h(*, \mathbf{Iso}(\mathcal{F})) \cong [*, \overline{W}\mathbf{Iso}(\mathcal{F})]$$
for all families of sheaves of groups $\mathcal{F}$ on the site $\mathcal{C}$.

There is a pointed version of Theorem 9.66. Suppose in the following that H is a presheaf of groupoids and that $\mathcal{F}$ is a family of sheaves of groups.

A base point of an object (p, j, α) of $\mathbf{Ext}(H, \mathcal{F})$ is a global section $x \in G$, where $p : G \to H$ is the groupoid morphism in the structure. A morphism $f : (p, j, \alpha) \to (p', j', \alpha')$ is pointed if the groupoid morphism $f : G \to G'$ preserves base points. Write $\mathbf{Ext}_*(H, \mathcal{F})$ for the corresponding pointed category.

A base point for a cocycle

$$H \xleftarrow{g} A \xrightarrow{F} \mathbf{Iso}(\mathcal{F})$$

is a choice of 0-cell x in the global sections of A, and a cocycle morphism $\theta : (g, F) \to (g', F')$ is pointed if the morphism $\theta : A \to A'$ of presheaves of groupoids preserves base points. Write $h_*(H, \mathbf{Iso}(\mathcal{F}))$ for the corresponding pointed cocycle category.

If $f : (p, j, \alpha) \to (p', j', \alpha')$ is a pointed map, then the base point $x \in G$ is also a 0-cell of the 2-groupoid $R(p)$, hence a base point, and the induced map $f_* : R(p) \to R(p')$ preserve base points. Similarly, if (g, F) is a cocycle, then the objects of the groupoid $E_A F$ are the 0-cells of A, so that we have a natural choice of base point for the object $\Psi(g, F) \in \mathbf{Ext}(H, \mathcal{F})$. It follows that the functors ϕ and ψ restrict to functors

$$\phi : \mathbf{Ext}_*(H, \mathcal{F}) \leftrightarrows h_*(H, \mathbf{Iso}(\mathcal{F})) : \psi$$

on the respective pointed categories. The morphisms in (9.26) and (9.27) preserve base points, so that the proof of Theorem 9.66 restricts to the pointed case, giving the following theorem:

Theorem 9.69 *Suppose that H is a presheaf of groupoids and that $\mathcal{F}$ is a family of sheaves of groups. Then the functions ϕ and ψ define a bijection*

$$\pi_0 \mathbf{Ext}_*(H, \mathcal{F}) \cong \pi_0 h_*(H, \mathbf{Iso}(\mathcal{F})).$$

Corollary 9.70 *Suppose that H is a presheaf of groupoids and that $\mathcal{F}$ is a family of sheaves of groups. Then there is a natural bijection*

$$\pi_0 \mathbf{Ext}_*(H, \mathcal{F}) \cong [BH, \overline{W} \mathbf{Iso}(\mathcal{F})]_*$$

relating $\pi_0 \mathbf{Ext}_*(H, \mathcal{F})$ to a set of morphisms in the homotopy category of pointed simplicial presheaves.

Suppose that H is a presheaf of groups and that L is a sheaf of groups. Suppose $x \in G$ is a base point of an object (p, j, α) of $\mathbf{Ext}_*(H, L)$, where L is the family of groups $L \to *$ consisting of L alone.

Let $p_x : G_x \to H$ be the local epimorphism of presheaves of groups which is associated to x, let $j_x : K_x \to G_x$ be the restricted kernel, and let α_x be the restricted family morphism

$$K_x \subset K \xrightarrow{\alpha} L.$$

9.5 Extensions and Gerbes

The map α_x is a fixed choice of local isomorphism $K_x \xrightarrow{\simeq} L$. The unique object of G_x is a natural choice of base point for the object (p_x, j_x, α_x), and there is a canonical pointed map $(p_x, j_x, \alpha_x) \to (p, j, \alpha)$, which is natural in pointed morphisms $f : (p, j, \alpha) \to (p', j', \alpha')$.

It follows that $\pi_0 \mathbf{Ext}_*(H, L)$ can be identified with isomorphism classes of the data

$$e \longrightarrow K \xrightarrow{j} G \xrightarrow{p} H$$
$$\alpha \downarrow \simeq$$
$$L$$

consisting of exact sequences of presheaves of groups as displayed, with fixed local isomorphisms $\alpha : K \to L$, and with p a local epimorphism. A morphism of these data is a local isomorphism $\theta : G \xrightarrow{\simeq} G'$ which respects all choices. In particular, the induced local isomorphism $\theta_* : K \to K'$ on kernels sits in a commutative diagram

The set $\pi_0 \mathbf{Ext}_*(H, L)$ can therefore be identified with the set of path components of the category of extensions

$$e \to L \to G \to \tilde{H} \to e.$$

in the sheaf category. A morphism in this category is a commutative diagram

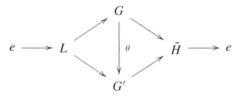

in the category of sheaves of groups, where the top and bottom sequences are exact. The morphism θ is necessarily an isomorphism.

We have proved the following:

Corollary 9.71 *Suppose that H is a presheaf of groups and that L is a sheaf of groups. Then the set $\pi_0 h_*(H, \mathbf{Iso}(L))$ of path components of the pointed cocycle category $h_*(H, \mathbf{Iso}(L))$ can be identified with the set of isomorphism classes of extensions*

$$e \to L \to G \to \tilde{H} \to e$$

in the category of sheaves of groups.

Corollary 9.72 *Suppose that H is a presheaf of groups and that L is a sheaf of groups. Then the set of isomorphism classes of extensions*

$$e \to L \to G \to \tilde{H} \to e$$

is naturally isomorphic to the set of morphisms $[BH, \overline{W}\mathbf{Iso}(L)]_*$ in the homotopy category of pointed simplicial presheaves.

Part IV
Stable Homotopy Theory

Chapter 10
Spectra and T-spectra

The local stable model structure for presheaves of spectra appeared in [51], soon after the introduction of the local homotopy theories for simplicial sheaves and presheaves.

A presheaf of spectra E on a site $\mathcal{C}$ is a diagram in ordinary spectra, in that it consists of pointed simplicial presheaves E^n, $n \geq 0$, and pointed maps

$$S^1 \wedge E^n \to E^{n+1},$$

where $S^1 = \Delta^1/\partial\Delta^1$ is the *simplicial circle*, interpreted as a constant simplicial presheaf. A morphism $E \to F$ of presheaves of spectra consists of pointed maps $E^n \to F^n$ which respect structure.

Elements of local stable homotopy theory were already in place before the model structure appeared, for example in Thomason's work on étale descent for Bott periodic algebraic K-theory [99]. Early outcomes of the theory included a diagram-theoretic definition of étale K-theory, the identification of sheaf cohomology groups as stable homotopy groups of spectra, and the Nisnevich descent theorem for algebraic K-theory [83].

The sheaf cohomology spectra, for example, are easily constructed with the methods of the first section of this chapter—see Example 10.2. If A is a sheaf of abelian groups on a site $\mathcal{C}$ and $j : H(A) \to QH(A)$ is a stable fibrant model of the Eilenberg–Mac Lane presheaf of spectra $H(A)$ (which is constructed in the way one would expect), then the global sections spectrum $\Gamma_* QH(A)$ for the stable fibrant object $QH(A)$ has stable homotopy groups given by the sheaf cohomology $H^*(\mathcal{C}, A)$ in negative degrees, in the sense that

$$\pi_s \Gamma_* QH(A) = \begin{cases} 0 & \text{if } s > 0, \\ H^{-s}(\mathcal{C}, A) & \text{if } s \leq 0. \end{cases}$$

This statement is the starting point for many calculations that one makes within the stable homotopy theories of presheaves of spectra, for example with descent spectral sequences. The moral is, quite generally, that the local stable homotopy type of a presheaf of spectra is determined by sheaf cohomology.

In the years following its introduction, the homotopy theory of presheaves of spectra has become a common tool in algebraic K-theory and in some areas of stable homotopy theory. The latter include elliptic cohomology theories and the theory of topological modular forms [34, 74], and, to a lesser extent, equivariant stable homotopy theory [40].

The stable model structure for presheaves of spectra is the subject of Sect. 10.1. The argument for its existence, which is presented here, is not the original (which followed the methods of Bousfield and Friedlander [13]), but instead focuses on cofibrant generation in a way that is consistent with the methods of other parts of this monograph. The reader will observe that results from the other sections of this chapter specialize to the basic results for presheaves of spectra in the first section, but the idea for this exposition is to start with a relatively simple and self-contained introduction to the early, "naive" form of the theory.

There are multiple forms of local stable homotopy theory, of varying degrees of complexity. These theories are largely modelled on and specialize to motivic stable homotopy theory [57]. The prominent formal aspects of motivic stable homotopy theory are discussed in a series of examples, which appear at various points in this chapter and in Chap. 11. The motivic stable category was introduced by Voevodsky [102], in part as a vehicle to represent motivic cohomology theory within an appropriate homotopy category.

In motivic stable homotopy theory, one requires that the affine line $\mathbb{A}^1$ should be a point, as in basic motivic homotopy theory (Examples 7.22, 7.28). In this context, the simplicial (or topological) circle S^1 is replaced by a geometric circle-like object $\mathbb{A}^1/(\mathbb{A}^1 - 0)$, which is a "mixed" 2-sphere in the sense that there is a pointed motivic weak equivalence

$$\mathbb{A}^1/(\mathbb{A}^1 - 0) \simeq S^1 \wedge (\mathbb{A}^1 - 0) = S^1 \wedge \mathbb{G}_m,$$

where $\mathbb{G}_m$ is the multiplicative group, with underlying scheme $\mathbb{A}^1 - 0$. There is also a motivic weak equivalence

$$\mathbb{A}^1/(\mathbb{A}^1 - 0) \simeq \mathbb{P}^1,$$

by a standard argument, where $\mathbb{P}^1$ is the projective line.

Motivic cohomology theory has the additional requirement that transfers should be built into the theory (Example 8.49), and this is done formally. The motive $\mathbb{Z}(n)$ is a motivic fibrant model of the free simplicial presheaf with transfers

$$\mathbb{Z}_{tr}((\mathbb{G}_m \wedge S^1)^{\wedge n})$$

that is associated to the simplicial presheaf $(\mathbb{G}_m \wedge S^1)^{\wedge n}$ on the smooth Nisnevich site. The motive $\mathbb{Z}(n)$ is the object at level n for a motivic Eilenberg–Mac Lane spectrum object $\mathbf{H}_\mathbb{Z}$ which has "bonding maps"

$$(S^1 \wedge \mathbb{G}_m) \wedge \mathbb{Z}(n) \to \mathbb{Z}(n+1),$$

which are defined by an adjunction map. This description of the motivic Eilenberg–Mac Lane spectrum thus involves smashing with a parameter object $S^1 \wedge \mathbb{G}_m$ which

Spectra and T-spectra

is certainly not a circle; it is the smash of the topological circle with a geometric circle, and is more like a twisted 2-sphere.

We begin a general discussion, in Sect. 10.2, by generalizing the parameter object S^1 to a pointed simplicial presheaf T on a site $\mathcal{C}$. A T-spectrum X consists of pointed simplicial presheaves X^n, $n \geq 0$, together with pointed maps

$$\sigma : T \wedge X^n \to X^{n+1},$$

called bonding maps. A morphism $X \to Y$ of T-spectra is a collection of pointed maps $X^n \to Y^n$ which respect bonding maps in the way one would expect from ordinary stable homotopy theory. The corresponding category of T-spectra is denoted by $\mathbf{Spt}_T(\mathcal{C})$. In this language, a presheaf of spectra is an S^1-spectrum.

As in the early constructions of stable model structures [13], there is a preliminary strict model structure on the T-spectrum category for which the fibrations and weak equivalences are defined levelwise in the injective model structure for pointed simplicial presheaves. All other model structures for T-spectra in this chapter are constructed by formally inverting a set of maps in the strict structure, by using the methods of Chap. 7. The main overall existence result is Theorem 10.20.

The stable model structure on $\mathbf{Spt}_T(\mathcal{C})$ is constructed in Example 10.22 by formally inverting the stablilization maps

$$(S_T \wedge T)[-1-n] \to S_T[-n], \ n \geq 0.$$

Here, S_T is the sphere T-spectrum

$$S^0, \ T, \ T \wedge T, \ \ldots T^{\wedge k}, \ \ldots,$$

which is composed of the smash powers of the parameter object T, and the displayed objects are shifted in the usual sense. The stabilization map $(S_T \wedge T)[-1] \to S_T$ is defined by the identities on the smash powers $T^{\wedge k}$ in levels $k \geq 1$.

In applications, one typically wants to invert some cofibration of simplicial presheaves $f : A \to B$ within the stable model structure, and one does this by *also* inverting a set of maps

$$(S_T \wedge C)[-n] \to (S_T \wedge D)[-n], \ n \geq 0,$$

which are induced by a set of generators $C \to D$ for the trivial cofibrations of the f-local model structure for pointed simplicial presheaves. The resulting localized model structure, which appears in Example 10.23, is the stable f-local model structure for the category of T-spectra.

The motivic stable category over a scheme S is the stable f-local model structure for T-spectra on the smooth Nisnevich site over S, where f is a rational point $* \to \mathbb{A}^1$ in the affine line $\mathbb{A}^1$ over S, and T is the smash $S^1 \wedge \mathbb{G}_m$. See Example 10.41. It is a consequence of Theorem 10.40 that we could use anything motivically equivalent to $S^1 \wedge \mathbb{G}_m$ (such as the projective line $\mathbb{P}^1$) as a parameter object T, and be assured that the localization process produces an equivalent model for the motivic stable category.

This localization-theoretic construction of the motivic stable model structure differs from that of [57], which uses the method of Bousfield and Friedlander. The approach that is used in this chapter more closely resembles a method which was displayed by Hovey in [45].

The overall localization construction for the f-local stable model structures for T-spectra is wildly general, and is not subject to any of the restrictions of the Bousfield–Friedlander method, such as properness for the underlying model structure or the assumption of the existence of a stable fibrant model $X \to QX$. One does, however, have to make some assumptions about both the parameter object T and the f-local model structure on the underlying category of simplicial presheaves in order for the f-local stable category of T-spectra to look anything like the ordinary stable category.

The first of these assumptions is a compactness requirement for the parameter object T, which makes the T-loops functor behave properly with respect to filtered colimits in the f-local model category.

This form of compactness is a bit technical; a pointed simplicial presheaf K is compact up to f-local equivalence if, for any inductive system $s \mapsto Z_s$ of pointed f-fibrant simplicial presheaves, the composite map

$$\varinjlim_s \Omega_K Z_s \to \Omega_K(\varinjlim_s Z_s) \to \Omega_K F(\varinjlim_s Z_s),$$

which is induced by an f-fibrant model

$$\varinjlim_s Z_s \to F(\varinjlim_s Z_s)$$

is an f-local equivalence. Here, the "K-loops" functor $Z \mapsto \Omega_K Z$ is the right adjoint of smashing with K.

It is not at all obvious that the K-loops functor preserves f-local equivalences, even for ordinary loop objects corresponding to $K = S^1$. This potential failure is counterintuitive and is one of the more difficult aspects of the theory to work around. Generally, something interesting must be afoot to show that this sort of compactness holds in a particular setting. For the motivic case (Example 10.29), one uses Nisnevich descent (Theorem 5.39) to show that all pointed S-schemes T are compact up to motivic equivalence, as are the objects $S^1 \wedge T$.

The specific assumption that we make is the following:

A1: The parameter object T is compact up to f-equivalence.

This assumption is powerful and is used almost everywhere, starting in Sect 10.3. The first major consequence is Theorem 10.32, which says that this assumption enables the construction

$$\eta : X \to Q_T X$$

of a natural stable f-fibrant model, by analogy with the stable fibrant model construction for ordinary spectra and presheaves of spectra.

Spectra and T-spectra

To go further, we need a second assumption, this time on the underlying f-local model structure. Say that the f-local model structure satisfies inductive colimit descent, if given an inductive system $s \mapsto Z_s$ of f-fibrant simplicial presheaves, then any f-fibrant model

$$\varinjlim_s Z_s \to F(\varinjlim_s Z_s) \tag{10.1}$$

must be a local weak equivalence, instead of just an f-local equivalence.

Here is the assumption:

A2: The f-local model structure satisfies inductive colimit descent.

In the presence of this assumption on the f-local model structure, any finite pointed simplicial set is compact up to f-equivalence (Lemma 10.35). The assumption **A2** holds in the motivic homotopy category, since maps of the form (10.1) are sectionwise weak equivalences in that case, by Nisnevich descent. More generally, if we assume that T is compact up to f-equivalence and that the f-local model structure satisfies inductive colimit descent, then the object $S^1 \wedge T$ is compact up to f-equivalence.

The special, pleasant features of $(S^1 \wedge T)$-spectra are discussed in Sects. 10.5 and 10.6. The f-local stable category of these objects looks most like the ordinary stable category, thanks to the fact that the parameter object is a suspension in the classical sense.

The assumptions **A1** and **A2** imply that fibre and cofibre sequences coincide in the f-local stable homotopy category of $(S^1 \wedge T)$-spectra (Corollary 10.59, Lemma 10.62), and that this stable homotopy category has the additivity property (Lemma 10.67). One can define bigraded sheaves of stable homotopy groups $\tilde{\pi}_{s,t}(X)$ for $(S^1 \wedge T)$-spectra X which reflect stable f-equivalence (Lemma 10.68), and there is a calculus of long exact sequences in sheaves of stable homotopy groups for fibre/cofibre sequences (Lemma 10.70) which extends known behaviour of the motivic stable category (Corollary 10.71)

The main technical tool in the proofs of these results is a theory of (S^1, T)-bispectra, which is discussed at the beginning of Sect. 10.5. Lemma 10.56 says that, if $X \to Y$ is a map of (S^1, T)-bispectra which is a stable f-equivalence in either the S^1-direction or the T-direction, then it induces a stable f-equivalence of diagonal $(S^1 \wedge T)$-spectra. This is the mechanism by which theorems from classical stable homotopy category imply corresponding statements for $(S^1 \wedge T)$-spectra.

The presence of long exact sequences for a fibration guarantees right properness (Theorem 10.64) for the f-local stable model structure on $(S^1 \wedge T)$-spectra, *without* any properness assumption on the underlying f-local model structure for simplicial presheaves. One can show that the f-local stable model structure of T-spectra is right proper for unsuspended parameter objects T, but only at the cost of assuming right properness of the f-local model structure for simplicial presheaves—this is the content of Theorem 10.36.

Section 10.6 contains a discussion of Postnikov towers and slice filtrations for $(S^1 \wedge T)$-spectra, again subject to the assumptions **A1** and **A2**. The formation of the qth Postnikov section, here denoted by $Z_{\mathbf{L}_q}$, in the T-direction for an object Z amounts to killing the stable homotopy group sheaves $\tilde{\pi}_{s,t} X$ with $t > q$ in the classical way. This process is a localization construction, for which the map $Z \to Z_{\mathbf{L}_q}$ is a fibrant model. The homotopy fibres of these maps form the slice filtration of Z, for T-connective objects Z. This general construction specializes to the Postnikov tower construction for spectra and presheaves of spectra, and to the slice filtration for the motivic stable category.

We say that the parameter object T is *cycle trivial* if the shuffle map

$$c_{1,2} : T^{\wedge 3} \to T^{\wedge 3}$$

which is defined by

$$x_1 \wedge x_2 \wedge x_3 \mapsto x_2 \wedge x_3 \wedge x_1$$

represents the identity map in the f-local homotopy category of pointed simplicial presheaves.

The simplicial circle S^1 is cycle trivial, by a standard argument on degree. The projective line $\mathbb{P}^1$ is cycle trivial in the motivic homotopy category since the shuffle $c_{1,2}$ is a product of elementary transformations in the special linear group $Sl_3(\mathbb{Z})$, and therefore has a homotopy along the affine line to the identity—see Example 10.45. The cycle triviality of the projective line $\mathbb{P}^1$ was first observed by Voevodsky [102].

Here is our last major assumption:

A3: The parameter object T is cycle trivial.

This assumption is used to show that the T-suspension functor $X \mapsto X \wedge T$ is invertible in the stable f-local category of T-spectra, with inverse given by T-loops (Theorem 10.50), and that T-suspension, "fake" T-suspension and shift by 1 are equivalent in the sense that there are natural f-local stable equivalences

$$X \wedge T \simeq \Sigma_T X \simeq X[1].$$

The fake suspension $\Sigma_T X$ is defined in levels by $(\Sigma_T X)^n = T \wedge X^n$ and has bonding maps of the form $T \wedge \sigma$. The equivalence between the T-suspension and the fake T-suspension functors is constructed in the category of $(T \wedge T)$-spectra in Proposition 10.53. The proofs of these results specialize to new demonstrations of some well known theorems about the ordinary stable category.

The parameter object $S^1 \wedge \mathbb{G}_m$ and the motivic model structure for pointed simplicial presheaves on the smooth Nisnevich site of a scheme together satisfy the assumptions **A1, A2** and **A3**. Thus, by the results of Sects. 10.5 and 10.6, the motivic stable category of $(S^1 \wedge \mathbb{G}_m)$-spectra has many of the calculational attributes of the ordinary stable category. The same can be said, more generally, for the f-local stable category of $(S^1 \wedge T)$-spectra on an arbitrary site $\mathcal{C}$, provided that the object T and the underlying f-local model structure of pointed simplicial presheaves satisfy these three requirements.

The stabilization philosophy can be imported into the context of simplicial R-modules, where R is a presheaf of commutative unitary rings, by using an analogy of the localization technique for simplicial R-modules of Sect. 8.5.

Again, suppose that the parameter object T is a pointed simplicial presheaf. If E is a simplicial R-module, we write

$$T \otimes E = R_\bullet(T) \otimes E,$$

where $R_\bullet(T)$ is the free reduced simplicial R-module which is associated to T.

A T-complex A is a T-spectrum object in simplicial R-modules. It consists of simplicial R-modules A^n, $n \geq 0$, together with bonding maps

$$\sigma : T \otimes A^n \to A^{n+1}.$$

A morphism $A \to B$ of T-complexes consists of simplicial R-module maps $A^n \to B^n$ which respect bonding maps in the obvious way, and we write $\mathbf{Spt}_T(s\mathbf{Mod}_R)$ for the resulting category.

The concept of T-complex is a generalization of S^1-spectrum objects in simplicial R-modules, or S^1-complexes. Derived categories of unbounded complexes can be recovered from categories of S^1-complexes, with appropriate stabilization maps inverted.

There is a strict model structure (Proposition 10.80) on the category of T-complexes, just as for T-spectra, for which the weak equivalences and fibrations are defined levelwise in the injective model structure for simplicial R-modules of Theorem 8.6. We proceed by formally inverting sets of maps of T-spectra in the strict model structure for the category of T-complexes, by following the method used for simplicial R-modules in Theorem 8.39. The corresponding general localization result for T-complexes is Theorem 10.84.

The same set of maps which are inverted to construct the f-local stable of T-spectra can be inverted in T-complexes to produce the f-local stable model structure for T-complexes. This construction implicitly produces (Remark 10.85) a Quillen adjunction

$$R_\bullet : \mathbf{Spt}_T(\mathcal{C}) \leftrightarrows \mathbf{Spt}_T(s\mathbf{Mod}_R) : u$$

between the respective f-local structures, which is defined by the free reduced simplicial R-module functor $R_\bullet$ and the forgetful functor u. Further, a T-complex Z is stable f-fibrant (i.e. fibrant for the f-local stable model structure) if and only if the underlying T-spectrum $u(Z)$ is stable f-fibrant.

The f-local stable model structure for T-complexes has the same general properties as does the f-local stable model structure for T-spectra, in the presence of the assumptions **A1**, **A2** and **A3**. For example, the usual stabilization construction $X \to Q_T X$ is a stable f-fibrant model if T is compact (Theorem 10.87), and the T-suspension functor $X \mapsto X \otimes T$ is invertible on the f-local stable category if T is compact and cycle trivial (Theorem 10.90).

If T is compact and the inductive colimit descent assumption holds, then the f-local stable category of $(S^1 \wedge T)$-complexes can be analysed with familiar tools from ordinary stable homotopy theory, by analogy with the behaviour of the f-local stable category for $(S^1 \wedge T)$-spectra.

In particular, if
$$H(R) = R_\bullet(S_{S^1 \wedge T})$$
is the sphere spectrum object for $(S^1 \wedge T)$-complexes and K is a pointed simplicial presheaf, then the canonical map
$$u(H(R)) \wedge K \to u(H(R) \otimes K)$$
is an f-local stable equivalence, by Lemma 10.91. This result is central to the proof of the assertion, in Proposition 10.94, that a map $E \to F$ of $(S^1 \wedge T)$-complexes is a stable f-equivalence if and only if the underlying map $u(E) \to u(F)$ is a stable f-equivalence of $(S^1 \wedge T)$-spectra, in the presence of assumptions **A1** and **A2**.

This general theory specializes to a motivic stable model structure for the category of $(S^1 \wedge \mathbb{G}_m)$-complexes in simplicial R-modules over a field k, as in Example 10.95. In particular, a map $E \to F$ of $(S^1 \wedge \mathbb{G}_m)$-complexes is a motivic stable equivalence if and only if the underlying map $u(E) \to u(F)$ is a motivic stable equivalence of $(S^1 \wedge T)$-spectra.

The Voevodsky cancellation theorem implies that the motivic cohomology spectrum $\mathbf{H}_\mathbb{Z}$ is a stable fibrant object in the category of $(S^1 \wedge \mathbb{G}_m)$-complexes in presheaves of simplicial abelian groups (or simplicial $\mathbb{Z}$-modules), and it represents motivic cohomology theory in the motivic stable category.

We can go further. The motives which make up the object $\mathbf{H}_\mathbb{Z}$ have a much richer structure, in that they are simplicial presheaves with transfers. The object $\mathbf{H}_\mathbb{Z}$ can be formed as an $(S^1 \wedge \mathbb{G}_m)$-complex in the category of simplicial presheaves with transfers over the field k. It is possible to put both a Nisnevich local and motivic stable model structure on the category of all such objects, by formally inverting suitable sets of cofibrations of $(S^1 \wedge \mathbb{G}_m)$-spectra in a sectionwise strict model structure on the category. The homotopy category associated to motivic stable model structure on the category of $(S^1 \wedge \mathbb{G}_m)$-complexes is Voevodsky's big category of motives $\mathbf{DM}_k$ over the field k.

The motivic construction is a special case of an f-local stable structure which can be imposed on a suitably enriched category of T-complexes in simplicial abelian presheaves. The existence of this model structure follows from Theorem 10.96.

Product theories for the stable categories which are discussed in this chapter form the subject of Chap. 11.

10.1 Presheaves of Spectra

This section gives a description of the standard model structures for the category $\mathbf{Spt}(\mathcal{C})$ of presheaves of spectra on a Grothendieck site $\mathcal{C}$.

We shall use the injective model structure for the category $s_*\mathbf{Pre}(\mathcal{C})$ of pointed simplicial presheaves. This model structure is inherited from the injective model structure on the category of simplicial presheaves of Theorem 5.8 — see Remark 6.16.

A *presheaf of spectra* X consists of pointed simplicial presheaves X^n, $n \geq 0$ together with *bonding maps*

$$\sigma : S^1 \wedge X^n \to X^{n+1}, \ n \geq 0.$$

Here, S^1 is identified with the constant pointed simplicial presheaf $U \mapsto S^1$ associated to the simplicial circle

$$S^1 = \Delta^1/\partial\Delta^1.$$

One says that the number n is the *level* of the pointed simplicial presheaf X^n for a presheaf of spectra X.

A map $f : X \to Y$ of presheaves of spectra consists of pointed simplicial presheaf maps $f : X^n \to Y^n$, $n \geq 0$, which respect structure in the obvious sense. Write **Spt**($\mathcal{C}$) for the category of presheaves of spectra on $\mathcal{C}$.

The ordinary category of spectra **Spt** is the category of presheaves of spectra on the one-object, one-morphism category, so that results about presheaves of spectra apply to spectra.

More generally, if I is a small category, then the category of I-diagrams $X : I \to$ **Spt** is a category of presheaves of spectra on I^{op}, where I^{op} has the trivial topology. Thus, results about presheaves of spectra apply to all categories of small diagrams of spectra.

Example 10.1 Any spectrum A determines an associated constant presheaf of spectra Γ^*A on $\mathcal{C}$, where

$$\Gamma^*A(U) = A,$$

and every morphism $\phi : V \to U$ induces the identity morphism $A \to A$. Write $A = \Gamma^*A$ when there is no possibility of confusion. The sphere spectrum S in **Spt**($\mathcal{C}$) is the constant object Γ^*S associated to the ordinary sphere spectrum

$$S : S^0, S^1, S^{\wedge 2}, \dots.$$

The functor $A \mapsto \Gamma^*A$ is left adjoint to the global sections functor $\Gamma_* : \mathbf{Spt}(\mathcal{C}) \to$ **Spt**, where

$$\Gamma_*X = \varprojlim_{U \in \mathcal{C}} X(U).$$

If K is a pointed simplicial set and A is a simplicial abelian group, write $K \otimes A$ for the simplicial abelian group which is defined by

$$K \otimes A = \mathbb{Z}_\bullet(K) \otimes A,$$

where $\mathbb{Z}_\bullet(K)$ is the reduced free simplicial abelian group associated to the pointed simplicial set K. Recall that this object is defined by the exact sequence

$$0 \to \mathbb{Z}(*) \xrightarrow{x} \mathbb{Z}(K) \to \mathbb{Z}_\bullet(K) \to 0,$$

where x is the base point of K.

Example 10.2 If A is a sheaf (or presheaf) of abelian groups, the Eilenberg–Mac Lane presheaf of spectra $H(A)$ is the presheaf of spectra underlying the suspension object

$$A, S^1 \otimes A, S^2 \otimes A, \dots$$

in the category of presheaves of spectra in simplicial abelian groups. As a simplicial presheaf, $S^n \otimes A = K(A, n)$, and if $j : K(A, n) \to FK(A, n)$ is an injective fibrant model of $K(A, n)$ then there are natural isomorphisms

$$\pi_j \Gamma_* FK(A, n) = \begin{cases} H^{n-j}(\mathcal{C}, A) & 0 \leq j \leq n \\ 0 & j > n, \end{cases}$$

by Proposition 8.32. These isomorphisms assemble to give an identification of the stable homotopy groups of global sections of a stable fibrant model for $H(A)$ with the cohomology of $\mathcal{C}$ with coefficients in A. In other words, all sheaf cohomology groups are stable homotopy groups.

More generally, every chain complex (bounded or unbounded) D determines a presheaf of spectra $H(D)$, which computes the hypercohomology of $\mathcal{C}$ with coefficients in D, via computing stable homotopy groups of global sections of a stable fibrant model. See Example 10.88 below, and [61].

Example 10.3 The algebraic K-theory presheaf of spectra **K** is defined on the category $\text{Sch}\,|_S$ of S-schemes. In sections corresponding to an S-scheme X, the group $\pi_j\mathbf{K}(X)$ is the jth algebraic K-group $K_j(X)$ of X. This object **K** is actually a presheaf of symmetric spectra—it is constructed explicitly in [68].

Say that a map $f : X \to Y$ of presheaves of spectra is a *strict weak equivalence* (respectively *strict fibration*) if all maps $f : X^n \to Y^n$ are local weak equivalences (respectively injective fibrations).

A *cofibration* $i : A \to B$ of **Spt**($\mathcal{C}$) is a map for which

1) the map $i : A^0 \to B^0$ is a cofibration, and
2) all maps

$$(S^1 \wedge B^n) \cup_{(S^1 \wedge A^n)} A^{n+1} \to B^{n+1}$$

are cofibrations.

The *function complex* **hom**(X, Y) for presheaves of spectra X, Y is defined in simplicial degree n by the assignment

$$\mathbf{hom}(X, Y)_n = \hom(X \wedge \Delta^n_+, Y).$$

10.1 Presheaves of Spectra

The pointed simplicial set $K_+ = K \sqcup \{*\}$ is constructed from a simplicial set K by freely adjoining a base point $*$. This construction is functorial, and therefore extends to simplicial presheaves.

Proposition 10.4 *The category* $\mathbf{Spt}(\mathcal{C})$ *of presheaves of spectra, together with the classes of strict weak equivalences, strict fibrations and cofibrations as defined above, satisfies the axioms for a proper closed simplicial model category.*

The model structure of Proposition 10.4 is typically called the *strict model structure* for presheaves of spectra. Its proof is an exercise.

A presheaf of spectra X has presheaves $\pi_n^s X$ of stable homotopy groups, defined by

$$U \mapsto \pi_n^s X(U).$$

Write $\tilde{\pi}_n^s X$ for the sheaf associated to the presheaf $\pi_n^s X$. The sheaves $\tilde{\pi}_n^s X$, $n \in \mathbb{Z}$, are the *sheaves of stable homotopy groups* of X.

Say that a map $f : X \to Y$ of presheaves of spectra is a *stable equivalence*, if it induces isomorphisms

$$\tilde{\pi}_n^s X \xrightarrow{\cong} \tilde{\pi}_n^s Y,$$

for all $n \in \mathbb{Z}$. Every strict weak equivalence is a stable equivalence.

Say that a map $p : Z \to W$ of presheaves of spectra is a *stable fibration* if it has the right lifting property with respect to all maps that are cofibrations and stable equivalences.

Theorem 10.5 *With the definitions of cofibration, stable equivalence and stable cofibration given above, the category* $\mathbf{Spt}(\mathcal{C})$ *of presheaves of spectra satisfies the axioms for a proper closed simplicial model category. This model structure is cofibrantly generated.*

The model structure of Theorem 10.5 is the *stable model structure* for the category of presheaves of spectra.

With the exception of the final sentence in the statement, Theorem 10.5 is the analogue of a well known result of Bousfield and Friedlander for ordinary spectra [13]. Theorem 10.5 first appeared in [51], and was proved by extending the methods of Bousfield and Friedlander to the context of presheaves of spectra. A different line of argument is presented here, which focuses on the cofibrant generation; this discussion is presented in a sequence of lemmas.

Lemma 10.6 *A map* $p : X \to Y$ *of presheaves of spectra is a stable fibration and a stable equivalence if and only if all level maps* $p : X^n \to Y^n$ *are trivial injective fibrations of simplicial presheaves.*

Proof If all maps $p : X^n \to Y^n$ are trivial injective fibrations, then p has the right lifting property with respect to all cofibrations, and is therefore a stable fibration. The map p is also a stable equivalence because it is a strict equivalence.

Suppose that p is a stable fibration and a stable equivalence. Then p has a factorization

where j is a cofibration and q is a trivial strict fibration. But then j is stable equivalence as well as a cofibration, so that the lifting exists in the diagram

and so p is a retract of q, and is therefore a trivial strict fibration.

Choose a regular cardinal α such that $|\operatorname{Mor}(\mathcal{C})| < \alpha$. Say that a presheaf of spectra A is α-bounded if all pointed simplicial sets $A^n(U)$, $n \geq 0$, $U \in \mathcal{C}$ are α-bounded. Every presheaf of spectra X is a union of its α-bounded subobjects.

Lemma 10.7 *Suppose given a cofibration $i : X \to Y$ which is a stable equivalence, and suppose that $A \subset Y$ is an α-bounded subobject. Then there is an α-bounded subobject $B \subset Y$ such that $A \subset B$ and the map $B \cap X \to B$ is a stable equivalence.*

Proof The sheaf $\tilde{\pi}_n^s Z$ is 0 if and only if for all $x \in \pi_n^s Z(U)$ there is a covering sieve $\phi : V \to U$ such that $\phi^*(x) = 0$ for all ϕ in the covering.

The sheaves $\tilde{\pi}_n^s(Y/X)$ are trivial (sheafify the natural long exact sequence for a cofibration), and

$$\tilde{\pi}_n^s(Y/X) = \varinjlim_C \tilde{\pi}_n^s(C/C \cap X)$$

where C varies over all α-bounded subobjects of Y. The list of elements of all $x \in \pi_n^s(A/A \cap X)(U)$ is α-bounded. For each such x there is an α-bounded subobject $B_x \subset X$ such that

$$x \mapsto 0 \in \tilde{\pi}_n(B_x/B_x \cap X).$$

It follows that there is an α-bounded subobject

$$B_1 = A \cup (\cup_x B_x)$$

such that all $x \mapsto 0 \in \tilde{\pi}_n(B_1/B_1 \cap X)$ for all n and all x.

10.1 Presheaves of Spectra

Write $A = B_0$. Then inductively, we can produce an ascending sequence

$$A = B_0 \subset B_1 \subset B_2 \subset \dots$$

of α-bounded subobjects of Y such that all sheaf homomorphisms

$$\pi_n^s(B_i/B_i \cap X) \to \tilde{\pi}_n^s(B_{i+1}/B_{i+1} \cap X)$$

are trivial. Set $B = \cup_i B_i$. Then the object B is α-bounded and all sheaves $\tilde{\pi}_n(B/B \cap X)$ are trivial.

One of the things which makes the argument for Lemma 10.7 work is that the maps $B \cap X \to B$ are monomorphisms for all subobjects B of Y (so that the long exact sequence for a cofibration can be used). It is an exercise to show that the maps $B \cap X \to B$ are cofibrations since $i : X \to Y$ is a cofibration.

The use of long exact sequences simplifies some of the arguments that appear in this section, but there are alternatives; see the proof of the corresponding result for T-spectra in Lemma 10.19.

Lemma 10.8 *The class of stable trivial cofibrations has a generating set, namely the set I of all α-bounded stable trivial cofibrations.*

Proof The proof of Lemma 5.44 is the prototype for this argument.

The class of cofibrations of $\mathbf{Spt}(\mathcal{C})$ is generated by the set J of cofibrations

$$\Sigma^\infty A[-n] \to \Sigma^\infty B[-n]$$

which are induced by α-bounded cofibrations $A \to B$ of pointed simplicial presheaves.

Suppose given a diagram

$$\begin{array}{ccc} A & \longrightarrow & X \\ {\scriptstyle j} \downarrow & & \downarrow {\scriptstyle f} \\ B & \longrightarrow & Y \end{array}$$

where j is a cofibration, B is α-bounded and f is a stable equivalence. Then f has a factorization $f = q \cdot i$ where i is a cofibration and q is a trivial strict fibration, hence a stable equivalence, and the lifting exists in the diagram

The cofibration $i : X \to Z$ is a stable equivalence, and the image $\theta(B) \subset Z$ is α-bounded, so there is an α-bounded subobject $D \subset Z$ with $\theta(B) \subset D$ such that the

map $D \cap X \to D$ is a stable equivalence, by Lemma 10.7. It follows that there is a factorization

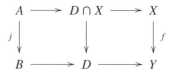

of the original diagram through an α-bounded stable trivial cofibration.

Now suppose that the map $i : C \to D$ is a cofibration and a stable equivalence. Then i has a factorization

where j is a cofibration in the saturation of the set of α-bounded stable trivial cofibrations and p has the right lifting property with respect to all α-bounded stable trivial cofibrations. The map j is a stable equivalence since the class of stable trivial cofibrations is closed under pushout (by a long exact sequence argument) and composition. It follows that p is a stable equivalence, and therefore has the right lifting property with respect to all α-bounded cofibrations by the previous paragraph. The map p therefore has the right lifting property with respect to all cofibrations, so p is a trivial strict fibration, and it follows that i is a retract of the map j.

Proof [Proof of Theorem 10.5] According to Lemma 10.8, a map is a stable fibration if and only if it has the right lifting property with respect to all α-bounded stable trivial cofibrations. A small object argument therefore implies that every map $f : X \to Y$ has a factorization

where j is a stable trivial cofibration and p is a stable fibration.

Lemma 10.6 says that a map is a stable fibration and a stable weak equivalence if and only if it is a strict fibration and a strict weak equivalence. There is a factorization

10.1 Presheaves of Spectra

for any map $f : X \to Y$, where i is a cofibration and p is a strict fibration and a strict equivalence—this gives the corresponding factorization for the stable structure.

We have proved **CM5**. The axiom **CM4** is a consequence of Lemma 10.6. The remaining closed model axioms are immediate.

The closed simplicial model structure is verified by showing that the cofibrations

$$(B \wedge \partial \Delta^n_+) \cup (A \wedge \Delta^n_+) \to B \wedge \Delta^n_+$$

induced by a cofibration $A \to B$ are stable equivalences if $A \to B$ is a stable equivalence. For this, one shows by induction on n (or otherwise) that the cofibrations

$$i \wedge \partial \Delta^n_+ : A \wedge \partial \Delta^n_+ \to B \wedge \partial \Delta^n_+$$

are stable equivalences.

Left and right properness are proved with long exact sequences in stable sheaves of stable groups.

Since the stable model structure on $\mathbf{Spt}(\mathcal{C})$ is cofibrantly generated there is a functorial stable fibrant model construction

$$j : X \to LX,$$

in which j is a stable trivial cofibration and LX is stable fibrant.

If X and Y are stable fibrant, any stable equivalence $f : X \to Y$ must be a strict equivalence. This is a consequence of Lemma 10.6. It follows that a map $f : X \to Y$ of arbitrary presheaves of spectra is a stable equivalence if and only if the induced map $LX \to LY$ is a strict equivalence.

We also have the following:

A4 The functor L preserves strict equivalences.
A5 The maps $j_{LX}, Lj_X : LX \to LLX$ are strict weak equivalences.
A6' Stable equivalences are preserved by pullback along stable fibrations.

We then have a formal consequence:

Lemma 10.9 *Suppose that the map $p : X \to Y$ of $\mathbf{Spt}(\mathcal{C})$ is a strict fibration. Then p is a stable fibration if and only if the diagram*

$$\begin{array}{ccc} X & \xrightarrow{j} & LX \\ {\scriptstyle p}\downarrow & & \downarrow{\scriptstyle Lp} \\ Y & \xrightarrow{j} & LY \end{array} \qquad (10.2)$$

is strictly homotopy cartesian.

Proof This result is proved in [13] by manipulating the statements **A4**, **A5** and **A6'**. The proof is organized here in a slightly different way, although the ideas are the same.

Suppose that p is a stable fibration, and find a factorization

of p_* such that q is a stable fibration and j is a stable equivalence. Then j is a stable equivalence of stably fibrant objects, so that j is a strict equivalence. The map q is also a strict fibration.

The map $Y \times_{LY} Z \to Z$ is a stable equivalence by right properness (**A6′**), so that the map $X \to Y \times_{LY} Z$ is a stable equivalence of stable fibrations, and is therefore a strict equivalence. It follows that the diagram (10.2) is strictly homotopy cartesian.

Suppose, conversely, that the diagram (10.2) is strictly homotopy cartesian, and find a factorization of p_* as above. Then the map j is a strict equivalence, and so the induced map $\theta : X \to Y \times_{LY} Z$ is a strict equivalence. But then the strict fibration p is strictly equivalent to a stable fibration q_*, and must be a stable fibration by a standard argument: factorize $\theta = \pi \cdot i$ where π is a trivial strict fibration, and i is a trivial strict cofibration, the map i defines p as a retract of the stable fibration $q_* \cdot \pi$.

We finish this section with a series of results that lead to a stable fibrant model construction $X \mapsto QX$ for presheaves of spectra X. This is the standard stable fibrant model construction from ordinary stable homotopy theory, but the following results show that it arises naturally from the stable model structure in a way that is useful in other contexts.

Lemma 10.10 *Suppose that the map $p : X \to Y$ is a stable fibration. Then the diagrams*

$$\begin{array}{ccc} X^n & \xrightarrow{\sigma_*} & \Omega X^{n+1} \\ p \downarrow & & \downarrow \Omega p \\ Y^n & \xrightarrow{\sigma_*} & \Omega Y^{n+1} \end{array} \qquad (10.3)$$

of simplicial presheaf maps are homotopy cartesian in the injective model structure.

Proof Since p is a stable fibration, any stable trivial cofibration $\gamma : A \to B$ between cofibrant objects induces a homotopy cartesian diagram of simplicial sets

$$\begin{array}{ccc} \hom(B, X) & \xrightarrow{\gamma^*} & \hom(A, X) \\ p_* \downarrow & & \downarrow p_* \\ \hom(B, Y) & \xrightarrow{\gamma^*} & \hom(A, Y) \end{array} \qquad (10.4)$$

10.1 Presheaves of Spectra

If $\gamma : A \to B$ is a stable equivalence between cofibrant objects, then the diagram above is still homotopy cartesian. In effect, γ has a factorization $\gamma = \pi \cdot j$ where j is a stable trivial cofibration and $\pi \cdot i = 1$ for some stable trivial cofibration i. It follows that the diagram (10.4) is a retract of a homotopy cartesian diagram, and is therefore homotopy cartesian.

The diagrams (10.3) arise in this way from the stable equivalences

$$\Sigma^\infty S^1[-1-n] \to S[-n]$$

associated to the sphere spectrum S.

The maps $\sigma_* : X^n \to \Omega X^{n+1}$ which are adjoint to the bonding maps $S^1 \wedge X^n \to X^{n+1}$ for a presheaf of spectra X are called the *adjoint bonding maps*.

Corollary 10.11 *If X is stable fibrant, then all constituent simplicial presheaves X^n are injective fibrant and all adjoint bonding maps $\sigma_* : X^n \to \Omega X^{n+1}$ are local weak equivalences.*

The converse of Corollary 10.11 has a more interesting proof.

Proposition 10.12 *A presheaf of spectra X is stable fibrant if and only if all X^n are injective fibrant and all adjoint bonding maps $\sigma_* : X^n \to \Omega X^{n+1}$ are local weak equivalences.*

Proof Suppose that all X^n are injective fibrant and all $\sigma_* : X^n \to \Omega X^{n+1}$ are local weak equivalences. It follows from Lemma 5.12 that all spaces $X^n(U)$ are fibrant and that all maps $\sigma_* : X^n(U) \to \Omega X^{n+1}(U)$ are weak equivalences of pointed simplicial sets. All maps

$$\pi_k X^n(U) \to \pi^s_{k-n} X(U)$$

are therefore isomorphisms.

Suppose that $j : X \to LX$ is a stable fibrant model for X. Then all spaces $LX^n(U)$ are fibrant and all maps $LX^n(U) \to \Omega LX^{n+1}(U)$ are weak equivalences, and so all maps

$$\pi_k LX^n(U) \to \pi^s_{k-n} LX(U)$$

are isomorphisms. The map j induces an isomorphism in all sheaves of stable homotopy groups, and so the maps $j : X^n \to LX^n$ induce isomorphisms

$$\tilde{\pi}_k X^n \to \tilde{\pi}_k LX^n$$

of sheaves of homotopy groups for $k \geq 0$. The level objects X^n and LX^n are presheaves of infinite loop spaces, and so the maps $X^n \to LX^n$ are local weak equivalences of simplicial presheaves. In particular, the map $j : X \to LX$ is a strict weak equivalence.

Now consider the lifting problem

$$\begin{array}{ccc} A & \xrightarrow{\alpha} & X \\ {\scriptstyle i}\downarrow & \nearrow & \\ B & & \end{array} \qquad (10.5)$$

where i is a stable trivial cofibration. Then the induced map $i_* : LA \to LB$ is a strict equivalence of stable fibrant objects, by assumption.

Take a factorization

$$\begin{array}{ccc} LA & \xrightarrow{L\alpha} & LX \\ {\scriptstyle j'}\searrow & & \nearrow {\scriptstyle p} \\ & Z & \end{array}$$

where j' is a cofibration and a strict weak equivalence and p is a strict fibration. The presheaf of spectra LB is strictly fibrant, so there is a map $\zeta : Z \to LB$ such that the diagram

$$\begin{array}{ccc} LA & \xrightarrow{j'} & Z \\ {\scriptstyle Li}\downarrow & \swarrow {\scriptstyle \zeta} & \\ LB & & \end{array}$$

commutes. The map ζ is therefore a strict equivalence. Form the pullback

$$\begin{array}{ccc} Z \times_{LX} X & \xrightarrow{p_*} & X \\ {\scriptstyle j_*}\downarrow & & \downarrow {\scriptstyle j} \\ Z & \xrightarrow{p} & LX \end{array}$$

and observe that the map j_* is a strict weak equivalence since j is a strict weak equivalence and p is a strict fibration. It follows that the solid arrow diagram (10.5) has a factorization

$$\begin{array}{ccccc} A & \longrightarrow & Z \times_{LX} X & \xrightarrow{p_*} & X \\ {\scriptstyle i}\downarrow & & \downarrow {\scriptstyle \zeta j_*} & & \\ B & \xrightarrow{j} & LB & & \end{array}$$

10.1 Presheaves of Spectra

in which the top composite is the map $\alpha : A \to X$. The vertical map ζj_* is a strict weak equivalence and X is strictly fibrant, and it is an exercise to show that the lifting problem can then be solved.

The strict model structure on **Spt**($\mathcal{C}$) is cofibrantly generated, so we are entitled to a natural strict fibrant model construction $X \to FX$ for presheaves of spectra X.

The bonding maps $\sigma : S^1 \wedge X^n \to X^{n+1}$ for a presheaf of spectra X can alternatively be described by their adjoints $\sigma_* : X^n \to \Omega X^{n+1}$. One defines the pointed simplicial presheaf $\Omega^\infty X^n$ for X by the colimit

$$\Omega^\infty X^n = \varinjlim_k \Omega^{n+k} X^n.$$

of the diagram

$$X^n \xrightarrow{\sigma_*} \Omega X^{n+1} \xrightarrow{\Omega \sigma_*} \Omega^2 X^{n+1} \xrightarrow{\Omega^2 \sigma_*} \cdots$$

The adjoint bonding maps also induce pointed maps

$$\Omega^\infty X^n \to \Omega(\Omega^\infty X^{n+1}),$$

which maps isomorphisms by a cofinality argument. The canonical maps $X^n \to \Omega^\infty X^n$ induce a natural map $X \to \Omega^\infty X$ of presheaves of spectra.

The construction $X \mapsto \Omega^\infty X$ makes little homotopy theoretic sense unless the presheaf of spectra X is at least strictly fibrant. If X is strictly fibrant, then the map $X \to \Omega^\infty X$ is a stable equivalence, by a sectionwise argument. Unlike the situation in standard stable homotopy theory, however, the object $\Omega^\infty X$ might not be strictly fibrant, even if X is strictly fibrant. We fix the problem, albeit somewhat brutally, in the following consequence of Proposition 10.12:

Corollary 10.13 *The presheaf of spectra*

$$QX = F\Omega^\infty FX$$

is stable fibrant, for any presheaf of spectra X. *The natural map* $\eta : X \to QX$ *which is defined by the composite*

$$X \to FX \to \Omega^\infty FX \to F\Omega^\infty FX$$

is a stable equivalence, so that $\eta : X \to QX$ *is a natural stable fibrant model for presheaves of spectra* X.

Example 10.14 Theorem 10.5 specializes to give a construction of stable homotopy theory for small diagrams $I \to$ **Spt** of spectra, in which the strict equivalences and strict fibrations, and hence the stable equivalences, are defined sectionwise. The topology on the index category I that one uses is the "chaotic" topology, for which every presheaf is a sheaf.

Suppose that G is a finite group. Write $\mathcal{B}G$ for the category whose objects are G-sets (i.e. sets with G-action) and the G-equivariant maps between them. This category satisfies the exactness conditions for Giraud's Theorem (Theorem 3.17), and is thus equivalent to a Grothendieck topos. The category $\mathcal{B}G$ is called the *classifying topos* for G. This is a special case of the general theme which is described in Example 3.19.

The method of proof of Giraud's Theorem, in this case, is to explicitly construct a site $\mathcal{C}_G$ such that the corresponding sheaf category $\mathbf{Shv}(\mathcal{C}_G)$ is equivalent to the category $\mathcal{B}G$. The category $\mathcal{B}G$ has a set of generators, which is usually described as the collection of finite G-sets X, and the category $\mathcal{C}$ is the full subcategory of $\mathcal{B}G$ on the generators X. The topology is defined by the families of maps $Y_i \to X$ such that the G-equivariant map $\sqcup\, Y_i \to X$ is surjective. The G-set E represents a sheaf on $\mathcal{B}G$, which is defined by the assignment

$$X \mapsto F(X) = \hom(X, E).$$

It follows that $E(G/H)$ is the set of H-fixed points E^H in E. One recovers the G-set E from the isomorphism $E \cong E(G)$.

A presheaf $F : \mathcal{C}_G^{op} \to \mathbf{Set}$ is a sheaf for this topology if and only if F is *additive* in the sense that it takes finite disjoint unions to products, and $F(G/H) = F(G)^H$ for all subgroups $H \subset G$. In general, the assignment $F \mapsto F(G)$ defines the associated sheaf functor.

The additivity property implies that sheaves, or G-sets, are completely determined by their restrictions to the full subcategory on the objects G/H of the site $\mathcal{C}_G$. This full subcategory is the *orbit category* $\mathcal{O}_G$ for G.

In equivariant homotopy theory (as in [39]), a G-equivariant weak equivalence (respectively G-fibration) $X \to Y$ of spaces with G-action is map which induces a weak equivalence (respectively fibration)

$$X^H = X(G/H) \to Y(G/H) = Y^H$$

for each subgroup H of G. The equivariant homotopy theory of G-spaces is therefore a subspecies of the sectionwise homotopy theory of presheaves of spaces which are defined on the orbit category $\mathcal{O}_G$, or equivalently on the site $\mathcal{C}_G$. The G-equivariant stable category is constructed from presheaves of spectra, interpreted as ordinary diagrams of spectra.

Abelian cohomology theories are represented by Eilenberg–Mac Lane spectrum objects in all stable model structures—see Example 10.88 below. Here are the basic G-equivariant examples:

1) The presheaves of stable homotopy groups which arise from the sectionwise theory for G-spectra are the Mackey functors, because transfers are formally defined for spectra. The Eilenberg–Mac Lane spectrum object $H(M)$, which is associated to a Mackey functor M, represents Bredon cohomology for G-spaces X with coefficients in M, in the G-equivariant stable category.
2) The sheaf theoretic cohomology associated to an abelian group A with G-action is the classical equivariant cohomology $H^*(G, A)$ for the group G with coefficients in A. It is represented by the Eilenberg–Mac Lane spectrum object $H(A)$ in the stable category which is associated to the classifying topos.

10.1 Presheaves of Spectra

The Bredon and sheaf theoretic cohomology theories are quite different; Bredon cohomology corresponds to the chaotic topology, while ordinary equivariant cohomology is defined by using the topology derived from the classifying topos. They are related by base change, or change of topology, via the geometric morphism from sheaves to presheaves which forgets the topology—see Sect. 5.3.

Galois cohomology theory is the sheaf cohomology theory associated to the classifying topos of a profinite group, and as such is a generalization of the ordinary cohomology theory for a finite group. The étale cohomology of a field is a special case. See [33, 56, 79, 92], as well as Example 3.20.

We say that a presheaf of spectra X satisfies *descent* (for the ambient topology) if some, hence any, stable fibrant model $j : X \to QX$ is a sectionwise stable equivalence, in the sense that the maps $j : X(U) \to QX(U)$ are stable equivalences for all objects U of the underlying site $\mathcal{C}$. All stably fibrant presheaves of spectra satisfy descent.

A descent problem for a given presheaf of spectra X is question of whether or not X satisfies descent in this sense.

Variants of descent problems have been major themes in algebraic K-theory. For example, suppose that k is a field and that ℓ is a prime not equal to the characteristic of k. Suppose that k has Galois cohomological dimension d with respect to ℓ-torsion sheaves. Then the Lichtenbaum–Quillen conjecture for the field k asserts that any stably fibrant model $j : K/\ell \to Q_{et}(K/\ell)$ *on the étale site* for k induces isomorphisms in stable homotopy groups

$$\pi_s K/\ell(L) \to \pi_s Q_{et}(K/\ell)(L)$$

for $s \geq d - 1$ and for all finite separable extensions L/k.

The fibrant model $Q_{et}(K/\ell)$ for K/ℓ is the *étale mod ℓ K-theory presheaf of spectra*. It is traditional to write

$$K_s(k, \mathbb{Z}/\ell) = \pi_s Q_{et}(K/\ell)(k),$$

and call these groups the *mod ℓ étale K-groups* of k—see [24].

The Nisnevich descent theorem [83] asserts that the algebraic K-theory presheaf of spectra K satisfies descent for the Nisnevich topology on the category $et|_S$, where S is a separated, regular Noetherian scheme. In other words, the Nisnevich fibrant model $j : K \to Q_{Nis}K$ induces a stable equivalence

$$K(S) \xrightarrow{\simeq} Q_{Nis}K(S)$$

for such schemes S. One also writes

$$K_s^{Nis}(S) = \pi_s Q_{Nis}K(S),$$

and calls these groups the *Nisnevich K-groups* of S. The displayed stable equivalence therefore gives an isomorphism between K-theory and Nisnevich K-theory for regular schemes S.

One proves the Nisnevich descent theorem (now) by using the localization sequence for algebraic K-theory to show that the level objects K^n satisfy the conditions for the Nisnevich descent theorem (Theorem 5.39). The original proof was more complicated.

The mod n K-theory presheaf of spectra also satisfies Nisnevich descent for separated, regular, Noetherian schemes. One uses the cofibre sequence

$$K \xrightarrow{\times n} K \to K/n$$

to prove this.

For more general schemes T, the Nisnevich K-theory presheaf

$$K^{Nis} = Q_{Nis} K$$

may diverge from the K-theory presheaf of spectra, but there is still a major positive statement. The comparison

$$K \to K^{TT}$$

with Thomason–Trobaugh K-theory induces isomorphisms

$$\pi_s K(A) \xrightarrow{\cong} \pi_s K^{TT}(A), \ s \geq 0,$$

for all affine schemes A, while Thomason–Trobaugh K-theory satisfies Nisnevich descent. By comparing simplicial presheaves at level 0, it follows that there are natural isomorphisms

$$f K_s^{Nis}(S) := \pi_s Q_{Nis} K(S) \cong \pi_s K^{TT}(S), \ s \geq 0,$$

for all schemes S. It follows that the Nisnevich K-theory and the Thomason–Trobaugh K-theory presheaves of spectra coincide on connective covers. See [68, 100].

The knowledge that a presheaf of spectra satisfies descent can have important calculational consequences, due to the fact that if Z is stable fibrant, then the functor

$$K \mapsto \mathbf{hom}(K, Z)$$

takes local weak equivalences of pointed simplicial presheaves to stable equivalences of spectra. Thus, for example, if the simplicial presheaf map $U \to *$ is a local weak equivalence, then there is an induced stable equivalence

$$\Gamma_* Z = \mathbf{hom}(*, Z) \to \mathbf{hom}(U, Z).$$

The simplicial structure of U defines a Bousfield–Kan type spectral sequence for the stable homotopy groups of the global sections spectrum of Z, which is otherwise known as a *descent spectral sequence*.

10.1 Presheaves of Spectra

Specializing further, suppose that L/k is a finite Galois extension of fields with Galois group G. Then the hypercover $EG \times_G \mathrm{Sp}(L) \to *$ on the étale site for the field k (see Example 4.17) determines a spectral sequence

$$H^s(G, \pi_t Q_{et} K/\ell(L)) \Rightarrow \pi_{t-s} Q_{et} K/\ell(k) = K^{et}_{t-s}(k, \mathbb{Z}/\ell). \qquad (10.6)$$

This spectral sequence is variously known as the *finite Galois descent* spectral sequence, or the *homotopy fixed points* spectral sequence for the étale K-theory of k. The term "homotopy fixed points" arises from the observation that the stable equivalence

$$Q_{et} K/\ell(k) \cong \mathbf{hom}(*, Q_{et} K/\ell) \xrightarrow{\simeq} \mathbf{hom}(EG \times_G \mathrm{Sp}(L), Q_{et} K/\ell)$$

identifies the étale K-theory spectrum $Q_{et} K/\ell(k)$ for the field k with the homotopy fixed points spectrum (i.e. the displayed function complex) for the action of the Galois group G on the spectrum $Q_{et} K/\ell(L)$.

The finite descent spectral sequence can also be constructed by applying the homotopy fixed points functor

$$\mathbf{hom}(EG \times_G \mathrm{Sp}(L), ?)$$

to the Postnikov tower $P_n Q_{et} K/\ell$ for the étale K-theory presheaf of spectra $Q_{et} K/\ell$, in the presence of a bound on Galois cohomological dimension for ℓ-torsion sheaves. The Postnikov tower of a presheaf of spectra is described in Sect. 10.6 below.

Taking a suitable fibrant model of the Postnikov tower $P_n Q_{et} K/\ell$ and evaluating in global sections construct the étale (or Galois) cohomological descent spectral sequence

$$H^s_{et}(k, \tilde{\pi}_t K/\ell) \Rightarrow \pi_{t-s} Q_{et} K/\ell(k) \qquad (10.7)$$

for the étale K-theory of the field k. For the construction, one uses the observations in Example 10.2, along with Proposition 8.32. See also Sect. 6.1 of [56].

The descent spectral sequences (10.6) and (10.7) for étale K-theory have straightforward derivations in the theory of presheaves of spectra, but the general principle that étale K-theory can be computed from étale cohomology has been known for quite some time. This principle was the motivation for the Lichtenbaum–Quillen conjecture—to derive an étale cohomological calculational device for algebraic K-theory—from the early days of the subject. It is now known that the Lichtenbaum–Quillen conjecture holds, as a consequence of the Bloch–Kato conjecture [96, 107].

Suppose that G is a finite group with subgroup H. Within the classifying topos for G of Example 10.14, the Čech resolution for the covering $G \to G/H$ can be identified up to isomorphism with the canonical map

$$G \times_H EH \to G/H.$$

This map is a hypercover, and induces a stable equivalence

$$Z(G/H) = \mathbf{hom}(G/H, Z) \xrightarrow{\simeq} \mathbf{hom}(G \times_H EH, Z)$$

for any stably fibrant object Z in the sheaf theoretic stable model structure for presheaves of spectra on the orbit category $\mathcal{O}_G$.

From the general definition, a G-spectrum E satisfies descent for the sheaf-theoretic stable model structure if some (hence any) stably fibrant model $j : E \to Z$ induces stable equivalences $E(G/H) \to Z(G/H)$ for all subgroups $H \subset G$.

The descent condition is independent of sectionwise stable equivalence, so we can suppose that, as a presheaf of spectra, E is sectionwise stably fibrant. In this case, the stably fibrant model $j : E \to Z$ induces a G-equivariant stable equivalence $E(G) \to Z(G)$ of stably fibrant spectra, and it follows that the induced maps

$$j_* : \mathbf{hom}(G \times_H EH, E) \to \mathbf{hom}(G \times_H EH, Z)$$

of homotopy fixed point spectra are stable equivalences. Thus, from the diagram

$$\begin{array}{ccc} E(G/H) & \longrightarrow & \mathbf{hom}(G \times_H EH, E) \\ j \downarrow & & j_* \downarrow \simeq \\ Z(G/H) & \xrightarrow{\simeq} & \mathbf{hom}(G \times_H EH, Z) \end{array}$$

one concludes that the H-fixed points of the G-spectrum E coincide with the H-homotopy fixed points of E for all subgroups H of G if E satisfies descent for the sheaf theoretic stable model structure.

10.2 T-spectra and Localization

The results and methods of ordinary stable homotopy theory admit substantial generalization, beyond even the case of presheaves of spectra.

Suppose that T is a pointed simplicial presheaf on a small Grothendieck site $\mathcal{C}$.

A T-*spectrum* X is a collection of pointed simplicial presheaves X^n, $n \geq 0$, with pointed maps

$$\sigma : T \wedge X^n \to X^{n+1},$$

called *bonding maps*. The simplicial presheaf X^n for a T-spectrum X is the object in *level n*.

One says that the simplicial presheaf T is the *parameter object*. Presheaves of spectra are S^1-*spectra* in the present language; the simplicial circle S^1 is the parameter object in that case.

10.2 T-spectra and Localization

A *map* $f : X \to Y$ *of* T-*spectra* consists of pointed simplicial presheaf maps $f : X^n \to Y^n$ which respect structure in the sense that the diagrams

$$\begin{array}{ccc} T \wedge X^n & \xrightarrow{\sigma} & X^{n+1} \\ {\scriptstyle T \wedge f} \downarrow & & \downarrow {\scriptstyle f} \\ T \wedge Y^n & \xrightarrow{\sigma} & Y^{n+1} \end{array}$$

commute. Write $\mathbf{Spt}_T(\mathcal{C})$ for the category of T-spectra for the site $\mathcal{C}$.

For notational convenience, for pointed simplicial presheaves X, write

$$\Omega_T X = \mathbf{Hom}(T, X),$$

where $\mathbf{Hom}(T, X)$ is the internal function complex in *pointed* simplicial presheaves. This construction is right adjoint to smashing with T, so that there is a natural bijection

$$\hom(K, \Omega_T X) \cong \hom(K \wedge T, X)$$

for pointed simplicial presheaves K and X. The adjoint of a map $f : K \to \Omega_T Y$ is the composite

$$K \wedge T \xrightarrow{f \wedge T} \Omega_T Y \wedge T \xrightarrow{ev} Y,$$

where ev is the evaluation map. The functor $X \mapsto \Omega_T X$ is often called the T-*loops functor*.

A T-spectrum X may therefore be defined to be a collection X^n, $n \geq 0$, of pointed simplicial presheaves, together with pointed maps $\sigma_* : X^n \to \Omega_T X^{n+1}$. The map σ_* is the "adjoint" of the bonding map $\sigma : T \wedge X^n \to X^{n+1}$, and is called an *adjoint bonding map*. The map σ_* is adjoint to the composite

$$X^n \wedge T \xrightarrow{\tau} T \wedge X^n \xrightarrow{\sigma} X^{n+1},$$

where τ is the isomorphism which flips smash factors.

Say that a map $f : X \to Y$ of T-spectra is a *strict weak equivalence* (respectively *strict fibration*) if all maps $f : X^n \to Y^n$ are local weak equivalences (respectively injective fibrations) of pointed simplicial presheaves. Another way of saying this is that a strict weak equivalence is a *level weak equivalence*, and a strict fibration is a *level fibration*.

A *cofibration* of T-spectra is a map $i : A \to B$ such that

a) $i : A^0 \to B^0$ is a cofibration of simplicial presheaves, and
b) all maps

$$(T \wedge B^n) \cup_{(T \wedge A^n)} A^{n+1} \to B^{n+1}$$

are cofibrations of simplicial presheaves.

If K is a pointed simplicial presheaf and X is a T-spectrum, then $X \wedge K$ has the obvious meaning:

$$(X \wedge K)^n = X^n \wedge K.$$

The function complex $\mathbf{hom}(X, Y)$ for T-spectra X and Y is the simplicial set with

$$\mathbf{hom}(X, Y)_n = \hom(X \wedge \Delta^n_+, Y).$$

Proposition 10.15 *The category of $\mathbf{Spt}_T(\mathcal{C})$ of T-spectra, together with the classes of strict weak equivalences, strict fibrations and cofibrations as defined above, satisfies the definitions for a proper closed simplicial model category. This model structure is cofibrantly generated.*

The proof is, like that of Proposition 10.4, an exercise—in fact, it is the same exercise. The model structure for the category of T-spectra of Proposition 10.15 is called the *strict model structure*.

The simplicial model structure is defined by the function complex $\mathbf{hom}(X, Y)$. The usual lemma about cofibrations that gives the simplicial model axiom can either be proved directly, or as a consequence of the following enriched version:

Lemma 10.16 *Suppose that $X \to Y$ is a cofibration of T-spectra and that $A \to B$ is a cofibration of pointed simplicial presheaves. Then the induced map*

$$(Y \wedge A) \cup (X \wedge B) \to Y \wedge B$$

is a cofibration of T-spectra, which is a strict equivalence if either $X \to Y$ is a strict equivalence or $A \to B$ is a local weak equivalence.

Proof There is a pointed version of Corollary 4.41: if $A \to B$ and $C \to D$ are cofibrations (i.e. monomorphisms) of pointed simplicial presheaves, then the map

$$(D \wedge A) \cup (C \wedge B) \to D \wedge B$$

is a cofibration, which is a local weak equivalence if either $A \to B$ or $C \to D$ is a local weak equivalence.

Thus, all that we have to prove is the assertion that the map

$$(Y \wedge A) \cup (X \wedge B) \to Y \wedge B$$

is a cofibration of T-spectra, but this is an exercise.

Suspension spectrum constructions and shifts have the same formal properties for T-spectra as for ordinary spectra, up to a point:

1) Given a pointed simplicial presheaf K, the *suspension T-spectrum* $\Sigma_T^\infty K$ is the T-spectrum

$$K, T \wedge K, T^{\wedge 2} \wedge K, \dots$$

10.2 T-spectra and Localization

where
$$T^{\wedge n} = T \wedge \cdots \wedge T$$

is the *n-fold smash power*. The functor $K \mapsto \Sigma_T^\infty K$ is left adjoint to the level 0 functor $X \mapsto X^0$.

The *sphere T-spectrum* S_T for the category of T-spectra is the suspension T-spectrum $\Sigma_T^\infty S^0$ of the 2-point simplicial set S^0, and it consists of the objects

$$S^0, T, T^{\wedge 2}, T^{\wedge 3}, \ldots.$$

2) Given a T-spectrum X and $n \in \mathbb{Z}$, define the *shifted T-spectrum* $X[n]$ by

$$X[n]^k = \begin{cases} X^{n+k} & n+k \geq 0 \\ * & n+k < 0 \end{cases}$$

There are natural isomorphisms

$$\hom(\Sigma_T^\infty A[-n], X) \cong \hom(\Sigma_T^\infty A, X[n]) \cong \hom(A, X^n),$$

for all pointed simplicial presheaves A and T-spectra X, and $n \geq 0$.

The generating sets I and J for the cofibrations, respectively trivial cofibrations of the strict structures are the sets of maps

$$\Sigma_T^\infty A[-n] \to \Sigma_T^\infty B[-n]$$

where $n \geq 0$ and $A \to B$ is an α-bounded cofibration (respectively trivial cofibration) of pointed simplicial presheaves. Here, as always, α is a regular cardinal such that $\alpha > |\operatorname{Mor}(\mathcal{C})|$.

Every cofibration of T-spectra is a monomorphism, but the converse assertion is false because not all T-spectra are cofibrant. The cofibrant T-spectra are those objects X for which the maps $T \wedge X^n \to X^{n+1}$ are cofibrations of simplicial presheaves.

One often says that a monomorphism of T-spectra is a *level cofibration*.

The *T-suspension* of a T-spectrum X, is the T-spectrum $X \wedge T$, with bonding maps

$$\sigma \wedge 1 : T \wedge X^n \wedge T \to X^{n+1} \wedge T.$$

The *fake T-suspension* $\Sigma_T X$ of X is, by contrast, the T-spectrum with the objects $T \wedge X^n$ in the various levels, and bonding maps

$$1 \wedge \sigma : T \wedge T \wedge X^n \to T \wedge X^{n+1}.$$

Up to natural isomorphism, the suspension and fake suspension differ from each other by a twist automorphism $\tau : T \wedge T \cong T \wedge T$ in the bonding maps. The distinction between T-suspension and fake T-suspension is important, and is encountered in calculations.

The T-suspension functor $X \mapsto X \wedge T$ is left adjoint to the T-loops functor, while the fake T-suspension functor is left adjoint to a fake T-loops functor.

We construct various model structures on the category of T-spectra, including all stable model structures, by localizing the strict structure according to the method presented in Sect. 7.1. As in the construction of localized theories for simplicial presheaves (Sect. 7.2) or simplicial modules (Sect. 8.5), we begin with the construction of a functor $X \mapsto L_{\mathcal{F}} X$, via a small object construction that starts with a set of monomorphisms $\mathcal{F}$.

Choose a regular cardinal β such that $\beta > \alpha$, $\beta > |T|$, $\beta > |\mathcal{F}|$ and $\beta > |B|$ for any cofibration $A \to B$ in the set $\mathcal{F}$. Choose a cardinal λ such that $\lambda > 2^\beta$.

By a small object argument (as in Sect. 7.2), every morphism $g : X \to Y$ of T-spectra has a functorial factorization

$$\begin{array}{c} X \xrightarrow{i} E_\lambda(g) \\ \searrow_{g} \downarrow p \\ Y \end{array}$$

such that i is in the saturation of the set of monomorphisms $\mathcal{F}$ and the map p has the right lifting property with respect to all members of $\mathcal{F}$. The small object argument has λ steps.

Write

$$L_{\mathcal{F}}(X) = E_\lambda(X \to *)$$

for the result of this construction when applied to the canonical map $X \to *$.

Then we have the following. Lemma 10.17 is the analogue for T-spectra of Lemma 7.16, which is the corresponding result for simplicial presheaves. It is also a consequence of that result. See also Lemma 8.38, which is the analogous result for simplicial R-modules.

Lemma 10.17

1) *Suppose that the assignment $t \mapsto X_t$ defines a diagram of monomorphisms, indexed by a cardinal $\gamma > 2^\beta$. Then the natural map*

$$\varinjlim_{t < \gamma} L_{\mathcal{F}}(X_t) \to L_{\mathcal{F}}(\varinjlim_{t < \gamma} X_t)$$

is an isomorphism.

2) *Suppose that ζ is a cardinal with $\zeta > \beta$, and let $B_\zeta(X)$ denote the filtered system of subobjects of X having cardinality less than ζ. Then the natural map*

$$\varinjlim_{Y \in B_\zeta(X)} L_{\mathcal{F}}(Y) \to L_{\mathcal{F}}(X)$$

is an isomorphism.

10.2 T-spectra and Localization

3) The functor $X \mapsto L_{\mathcal{F}}(X)$ preserves monomorphisms.
4) Suppose that U, V are subobjects of a T-spectrum X. Then the natural map

$$L_{\mathcal{F}}(U \cap V) \to L_{\mathcal{F}}(U) \cap L_{\mathcal{F}}(V)$$

is an isomorphism.
5) If $|X| \leq 2^{\mu}$ where $\mu > \beta$, then $|L_{\mathcal{F}}(X)| \leq 2^{\mu}$.

Say that a map $f : X \to Y$ of $\mathbf{Spt}_T(\mathcal{C})$ is an $L_{\mathcal{F}}$-*equivalence* if it induces a strict weak equivalence

$$f_* : L_{\mathcal{F}} X \to L_{\mathcal{F}} Y.$$

Let κ be the successor cardinal for 2^{μ}, where μ is cardinal of statement 5) of Lemma 10.17. Then κ is a regular cardinal, and Lemma 10.17 implies that if a T-spectrum X is κ-bounded then $L_{\mathcal{F}}(X)$ is κ-bounded. A T-spectrum X is said to be κ-*bounded* if $|X_m^n(U)| < \kappa$ in all sections, levels and simplicial degrees.

The following Lemma establishes the bounded monomorphism property for the strict model structure on T-spectra:

Lemma 10.18 *Suppose given a diagram of monomorphisms*

such that the map i is a strict equivalence and A is β-bounded. Then there is a factorization of j by monomorphisms $A \to B \to Y$ such that B is β-bounded and the map $B \cap X \to B$ is a strict equivalence.

Proof Start with the diagram of cofibrations

$$\begin{array}{c} X^0 \\ \downarrow i \\ A^0 \longrightarrow Y^0 \end{array}$$

of pointed simplicial presheaves. Lemma 5.2 implies that there is a subobject $B^0 \subset Y^0$ such that B^0 is β-bounded, $A^0 \subset B^0$ and $B^0 \cap X^0 \to B^0$ is a local weak equivalence.

Form the diagram

$$\begin{array}{ccccc} T \wedge A^0 & \longrightarrow & T \wedge B^0 & \longrightarrow & T \wedge Y^0 \\ \sigma \downarrow & & & & \downarrow \sigma \\ A^1 & & \longrightarrow & & Y^1 \end{array}$$

Then the induced map

$$A^1 \cup_{T \wedge A^0} T \wedge B^0 \to Y^1$$

factors through a κ-bounded subobject $C^1 \subset Y^1$. There is an β-bounded subobject $B^1 \subset Y^1$ such that $C^1 \subset B^1$ and $B^1 \cap X^1 \to B^1$ is a local weak equivalence. The composite

$$T \wedge B^0 \to A^1 \cup_{T \wedge A^0} T \wedge B^0 \to C^1 \subset B^1$$

is the bonding map up to level 1 for the object B. Continue inductively to construct all B^n, $n \geq 1$.

The following result establishes the bounded monomorphism property for $L_\mathcal{F}$-equivalences of T-spectra.

Lemma 10.19 *Suppose given a diagram of monomorphisms*

$$\begin{array}{ccc} & & X \\ & & \downarrow i \\ A & \xrightarrow{j} & Y \end{array}$$

such that the map i is an $L_\mathcal{F}$-equivalence and A is κ-bounded. Then there is a factorization of j by monomorphisms $A \to B \to Y$ such that B is κ-bounded and the map $B \cap X \to B$ is an $L_\mathcal{F}$-equivalence.

Proof In the presence of the bounded cofibration condition for the strict structure (Lemma 10.18) and the list of properties for the functor $L_\mathcal{F}$ given by Lemma 10.17, the proof is formally the same as that of Lemma 7.17.

Now suppose that $\mathcal{F}$ is a set of cofibrations of T-spectra which satisfies the following conditions:

C1 The T-spectrum A is cofibrant for all maps $i : A \to B$ in $\mathcal{F}$.
C2 The set $\mathcal{F}$ includes the set J of generators for the trivial cofibrations for the strict model structure on $\mathbf{Spt}_T(\mathcal{C})$.
C3 If the map $i : A \to B$ is a member of the set $\mathcal{F}$, then the cofibrations

$$(A \wedge D) \cup (B \wedge C) \to B \wedge D$$

induced by i and all α-bounded cofibrations $C \to D$ of pointed simplicial presheaves are in $\mathcal{F}$.

Smashing with pointed simplicial presheaves preserves cofibrations of T-spectra, by Lemma 10.16. In particular, the T-spectrum $A \wedge D$ is cofibrant if A is cofibrant, for all pointed simplicial presheaves D. It follows that the objects

10.2 T-spectra and Localization

$$(A \wedge D) \cup (B \wedge C)$$

appearing in **C3** are cofibrant T-spectra.

If the cofibration $A \to B$ is in $\mathcal{F}$, then $\mathcal{F}$ also contains the cofibrations

$$(A \wedge \Delta_+^m) \cup (B \wedge \partial \Delta_+^m) \to B \wedge \Delta_+^m, \ m \geq 0.$$

The set $\mathcal{F}$ therefore satisfies the requirement **F3** for the general localization setup of Chap. 7.

In the language of Sect. 7.1, a map $p : X \to Y$ is said to be $\mathcal{F}$-*injective* if it has the right lifting property with respect to all maps of $\mathcal{F}$. An object X is $\mathcal{F}$-*injective* if the map $X \to *$ is $\mathcal{F}$-injective.

The object $L_\mathcal{F} X$ is $\mathcal{F}$-injective for every object X. Every $\mathcal{F}$-injective T-spectrum is strictly fibrant.

A map of T-spectra $X \to Y$ is an $\mathcal{F}$-*equivalence* if and only if the map

$$\mathbf{hom}(Y_c, Z) \to \mathbf{hom}(X_c, Z)$$

is a weak equivalence of simplicial sets for every $\mathcal{F}$-injective object Z, and where $X_c \to X$ is a natural cofibrant model construction in the strict model category. It is a consequence of Lemma 7.6 that a map $X \to Y$ is an $\mathcal{F}$-equivalence if and only if it is an $L_\mathcal{F}$-equivalence.

An $\mathcal{F}$-*fibration* is a map that has the right lifting property with respect to all cofibrations which are $\mathcal{F}$-equivalences. The $\mathcal{F}$-fibrations coincide with the $L_\mathcal{F}$-fibrations of Theorem 7.10.

The methods in Sect. 7.1 (specifically, Theorem 7.10 with Lemma 10.19) then give the following:

Theorem 10.20 *Suppose that $\mathcal{C}$ is a small Grothendieck site, and that $\mathcal{F}$ is a set of cofibrations of T-spectra that satisfies the conditions* **C1**, **C2** *and* **C3**. *Then the category* $\mathbf{Spt}_T(\mathcal{C})$ *of T-spectra, together with the classes of cofibrations, $\mathcal{F}$-equivalences and $\mathcal{F}$-fibrations, satisfies the axioms for a cofibrantly generated closed simplicial model category. This model structure is left proper.*

The model structure of Theorem 10.20 is the $\mathcal{F}$-*local structure* for the category of T-spectra. The left properness of this model structure is a consequence of Lemma 7.8.

Lemma 7.11 implies that if $p : X \to Y$ is an $\mathcal{F}$-injective map and Y is an $\mathcal{F}$-injective object, then the map p is an $\mathcal{F}$-fibration. It follows in particular (Corollary 7.12) that a T-spectrum Z is $\mathcal{F}$-injective if and only if it is $\mathcal{F}$-fibrant.

The *internal function complex* $\mathbf{Hom}(A, X)$, for T-spectra A and X, is the pointed simplicial presheaf which is defined in sections by

$$\mathbf{Hom}(A, X)(U) = \mathbf{hom}(A|_U, X|_U) = \mathbf{hom}(A \wedge U_+, Y),$$

for $U \in \mathcal{C}$ (see (5.4)). The functor $X \mapsto \mathbf{Hom}(A, X)$ is right adjoint to the functor

$$s\mathbf{Pre}_*(\mathcal{C}) \to \mathbf{Spt}_T(\mathcal{C})$$

which is defined for pointed simplicial presheaves K by $K \mapsto A \wedge K$.

Suppose given a subset S of $\mathcal{F}$, and that S contains the generating set J for the strict trivial cofibrations. Say that S *generates* the set $\mathcal{F}$ if $\mathcal{F}$ is the smallest subset which contains S and has the closure property **C3**.

The following result will be needed later.

Lemma 10.21 *Suppose that the subset $S \subset \mathcal{F}$ generates $\mathcal{F}$. Then a map $p : X \to Y$ is $\mathcal{F}$-injective if and only if all maps*

$$(i^*, p_*) : \mathbf{Hom}(B, X) \to \mathbf{Hom}(A, X) \times_{\mathbf{Hom}(A,Y)} \mathbf{Hom}(B, Y)$$

are trivial injective fibrations of simplicial presheaves for all members $i : A \to B$ of the set S.

Proof Suppose that the maps (i^*, p_*) are trivial injective fibrations for all $i \in S$. Suppose that $\mathcal{F}' \subset \mathcal{F}$ is the collection of all cofibrations $j : A \to B$ in $\mathcal{F}$ such that the map (j^*, p_*) is a trivial fibration. This is the collection of cofibrations $j : A \to B$ in $\mathcal{F}$ such that the map p has the right lifting property with respect to all maps

$$(j, i) : (A \wedge D) \cup (B \wedge C) \to B \wedge D$$

which are induced by cofibrations $i : C \to D$ of pointed simplicial presheaves.

If $j : A \to B$ is a member of $\mathcal{F}'$, then the cofibration (j, i') induced by j and an α-bounded cofibration $i' : C' \to D'$ is also in $\mathcal{F}'$. The subset $\mathcal{F}'$ contains S by assumption, and S generates $\mathcal{F}$, so that $\mathcal{F}' = \mathcal{F}$. It follows that p has the right lifting property with respect to all morphisms of $\mathcal{F}$, and is therefore an $\mathcal{F}$-injective morphism.

Conversely, if the morphism p is $\mathcal{F}$-injective then the simplicial presheaf map (i^*, p_*) is a trivial fibration for all morphisms i of S, because the set $\mathcal{F}$ satisfies the condition **C3**.

10.3 Stable Model Structures for T-spectra

Theorem 10.20 gives an $\mathcal{F}$-model structure for the category $\mathbf{Spt}_T(\mathcal{C})$ in which the cofibrations in a set $\mathcal{F}$ are formally inverted. Recall that this set $\mathcal{F}$ is required to satisfy the following conditions:

C1 The T-spectrum A is cofibrant for all maps $i : A \to B$ in $\mathcal{F}$.

C2 The set $\mathcal{F}$ includes the set J of generators for the trivial cofibrations for the strict model structure on $\mathbf{Spt}_T(\mathcal{C})$.

C3 If $i : A \to B$ is in the set $\mathcal{F}$, then the cofibrations

$$(A \wedge D) \cup (B \wedge C) \to B \wedge D$$

10.3 Stable Model Structures for T-spectra

induced by i and all α-bounded cofibrations $C \to D$ of pointed simplicial presheaves are also in $\mathcal{F}$.

This set $\mathcal{F}$ is generated by a subset S containing J if $\mathcal{F}$ is the smallest subset which contains S and is closed under condition **C3**.

The *stabilization maps*

$$\Sigma_T^\infty T[-n-1] = (S_T \wedge T)[-n-1] \to S_T[-n]$$

are shifts of the map

$$(S_T \wedge T)[-1] \to S_T$$

which consists of canonical isomorphisms

$$T^{\wedge(m-1)} \wedge T \to T^{\wedge m}$$

in levels $m \geq 1$.

Say that the $\mathcal{F}$-local structure on $\mathbf{Spt}_T(\mathcal{C})$ is *stable* if the stabilization maps are $\mathcal{F}$-equivalences. One forces this by insisting that the set $\mathcal{F}$ contains cofibrant replacements of these maps.

Example 10.22 If the set $\mathcal{F}$ is generated over the generating set J for the trivial cofibrations of simplicial presheaves by cofibrant replacements of the stabilization maps alone, then the corresponding $\mathcal{F}$-local structure is the *stable model structure* for the category of T-spectra. The weak equivalences for this model structure are called *stable equivalences*, and the fibrations are *stable fibrations*.

This case specializes to classical stable homotopy theory of spectra, or S^1-spectra.

Example 10.23 Suppose that the map $f : A \to B$ is a cofibration of pointed simplicial presheaves.

The f-local model structure for simplicial presheaves of Theorem 7.18 is cofibrantly generated. Let J_f be the collection of maps

$$\Sigma_T^\infty C_+[-n] \to \Sigma_T^\infty D_+[-n]$$

of T-spectra which are induced by a fixed set of generators $C \to D$ for the trivial cofibrations of the f-local model structure on the category $s\mathbf{Pre}(\mathcal{C})$ of simplicial presheaves (Theorem 7.18). We assume that this set of generators $C \to D$ for the f-local structure contains a set of generators for the trivial cofibrations for the injective model structure on simplicial presheaves, so that $J \subset J_f$.

Suppose that the set of cofibrations S_f is generated by the set J_f, together with cofibrant replacements of the stabilization maps

$$(S_T \wedge T)[-n-1] \to S_T[-n], \ n \geq 0.$$

The model structure on $\mathbf{Spt}_T(\mathcal{C})$ given by Theorem 10.20 for the set S_f is the *f-local stable model structure*, and the corresponding L-equivalences and L-fibrations are called *stable f-equivalences* and *stable f-fibrations*, respectively.

A map $p : X \to Y$ is f-*injective* if it has the right lifting property with respect to all members of the set S_f, and a T-spectrum Z is f-injective if the map $Z \to *$ is f-injective.

It is a formal result (Corollary 7.12) that the f-injective T-spectra coincide with the stable f-fibrant T-spectra.

This remainder of this section is a study of f-local stable model structures for T-spectra. We begin with general theory, but then we make some assumptions on both the underlying site and the object T to recover some traditional aspects of stable homotopy theory in the $\mathcal{F}$-local structures. These include an infinite loop space model for stable fibrant replacement (Theorem 10.32), and properness of the stable model structure (Theorem 10.36).

Remark 10.24 A *strict f-fibration* (respectively *strict f-equivalence*) is a map $X \to Y$ of T-spectra which consists of f-fibrations (respectively f-equivalences) $X^n \to Y^n$ of simplicial presheaves in all levels. The cofibrations, strict f-fibrations and strict f-equivalences together form a closed simplicial model structure on the category of T-spectra, called the *strict f-local model structure*—see also Proposition 10.15. This model structure is cofibrantly generated, and therefore has natural fibrant models. It is also left proper. The strict f-local model structure is right proper if the f-local model structure on simplicial presheaves is right proper.

It is an exercise to show that every strict f-equivalence of T-spectra is a stable f-equivalence.

We shall use the notation $X \to FX$ to denote a (natural) strict f-fibrant model for a T-spectrum X.

We have an analogue of Lemma 10.10 for the stable f-local theory:

Lemma 10.25 *Suppose that the T-spectrum Y is strictly f-fibrant. Then a map $p : X \to Y$ is f-injective if and only if p is a strict f-fibration and all diagrams of pointed simplicial presheaf maps*

$$\begin{array}{ccc} X^n & \xrightarrow{\sigma_*} & \Omega_T X^{n+1} \\ \downarrow & & \downarrow \\ Y^n & \xrightarrow{\sigma_*} & \Omega_T Y^{n+1} \end{array}$$

are homotopy cartesian in the f-local model structure.

Proof Lemma 10.21 says that the map p is an f-injective fibration if and only if p induces trivial fibrations of simplicial presheaves

$$\mathbf{Hom}(B, X) \to \mathbf{Hom}(A, X) \times_{\mathbf{Hom}(A,Y)} \mathbf{Hom}(B, Y) \qquad (10.8)$$

for all members $A \to B$ of the set S_f.

In particular, if p is f-injective then the maps (10.8) are surjective in global sections so that p has the right lifting property with respect to all members of S_f.

10.3 Stable Model Structures for T-spectra

This set includes all cofibrations

$$\Sigma_T^\infty C_+[-n] \to \Sigma_T^\infty D_+[-n]$$

arising from a set of generators $C \to D$ for the f-local structure on simplicial presheaves, so that $p : X \to Y$ must be a strict f-fibration. The other members of the generating set for S_f are cofibrant replacements for the stabilization maps

$$\Sigma_T^\infty T[-n-1] \to S_T[-n].$$

The objects $\Sigma_T^\infty T[-n-1]$ and $S_T[-n]$ are cofibrant, and it follows that the diagram

$$\begin{array}{ccc} \mathbf{Hom}(S_T[-n], X) & \longrightarrow & \mathbf{Hom}(\Sigma_T^\infty T[-n-1], X) \\ \downarrow & & \downarrow \\ \mathbf{Hom}(S_T[-n], Y) & \longrightarrow & \mathbf{Hom}(\Sigma_T^\infty T[-n-1], Y) \end{array}$$

is homotopy cartesian since X and Y are strictly f-fibrant. The map

$$X^n \to Y^n \times_{\Omega_T Y^{n+1}} \Omega_T X^{n+1}$$

is therefore a local weak equivalence.

The proof of the converse assertion is an exercise. □

Corollary 10.26 *A T-spectrum X is stable f-fibrant if and only if it is strictly f-fibrant and all maps $\sigma_* : X^n \to \Omega_T X^{n+1}$ are sectionwise equivalences of pointed simplicial presheaves.*

Proof The stable f-fibrant objects coincide with the f-injective objects. Lemma 10.25 implies that these are the objects X that are strictly f-fibrant and such that the maps $\sigma_* : X^n \to \Omega_T X^{n+1}$ (of f-fibrant simplicial presheaves) are f-local weak equivalences, and hence sectionwise weak equivalences. See the preamble to Theorem 7.18 for the claim about sectionwise weak equivalences. □

Corollary 10.27 *Every stable f-fibrant T-spectrum is stable fibrant. Every stable equivalence of T-spectra is a stable f-equivalence.*

Proof If a T-spectrum Z is stable f-fibrant, then all component presheaves Z^n are f-fibrant, hence injective fibrant, and all maps $Z^n \to \Omega_T Z^{n+1}$ are sectionwise equivalences. It follows from Corollary 10.26 that Z is stable fibrant.

If the map $X \to Y$ is a stable equivalence, then it has a cofibrant replacement $X_c \to Y_c$ in the strict model structure such that the simplicial set map

$$\mathbf{hom}(Y_c, W) \to \mathbf{hom}(X_c, W)$$

is a weak equivalence for all stable fibrant T-spectra W. It follows that the map

$$\mathbf{hom}(Y_c, Z) \to \mathbf{hom}(X_c, Z)$$

is a weak equivalence for all stable f-fibrant T-spectra Z, and so the map $X \to Y$ is a stable f-equivalence.

An *inductive system* (or an *inductive diagram*) is a diagram $s \mapsto X_s$ which is indexed by $s \in \beta$, where β is some limit ordinal. An *inductive colimit* is a colimit for an inductive system. Inductive colimits are necessarily filtered colimits.

Say that the pointed simplicial presheaf T is *compact up to f-equivalence* if, for any inductive system $s \mapsto X_s$ of f-fibrant pointed simplicial presheaves, the composite

$$\varinjlim_s \Omega_T X_s \xrightarrow{c} \Omega_T(\varinjlim_s X_s) \xrightarrow{\Omega_T j} \Omega_T F(\varinjlim_s X_s)$$

is an f-equivalence, where c is the canonical map.

Lemma 10.28

1) If T and T' are compact up to f-equivalence, then the smash product $T \wedge T'$ is compact up to f-equivalence.
2) Suppose that the map $T \to T'$ is an f-equivalence. If either T or T' is compact up to equivalence, then so is the other.

Proof We prove statement 1). Statement 2) is an exercise.

We have a natural isomorphism

$$\Omega_{T \wedge T'} Y \cong \Omega_T(\Omega_{T'} Y),$$

and the canonical map

$$\varinjlim_s \Omega_T \Omega_{T'} X_s \to \Omega_T \Omega_{T'}(\varinjlim_s X_s)$$

may be identified with the composite

$$\varinjlim_s \Omega_T \Omega_{T'} X_s \to \Omega_T(\varinjlim_s \Omega_{T'} X_s) \to \Omega_T(\Omega_{T'}(\varinjlim_s X_s)).$$

The composite

$$\varinjlim_s \Omega_{T'} X_s \to \Omega_{T'}(\varinjlim_s X_s) \to \Omega_{T'} F(\varinjlim_s X_s)$$

is a stable fibrant model for $\varinjlim_s \Omega_{T'} X_s$ since T' is compact up to f-equivalence. It follows that the composite

$$\varinjlim_s \Omega_T \Omega_{T'} X_s \to \Omega_T(\varinjlim_s \Omega_{T'} X_s) \to \Omega_T(\Omega_{T'}(\varinjlim_s X_s)) \to \Omega_T \Omega_{T'} F(\varinjlim_s X_s)$$

is an f-equivalence since T is compact up to f-equivalence.

10.3 Stable Model Structures for T-spectra

Example 10.29 In the motivic model structure (defined over a base scheme S—see Example 7.20), a filtered colimit $\varinjlim_s X_s$ of motivic fibrant simplicial presheaves X_s satisfies motivic descent in the sense that any motivic fibrant model

$$j : \varinjlim_s X_s \to F(\varinjlim_s X_s)$$

is a sectionwise weak equivalence.

We have the following observations:

1) If K is a finite pointed simplicial set, then K is compact up to motivic equivalence. In the composite

$$\varinjlim_s \Omega_K X_s \xrightarrow{c} \Omega_K(\varinjlim_s X_s) \xrightarrow{\Omega_K j} \Omega_K F(\varinjlim_s X_s),$$

the map c is an isomorphism since K is finite, and the map $j : \varinjlim_s X_s \to F(\varinjlim_s X_s)$ is a sectionwise equivalence of presheaves of pointed simplicial sets by motivic descent, so that $\Omega_K j$ is a sectionwise equivalence.

2) If $x : * \to U$ is a pointed S-scheme, then U is compact up to motivic equivalence. In sections, $\Omega_U(V)$ is the fibre of the map $X(V \times U) \to X(V)$ which is induced by the map $(1_V, x) : V \to V \times U$ that is determined by the base point x of U.

In the composite

$$\varinjlim_s \Omega_U X_s \xrightarrow{c} \Omega_U(\varinjlim_s X_s) \xrightarrow{\Omega_U j} \Omega_U F(\varinjlim_s X_s),$$

the map c is an isomorphism, and the map $\Omega_U j$ is a sectionwise equivalence since the colimit satisfies motivic descent.

Example 10.30 Suppose that K is a finite pointed simplicial set. Then K is compact up to equivalence for the injective model structure on pointed simplicial presheaves.

If X is a pointed presheaf of Kan complexes, then there is a pullback square

$$\begin{array}{ccc} \Omega_K X & \longrightarrow & \mathbf{hom}(K, X) \\ \downarrow & & \downarrow p \\ * & \longrightarrow & X \end{array}$$

The map p is a sectionwise Kan fibration, so that the diagram is homotopy cartesian in the injective model structure by Lemma 5.20. Properness for the injective model structure implies that any local weak equivalence $X \to Y$ of presheaves of pointed Kan complexes induces a local weak equivalence $\Omega_K X \to \Omega_K Y$.

Thus, if $s \mapsto X_s$ is an inductive system of injective fibrant simplicial presheaves, then in the composite

$$\varinjlim_s \Omega_K X_s \xrightarrow{c} \Omega_K(\varinjlim_s X_s) \xrightarrow{\Omega_K j} \Omega_K F(\varinjlim_s X_s)$$

then the map c is an isomorphism and the map $\Omega_K j$ is a local weak equivalence.

The *fake T-loop object* $\Omega_T X$ associated with a T-spectrum X is the T-spectrum with

$$(\Omega_T X)^n = \Omega_T(X^n),$$

with the maps

$$\Omega_T \sigma_* : \Omega_T X^n \to \Omega_T \Omega_T X^{n+1}$$

as adjoint bonding maps. There is a natural map of T-spectra

$$\gamma : X \to \Omega_T X[1],$$

which is defined by the adjoint bonding map

$$\sigma_* : X^n \to \Omega_T X^{n+1}$$

in each level n.

The "real" T-loop spectrum is the object $\mathbf{Hom}(T, X)$ given by the pointed internal function complex. Generally, if A is a pointed simplicial presheaf and X is a T-spectrum, then there is a T-spectrum $\mathbf{Hom}(A, X)$ with

$$\mathbf{Hom}(A, X)^n := \mathbf{Hom}(A, X^n),$$

given by the pointed internal function complex, with bonding map

$$T \wedge \mathbf{Hom}(A, X^n) \to \mathbf{Hom}(A, X^{n+1})$$

which is adjoint to the composite

$$T \wedge \mathbf{Hom}(A, X^n) \wedge A \xrightarrow{T \wedge ev} T \wedge X^n \xrightarrow{\sigma} X^{n+1}.$$

It is an exercise to show that the bonding maps for the real and fake T-loop spectra, $\mathbf{Hom}(T, X)$ and $\Omega_T X$ respectively, have adjoints defined in levels by maps

$$\Omega_T X^n \wedge T \wedge T \to X^{n+1}$$

which differ by a twist $\tau : T \wedge T \to T \wedge T$ of the smash factor $T \wedge T$.

There is a natural directed system of maps

$$X \xrightarrow{\gamma} \Omega_T X[1] \xrightarrow{\Omega_T \gamma[1]} \Omega_T^2 X[2] \xrightarrow{\Omega_T^2 \gamma[2]} \Omega_T^3 X[3] \xrightarrow{\Omega_T^3 \gamma[3]} \ldots$$

We define the T-spectrum $Q_T X$ by

$$Q_T X = F(\varinjlim_n \Omega_T^n F X[n]),$$

10.3 Stable Model Structures for T-spectra

where $j : Y \to FY$ is the natural strictly f-fibrant model for a T-spectrum Y. The natural map $\eta : X \to Q_T X$ is the composite

$$X \xrightarrow{j} FX \to \varinjlim_n \Omega_T^n FX[n] \xrightarrow{j} F(\varinjlim_n \Omega_T^n FX[n]).$$

The functor $X \mapsto Q_T X$ preserves strict f-equivalences. In effect, if the map $f : X \to Y$ is a strict f-equivalence, then all induced maps

$$\Omega_T^k FX^{n+k} \to \Omega_T^k FY^{n+k}$$

are f-equivalences (even sectionwise equivalences), as is the induced map

$$\varinjlim_k \Omega_T^k FX^{n+k} \to \varinjlim_k \Omega_T^k FY^{n+k},$$

and the claim follows.

More generally, Lemma 7.7 implies that inductive colimits preserve f-equivalences in the category of simplicial presheaves, in the sense that if κ is a limit ordinal and $X_s \to Y_s$ is a natural transformation of f-equivalences defined for $s < \kappa$, then the induced map

$$\varinjlim_{s<\kappa} X_s \to \varinjlim_{s<\kappa} Y_s$$

is an f-equivalence of simplicial presheaves. It follows that strict f-equivalences of T-spectra are preserved by inductive colimits, in the same sense.

Lemma 7.7 also implies that stable f-equivalences are preserved by inductive colimits.

Lemma 10.31 *Suppose that T is compact up to f-equivalence. Then the canonical map*

$$\varinjlim_{s<\kappa} Q_T X_s \to Q_T(\varinjlim_{s<\kappa} X_s) \tag{10.9}$$

is a strict f-equivalence.

Proof We can assume that all of the objects X_s are strictly f-fibrant. Then the composites

$$\varinjlim_{s<\kappa} \Omega_T^n X_s[n] \to \Omega_T^n(\varinjlim_{s<\kappa} X_s)[n] \to \Omega_T^n F(\varinjlim_{s<\kappa} X_s)[n]$$

are strict f-equivalences by the compactness assumption, and it follows that the map (10.9) is a strict f-equivalence.

Theorem 10.32 *Suppose that T is compact up to f-equivalence. Then the map*

$$\eta : X \to Q_T X$$

is a natural stable f-fibrant model for all T-spectra X.

Proof The T-spectrum $Q_T X$ is stably f-fibrant by construction, the compactness of T, and Corollary 10.26.

Suppose that the map $j : X \to LX$ is a natural injective (equivalently, stable f-fibrant) model. There is a commutative diagram

$$\begin{array}{ccc} X & \xrightarrow{j} & LX \\ \eta \downarrow & & \downarrow \eta \\ Q_T X & \xrightarrow{j_*} & Q_T LX \end{array}$$

We show that the maps $\eta : LX \to Q_T LX$ and $j_* : Q_T X \to Q_T LX$ are strict f-equivalences. It then follows that the map $\eta : X \to Q_T X$ is a stable f-equivalence.

If Z is stable f-fibrant, the map $\eta : Z \to Q_T Z$ consists of the maps

$$Z^n \xrightarrow{\sigma_*} \Omega_T Z^{n+1} \xrightarrow{\Omega_T \sigma_*} \Omega_T^2 Z^{n+2} \xrightarrow{\Omega_T^2 \sigma_*} \cdots \to \varinjlim \Omega_t^k Z^{n+k} \xrightarrow{j} F(\varinjlim \Omega_t^k Z^{n+k}),$$

all of which are f-equivalences, so that the map $\eta : Z \to Q_T Z$ is a strict f-equivalence. In particular, the map $\eta : LX \to Q_T LX$ is a strict f-equivalence.

Say that a map $f : X \to Y$ of T-spectra is an *equivalence after level N* if the component maps $X^k \to Y^k$ are f-equivalences for $k > N$. If $f : X \to Y$ is an equivalence after level N, then the map $\Omega_T^k FX[k] \to \Omega_T^k FY[k]$ is a (sectionwise) strict equivalence for $k > N$, and so the map $Q_T X \to Q_T Y$ is a strict f-equivalence.

Each generating cofibration for the set S_f is an equivalence after level N for some N, as are all of its pushouts. It follows that the stable f-fibrant model $j : X \to LX$ can be written as the composite

$$X = X_0 \to X_1 \to \cdots$$

of maps between objects X_s, $s \in \beta$ for some regular cardinal β such that each map $X_s \to X_{s+1}$ induces a strict equivalence $Q_T X_s \to Q_T X_{s+1}$, and where $X_t = \varinjlim_{s<t} X_s$ for all limit ordinals $t < \beta$. It follows from Lemma 10.31 that the map

$$X = X_0 \to \varinjlim_{s<\beta} X_s = LX$$

induces a strict equivalence

$$j_* : Q_T X \to Q_T LX.$$

Corollary 10.33 *Suppose that T is compact up to f-equivalence. Then a map $\alpha : X \to Y$ of T-spectra is a stable f-equivalence if and only if the induced map $\alpha_* : Q_T X \to Q_T Y$ is a strict f-equivalence.*

10.3 Stable Model Structures for T-spectra

Say that the f-local model structure on the category of simplicial presheaves satisfies *inductive colimit descent* if, given an inductive system $s \mapsto X_s$ of f-fibrant pointed simplicial presheaves, and an f-fibrant model

$$j : \varinjlim_s X_s \to F(\varinjlim_s X_s),$$

then the map j is a *local* weak equivalence.

Example 10.34 The motivic model structure (Example 7.20) satisfies inductive colimit descent, since an inductive colimit of motivic fibrant simplicial presheaves satisfies motivic descent.

The injective model structure satisfies inductive colimit descent automatically.

The following result generalizes observations of Examples 10.29 and 10.30:

Lemma 10.35 *Suppose that the f-local model structure on the category of simplicial presheaves satisfies inductive colimit descent. Then all finite pointed simplicial sets K are compact up to f-equivalence.*

Proof Suppose that $s \mapsto X_s$ is an inductive system of f-fibrant pointed simplicial presheaves. Suppose that the map $j : \varinjlim_s X_s \to F(\varinjlim_s X_s)$ is an f-fibrant model, and consider the composite

$$\varinjlim_s \Omega_K(X_s) \xrightarrow{c} \Omega_K(\varinjlim_s X_s) \xrightarrow{\Omega_K(j)} \Omega_K(F(\varinjlim_s X_s)).$$

Here, the map c is an isomorphism since K is finite. The map j is a local weak equivalence by the inductive colimit descent assumption, and it follows from Lemma 5.20 that the map $\Omega_K(j)$ is a local weak equivalence.

Theorem 10.36 *Suppose that T is compact up to f-equivalence, and that the f-local model structure for simplicial presheaves satisfies inductive colimit descent. Suppose also that the f-local structure for simplicial presheaves is right proper. Then the f-local stable model structure on the category of T-spectra is right proper.*

Proof Suppose given a pullback diagram

$$\begin{array}{ccc} Z \times_Y X & \xrightarrow{g_*} & X \\ \downarrow & & \downarrow p \\ Z & \xrightarrow{g} & Y \end{array} \qquad (10.10)$$

of T-spectra such that p is a strict f-fibration and g is a stable f-equivalence. We show that the map g_* is a stable f-equivalence.

The strict f-local model structure is right proper, on account of the assumption that the f-local structure on the category of simplicial presheaves is right proper. We can therefore assume that all objects in the diagram (10.10) are strictly f-fibrant.

Write
$$\overline{Q}_T X = \varinjlim_n \Omega_T^n X[n].$$

The induced diagram

$$\begin{array}{ccc} \overline{Q}_T(Z \times_Y X) & \xrightarrow{\overline{Q}_T f_*} & \overline{Q}_T X \\ \downarrow & & \downarrow \overline{Q}_T p \\ \overline{Q}_T Z & \xrightarrow{\overline{Q}_T f} & \overline{Q}_T Y \end{array}$$

is a pullback. The map $\overline{Q}_T p$ is a sectionwise Kan fibration and hence a local fibration in all levels, and the map $\overline{Q}_T f$ is a local weak equivalence in all levels by the inductive colimit descent assumption and Theorem 10.32. It follows from Lemma 5.20 that the map $\overline{Q}_T f_*$ is a local weak equivalence in all levels, and so the map $Q_T f_*$ is a local weak equivalence in all levels, again by the inductive colimit descent assumption. The map $f_* : Z \times_Y X \to X$ is therefore a stable f-equivalence.

Corollary 10.37 *Suppose that T is compact up to equivalence and that the f-local model structure for simplicial presheaves satisfies inductive colimit descent. Suppose that the f-local structure for simplicial presheaves is proper. Suppose that $p : X \to Y$ is a strict f-fibration. Then p is a stable f-fibration if and only if the diagram*

$$\begin{array}{ccc} X & \xrightarrow{i} & LX \\ p \downarrow & & \downarrow Lp \\ Y & \xrightarrow{i} & LY \end{array}$$

is strictly homotopy cartesian.

Proof This result is a formal consequence of the right properness given by Theorem 10.36. See the proof of Lemma 10.9.

Example 10.38 The motivic stable category

Recall the setup for the motivic model structure for simplicial presheaves, over a base scheme S, from Example 7.20. The motivic model structure is right proper by Theorem 7.27.

It is common to write T for the projective line $\mathbb{P}^1$ over S, pointed by some global section. The scheme $\mathbb{P}^1$ (or rather, the sheaf represented by the scheme) is compact up to f-equivalence, as in Example 10.29.

10.3 Stable Model Structures for T-spectra

The resulting stable category of T-spectra is the motivic stable category of Morel and Voevodsky [57, 82].

The natural map $\eta : X \to Q_T X$ defines a natural stable motivic fibrant model for T-spectra X by Theorem 10.32, and the motivic stable model structure of T-spectra is proper by Theorem 10.36.

Motivic stable model structures of S^1-spectra and $\mathbb{G}_m$-spectra are also in common use. Both model structures have "standard" natural stable motivic fibrant models and are proper, by Theorem 10.32 and Theorem 10.36, respectively.

Example 10.39 Presheaves of Spectra

In this example, the underlying model structure for simplicial presheaves is the injective model structure, and T is the simplicial circle S^1. The object S^1 is a finite pointed simplicial set, and is therefore compact up to equivalence as in Example 10.30.

The category of S^1-spectra is the category of presheaves of spectra, and the stable model structure for the category of S^1-spectra arising from Theorem 10.20 coincides with the standard model structure on presheaves of spectra given by Theorem 10.5.

To see this, observe that the two model structures have the same cofibrations, by definition. Also, from Theorem 10.32, a map $f : X \to Y$ of S^1-spectra is a stable equivalence for Theorem 10.20 if and only if the map

$$f_* : Q_{S^1} X \to Q_{S^1} Y$$

is a levelwise equivalence, and this is true if and only if f induces isomorphisms $\tilde{\pi}_k X \to \tilde{\pi}_k Y$ in all sheaves of stable homotopy groups.

For this last point, there are natural sheaf isomorphisms

$$\tilde{\pi}_k(Q_{S^1} X^n, *) \cong \tilde{\pi}_{k-n} X.$$

There is a potential issue, in that the sheaves of homotopy groups $\tilde{\pi}_k(Q_{S^1} X^n, *)$ involve only a single choice of base point instead of all local choices, but one gets around this by using the fact that the simplicial presheaves $Q_{S^1} X^n$ are presheaves of H-spaces, so that the choice of base point does not matter.

The ordinary stable model structure on presheaves of spectra is proper from two points of view; one can either use a long exact sequence comparison, or Theorem 10.36 to prove right properness.

The standard recognition principle for a stable fibration of presheaves of spectra says that $p : X \to Y$ is a stable fibration if and only if it is a strict fibration and the diagram

$$\begin{array}{ccc} X & \longrightarrow & Q_{S^1} X \\ {\scriptstyle p} \downarrow & & \downarrow {\scriptstyle p_*} \\ Y & \longrightarrow & Q_{S^1} Y \end{array}$$

is strictly homotopy cartesian. This is proved in both Lemma 10.9 and Corollary 10.37, by using essentially the same argument based on right properness of the model structure.

More generally, suppose L is a finite pointed simplicial set. Then L is compact up to equivalence, and there is a stable model structure on L-spectra for which a map $f : X \to Y$ is a stable equivalence if and only if the induced map $Q_L X \to Q_L Y$ is a levelwise equivalence, by Corollary 10.33. It is, however, rather hard to say more than this without knowing that each simplicial presheaf $Q_L X^n$ has something like an H-space structure.

There would be such an H-space structure if L is an S^1-suspension $L = S^1 \wedge K$ for some finite pointed simplicial presheaf K. In that case, there is a functorial bigraded system of sheaves of stable homotopy groups which describes stable equivalence. See Sect. 10.5.

Every pointed simplicial presheaf map $\alpha : T' \to T$ induces a functor

$$\alpha^* : \mathbf{Spt}_T(\mathcal{C}) \to \mathbf{Spt}_{T'}(\mathcal{C}).$$

If X is a T-spectrum, then $\alpha^* X$ is the T'-spectrum with

$$\alpha^* X^n = X^n$$

for $n \geq 0$, and with bonding maps given by the composites

$$T' \wedge X^n \xrightarrow{\alpha \wedge 1} T \wedge X^n \xrightarrow{\sigma} X^{n+1},$$

where the morphisms σ are the bonding maps of the T-spectrum X.

The functor α^* preserves strict f-equivalences and strict f-fibrations in the sense of Remark 10.24.

Theorem 10.40 *Suppose that the map $\alpha : T' \to T$ is an f-equivalence of pointed simplicial presheaves. Suppose that the objects T and T' are compact up to f-equivalence. Then the functor*

$$\alpha^* : \mathbf{Spt}_T(\mathcal{C}) \to \mathbf{Spt}_{T'}(\mathcal{C})$$

preserves stable f-equivalences, and the resulting functor

$$\alpha^* : \mathrm{Ho}\,(\mathbf{Spt}_T(\mathcal{C})) \to \mathrm{Ho}\,(\mathbf{Spt}_{T'}(\mathcal{C}))$$

is an equivalence of f-local stable homotopy categories.

Proof It suffices, by Lemma 10.28, to assume that the map α is a cofibration. In this case, the functor α^* preserves cofibrations.

The natural weak equivalences

$$\Omega_T Y \xrightarrow{\alpha^*} \Omega_{T'} Y$$

10.3 Stable Model Structures for T-spectra

for f-fibrant simplicial presheaves Y induce natural f-equivalences

$$\alpha^{*'} : Q_T X^n \xrightarrow{\simeq} Q_{T'}(\alpha^* X)^n$$

for T-spectra X. It follows that the functor α^* preserves and reflects stable f-equivalences. The functor α^* also preserves stably f-fibrant objects.

Every stable f-fibrant T'-spectrum X is of the form $\alpha^* \overline{X}$ for some stable f-fibrant T-spectrum $\overline{X}$ with $\overline{X}^n = X^n$. The bonding maps $\overline{\sigma} : T \wedge X^n \to X^{n+1}$ are found by solving lifting problems

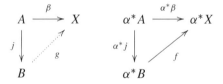

One finishes the proof by showing that the function

$$\alpha^* : \pi(A, X) \to \pi(\alpha^* A, \alpha^* X)$$

in simplicial homotopy classes of maps is a bijection for all stably f-fibrant T-spectra X and cofibrant T-spectra A. This is shown by verifying the following; given solid arrow commutative diagrams

$$\begin{array}{ccc} A & \xrightarrow{\beta} & X \\ {\scriptstyle j}\downarrow & \nearrow{\scriptstyle g} & \\ B & & \end{array} \qquad \begin{array}{ccc} \alpha^* A & \xrightarrow{\alpha^*\beta} & \alpha^* X \\ {\scriptstyle \alpha^* j}\downarrow & \nearrow{\scriptstyle f} & \\ \alpha^* B & & \end{array}$$

with A as cofibrant, j a cofibration and X stable f-fibrant, then the dotted arrow g exists making the diagram of T-spectra commute, and there is a homotopy $\alpha^* g \simeq f$, which is constant at $\alpha^* \beta$ on $\alpha^* A$. This is proved by induction on levels.

Example 10.41 In motivic homotopy theory (Examples 7.20 and 10.38), the parameter object T is the projective line $\mathbb{P}^1$ or any of its weakly equivalent compact models.

The standard Zariski covering of $\mathbb{P}^1$ given by the pushout (and pullback)

$$\begin{array}{ccc} \mathbb{A}^1 - 0 & \longrightarrow & \mathbb{A}^1 \\ \downarrow & & \downarrow \\ \mathbb{A}^1 & \longrightarrow & \mathbb{P}^1 \end{array}$$

defines a homotopy cofibre square in the motivic model structure on the category of simplicial presheaves on the Nisnevich site. The object $\mathbb{A}^1 - 0$ is the scheme underlying the multiplicative group-scheme $\mathbb{G}_m$. The contractibility of the affine line

$\mathbb{A}^1$ implies that $\mathbb{P}^1$ can be identified with either an S^1-suspension or a quotient in the motivic homotopy category, in the sense that there are (zig-zags of) motivic weak equivalences

$$\mathbb{P}^1 \simeq S^1 \wedge \mathbb{G}_m \simeq \mathbb{A}^1/(\mathbb{A}^1 - 0)$$

through compact objects, and all of these interpretations are useful. Theorem 10.40 implies that there is an equivalence of motivic stable categories

$$\text{Ho}(\mathbf{Spt}_{\mathbb{P}^1}(Sm/S)_{Nis})) \simeq \text{Ho}(\mathbf{Spt}_{S^1 \wedge \mathbb{G}_m}(Sm/S)_{Nis})).$$

This equivalence is used to generate both the motivic stable homotopy groups and the exact sequences relating them, which are found in Sect. 10.5—see Remark 10.73.

The simplicial presheaf $S^1 \wedge \mathbb{G}_m$ is a nonabelian version of the motive $\mathbb{Z}(1)$ of [78] and Example 8.49. The S^1-suspension functor $X \mapsto S^1 \wedge X$ corresponds to the shift $C \mapsto C[-1]$ of chain complexes.

Corollary 10.42 *Suppose that T is compact up to f-equivalence, and suppose given pointed simplicial presheaves Y^n, together with maps $\sigma, \sigma' : T \wedge Y^n \to Y^{n+1}$, $n \geq 0$ such that all maps σ and σ' coincide in the f-local homotopy category of pointed simplicial presheaves.*

Let Y_0 be the T-spectrum with spaces Y^n and bonding maps $\sigma : T \wedge Y^n \to Y^{n+1}$, and let Y_1 be the T-spectrum with spaces Y^n and bonding maps $\sigma' : T \wedge Y^n \to Y^{n+1}$. Then the T-spectra Y_0 and Y_1 are stable f-equivalent.

Proof We can assume that all simplicial presheaves Y^n are f-fibrant. All simplicial presheaves are cofibrant, so there is a collection of naive homotopies

$$h_n : T \wedge \Delta^1_+ \wedge Y^n \cong T \wedge Y^n \wedge \Delta^1_+ \to Y^{n+1}$$

between the maps $\sigma, \sigma' : T \wedge Y^n \to Y^{n+1}$. These homotopies determine a $(T \wedge \Delta^1_+)$-spectrum Z which restricts to the T-spectra Y_0 and Y_1 by composing with the inclusions i_0, i_1.

The finite pointed simplicial set Δ^1_+ is compact up to f-equivalence, since there is a weak

$$\Omega_{\Delta^1_+}(X) \simeq \mathbf{hom}(\Delta^1, X)$$

with the unpointed function complex, which is locally weakly equivalent to X for all locally fibrant pointed simplicial presheaves X. It follows that the smash product $T \wedge \Delta^1_+$ is compact up to f-equivalence, by Lemma 10.28.

The inclusions $i_0, i_1 : T \wedge \Delta^1_+$ are both split by the projection $s : T \wedge \Delta^1_+ \to T$, and all of these maps are f-equivalences.

It is a consequence of Theorem 10.40 that the $(T \wedge \Delta^1_+)$-spectrum Z is stable equivalent to an object s^*V, for some T-spectrum V and so there are stable equivalences

$$Y_0 = i_0^* Z \simeq i_0^* s^* V = V = i_1^* s^* V \simeq i_1^* Z = Y_1.$$

Remark 10.43 Compare the proof of Corollary 10.42 with (for example) the proof of Corollary 1.15 of [56], which is a similar result for presheaves of spectra. The latter is far more traditional, in that its proof uses the classical telescope construction.

10.4 Shifts and Suspensions

Suppose, as before, that $f : A \to B$ is a cofibraton of pointed simplicial presheaves.

Recall that if X is a T-spectrum and n is an integer, then the shifted T-spectrum $X[n]$ is defined by

$$X[n]^k = \begin{cases} X^{n+k} & \text{if } n + k \geq 0, \text{ and} \\ * & \text{if } n + k < 0. \end{cases}$$

Proposition 10.44

1) The shift functors

$$[-1] : \mathbf{Spt}_T(\mathcal{C}) \leftrightarrows \mathbf{Spt}_T(\mathcal{C}) : [1]$$

define a Quillen self-equivalence of the f-local stable model structure on the category of T-spectra.

2) The shift functors $X \mapsto X[n]$ preserve and reflect stable f-equivalences.

Proof It is a consequence of Corollary 10.26 that if the T-spectrum Z is stable f-fibrant, then the shifted T-spectrum $Z[1]$ is stable f-fibrant. It follows that the left adjoint functor $X \mapsto X[-1]$ preserves stable equivalences. This functor also preserves cofibrations.

The functor $X \mapsto X[1]$ preserves the generators of the set of cofibrations S_f, and preserves colimits. It follows that if $j : X \to LX$ is the stable f-fibrant model (or f-injective model) for a T-spectrum X, then the shifted map $j[1] : X[1] \to LX[1]$ is a stable f-fibrant model for $X[1]$. It follows that the functor $X \mapsto X[1]$ preserves stable f-equivalences.

The canonical map

$$\epsilon : X[-1][1] \to X$$

is a natural isomorphism.

Suppose that the T-spectrum Z is stable f-fibrant, and that $j : Z[1][-1] \to W$ is a stable f-fibrant model for the T-spectrum $Z[1][-1]$. The shifted map $(j \cdot \eta)[1]$ is a stable f-equivalence of stable f-fibrant T-spectra, and is therefore a level equivalence. It follows that the map $j \cdot \eta$ is a stable f-equivalence, so that the map $\eta : Z \to Z[1][-1]$ is a stable f-equivalence. Shifts preserve stable f-equivalences, so the map

$$\eta : X \to X[1][-1]$$

is a stable f-equivalence for all T-spectra X.

It follows that the shift functors $X \mapsto X[n]$ preserve and reflect stable f-equivalences for all $n = \pm 1$. The claim for general n also follows.

Say that the parameter object T is *cycle trivial* if the automorphism

$$c_{1,2} : T^{\wedge 3} \to T^{\wedge 3},$$

which permutes smash factors by the shuffle

$$x_1 \wedge x_2 \wedge x_3 \mapsto x_2 \wedge x_3 \wedge x_1$$

is the identity on $T^{\wedge 3}$ in the f-local pointed homotopy category for simplicial presheaves.

Example 10.45

1) The circle S^1 is cycle trivial for all local homotopy categories of pointed simplicial presheaves, since the permutation $c_{1,2}$ has trivial degree on S^3.
2) The projective line $T \simeq \mathbb{P}^1$ is cycle trivial in the motivic model structure for pointed simplicial presheaves on the smooth Nisnevich site of a scheme S. In effect, there is a motivic weak equivalence

$$\mathbb{P}^1 \simeq \mathbb{A}^1/(\mathbb{A}^1 - 0),$$

and an isomorphism

$$(\mathbb{A}^1/(\mathbb{A}^1 - 0))^{\wedge 3} \cong \mathbb{A}^3/(\mathbb{A}^3 - 0).$$

The action of $c_{1,2}$ on $\mathbb{A}^3/(\mathbb{A}^3 - 0)$ is induced by the action

$$Gl_3 \times \mathbb{A}^3 \to \mathbb{A}^3,$$

of the general linear group Gl_3. The element $c_{1,2} \in Sl_3 \subset Gl_3$ is a product of elementary transformation matrices, and so there is a path $\mathbb{A}^1 \to Gl_3$ from $c_{1,2}$ to the identity matrix. This path induces an algebraic homotopy

$$\mathbb{A}^3/(\mathbb{A}^3 - 0) \times \mathbb{A}^1 \to \mathbb{A}^3/(\mathbb{A}^3 - 0)$$

from $c_{1,2}$ to the identity on $\mathbb{A}^3/(\mathbb{A}^3 - 0)$. See also Lemma 3.13 of [57].

3) If the objects T and T' are cycle trivial then the smash product $T \wedge T'$ is cycle trivial. The proof is an exercise.

We show that the T-suspension functor $X \mapsto X \wedge T$ and the T-loops functor $Y \mapsto \mathbf{hom}(T, F(Y))$ are inverse to each other in the stable homotopy category, subject to the assumptions that the parameter object T is compact up to f-equivalence and is cycle trivial. This is the substance of Theorem 10.46 below.

The proof of Theorem 10.46 uses the *layer filtration* $L_n X$, $n \geq 0$, for a T-spectrum X.

10.4 Shifts and Suspensions

The *nth layer* $L_n X$ of X is defined by the sequence of pointed simplicial presheaves

$$X^0, \ldots, X^n, T \wedge X^n, T \wedge T \wedge X^n, \ldots \tag{10.11}$$

There are natural maps $L_n X \to L_{n+1} X$ and maps $L_n X \to X$, which together induce an isomorphism

$$\varinjlim_n L_n X \xrightarrow{\cong} X.$$

There is a sequence of pushouts

$$\begin{array}{ccc} \Sigma_T^\infty (T \wedge X^n)[-n-1] & \longrightarrow & L_n X \\ \downarrow & & \downarrow \\ \Sigma_T^\infty X^{n+1}[-n-1] & \longrightarrow & L_{n+1} X \end{array} \tag{10.12}$$

The maps $L_n X \to L_{n+1} X$ are level cofibrations if the T-spectrum X is cofibrant. There is a natural stable equivalence

$$\Sigma_T^\infty X^n [-n] \xrightarrow{\simeq} L_n X$$

for each n and T-spectrum X.

Theorem 10.46 *Suppose that T is compact up to f-equivalence and is cycle trivial. Then the composite*

$$X \to \mathbf{Hom}(T, X \wedge T) \xrightarrow{\eta_*} \mathbf{Hom}(T, Q_T(X \wedge T)),$$

which is defined by the adjunction map and the stable fibrant model $\eta : X \wedge T \to Q_T(X \wedge T)$, is a stable f-equivalence for all T-spectra X.

Proof The map $\eta : X \wedge T \to Q_T(X \wedge T)$ factors as a composite

$$X \wedge T \xrightarrow{j} F(X \wedge T) \xrightarrow{\tilde{\eta}} Q_T(X \wedge T),$$

where the map j is the strictly f-fibrant model and $\tilde{\eta}$ is a stable equivalence of strictly f-fibrant T-spectra. The induced map

$$\tilde{\eta}_* : \mathbf{Hom}(T, F(X \wedge T)) \to \mathbf{Hom}(T, Q_T(X \wedge T))$$

is a stable f-equivalence by Lemma 10.48 below, so it suffices to show that the composite

$$X \to \mathbf{Hom}(T, X \wedge T) \xrightarrow{j_*} \mathbf{Hom}(T, F(X \wedge T)) \tag{10.13}$$

is a stable f-equivalence

The question of whether or not the map (10.13) is a stable f-equivalence is independent of the choice of representative for the stable homotopy type of X and is insensitive to shifts. This map also respects filtered colimits in X, by a compactness argument. A layer filtration argument then implies that it is enough to assume that X is a suspension spectrum $\Sigma_T^\infty K$ for some pointed simplicial presheaf K. This claim is proved in Lemma 10.47.

In all that follows, we shall write T^p for the p-fold wedge $T^{\wedge p}$, for all pointed simplicial presheaves T. We shall also write S^q for the q-fold wedge $(S^1)^{\wedge q}$ of the circle S^1, when such an object appears.

Lemma 10.47 *Suppose given the conditions of Theorem 10.46, and that K is a pointed simplicial presheaf. Then the composite map*

$$\Sigma_T^\infty K \to \mathbf{Hom}(T, (\Sigma_T^\infty K) \wedge T)) \xrightarrow{j_*} \mathbf{Hom}(T, F((\Sigma_T^\infty K) \wedge T))$$

is a stable equivalence.

Proof If Y is a T-spectrum, then the homotopy group presheaves $\pi_r(Q_T Y)^n(U)$ of the stable fibrant model $Q_T Y$ are computed by the filtered colimits

$$[S^r, Y^n]_U \xrightarrow{\Sigma} [T \wedge S^r, Y^{n+1}]_U \xrightarrow{\Sigma} \cdots$$

where $[K, X]_U = [K|_U, X|_U]$ means homotopy classes of maps of the restrictions to the site over U. The suspension homomorphism Σ is induced by smashing with T on the left; it takes a morphism $\theta : T^k \wedge S^r \to Y^{n+k}$ to the composite

$$T \wedge T^k \wedge S^r \xrightarrow{T \wedge \theta} T \wedge Y^{n+k} \xrightarrow{\sigma} Y^{n+k+1}.$$

If Y is level f-fibrant, then the adjunction isomorphisms

$$[T^k \wedge S^r, \Omega_T Y^{n+k}]_U \cong [T^k \wedge S^r \wedge T, Y^{n+k}]_U$$

fit into commutative diagrams

$$\begin{array}{ccc} [T^k \wedge S^r, \Omega_T Y^{n+k}]_U & \xrightarrow{\cong} & [T^k \wedge S^r \wedge T, Y^{n+k}]_U \\ \Sigma \downarrow & & \downarrow \Sigma \\ [T^{k+1} \wedge S^r, \Omega_T Y^{n+k+1}]_U & \xrightarrow{\cong} & [T^{k+1} \wedge S^r \wedge T, Y^{n+k+1}]_U \end{array}$$

It follows that the map in presheaves of stable homotopy groups which is induced by the composite

$$\Sigma_T^\infty K \to \mathbf{Hom}(T, (\Sigma_T^\infty K) \wedge T) \xrightarrow{j_*} \mathbf{Hom}(T, F((\Sigma_T^\infty K) \wedge T))$$

is isomorphic to the filtered colimit of the maps

10.4 Shifts and Suspensions

$$[T^k \wedge S^r, T^{n+k} \wedge K]_U \xrightarrow{\wedge T} [T^k \wedge S^r \wedge T, T^{n+k} \wedge K \wedge T]_U$$

which are induced by smashing with T on the right.

Suppose that $\phi : K \wedge T \to X \wedge T$ is a map of pointed simplicial presheaves, and write $c_t(\phi)$ for the map $T \wedge K \to T \wedge X$ arises from ϕ by conjugation with the twist of smash factors. There is a commutative diagram

$$\begin{array}{ccccc}
T \wedge (T^2 \wedge K) & \xleftarrow{t} & T^2 \wedge K \wedge T & \xrightarrow{T^2 \wedge t} & T^2 \wedge T \wedge K \\
{\scriptstyle c_t(T^2 \wedge \phi)} \downarrow & & \downarrow {\scriptstyle T^2 \wedge \phi} & & \downarrow {\scriptstyle T^2 \wedge c_t(\phi)} \\
T \wedge (T^2 \wedge X) & \xleftarrow{t} & T^2 \wedge X \wedge T & \xrightarrow{T^2 \wedge t} & T^2 \wedge T \wedge X
\end{array}$$

and hence a diagram

$$\begin{array}{ccc}
T^3 \wedge K & \xrightarrow{c_{1,2} \wedge K} & T^3 \wedge K \\
{\scriptstyle c_t(T^2 \wedge \phi)} \downarrow & & \downarrow {\scriptstyle T^2 \wedge c_t(\phi)} \\
T^3 \wedge X & \xrightarrow{c_{1,2} \wedge X} & T^3 \wedge X
\end{array}$$

We are assuming that the map $c_{1,2}$ is the identity in the homotopy category, and it follows that the maps in the homotopy category represented by $T^2 \wedge c_t(\phi)$ and $c_t(T^2 \wedge \phi)$ coincide.

We therefore have commutative diagrams

$$\begin{array}{ccc}
[T^k \wedge S^r, T^{n+k} \wedge K]_U & \xrightarrow{T^2 \wedge} & [T^2 \wedge T^k \wedge S^r, T^2 \wedge T^{n+k} \wedge K]_U \\
{\scriptstyle \wedge T} \downarrow & & \downarrow {\scriptstyle \wedge T} \\
[T^k \wedge S^r \wedge T, T^{n+k} \wedge K \wedge T]_U & \xrightarrow{T^2 \wedge} & [T^2 \wedge T^k \wedge S^r \wedge T, T^2 \wedge T^{n+k} \wedge K \wedge T]_U \\
{\scriptstyle c_t} \downarrow \cong & & \cong \downarrow {\scriptstyle c_t} \\
[T \wedge T^k \wedge S^r, T \wedge T^{n+k} \wedge K]_U & \xrightarrow{T^2 \wedge} & [T^3 \wedge T^k \wedge S^r, T^3 \wedge T^{n+k} \wedge K]_U
\end{array}$$

The vertical composites coincide with the map $T \wedge$ induced by smashing on the left with T. It follows from a cofinality argument that the induced map on the filtered colimits is an isomorphism.

Lemma 10.48 *Suppose that T is compact up to f-equivalence, and that X and Y are strictly f-fibrant T-spectra. Then the map $\alpha : X \to Y$ is a stable f-equivalence if and only if the induced map*

$$\alpha_* : \mathbf{Hom}(T, X) \to \mathbf{Hom}(T, Y)$$

is a stable f-equivalence.

Proof There is a commutative diagram

$$\begin{array}{ccc} \Omega_T X^n & \xrightarrow{\sigma_*} & \Omega_T \Omega_T X^{n+1} \\ & \searrow{\Omega_T \sigma_*} & \cong \downarrow \tau \\ & & \Omega_T \Omega_T X^{n+1} \end{array}$$

where τ flips loop factors. The horizontal map σ_* is the adjoint bonding map

$$\sigma_* : \mathbf{Hom}(T, X^n) \to \Omega_T \mathbf{Hom}(T, X^{n+1})$$

for the T-spectrum $\mathbf{Hom}(T, X)$.

This diagram is natural in strictly fibrant T-spectra X, and it follows that there are natural f-equivalences

$$Q_T(\mathbf{Hom}(T, X))^n \xrightarrow{\simeq} \Omega_T(Q_T X^n) \xleftarrow{\simeq} Q_T X^{n-1}$$

for all such X. The claim follows.

Corollary 10.49 *Suppose that T is compact up to f-equivalence and is cycle trivial. Suppose that the T-spectrum Y is strictly f-fibrant. Then the evaluation map*

$$ev : \mathbf{Hom}(T, Y) \wedge T \to Y$$

is a stable f-equivalence.

Proof Take a strictly f-fibrant model $j : \mathbf{Hom}(T, Y) \wedge T \to F(\mathbf{Hom}(T, Y) \wedge T)$ and form the diagram

$$\begin{array}{ccc} \mathbf{Hom}(T, Y) \wedge T & \xrightarrow{j} & F(\mathbf{Hom}(T, Y) \wedge T) \\ ev \downarrow & \swarrow \epsilon & \\ Y & & \end{array}$$

in the strict homotopy category. We show that the map $\epsilon_* = \mathbf{Hom}(T, \epsilon)$ is a stable f-equivalence, and then apply Lemma 10.48.

There is a commutative diagram

$$\begin{array}{ccccc} \mathbf{Hom}(T, Y) & \xrightarrow{\eta} & \mathbf{Hom}(T, \mathbf{Hom}(T, Y) \wedge T) & \xrightarrow{j_*} & \mathbf{Hom}(T, F(\mathbf{Hom}(T, \mathbf{Hom}(T, Y) \wedge T)) \\ & \searrow{1} & \downarrow{ev_*} & \swarrow{\epsilon_*} & \\ & & \mathbf{Hom}(T, Y) & & \end{array}$$

10.4 Shifts and Suspensions

in which the horizontal composite is a stable f-equivalence by Theorem 10.46, so that ϵ_* is a stable f-equivalence.

The following result now has an easy proof:

Theorem 10.50 *Suppose that T is compact up to f-equivalence and is cycle trivial. Then the T-suspension and T-loops functors induce a Quillen equivalence*

$$\wedge T : \mathbf{Spt}_T(\mathcal{C}) \rightleftarrows \mathbf{Spt}_T(\mathcal{C}) : \mathbf{Hom}(T, \ )$$

for the f-local stable model structure on the category of T-spectra.

Every T-spectrum X determines a $(T \wedge T)$-spectrum $R(X)$ with

$$R(X)^n = X^{2n},$$

and bonding maps

$$T \wedge T \wedge X^{2n} \xrightarrow{T \wedge \sigma} T \wedge X^{2n+1} \xrightarrow{\sigma} X^{2n+2}.$$

The assignment $X \mapsto R(X)$ is functorial.

Every $(T \wedge T)$-spectrum Y determines a T-spectrum $L(Y)$ with $L(Y)^{2n} = Y^n$, $L(Y)^{2n+1} = T \wedge Y^n$, with the bonding maps

$$T \wedge L(Y)^{2n} \to L(Y)^{2n+1}$$

defined by identity maps (as in a suspension spectrum), and with the map

$$T \wedge L(Y)^{2n+1} \to L(Y)^{2n+2}$$

defined to be

$$T \wedge L(Y)^{2n+1} = T \wedge T \wedge Y^n \xrightarrow{\sigma} Y^{n+1}$$

where σ is the bonding map if the $(T \wedge T)$-spectrum Y. The assignment $Y \mapsto L(Y)$ is again functorial.

The functor L is left adjoint to the functor R. The unit map

$$\eta : Y \cong RL(Y)$$

for the adjunction is a natural isomorphism. The counit map

$$\epsilon : LR(X) \to X$$

is the identity in evenly indexed levels, and is the map $\sigma : T \wedge X^{2k} \to X^{2k+1}$ in odd levels.

The following result is a consequence of Theorem 10.32:

Proposition 10.51 *Suppose that T is compact up to f-equivalence. Then the natural map*

$$\epsilon : LR(X) \to X$$

is a stable f-equivalence for all T-spectra X, and the functors L and R determine a Quillen equivalence

$$L : \mathbf{Spt}_{T \wedge T}(\mathcal{C}) \leftrightarrows \mathbf{Spt}_T(\mathcal{C}) : R.$$

Proof The object $T \wedge T$ is compact up to f-equivalence, by Lemma 10.28.

The map $\epsilon : LR(X) \to X$ is an isomorphism in evenly indexed levels and therefore induces f-equivalences

$$\epsilon_* : Q_T(LR(X))^n \to Q_T X^n$$

by a cofinality argument. It follows that the map ϵ is a stable f-equivalence.

The functor R preserves stable f-fibrant objects. The functor L preserves cofibrations, and therefore L preserves stable f-trivial cofibrations. The functors L and R therefore form a Quillen adjunction.

Recall that the fake T-suspension $\Sigma_T X$ of a T-spectrum X is defined by

$$(\Sigma_T X)^n = T \wedge X^n,$$

and has bonding maps

$$T \wedge T \wedge X^n \xrightarrow{T \wedge \sigma} T \wedge X^{n+1}$$

where the maps $\sigma : T \wedge X^n \to X^{n+1}$ are the bonding maps of X.

Lemma 10.52 *The bonding maps for X define a natural stable equivalence*

$$\sigma : \Sigma_T X \to X[1]$$

for all T-spectra X.

Proof The functors $X \mapsto \Sigma_T X$ and $X \mapsto X[1]$ both preserve stable equivalences and filtered colimits. They also preserve shifts of the form $Y \mapsto Y[n]$ with $n \geq 0$. Thus, by replacing X by its layers, it is enough to show that the map

$$\sigma : \Sigma_T(\Sigma_T^\infty K) \to \Sigma_T^\infty K[1]$$

is a stable equivalence for each pointed simplicial presheaf K. The map σ is an isomorphism for all suspension T-spectra.

Lemma 10.52 holds in great generality—there are no conditions on the object T. The following result is a little more subtle:

Proposition 10.53 *Suppose that T is compact up to f-equivalence and is cycle trivial. Then there is a natural stable equivalence*

$$\Sigma_T X \simeq X \wedge T$$

for all T-spectra X.

10.5 Fibre and Cofibre Sequences

Proof The twist isomorphisms

$$\tau : X^n \wedge T \xrightarrow{\cong} T \wedge X^n$$

define bonding maps

$$T \wedge T \wedge X^n \xrightarrow{T \wedge \tau} T \wedge X^n \wedge T \xrightarrow{\sigma \wedge T} X^{n+1} \wedge T \xrightarrow{\tau} T \wedge X^{n+1},$$

which differ from the bonding maps on the fake T-suspension $\Sigma_T X$ by composition with the twist automorphism $\tau : T \wedge T \to T \wedge T$. The assumption that the permutation

$$c_{1,2} : T^{\wedge 3} \to T^{\wedge 3}$$

is the identity map in the f-local homotopy category implies that the $(T \wedge T)$-spectra $R(\Sigma_T X)$ and $R(X \wedge T)$ are naturally stable equivalent, by Corollary 10.42. It follows from Proposition 10.51 that the T-spectra $\Sigma_T X$ and $X \wedge T$ are naturally stable equivalent.

Corollary 10.54 *Suppose that T is compact up to f-equivalence and is cycle trivial. Then we have the following:*

1) *There are natural stable f-equivalences*

$$X \wedge T \simeq \Sigma_T X \simeq X[1]$$

for all T-spectra X.

2) *There are natural stable f-equivalences*

$$\mathbf{Hom}(T, X) \simeq \Omega_T X \simeq X[-1]$$

for all strictly f-fibrant T-spectra X.

Proof Statement 1) is proved in Lemma 10.52 and Proposition 10.53. It follows that the derived adjoints of the functors in statement 1) are also naturally stable equivalent, giving statement 2).

10.5 Fibre and Cofibre Sequences

We continue to suppose that the map $f : A \to B$ is a cofibration of simplicial presheaves.

We shall need a suitable theory of bispectra. We begin with an example.

Suppose that S and T are arbitrary pointed simplicial presheaves, and suppose that X is an $(S \wedge T)$-spectrum with bonding maps

$$\sigma : S \wedge T \wedge X^n \to X^{n+1}.$$

In all that follows, let $\tau : X \wedge Y \xrightarrow{\cong} Y \wedge X$ be the automorphism which flips smash factors.

The $(S \wedge T)$-spectrum X determines a T-spectrum object $X^{*,*}$ in S-spectra, which at T-level n is the S-spectrum

$$X^{*,n} : \ X^0 \wedge T^{\wedge n}, X^1 \wedge T^{\wedge(n-1)}, \ldots,$$
$$X^{n-1} \wedge T, X^n, S \wedge X^n, S^{\wedge 2} \wedge X^n, \ldots$$

The bonding maps of the S-spectrum $X^{*,n}$ are the maps σ_S which are defined by the composites

$$S \wedge X^{n-j} \wedge T^{\wedge j} \xrightarrow{S \wedge \tau \wedge T^{\wedge(n-1)}} S \wedge T \wedge X^{n-j} \wedge T^{\wedge(j-1)}$$
$$\xrightarrow{\sigma \wedge T^{\wedge(n-1)}} X^{n-(j-1)} \wedge T^{\wedge(j-1)}$$

up to level $n - 1$ (i.e. for $j \geq 1$) and are the identities

$$S \wedge S^{\wedge k} \wedge X^n = S^{\wedge(k+1)} \wedge X^n$$

for $k \geq 0$. There is a map

$$\sigma_T : X^{*,n} \wedge T \to X^{*,n+1}$$

of S-spectra which is the identity map up to level n, and consists of the composites

$$S^{\wedge k} \wedge X^n \wedge T \xrightarrow{S^{\wedge k} \wedge \tau} S^{\wedge(k-1)} \wedge S \wedge T \wedge X^n \xrightarrow{S^{\wedge(k-1)} \wedge \sigma} S^{\wedge(k-1)} \wedge X^{n+1}$$

in higher levels.

Formally, a T-spectrum object Y in S-spectra consists of S-spectra Y^n and maps of S-spectra

$$\sigma_T : Y^n \wedge T \to Y^{n+1}.$$

The object $X^{*,*}$ which arises from an $(S \wedge T)$-spectrum X as above is a functorially defined example.

We also say that a T-spectrum object Y in S-spectra is an (S, T)-*bispectrum*. One can specify the (S, T)-bispectrum Y by a list $Y^{p,q}$ of pointed simplicial presheaves with $p, q \geq 0$, together with bonding maps $\sigma_S : S \wedge Y^{p,q} \to Y^{p+1,q}$ and $\sigma_T : Y^{p,q} \wedge T \to Y^{p,q+1}$ such that the diagrams

$$\begin{array}{ccc} S \wedge X^{p,q} \wedge T & \xrightarrow{\sigma_S \wedge T} & X^{p+1,q} \wedge T \\ {\scriptstyle S \wedge \sigma_T} \downarrow & & \downarrow {\scriptstyle \sigma_T} \\ S \wedge X^{p,q+1} & \xrightarrow{\sigma_S} & X^{p+1,q+1} \end{array}$$

commute.

10.5 Fibre and Cofibre Sequences

A morphism $X \to Y$ of (S,T)-bispectra consists of a family of pointed simplicial presheaf maps $X^{p,q} \to Y^{p,q}$ which respects bonding maps in both the S and T directions. Write $\mathbf{Spt}_{T,S}(\mathcal{C})$ for the corresponding category.

We say that a (T,T)-bispectrum is a T-*bispectrum*.

Each (S,T)-bispectrum Z has a functorially associated *diagonal* $(S \wedge T)$-spectrum dZ with

$$(dZ)^n = Z^{n,n},$$

and with bonding maps given by the composites

$$S \wedge T \wedge Z^{n,n} \xrightarrow{S \wedge \tau} S \wedge Z^{n,n} \wedge T \xrightarrow{S \wedge \sigma_T} S \wedge Z^{n+1,n} \xrightarrow{\sigma_S} Z^{n+1,n+1}.$$

There is a natural isomorphism

$$d(X^{*,*}) \cong X,$$

which is defined for all $(S \wedge T)$-spectra X.

Remark 10.55 Say that a map $X \to Y$ is a *strict f-equivalence* of (S,T)-spectra if all maps $X^{p,q} \to Y^{p,q}$ are f-equivalences. Any strict f-equivalence $X \to Y$ induces a strict f-equivalence $dX \to dY$ of $(S \wedge T)$-spectra.

Say that a map $X \to Y$ of (S,T)-bispectra is a *strict f-fibration* if all maps $X^{p,q} \to Y^{p,q}$ are f-fibrations. By analogy with Propositions 10.15 and 10.4, there is a strict model structure on the category of (S,T)-bispectra, for which the weak equivalences are the strict f-equivalences and the fibrations are the strict f-fibrations.

The cofibrations can be defined within either S-spectra or T-spectra. It is an exercise to show that an (S,T)-bispectrum A is cofibrant if and only if the bonding maps $\sigma_S : S \wedge A^{p,q} \to A^{p+1,q}$ and $\sigma_T : A^{p,q} \wedge T \to A^{p,q+1}$ and all maps

$$(S \wedge A^{p,q+1}) \cup_{(S \wedge A^{p,q} \wedge T)} (A^{p+1,q} \wedge T) \to A^{p+1,q+1}$$

are cofibrations of pointed simplicial presheaves.

Lemma 10.56 *Suppose that the parameter objects S and T are compact up to f-equivalence. Suppose that a map $g : X \to Y$ of (S,T)-bispectra consists of stable f-equivalences of S-spectra $X^{*,n} \to Y^{*,n}$ for all $n \geq 0$. Then the induced map $dX \to dY$ is a stable f-equivalence of $(S \wedge T)$-spectra.*

Proof We can assume that the (S,T)-bispectra X and Y are strictly f-fibrant in the sense that all simplicial presheaves $X^{p,q}$ and $Y^{p,q}$ are f-fibrant.

Consider the natural diagram

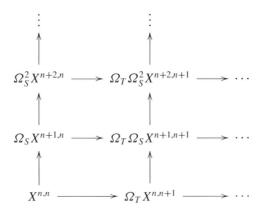

which is associated to a strictly f-fibrant (S,T)-spectrum X. The filtered colimit of this diagram coincides with the object $Q_{S \wedge T}(dX)^n$ up to f-equivalence. The map g induces f-equivalences

$$\varinjlim \Omega_S^p X^{n+p,m} \to \varinjlim \Omega_S^p Y^{n+p,m}$$

by assumption, and g therefore induces f-equivalences

$$\varinjlim \Omega_T^q \Omega_S^p X^{n+p,m} \to \varinjlim \Omega_T^q \Omega_S^p Y^{n+p,m}$$

by the compactness of T. It follows that g induces f-equivalences

$$Q_{S \wedge T}(dX)^n \to Q_{S \wedge T}(dY)^n$$

for all $n \geq 0$, so that the induced map $dX \to dY$ is a stable f-equivalence of $(S \wedge T)$-spectra.

We say that a map $g : X \to Y$ of (S,T)-bispectra is a *stable f-equivalence* if the induced map $g_* : d(X) \to d(Y)$ is a stable f-equivalence of $(S \wedge T)$-spectra.

From this point of view, Lemma 10.56 says that g is a stable f-equivalence of (S,T)-bispectra if all maps $X^{*,q} \to Y^{*,q}$ are stable f-equivalences of S-spectra. Similarly, the map g is a stable f-equivalence if all maps $X^{p,*} \to Y^{p,*}$ are stable f-equivalences of T-spectra.

We now specialize to $(S^1 \wedge T)$-spectra.

We shall also assume that the f-local model for simplicial presheaves satisfies inductive colimit descent. This assumption implies that finite pointed simplicial sets are compact up to f-equivalence, by Lemma 10.35.

It follows in particular that the simplicial circle S^1 is compact up to f-equivalence. The class of pointed simplicial presheaves which are compact up to f-equivalence is closed under finite smash products by Lemma 10.28, and so the simplicial presheaf $S^1 \wedge T$ is compact up to f-equivalence.

10.5 Fibre and Cofibre Sequences

The results of Sect. 10.3, in particular the identification of the object $Q_{S^1 \wedge T} X$ as a stable fibrant model of an object X of Theorem 10.32, apply to $(S^1 \wedge T)$-spectra in this case. We shall make aggressive use of Lemma 10.56.

Suppose that the maps

$$F \xrightarrow{i} X \xrightarrow{p} Y$$

form a strict fibre sequence of $(S^1 \wedge T)$-spectra, in the sense that p is a strict fibration (for the *unlocalized* strict model structure of Proposition 10.15) and F is the pullback of the canonical base point along p.

Every map $g : Z \to W$ of (S^1, T)-bispectra has a factorization

where q is a strict fibration and j is a cofibration and a strict weak equivalence, again unlocalized. Take such a factorization

$$X^{*,*} \xrightarrow{j} V$$
$$\searrow_{p} \quad \downarrow_{q}$$
$$Y^{*,*}$$

for the map of (S^1, T)-bispectra which is induced by the $(S^1 \wedge T)$-spectrum map $p : X \to Y$, and let $\overline{F}$ be the fibre of q. Then there are induced comparisons of fibre sequences of simplicial presheaves

$$\begin{array}{ccccc}
F^n & \xrightarrow{i} & X^n & \xrightarrow{p} & Y^n \\
\downarrow & & \downarrow \simeq & & \downarrow \simeq \\
\overline{F}^{n,n} & \longrightarrow & X^{n,n} & \longrightarrow & Y^{n,n}
\end{array}$$

for each $n \geq 0$, and it follows (by properness for pointed simplicial presheaves) that the induced map $F \to d\overline{F}$ is a strict weak equivalence.

Recall the following:

Lemma 10.57 *Suppose that $p : X \to Y$ is a strict fibration of (ordinary) S^1-spectra with fibre F. Then the canonical map $X/F \to Y$ is a stable equivalence.*

Lemma 10.57 is part of the general, well known yoga which says that fibre and cofibre sequences coincide in ordinary stable homotopy theory. It appears, for example, as Corollary 4.4 of [56].

We now show that strict fibre and cofibre sequences coincide up to natural stable f-equivalence in $(S^1 \wedge T)$-spectra. This claim is the colloquial version of Lemmas 10.58 and 10.62, taken together.

Lemma 10.58 *Suppose that $p : X \to Y$ is a strict fibration of $(S^1 \wedge T)$-spectra, with fibre F. Then the canonical map*

$$X/F \to Y$$

is a stable equivalence.

Proof Take a factorization

of the corresponding map of bispectrum objects, where q is a strict fibration and j is a strict trivial cofibration. Let $\overline{F}$ be the fibre of q in the bispectrum category. Then the map

$$V/\overline{F} \to Y^{*,*}$$

is a stable equivalence of S^1-spectra in all T-levels (by Lemma 10.57), and so the induced map

$$d(V/\overline{F}) \to d(Y^{*,*}) = Y$$

is a stable equivalence of $(S^1 \wedge T)$-spectra by Lemma 10.56. The map

$$X/F = d(X^{*,*}/F^{*,*}) \to d(V/\overline{F})$$

is a strict equivalence of $(S^1 \wedge T)$-spectra since the map

$$F = d(F^{*,*}) \to d\overline{F}$$

is a strict equivalence.

Corollary 10.59 *Suppose that $p : X \to Y$ is a strict f-fibration of $(S^1 \wedge T)$-spectra, with fibre F. Then the canonical map*

$$X/F \to Y$$

is a stable f-equivalence.

Proof The map $X/F \to Y$ is a stable equivalence by Lemma 10.58, and is therefore a stable f-equivalence by Corollary 10.27.

We have, so far, made a project of incorporating standard results from stable homotopy theory into the f-local stable homotopy theory for $(S^1 \wedge T)$-spectra.

10.5 Fibre and Cofibre Sequences

The following result must be proved in the f-local stable category. It expresses the fundamental relation between f-local stable equivalences and cofibre sequences of $(S^1 \wedge T)$-spectra.

Lemma 10.60 *Suppose that the diagram*

$$\begin{array}{ccccc} A_1 & \longrightarrow & A_2 & \longrightarrow & A_3 \\ {\scriptstyle g_1}\downarrow & & {\scriptstyle g_2}\downarrow & & {\scriptstyle g_3}\downarrow \\ B_1 & \longrightarrow & B_2 & \longrightarrow & B_3 \end{array}$$

is a comparison of level cofibre sequences of $(S^1 \wedge T)$-spectra. If any two of the maps g_1, g_2 and g_3 are stable f-equivalences, then so is the third.

Proof Suppose that A is a cofibrant $(S^1 \wedge T)$-spectrum and that $\Sigma_{S^1 \wedge T} A$ is the fake $(S^1 \wedge T)$-suspension of A. Then the canonical map $\Sigma_{S^1 \wedge T} A \to A[1]$ is a stable equivalence, by Lemma 10.52. Shifting preserves stable equivalences (Proposition 10.44), so the map $\Sigma_{S^1 \wedge T} A[-1] \to A$ is also a stable equivalence.

The bispectrum $(\Sigma_{S^1 \wedge T} A)^{*,*}$ is naturally isomorphic to the fake S^1-suspension $\Sigma_{S^1} \Sigma_T A^{*,*}$ of the fake T-suspension of $A^{*,*}$. The object $\Sigma_T A^{*,*}$ is a T-spectrum object in S^1-spectra, and it follows from classical stable homotopy theory (or Proposition 10.53) that there are natural S^1-stable equivalences

$$\Sigma_T A^{*,*} \wedge S^1 \simeq \Sigma_{S^1} \Sigma_T A^{*,*},$$

defined levelwise with respect to T. Lemma 10.56 implies that there is a natural stable equivalence

$$d(\Sigma_T A^{*,*}) \wedge S^1 \simeq d(\Sigma_{S^1} \Sigma_T A^{*,*}) \cong \Sigma_{S^1 \wedge T} A,$$

and so there are natural stable equivalences

$$d(\Sigma_T A^{*,*})[-1] \wedge S^1 \simeq \Sigma_{S^1 \wedge T} A[-1] \simeq A. \tag{10.14}$$

To complete the proof of the Lemma, it suffices to assume that all objects A_i and B_i are cofibrant.

The natural equivalence (10.14) implies that the comparison of cofibre sequences in the statement of the Lemma induces a comparison of fibre sequences

$$\begin{array}{ccccc} \mathbf{hom}(B_3, Z) & \longrightarrow & \mathbf{hom}(B_2, Z) & \longrightarrow & \mathbf{hom}(B_1, Z) \\ {\scriptstyle f_3^*}\downarrow & & {\scriptstyle f_2^*}\downarrow & & {\scriptstyle f_1^*}\downarrow \\ \mathbf{hom}(A_3, Z) & \longrightarrow & \mathbf{hom}(A_2, Z) & \longrightarrow & \mathbf{hom}(A_1, Z) \end{array}$$

of infinite loop spaces, for all stable f-fibrant objects Z. If any two of the vertical maps in this diagram are weak equivalences, then so is the third.

Remark 10.61 A simpler proof of Lemma 10.60 is available if we assume that the object T is cycle trivial. Then the object $S^1 \wedge T$ is also cycle trivial, and there are natural stable equivalences

$$A \simeq \Sigma_{S^1 \wedge T} A[-1] \simeq (A \wedge S^1 \wedge T)[-1] \cong (A \wedge T)[-1] \wedge S^1,$$

by Proposition 10.53.

Lemma 10.62 *Suppose that the map $i : A \to B$ is a level cofibration of $(S^1 \wedge T)$-spectra, and take a factorization*

of the quotient map $\pi : B \to B/A$, where j is a strict f-trivial cofibration and p is a strict f-fibration. Let F be the fibre of p. Then the induced map $A \to F$ is a stable equivalence.

Proof The canonical map $p_* : Z/F \to B/A$ associated to the fibration $p : Z \to B/A$ is a stable equivalence by Lemma 10.58. There is also a commutative diagram

$$\begin{array}{ccccc} A & \longrightarrow & B & \longrightarrow & B/A \\ \downarrow & & \simeq \downarrow j & & \downarrow j_* \\ F & \longrightarrow & Z & \longrightarrow & Z/F \end{array}$$

and $p_* j_* = 1$ so that j_* is a stable equivalence. It follows that the map $A \to F$ in the diagram is a stable equivalence, by the unlocalized version of Lemma 10.60.

Corollary 10.63 *Suppose that the diagram*

$$\begin{array}{ccccc} F_1 & \longrightarrow & X_1 & \longrightarrow & Y_1 \\ g_1 \downarrow & & g_2 \downarrow & & g_3 \downarrow \\ F_2 & \longrightarrow & X_2 & \longrightarrow & Y_2 \end{array}$$

is a comparison of level f-fibre sequences of $(S^1 \wedge T)$-spectra. If any two of the maps g_1, g_2 and g_3 are stable f-equivalences, then so is the third.

Proof Use Corollary 10.59 and Lemma 10.60.

The following result is an immediate consequence of Corollary 10.63:

10.5 Fibre and Cofibre Sequences

Theorem 10.64 *Supppose that T is compact up to f-equivalence and that the f-local model structure on simplicial presheaves satisfies inductive colimit descent.*

Then the f-local stable model structure on the category of $(S^1 \wedge T)$-spectra is right proper.

We already know that the f-local stable model structure is left proper, by Theorem 7.10.

Properness for the stable model structure has a standard consequence in the ordinary stable category, which appears above as Lemma 10.9 with the traditional proof. That result is also a consequence of Lemma 7.14.

The following is the analogue of Lemma 10.9 for $(S^1 \wedge T)$-spectra, in the f-local case:

Theorem 10.65 *Supppose that T is compact up to f-equivalence and that the f-local model structure on simplicial presheaves satisfies inductive colimit descent.*

Suppose that the map $p : X \to Y$ is a strict fibration of $(S^1 \wedge T)$-spectra. Then p is a stable f-fibration if and only if the square

$$\begin{array}{ccc} X & \longrightarrow & Q_{S^1 \wedge T} X \\ p \downarrow & & \downarrow p_* \\ Y & \longrightarrow & Q_{S^1 \wedge T} Y \end{array} \qquad (10.15)$$

is homotopy cartesian in the strict model structure.

Proof Suppose that the diagram (10.15) is homotopy cartesian in the strict model structure. Find a factorization

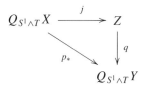

such that q is a stable f-fibration and j is a stable f-equivalence. Then the map j is a stable f-equivalence of stable f-fibrant objects, and is therefore a strict equivalence. The map q is a strict fibration. The assumption that the diagram (10.15) is homotopy cartesian in the strict model structure implies that the induced map

$$\theta : X \to Y \times_{Q_{S^1 \wedge T} Y} Z$$

is a strict equivalence. The map θ defines a strict weak equivalence from the strict fibration p to the stable f-fibration

$$q_* : Y \times_{Q_{S^1 \wedge T} Y} Z \to Y.$$

It follows, by the usual argument (see the proof of Lemma 7.13), that the map p is a retract of a stable f-fibration, and is therefore a stable f-fibration.

Suppose that the map p is a stable f-fibration, and use the factorization $p_* = q \cdot j$ of the previous paragraph. Then again, j is a strict equivalence and q is a strict fibration. The map

$$Y \times_{Q_{S^1 \wedge T} Y} Z \to Z$$

is a stable f-equivalence by properness of the f-local stable model structure (Theorem 10.64), and it follows that the map θ is a stable f-equivalence. This map θ defines a stable f-equivalence between stable f-fibrations, and must therefore be a strict equivalence. It follows that the diagram 10.15 is homotopy cartesian for the strict model structure.

Suppose that the map $p : X \to Y$ is a strict f-fibration. Then one can show that the diagram (10.15) is homotopy cartesian in the strict model structure if and only if it is homotopy cartesian in the strict f-local model structure. This is done by replacing the map p_* by a stable f-fibration q, as in the proof of Theorem 10.65. We therefore have the following corollary:

Corollary 10.66 *Supppose that T is compact up to f-equivalence and that the f-local model structure on simplicial presheaves satisfies inductive colimit descent.*

Suppose that the map $p : X \to Y$ is a strict f-fibration of $(S^1 \wedge T)$-spectra. Then p is a stable f-fibration if and only if the square

$$\begin{array}{ccc} X & \longrightarrow & Q_{S^1 \wedge T} X \\ p \downarrow & & \downarrow p_* \\ Y & \longrightarrow & Q_{S^1 \wedge T} Y \end{array}$$

is homotopy cartesian in the strict f-local model structure.

The additivity property is also inherited from the ordinary stable category:

Lemma 10.67 *[additivity] Supppose that T is compact up to f-equivalence and that the f-local model structure on simplicial presheaves satisfies inductive colimit descent.*

Suppose that X and Y are $(S^1 \wedge T)$-spectra. Then the canonical map

$$c : X \vee Y \to X \times Y$$

is a stable equivalence.

Proof Suppose that X' and Y' are (S^1, T)-bispectra. The canonical map

$$c : X' \vee Y' \to X' \times Y'$$

is a stable equivalence in all T-levels by the additivity property for the ordinary stable category. Lemma 10.56 implies that the associated map

$$d(X') \vee d(Y') = d(X' \vee Y') \xrightarrow{c} d(X' \times Y') = d(X') \times d(Y')$$

10.5 Fibre and Cofibre Sequences

is a stable equivalence of $(S^1 \wedge T)$-spectra. The desired result follows by setting $X' = X^{*,*}$ and $Y' = Y^{*,*}$.

Generally, an $(S^1 \wedge T)$-spectrum X has *bigraded presheaves of stable homotopy groups* $\pi_{s,t}X$, which are defined for objects U of the site $\mathcal{C}$ by setting

$$\pi_{s,t}X(U) = \varinjlim_{n \geq 0} [S^{n+s} \wedge T^{n+t} \wedge U_+, X^n],$$

where the homotopy classes of maps are computed with respect to the f-local model structure for pointed simplicial presheaves. The transition maps are defined by suspension with $S^1 \wedge T$ in the way one expects: a representing map

$$\alpha : S^{n+s} \wedge T^{n+t} \wedge U_+ \to X^n$$

is sent to the composite

$$S^{n+1+s} \wedge T^{n+1+t} \wedge U_+ \xrightarrow{S^1 \wedge \tau \wedge T^{n+t}} S^1 \wedge T \wedge S^{n+s} \wedge T^{n+t} \wedge U_+$$
$$\xrightarrow{1 \wedge \alpha} S^1 \wedge T \wedge X^n \xrightarrow{\sigma} X^{n+1},$$

provided that X is strictly f-fibrant.

These stable homotopy group presheaves are specializations of bigraded presheaves of stable homotopy groups $\pi_{s,t}Y$ which are defined for (S^1, T)-bispectra Y as filtered colimits of morphisms in the f-local homotopy category:

$$\pi_{s,t}Y(U) = \varinjlim_{k,l} [S^{k+s} \wedge T^{l+t} \wedge U_+, Y^{k,l}].$$

In particular, there are natural isomorphisms of presheaves

$$\pi_{s,t}Y \cong \pi_{s,t}dY$$

for (S^1, T)-bispectra Y. It follows that there are natural isomorphisms of presheaves

$$\pi_{s,t}X^{*,*} \cong \pi_{s,t}X \tag{10.16}$$

for all $(S^1 \wedge T)$-spectra X.

Write $\tilde{\pi}_{s,t}X$ for the sheaf associated to the presheaf $\pi_{s,t}X$, for both $(S^1 \wedge T)$-spectra and (S^1, T)-bispectra. These are the *bigraded sheaves of stable homotopy groups* for X.

The bonding maps $Y^n \wedge T \to Y^{n+1}$ of S^1-spectra in an (S^1, T)-bispectrum Y induce homomorphisms of stable homotopy groups

$$[S[s] \wedge T^{n+t} \wedge U_+, Y^n] \to [S[s] \wedge T^{n+t+1} \wedge U_+, Y^{n+1}]$$
$$\to [S[s] \wedge T^{n+t+2} \wedge U_+, Y^{n+2}] \to \dots$$

for all $U \in \mathcal{C}$, and the filtered colimit of this system is naturally isomorphic to $\pi_{s,t}Y(U)$.

There are natural isomorphisms of sheaves

$$\tilde{\pi}_{k-n,-n} X \cong \tilde{\pi}_k Q_{S^1 \wedge T} X^n$$

for all $(S^1 \wedge T)$-spectra X, since the f-local model structure satisfies filtered colimit descent. In effect, the defining strict f-equivalence

$$\varinjlim_n \Omega^n_{S^1 \wedge T} F X[n] \to Q_{S^1 \wedge T} X$$

is a strict local weak equivalence, by this assumption.

Lemma 10.68 *Supppose that T is compact up to f-equivalence and that the f-local model structure on simplicial presheaves satisfies inductive colimit descent.*

Then a map $X \to Y$ of $(S^1 \wedge T)$-spectra is a stable f-equivalence if and only if it induces an isomorphism $\tilde{\pi}_{s,t} X \xrightarrow{\cong} \tilde{\pi}_{s,t} Y$ of sheaves of stable homotopy groups for all $s, t \in \mathbb{Z}$.

Proof The assumption that g induces an isomorphism in all bigraded sheaves of stable homotopy groups implies that the maps $(Q_{S^1 \wedge T} X)^n \to (Q_{S^1 \wedge T} Y)^n$ induce isomorphisms

$$\tilde{\pi}_k((Q_{S^1 \wedge T} X)^n) \xrightarrow{\cong} \tilde{\pi}_k((Q_{S^1 \wedge T} Y)^n)$$

in all sheaves of homotopy groups pointed by the canonical base point. The objects $(Q_{S^1 \wedge T} X)^n$ and $(Q_{S^1 \wedge T} Y)^n$ are presheaves of H-spaces, so that the maps are local weak equivalences of simplicial presheaves, for all n, and so g is a stable equivalence.

Conversely, there is an isomorphism

$$\tilde{\pi}_{s,t} X \cong \tilde{\pi}_{n+s}(\Omega_T^{n+t} Q_{S^1 \wedge T} X^n)$$

provided that $s + n, t + n \geq 0$. Thus, any map $X \to Y$ of $(S^1 \wedge T)$-spectra which induces a level weak equivalence $Q_{S^1 \wedge T} X \to Q_{S^1 \wedge T} Y$ also induces isomorphisms $\tilde{\pi}_{s,t} X \xrightarrow{\cong} \tilde{\pi}_{s,t} Y$ for all s and t.

Remark 10.69 The motivic stable category is the primary example for the discussion of bigraded stable homotopy groups, but it is also quite special thanks to the Nisnevich descent theorem (Theorem 5.39).

Specifically, given a strictly motivic fibrant $(S^1 \wedge \mathbb{G}_m)$-spectrum X, the Nisnevich fibrant model

$$j : \varinjlim_k \Omega^{n+k}_{S^1 \wedge \mathbb{G}_m} X^{n+k} \to F(\varinjlim_k \Omega^{n+k}_{S^1 \wedge \mathbb{G}_m} X^{n+k})$$

is a sectionwise equivalence by Nisnevich descent, with a target that is motivic fibrant—see Example 7.20.

It follows that a map $X \to Y$ of $(S^1 \wedge \mathbb{G}_m)$-spectra is a motivic stable equivalence if and only if all *presheaf* maps

$$\pi_{s,t} X \to \pi_{s,t} Y$$

are isomorphisms.

10.5 Fibre and Cofibre Sequences

Suppose given a strict f-fibre sequence

$$F \xrightarrow{i} X \xrightarrow{p} Y$$

of (S^1, T)-bispectra and that Y is strictly f-fibrant in all T-levels. This means in particular that each map $p : X^{r,s} \to Y^{r,s}$ is an f-fibration between f-fibrant simplicial presheaves, with fibre $F^{r,s}$.

Then all induced sequences

$$\Omega_T^{t+n} F^n \to \Omega_T^{t+n} X^n \to \Omega_T^{t+n} Y^n$$

are strict f-fibre sequences of S^1-spectra, and all S^1-spectra $\Omega_T^{t+n} Y^n$ are strictly f-fibrant. It follows that there is a long exact sequence in presheaves of stable homotopy groups of the form

$$\cdots \to \pi_s \Omega_T^{t+n} F^n \to \pi_s \Omega_T^{t+n} X^n \to \pi_s \Omega_T^{t+n} Y^n \xrightarrow{\partial} \pi_{s-1} \Omega_T^{t+n} F^n \to \cdots$$

There are, as well, comparisons of f-fibre sequences

$$\begin{array}{ccccc} \Omega_T^{t+n} F^n & \to & \Omega_T^{t+n} X^n & \to & \Omega_T^{t+n} Y^n \\ \downarrow & & \downarrow & & \downarrow \\ \Omega_T^{t+n+1} F^{n+1} & \to & \Omega_T^{t+n+1} X^{n+1} & \to & \Omega_T^{t+n+1} Y^{n+1} \end{array}$$

Taking a filtered colimit in n gives a long exact sequence of presheaves

$$\cdots \to \pi_{s,t} F \xrightarrow{i_*} \pi_{s,t} X \xrightarrow{p_*} \pi_{s,t} Y \xrightarrow{\partial} \pi_{s-1,t} F \to \cdots \qquad (10.17)$$

It follows that there is an induced long exact sequence

$$\cdots \to \tilde{\pi}_{s,t} F \xrightarrow{i_*} \tilde{\pi}_{s,t} X \xrightarrow{p_*} \tilde{\pi}_{s,t} Y \xrightarrow{\partial} \tilde{\pi}_{s-1,t} F \to \cdots \qquad (10.18)$$

in sheaves of bigraded stable homotopy groups.

Observe that, in the long exact sequence (10.18), the "degree" s changes with the boundary map while the "weight" t does not. There is one of these exact sequences for each $t \in \mathbb{Z}$.

Suppose that $p : X \to Y$ is a strict f-fibration of (S^1, T)-bispectra with fibre F. The map p has a strictly f-fibrant replacement $p' : X' \to Y'$ up to strict f-equivalence, where Y' is strictly f-fibrant in all T-levels, and p' is a strict f-fibration in all T-levels. Let F' be the fibre of p'. The properness of the stable f-local model structure on $(S^1 \wedge T)$-spectra (Theorem 10.64) implies that the induced map $dF \to dF'$ is a stable f-equivalence, and so the maps

$$\tilde{\pi}_{s,t} F \to \tilde{\pi}_{s,t} F'$$

are isomorphisms, via (10.16). Any two such fibrant replacements are strictly equivalent, and hence equivalent in all sections and all levels.

It follows that there is a long exact sequence of the form (10.18) for any strict fibre sequence

$$F \xrightarrow{i} X \xrightarrow{p} Y$$

of (S^1, T)-bispectra.

We have proved:

Lemma 10.70 *Supppose that T is compact up to f-equivalence and that the f-local model structure on simplicial presheaves satisfies inductive colimit descent.*

Then every strict f-fibre sequence

$$F \xrightarrow{i} X \xrightarrow{p} Y$$

of $(S^1 \wedge T)$-spectra induces a natural long exact sequence

$$\cdots \to \tilde{\pi}_{s,t} F \xrightarrow{i_*} \tilde{\pi}_{s,t} X \xrightarrow{p_*} \tilde{\pi}_{s,t} Y \xrightarrow{\partial} \tilde{\pi}_{s-1,t} F \to \cdots$$

in sheaves of bigraded stable homotopy groups, for every $t \in \mathbb{Z}$.

Proof Use the discussion preceding Lemma 10.57 to replace the fibre sequence of $(S^1 \wedge T)$-spectra by a fibre sequence of (S^1, T)-bispectra. Then the desired long exact sequence is an instance of the long exact sequence (10.18).

Every cofibre sequence is a fibre sequence by Lemma 10.62, so we have the following:

Corollary 10.71 *Supppose that T is compact up to f-equivalence and that the f-local model structure on simplicial presheaves satisfies inductive colimit descent.*

Then every level cofibre sequence

$$A \to B \to B/A$$

of $(S^1 \wedge T)$-spectra has a naturally associated long exact sequence

$$\cdots \to \tilde{\pi}_{s,t} A \to \tilde{\pi}_{s,t} B \to \tilde{\pi}_{s,t}(B/A) \xrightarrow{\partial} \tilde{\pi}_{s-1,t} A \to \cdots.$$

of sheaves of stable homotopy groups.

Corollary 10.72 *Supppose that T is compact up to f-equivalence and that the f-local model structure on simplicial presheaves satisfies inductive colimit descent.*

Then there are natural isomorphisms

$$\tilde{\pi}_{s+1,t}(Y \wedge S^1) \cong \tilde{\pi}_{s,t} Y$$

for all $(S^1 \wedge T)$-spectra Y.

Remark 10.73 As noted in Remark 10.69, the presheaves $\pi_{s,t}(E)$ completely determine the motivic stable homotopy types of $(S^1 \wedge \mathbb{G}_m)$-spectra E.

In this case, the long exact sequence (10.17) for a strict fibre sequence determines a corresponding long exact sequence of bigraded presheaves

$$\cdots \to \pi_{s,t} A \to \pi_{s,t} B \to \pi_{s,t}(B/A) \xrightarrow{\partial} \pi_{s-1,t} A \to \cdots$$

for a level cofibre sequence

$$A \to B \to B/A.$$

Finally, there are isomorphisms of presheaves

$$\pi_{s+1,t}(E \wedge S^1) \cong \pi_{s,t} E$$

for all $(S^1 \wedge \mathbb{G}_m)$-spectra E.

10.6 Postnikov Sections and Slice Filtrations

Suppose that E is an ordinary spectrum, or S^1-spectrum, in pointed simplicial sets. The *nth Postnikov section* $P_n E$ of E is a functorially assigned spectrum together with natural map $E \to P_n E$ such that the map $\pi_s E \to \pi_s P_n E$ in stable homotopy groups is an isomorphism for $s \leq n$, and $\pi_s P_n E = 0$ for $s > n$. The diagram

$$\begin{array}{ccc} E & \longrightarrow & P_n E \\ \downarrow & & \downarrow \\ P_{n-1} E & \xrightarrow{\simeq} & P_{n-1} P_n E \end{array}$$

defines a natural map $P_n E \to P_{n-1} E$ in the stable category.

The homotopy fibre $f_{n+1} E$ of the map $E \to P_n E$ is the *n-connected cover* of the spectrum E. There is a natural diagram in the stable category of the form

$$\begin{array}{ccc} f_{n+1} E & \longrightarrow & E & \longrightarrow & P_n E \\ \downarrow & & \downarrow {\scriptstyle 1} & & \downarrow \\ f_n E & \longrightarrow & E & \longrightarrow & P_{n-1} E \end{array}$$

We say that E is *connective* if, equivalently, $P_{-1} E$ is contractible or the map $f_0 E \to E$ is a stable equivalence.

In newer language, the family of maps

$$\cdots \to f_2 E \to f_1 E \to f_0 E = E$$

of connected covers is the *slice filtration* of a connective spectrum E. We write $s_n E$ for the homotopy cofibre of the map $f_{n+1} E \to f_n E$, and say that $s_n E$ is the *nth slice* of the spectrum E.

Of course, the slice $s_n E$ is pretty simple in this context; it is a copy of the shifted Eilenberg–Mac Lane spectrum $H(\pi_n E)[-n]$ that has one nontrivial stable homotopy group, namely $\pi_n E$ in degree n.

There are various constructions of the Postnikov section functors for spectra. One can use the Moore–Postnikov section construction $X \mapsto P_n X$ for Kan pointed complexes X as in [56, Sect. 4.7], or coskeleta, but the oldest, least awkward, and theoretically most powerful method is to kill stable homotopy groups, in conjunction with localization techniques.

One formally inverts the cofibrations

$$* \to \Sigma^\infty(S^q)[-r], \quad q-r > n,$$

to form a new model structure from the stable model structure for which a spectrum Z is fibrant, here called *n-fibrant*, if and only if it is stable fibrant and also satisfies the requirement that all pointed function complexes

$$\mathbf{hom}(\Sigma^\infty(S^q)[-r], Z) \simeq \Omega^q Z^r$$

are contractible for $q - r > n$. Thus, a spectrum Z is n-fibrant if and only if Z is stable fibrant and the stable homotopy groups $\pi_s Z$ are trivial for $s > n$.

Say that a map $E \to F$ of spectra is an *n-equivalence* if the induced map $\pi_p E \to \pi_p F$ in stable homotopy groups is an isomorphism for $p \leq n$.

To kill a stable homotopy group element $\alpha : \Sigma^\infty(S^q)[-r] \to E$, $q - r > n$, in a stable fibrant spectrum E, one forms the homotopy cocartesian diagram

$$\begin{array}{ccc} \Sigma^\infty(S^q)[-r] & \longrightarrow & * \\ \downarrow & & \downarrow \\ E & \xrightarrow{f} & E' \end{array}$$

in which the spectrum E' is stable fibrant. Then the map f induces isomorphisms $\pi_p E \cong \pi_p E'$ for $p \leq n \leq (q-r) - 1$ by a long exact sequence argument, so the map f is an n-equivalence. The map f is also a weak equivalence of the localized model structure. Repeating this process inductively produces a map of spectra

$$j : E \to Z,$$

which is an n-equivalence and a weak equivalence of the localized structure, and where Z is stable fibrant and $\pi_p Z = 0$ for $p > n$. In particular, the map f is a fibrant model for the localized model structure. This map f is the natural model for the *nth* Postnikov section

$$E \to P_n E.$$

10.6 Postnikov Sections and Slice Filtrations

It follows, as well, that a map $E \to F$ is a weak equivalence for the localized model structure if and only if it is an n-equivalence, which might therefore be called the *n-equivalence model structure*.

The naturality of this construction for spectra immediately leads to a natural Postnikov section construction $X \to P_n X$ for presheaves of spectra X on a site $\mathcal{C}$.

We return to the setting and basic assumptions from Sect. 10.5, and approximate the Postnikov section construction in the category of $(S^1 \wedge T)$-spectra.

Suppose that T is compact up to f-equivalence, where $f: A \to B$ is a fixed choice of cofibration of simplicial presheaves. Suppose that the f-local model structure satisfies filtered colimit descent. We also assume that the site $\mathcal{C}$ has finite products.

To compress notation, write

$$F_n(L) = \Sigma^\infty_{S^1 \wedge T}(L)[-n]$$

for a pointed simplicial presheaf L.

Suppose that $\mathbf{L}$ is a set of pointed simplicial presheaves L which are compact up to f-equivalence, and consider a corresponding set of $(S^1 \wedge T)$-spectra

$$F_{n_L}(L \wedge U_+) = \Sigma^\infty_{S^1 \wedge T}(L \wedge U_+)[-n_L],$$

where $n_L \geq 0$ depends only on L, and where U varies through the objects of $\mathcal{C}$. Let F be the set of cofibrations

$$* \to F_{n_L}(L \wedge U_+)$$

We formally invert these cofibrations within the f-local stable model structure. To do so, we let the set F, a generating set of trivial cofibrations for the f-local model structure, and cofibrant replacements of the stabilization maps

$$F_{n+1}(T) \to F_n(S^0)$$

together generate a set $\mathcal{F}$ of cofibrations in the strict model structure for $(S^1 \wedge T)$-spectra in the sense of Example 10.23. This gives a model structure on the category $\mathbf{Spt}_{S^1 \wedge T}(\mathcal{C})$ of $(S^1 \wedge T)$-spectra, via Theorem 10.20. The resulting model structure is a localization of the f-local stable model structure, and will be called the $\mathbf{L}$-*local model structure*.

One says that the fibrant objects for the $\mathbf{L}$-local model structure are the $\mathbf{L}$-*fibrant* objects. The $\mathbf{L}$-*equivalences* are the weak equivalences for this structure.

It is a consequence of Lemma 10.21 that an $(S^1 \wedge T)$-spectrum Z is $\mathbf{L}$-fibrant if and only if it is stable f-fibrant, and all induced pointed simplicial set maps

$$\mathbf{hom}(L \wedge U_+, Z^{n_L}) \cong \mathbf{hom}(F_{n_L}(L \wedge U_+), Z) \to *$$

are trivial fibrations. It follows that the object Z is $\mathbf{L}$-fibrant if and only if Z is stable f-fibrant and all simplicial presheaf maps

$$\mathbf{Hom}(L, Z^{n_L}) \to *$$

are sectionwise (or equivalently local) weak equivalences.

Each simplicial set

$$\mathbf{Hom}(L, Z^{n_L})(U) = \mathbf{hom}(L \wedge U_+, Z^{n_L})$$

is an infinite loop space since the map

$$Z^m \to \Omega_{S^1 \wedge T} Z^{m+1} \cong \Omega_{S^1} \Omega_T Z^{m+1}$$

is an f-local weak equivalence of f-fibrant simplicial presheaves. It follows that the Kan complex

$$\mathbf{Hom}(L, Z^{n_L})(U) = \mathbf{hom}(F_{n_L}(L \wedge U_+), Z)$$

is contractible if and only if all maps

$$\alpha : F_{n_L}(L \wedge U_+ \wedge S^p) \to Z$$

are trivial in the f-local stable category.

We construct a sequence of cofibrations

$$Z = Z_0 \to Z_1 \to Z_2 \to \dots$$

where each Z_i is stable f-fibrant, and Z_{i+1} is constructed from Z_i by taking a cofibre sequence

$$\bigvee_\alpha F_{n_L}(L \wedge U_+ \wedge S^p) \to Z_i \to Z_{i+1} \qquad (10.19)$$

which kills all elements $\alpha : F_{n_L}(L \wedge U_+ \wedge S^p) \to Z_i$ in Z_i.

Suppose that the map

$$j : \varinjlim_i Z_i \to Z_\mathbf{L}$$

is a strict f-fibrant model. Then the object $Z_\mathbf{L}$ is stable f-fibrant by the compactness of $S^1 \wedge T$.

Consider the induced map

$$\varinjlim_i \mathbf{Hom}(F_{n_L}(L), Z_i) \to \mathbf{Hom}(F_{n_L}(L), Z_\mathbf{L}),$$

or equivalently the map

$$\varinjlim_i \mathbf{Hom}(L, Z_i^{n_L}) \to \mathbf{Hom}(L, Z_\mathbf{L}^{n_L}).$$

10.6 Postnikov Sections and Slice Filtrations

This map is an f-equivalence of simplicial presheaves by the compactness of L (and Lemma 10.28). It is therefore a local weak equivalence by the descent condition for the f-local model structure. The filtered colimit

$$\varinjlim_i \mathbf{Hom}(L, Z_i^{n_L})$$

has trivial presheaves of homotopy groups by construction and the finite products assumption on the site $\mathcal{C}$, so the simplicial presheaf $\mathbf{Hom}(L, Z_\mathbf{L}^{n_L})$ has trivial sheaves of homotopy groups, and is therefore locally (hence sectionwise) weakly equivalent to a point.

All maps $Z_i \to Z_{i+1}$ in the cofibre sequences (10.19) are **L**-equivalences by construction. It follows that the composite map

$$p : Z \to Z_\mathbf{L}$$

is an **L**-equivalence, taking values in an **L**-fibrant object, and is therefore an **L**-fibrant model of Z.

What comes next can be expressed in terms of the modern theory of colocalizations [41], but for simplicity we use the language of Bousfield's original paper [11] on homology localizations.

Write $_\mathbf{L}Z$ for the (strict) f-local homotopy fibre of the map $p : Z \to Z_\mathbf{L}$. By Corollary 10.59, the sequence

$$_\mathbf{L}Z \xrightarrow{i} Z \xrightarrow{p} Z_\mathbf{L} \qquad (10.20)$$

is also a cofibre sequence.

Say that a map $X \to Y$ of $(S^1 \wedge T)$-spectra is an **L**-*coequivalence* if all induced simplicial presheaf maps

$$\mathbf{Hom}(F_{n_L}(L), F(X)) \to \mathbf{Hom}(F_{n_L}(L), F(Y))$$

induced by the map $F(X) \to F(Y)$ of stable f-fibrant models are sectionwise weak equivalences.

The map $i : {}_\mathbf{L}Z \to Z$ is an **L**-coequivalence. In effect, all simplicial presheaves $\mathbf{Hom}(F_{n_L}(L), Z_\mathbf{L})$ are sectionwise contractible by construction, and we have fibre sequences

$$\mathbf{Hom}(F_{n_L}(L), {}_\mathbf{L}Z) \xrightarrow{i_*} \mathbf{Hom}(F_{n_L}(L), Z) \to \mathbf{Hom}(F_{n_L}(L), Z_\mathbf{L}).$$

Say that an $(S^1 \wedge T)$-spectrum E is **L**-*colocal* if E has a cofibrant model E_c such that the simplicial presheaf map

$$\mathbf{Hom}(E_c, F(X)) \to \mathbf{Hom}(E_c, F(Y))$$

is a sectionwise weak equivalence for all **L**-coequivalences $X \to Y$.

All objects $F_{n_L}(L \wedge U_+)$ are **L**-colocal, since we assume that the site $\mathcal{C}$ has finite products. The class of **L**-colocal objects is closed under wedge sums and filtered colimits. If
$$A \to B \to C$$
is a cofibre sequence, then if any two of the objects in the sequence are **L**-colocal, so is the third. It follows that the object $_L Z$ in the cofibre sequence (10.20) is **L**-colocal.

In effect, if $g : X \to Y$ is an **L**-coequivalence of stable f-fibrant objects, then the induced map
$$\mathbf{hom}(S^1, X) \to \mathbf{hom}(S^1, Y)$$
is also an **L**-coequivalence. For this, observe that all maps
$$\mathbf{Hom}(L, X^{n_L}) \to \mathbf{Hom}(L, Y^{n_L})$$
are sectionwise equivalences of f-fibrant simplicial presheaves, and so the maps
$$\Omega_{S^1} \mathbf{Hom}(L, X^{n_L}) \to \Omega_{S^1} \mathbf{Hom}(L.Y^{n_L})$$
are also sectionwise weak equivalences. It follows that if E is **L**-colocal, then the object $E \wedge S^1$ is **L**-colocal.

It is an exercise to show that, if the map $\phi : E \to E'$ is an **L**-coequivalence of **L**-colocal objects, then the map $\phi : E \to E'$ is a stable f-equivalence.

We therefore have the following result:

Proposition 10.74 *In the presence of the assumptions listed above, there is natural cofibre sequence*
$$_L Z \xrightarrow{i} Z \xrightarrow{p} Z_L$$
*in $(S^1 \wedge T)$-spectra such that Z_L is **L**-local and the map p is an **L**-equivalence, and the object $_L Z$ is **L**-colocal and the map i is an **L**-coequivalence.*

Example 10.75 Suppose that the family $\mathbf{L}_q$ consists of the cofibrations
$$* \to F_n(S^s \wedge T^{\wedge t}),$$
where $s, t \geq n$ and $t - n > q$, and that Z is stable f-fibrant. The presheaf of homotopy groups of the space
$$\mathbf{Hom}(F_n(S^s \wedge T^{\wedge t}), Z)$$
consists of the groups
$$\pi_{s-n+r, t-n} Z(U),$$
where $r \geq 0$ and $t - n > q$. It follows that killing all maps
$$F_n(S^{s+r} \wedge T^{\wedge t} \wedge U_+) \to Z$$
kills all groups $\pi_{s,t} Z(U)$ for $t > q$.

10.6 Postnikov Sections and Slice Filtrations

The corresponding cofibre sequence

$$\mathbf{L}_q Z \xrightarrow{i} Z \xrightarrow{p} Z_{\mathbf{L}_q}$$

therefore consists of a stable f-fibrant object $Z_{\mathbf{L}_q}$ with $\pi_{s,t}(Z_{\mathbf{L}_q}) = 0$ for $t > q$ with an $\mathbf{L}_q$-equivalence p, and an $\mathbf{L}_q$-colocal object $\mathbf{L}_q Z$ such that the map i induces isomorphisms

$$\pi_{s,t}(\mathbf{L}_q Z) \cong \pi_{s,t} Z$$

for $t > q$.

For alternate notation, as in [85], set $s_{<q} Z := Z_{\mathbf{L}_q}$ and $f_{q+1} Z = \mathbf{L}_q Z$.

Say that the $(S^1 \wedge T)$-spectrum Z is T-*connective* if the map $f_0 Z \to Z$ is a stable f-equivalence. The resulting sequence of maps

$$\ldots f_2 Z \to f_1 Z \to f_0 Z \simeq Z$$

is the *slice filtration* for a T-connective object Z. The cofibre sequences

$$f_{q+1} Z \to f_q Z \to s_q Z$$

define the *slices* $s_q Z$ of the object Z.

Example 10.76 The motivic stable category is the category of f-local $(S^1 \wedge \mathbb{G}_m)$-spectra on the smooth Nisnevich site of a scheme S, where $f : * \to \mathbb{A}^1$ is the 0-section of the affine line over S.

All assumptions leading to the constructions of Example 10.75 hold for this category of $(S^1 \wedge \mathbb{G}_m)$-spectra, and so we are entitled to a slice filtration

$$\ldots f_2 Z \to f_1 Z \to f_0 Z \simeq Z$$

for objects Z which are $\mathbb{G}_m$-connective, along with cofibre sequences

$$f_{q+1} Z \to f_q Z \to s_q Z$$

which define the slices.

The slice filtration for the motivic stable category is due to Voevodsky, and first appeared in [103]. Pelaez' thesis [85] contains a more homotopy theoretic description, which he imports into the symmetric spectrum context to study its multiplicative properties. This passage to symmetric spectra is also possible in the generality of Example 10.75.

Example 10.77 Suppose that G is a finite group. The slice filtration for G-equivariant stable homotopy theory is defined by killing elements in a G-spectrum Z which are parameterized by real regular representations on subgroups H of G, in an appropriate range; see [38, 39]. The ideas behind the construction of the G-equivariant slice filtration are similar to those displayed above, but are implemented for presheaves of spectra, without the extra complications of $(S^1 \wedge T)$-spectra. See Example 10.14.

10.7 T-Complexes

Suppose that R is a presheaf of commutative unitary rings on a small Grothendieck site $\mathcal{C}$ as in Sect. 8.1, and let T be a pointed simplicial presheaf on $\mathcal{C}$. We present, in this section, a general approach to T-spectrum objects in simplicial R-modules, their stable homotopy types, and localizations.

In general, if K is a pointed simplicial presheaf, we write $R_\bullet(K)$ for the quotient $R(K)/R(*)$ that is defined by the base point $* \to K$ of T, as in Chap. 8 (Remark 8.1). The object $R_\bullet(K)$ is the *reduced free simplicial R-module* which is associated to K.

Suppose that A is a simplicial R-module. With a small risk of confusion (compare with (8.1)), we write

$$K \otimes A = R_\bullet(K) \otimes A \text{ and } A \otimes K = A \otimes R_\bullet(K).$$

for pointed simplicial presheaves K. These complexes are naturally isomorphic.

In this notation, there is a natural isomorphism

$$A \otimes (L_+) \cong A \otimes R(L)$$

for all simplicial presheaves L, where

$$L_+ = L \sqcup \{*\}$$

is L with a disjoint base point $*$ formally attached.

If L' is a second pointed simplicial presheaf, then there is an isomorphism

$$R_\bullet(L \wedge L') \cong R_\bullet(L) \otimes R_\bullet(L')$$

which is natural in L and L'. We shall denote this object by $L \otimes L'$.

Write $u(A)$ for the simplicial presheaf underlying a simplicial R-module A. This object is canonically pointed by 0. There is a natural pointed map $K \to u(R_\bullet(K))$ for pointed simplicial presheaves K, which is given by the composite

$$K \to u(R(K)) \to u(R(K)/R(*)),$$

and is initial among all pointed maps $K \to u(A)$ for simplicial R-modules A. We therefore have an adjoint pair of functors

$$R_\bullet : s_*\mathbf{Pre}(\mathcal{C}) \leftrightarrows s\mathbf{Pre}_R : u,$$

relating pointed simplicial presheaves and simplicial R-modules.

There is a pointed map

$$\gamma : K \wedge u(A) \to u(K \otimes A) \qquad (10.21)$$

that is defined in all degrees and sections by $x \wedge a \mapsto x \otimes a$. This map is natural in pointed simplicial presheaves K and simplicial R-modules A.

10.7 T-Complexes

Theorem 8.6 says that there is a model structure on the category $s\mathbf{Pre}_R$ of simplicial modules for which the weak equivalences, respectively fibrations, are those maps $X \to Y$ whose underlying maps of simplicial presheaves are local weak equivalences, respectively injective fibrations.

The simplicial R-module $R_\bullet(K)$ is a direct summand of $R(K)$, since the base point $* \to K$ splits the canonical simplicial presheaf map $K \to *$. It follows from Lemma 8.2 that the functor $K \mapsto R_\bullet(K)$ preserves local weak equivalences of pointed simplicial presheaves.

Lemma 10.78 *Suppose that $K \to L$ is a cofibration of pointed simplicial presheaves and that $A \to B$ is a cofibration of simplicial R-modules. Then the map*

$$(L \otimes A) \cup (K \otimes B) \to L \otimes B$$

is a cofibration of simplicial R-modules which is a local weak equivalence if either $K \to L$ or $A \to B$ is a local weak equivalence.

Proof The unpointed version of this result is a special case of Lemma 8.9. The present result follows, by a retraction argument.

It is a consequence of Lemma 10.78 that the functor $A \mapsto K \otimes A$ preserves cofibrations and trivial cofibrations of simplicial R-modules, for any pointed simplicial presheaf K.

It is observed in Corollary 8.10 that the functor $A \mapsto R(K) \otimes A$ preserves local weak equivalences in simplicial R-modules A. It follows that the functor $A \mapsto K \otimes A$ also preserves local weak equivalences in A.

A *T-complex* A is a collection of simplicial R-modules A^n, $n \geq 0$, together with simplicial R-module homomorphisms

$$\sigma : T \otimes A^n \to A^{n+1},$$

called *bonding maps*. A morphism $g : A \to B$ of T-complexes consists of simplicial R-module homomorphisms $g : A^n \to B^n$ that preserve structure in the sense that the diagrams

$$\begin{array}{ccc} T \otimes A^n & \xrightarrow{\sigma} & A^{n+1} \\ {\scriptstyle T \otimes g} \downarrow & & \downarrow {\scriptstyle g} \\ T \otimes B^n & \xrightarrow{\sigma} & B^{n+1} \end{array}$$

commute.

A T-complex A is what one would call a *T-spectrum object* in the category $s\mathbf{Pre}_R$ of simplicial R-modules. Write $\mathbf{Spt}_T(s\mathbf{Pre}_R)$ for the resulting category of T-complexes and their morphisms.

Example 10.79 Write

$$H(R) = R_\bullet(S_T),$$

and call it the *sphere T-spectrum object* for the category of T-complexes in simplicial R-modules.

There is a functorial *suspension T-spectrum object*

$$H(A) = H(R) \otimes A$$

for a simplicial R-module A. This object is defined in levels by the simplicial R-modules

$$A, \; T \otimes A, \; T \otimes T \otimes A, \; \dots.$$

There is a natural bijection

$$\hom(H(A), B) = \hom(H(R) \otimes A, B) \cong \hom(A, B^0)$$

for all simplicial R-modules A and T-complexes B.

The construction $A \mapsto H(R) \otimes A$ is the natural source of Eilenberg–Mac Lane T-spectra, and one says that $H(R) \otimes A$ is the *Eilenberg–Mac Lane T-complex* which is associated to the simplicial R-module A.

The shift $X[n]$ of a T-complex X is defined by

$$X[n]^k = \begin{cases} X^{n+k} & \text{if } n+k \geq 0, \text{ and} \\ 0 & \text{if } n+k < 0. \end{cases}$$

Say that a map $X \to Y$ of T-complexes is a *strict weak equivalence* (respectively *strict fibration*) if all of the maps $X^n \to Y^n$ are local weak equivalences (respectively injective fibrations) for the model structure on the category of simplicial R-modules given by Theorem 8.6. A *cofibration* of T-complexes is a map $A \to B$ such that the map $A^0 \to B^0$ and all induced maps

$$(T \otimes B^n) \cup A^{n+1} \to B^{n+1}$$

are cofibrations of simplicial R-modules.

Proposition 10.80 *The category $\mathbf{Spt}_T(s\mathbf{Pre}_R)$ of T-complexes, together with the classes of strict weak equivalences, strict fibrations and cofibrations, satisfies the axioms for a proper, cofibrantly generated closed simplicial model category.*

Proof The proof of this result is an exercise—see also Propositions 10.4 and 10.15.

The key to the proofs of both the factorization axiom **CM5** and the lifting axiom **CM4** is that the functor $A \mapsto T \otimes A$ preserves cofibrations and trivial cofibrations in simplicial R-modules A.

The function complex $\mathbf{hom}(X, Y)$ for T-complexes X and Y has n-simplices given by the T-complex morphisms $X \otimes (\Delta^n_+) \to Y$.

10.7 T-Complexes

Suppose that the set I, respectively J, consists of maps of the form

$$H(A)[-n] \to H(B)[-n]$$

which are induced by members $A \to B$ of generating sets for the cofibrations, respectively the trivial cofibrations, of simplicial R-modules. The sets I and J give generating sets for the classes of cofibrations and trivial cofibrations for this model structure.

The properness claim is a consequence of properness for the injective model structure on the category of simplicial R-modules.

The model structure of Proposition 10.80 is the *strict model structure* for the category of T-complexes.

Every T-spectrum X has an associated T-complex $R_\bullet X$, which is defined in levels by the assignments

$$(R_\bullet X)^n = R_\bullet(X^n), \quad n \geq 0.$$

The bonding maps for $R_\bullet(X)$ are the composite maps

$$T \otimes R_\bullet(X^n) \cong R_\bullet(T \wedge X^n) \xrightarrow{\sigma_*} R_\bullet(X^{n+1}).$$

If A is a T-complex, then the natural composites

$$T \wedge u(A^n) \xrightarrow{\gamma} u(T \otimes A^n) \xrightarrow{u(\sigma)} u(A^{n+1}),$$

which are induced by the bonding maps of A and the natural map γ of (10.21), give the list of pointed simplicial presheaves $u(A^n)$ the structure of a T-spectrum. This T-spectrum is denoted by $u(A)$ and is called the *underlying T-spectrum* for the T-complex A. The functor $X \mapsto R_\bullet X$ is left adjoint to the underlying T-spectrum functor $A \mapsto u(A)$, and the functors

$$R_\bullet : \mathbf{Spt}_T(\mathcal{C}) \leftrightarrows \mathbf{Spt}_T(s\mathbf{Pre}_R) : u$$

form a Quillen adjunction between the respective strict structures.

In particular, the functor $X \mapsto R_\bullet X$ preserves cofibrations and trivial cofibrations for the strict model structure on T-spectra. The functor $R_\bullet$ also commutes with the shift functors and preserves suspension spectrum objects.

There is a natural isomorphism

$$R_\bullet(X \wedge K) \cong R_\bullet(X) \otimes K,$$

for all T-spectra X and pointed simplicial presheaves K. It follows that the functor $A \mapsto A \otimes K$ preserves cofibrations of T-complexes. In particular, the T-complex $A \otimes K$ is cofibrant if A is cofibrant.

Suppose (as in Sect. 10.2) that $\mathcal{F}$ is a set of cofibrations of T-spectra that satisfies the following:

C1 The T-spectrum A is cofibrant for all maps $i : A \to B$ in $\mathcal{F}$.
C2 The set $\mathcal{F}$ includes the set J of generators for the trivial cofibrations for the strict model structure on $\mathbf{Spt}_T(\mathcal{C})$.
C3 If the map $i : A \to B$ is in $\mathcal{F}$, then the cofibrations

$$(A \wedge D) \cup (B \wedge C) \to B \wedge D$$

induced by i and all α-bounded cofibrations $C \to D$ of pointed simplicial presheaves are also in $\mathcal{F}$.

Write $R_\bullet(\mathcal{F})$ for the set of morphisms $R_\bullet(A) \to R_\bullet(B)$ of T-complexes which are induced by morphisms $A \to B$ of $\mathcal{F}$.

Choose a regular cardinal β such that $\beta > \alpha$, $\beta > |\mathcal{F}|$, $\beta > |B|$ for any cofibration $A \to B$ of $\mathcal{F}$. Finally, suppose that $\beta > |R|$. Choose a cardinal λ such that $\lambda > 2^\beta$.

Suppose that $g : X \to Y$ is a morphism of T-complexes. There is a functorial factorization

of g such that the map i is a cofibration which is in the saturation of $R_\bullet(\mathcal{F})$ and the map p has the right lifting property with respect to all members of $R_\bullet(\mathcal{F})$. The construction of this factorization uses a small object argument, which terminates after λ steps.

Set

$$L_\mathcal{F}(X) = E_\lambda(X \to 0)$$

for all T-complexes X. Say that a map $X \to Y$ of T-complexes is an $L_\mathcal{F}$-*equivalence* if the induced map $L_\mathcal{F} X \to L_\mathcal{F} Y$ is a strict equivalence of T-complexes.

We have the following:

Lemma 10.81

1) Suppose that the assignment $t \mapsto X_t$ defines a diagram of monomorphisms indexed by $t < \gamma$ where γ is a cardinal such that $\gamma > 2^\beta$. Then the natural map

$$\varinjlim_{t<\gamma} L_\mathcal{F}(X_t) \to L_\mathcal{F}(\varinjlim_{t<\gamma} X_t)$$

is an isomorphism.

10.7 T-Complexes

2) *Suppose that ζ is a cardinal with $\zeta > \beta$, and let $B_\zeta(X)$ denote the filtered system of subobjects of X having cardinality less than ζ. Then the natural map*

$$\varinjlim_{Y \in B_\zeta(X)} L_{\mathcal{F}}(Y) \to L_{\mathcal{F}}(X)$$

is an isomorphism.
3) *The functor $X \mapsto L_{\mathcal{F}}(X)$ preserves monomorphisms.*
4) *Suppose that U, V are subobjects of a T-complex X. Then the natural map*

$$L_{\mathcal{F}}(U \cap V) \to L_{\mathcal{F}}(U) \cap L_{\mathcal{F}}(V)$$

is an isomorphism.
5) *If $|X| \leq 2^\mu$ where $\mu \geq \lambda$ then $|L_{\mathcal{F}}(X)| \leq 2^\mu$.*

Proof This result is a consequence of Lemma 8.38. □

Let κ be the successor cardinal for 2^μ, where μ is cardinal of statement 5) of Lemma 10.81. Then κ is a regular cardinal, and Lemma 10.81 implies that, if a T-complex X is κ-bounded, then the T-complex $L_{\mathcal{F}}(X)$ is κ-bounded.

The following is the bounded monomorphism statement for the strict model structure on the category of T-complexes:

Lemma 10.82 *Suppose given a diagram of monomorphisms*

of T-complexes such that the map i is a strict equivalence and A is κ-bounded. Then there is a factorization of j by monomorphisms $A \to B \to Y$ such that B is κ-bounded and the map $B \cap X \to B$ is a strict equivalence.

Proof One can prove this statement by following the method of Lemma 10.18, by using the bounded subobject statement Lemma 8.11 for simplicial R-modules in place of Lemma 5.2. A variant of the argument for Lemma 8.11 can also be applied directly. □

We now have the bounded monomorphism property for $L_{\mathcal{F}}$-equivalences of T-complexes.

Lemma 10.83 *Suppose given a diagram of monomorphisms*

of T-complexes such that the map i is an $L_{\mathcal{F}}$-equivalence and A is κ-bounded. Then there is a factorization of j by monomorphisms $A \to B \to Y$ such that B is κ-bounded and the map $B \cap X \to B$ is an $L_{\mathcal{F}}$-equivalence.

Proof The proof of this result is formally the same as that of Lemma 7.17. It uses the bounded cofibration statement for strict equivalences of Lemma 10.82.

Following the methods of Sect. 7.2, a map of T-complexes $X \to Y$ is an $\mathcal{F}$-*equivalence* if and only if the map

$$\mathbf{hom}(Y_c, Z) \to \mathbf{hom}(X_c, Z)$$

is a weak equivalence of simplicial sets for every $\mathcal{F}$-injective object Z, and where $X_c \to X$ is a natural cofibrant model construction in the strict model category of Proposition 10.80. It is a consequence of Lemma 7.6 that a map $X \to Y$ is an $\mathcal{F}$-equivalence if and only if it is an $L_{\mathcal{F}}$-equivalence.

An $\mathcal{F}$-*fibration* is a map that has the right lifting property with respect to all cofibrations which are $\mathcal{F}$-equivalences.

Theorem 7.10 then implies the following:

Theorem 10.84 *Suppose that $\mathcal{C}$ is a small Grothendieck site, and that $\mathcal{F}$ is a set of cofibrations of T-spectra which satisfies the conditions* **C1**, **C2** *and* **C3**. *Then the category* $\mathbf{Spt}_T(s\mathbf{Pre}_R(\mathcal{C}))$ *of T-complexes, together with the classes of cofibrations, $\mathcal{F}$-equivalences and $\mathcal{F}$-fibrations, satisfies the axioms for a cofibrantly generated closed simplicial model category. This model structure is left proper.*

The model structure of Theorem 10.84 for the category of T-complexes is the $\mathcal{F}$-*local structure*. The left properness of this model structure is a consequence of Lemma 7.8.

Corollary 7.12 implies that a T-complex Z is $\mathcal{F}$-fibrant if and only it is $\mathcal{F}$-injective in the sense that the map $Z \to 0$ has the right lifting property with respect to all members of $R_\bullet(\mathcal{F})$. It follows that the T-complex Z is $\mathcal{F}$-fibrant if and only if the underlying T-spectrum $u(Z)$ is $\mathcal{F}$-fibrant, meaning fibrant for the model structure of Theorem 10.20.

Remark 10.85 The functors

$$R_\bullet : \mathbf{Spt}_T(\mathcal{C}) \leftrightarrows \mathbf{Spt}_T(s\mathbf{Pre}_R(\mathcal{C})) : u$$

define a Quillen adjunction between the respective $\mathcal{F}$-local structures on the categories of T-spectra and T-complexes, since the functor $R_\bullet$ preserves $\mathcal{F}$-equivalences as well as cofibrations. To verify the claim about $\mathcal{F}$-equivalences, observe that the functor $R_\bullet$ preserves strictly cofibrant models, so it suffices to show that $R_\bullet$ preserves $\mathcal{F}$-equivalences $X \to Y$ between cofibrant T-spectra.

For this last claim, if Z is an $\mathcal{F}$-fibrant T-complex, then the underlying T-spectrum $u(Z)$ is $\mathcal{F}$-fibrant, and the simplicial set map

$$\mathbf{hom}(R_\bullet(Y), Z) \to \mathbf{hom}(R_\bullet(X), Z)$$

is isomorphic to the map

$$\mathbf{hom}(Y, u(Z)) \to \mathbf{hom}(X, u(Z)),$$

which is a weak equivalence.

The cofibrations in the set $R_\bullet(\mathcal{F})$ are formally inverted in the $\mathcal{F}$-local model structure of Theorem 10.84 for T-complexes. Recall that the set $\mathcal{F}$ is generated (over J) by a subset S that contains J if $\mathcal{F}$ is the smallest subset which contains S and is closed under condition **C3**.

Recall that the stabilization maps

$$(S_T \wedge T)[-n-1] \to S_T[-n], \ n \geq 0,$$

are shifts of the map

$$(S_T \wedge T)[-1] \to S_T$$

which consists of the canonical isomorphisms

$$T^{\wedge(m-1)} \wedge T \to T^{\wedge m}$$

in levels $m \geq 1$.

Say that the $\mathcal{F}$-local structure on $\mathbf{Spt}_T(s\mathbf{Pre}_R)$ is *stable* if the stabilization maps induce $\mathcal{F}$-equivalences

$$(H(R) \otimes T)[-n-1] \to H(R)[-n].$$

One forces this by insisting that the set $\mathcal{F}$ contains cofibrant replacements of these maps.

If the set $\mathcal{F}$ is generated over J by the cofibrant replacements of the stabilization maps alone, then the corresponding $\mathcal{F}$-local structure is the *stable model structure* for the category of T-complexes. The weak equivalences for this theory are the *stable equivalences*, and the fibrations are the *stable fibrations*.

More generally, given a cofibration $f : A \to B$ of pointed simplicial presheaves, one constructs the f-local stable structure on the category of T-complexes by forming the $\mathcal{F}$-local structure for the set $\mathcal{F} = S_f$ which is defined in Example 10.23. This set of cofibrations is generated by the stabilization maps and all maps

$$(H(R) \otimes A_+)[-n] \to (H(R) \otimes B_+)[-n]$$

induced by members $A \to B$ of a generating set of trivial cofibrations for the f-local model structure on simplicial presheaves.

The model structure on the category $\mathbf{Spt}_T(s\mathbf{Pre}_R)$ which is given by Theorem 10.84 for the set $\mathcal{F} = S_f$ is the f-*local stable model structure*, on the category of T-complexes, and the corresponding $\mathcal{F}$-equivalences and $\mathcal{F}$-fibrations are called *stable f-equivalences* and *stable f-fibrations*, respectively.

A T-complex Z is stable f-fibrant if and only if its underlying T-spectrum $u(Z)$ is stable f-fibrant (see Remark 10.85). This observation has various consequences, which follow from the results of Sect. 10.3:

Corollary 10.86 *A T-complex Z is stable f-fibrant if and only if all level objects Z^n are f-fibrant simplicial R-modules and all maps $Z^n \to \Omega_T Z^{n+1}$ are sectionwise equivalences of simplicial R-modules.*

Proof This result follows from Corollary 10.26.

There is a fake T-loop spectrum functor $X \mapsto \Omega_T X$ for T-complexes, just as there is for T-spectra, along with a natural map

$$X \to \Omega_T X[1]$$

that is defined by adjoint bonding maps. There is also a natural directed system

$$X \to \Omega_T X[1] \to \Omega_T^2 X[2] \to \cdots$$

of T-complexes. Define

$$Q_T X = F(\varinjlim_n \Omega_T^n(FX)[n]),$$

where $j : Y \to FY$ is the natural strictly f-fibrant model for a T-complex Y (see Remark 10.24). Then the composite

$$X \xrightarrow{j} FX \to \varinjlim \Omega_T^n(FX)[n] \xrightarrow{j} F(\varinjlim \Omega_T^n(FX)[n])$$

defines a natural map $\eta : X \to Q_T X$ in T-complexes.

We have the following analogue of Theorem 10.32:

Theorem 10.87 *Suppose that T is compact up to f-equivalence. Then the natural map*

$$\eta : X \to Q_T X$$

is a stable f-fibrant model for each T-complex X.

Proof All steps in the proof of Theorem 10.32 have analogues in the category of T-complexes.

Example 10.88 [Stable homotopy theory of S^1-complexes] An S^1-complex X is a collection of simplicial R-modules X^n, together with bonding maps $S^1 \otimes X^n \to X^n$. Alternatively, one says that an S^1-complex is a spectrum object in simplicial R-modules.

The simplicial circle S^1 is compact, and the filtered colimit descent is automatic in the injective model structure for simplicial presheaves. It follows that the map $X \to Q_{S^1} X$ is a fibrant model for this theory.

10.7 T-Complexes

A map $X \to Y$ is a weak equivalence for the stable model structure on this category if and only if the underlying map $u(X) \to u(Y)$ is a stable equivalence of presheaves of spectra in the usual sense, so that the stable model structure on S^1-complexes that one obtains by formally inverting the stabilization maps

$$(S_{S^1} \wedge S^1)[-n-1] \to S_{S^1}[-n]$$

in the strict model structure for S^1-complexes is the "standard" stable model structure. This model structure is also proper.

The Dold–Kan correpondence extends to an equivalence of categories

$$N : \mathbf{Spt}_{S^1}(s\mathbf{Pre}_R(\mathcal{C})) \leftrightarrows \mathbf{Spt}(\mathrm{Ch}_+(\mathbf{Pre}_R(\mathcal{C}))) : \Gamma.$$

The functors N and Γ are defined levelwise by the normalized chains functor N and its inverse Γ, respectively. One uses the natural homotopy equivalence

$$N(S^1 \otimes A) \xrightarrow{\simeq} NA[-1]$$

and its natural homotopy inverse

$$NA[-1] \xrightarrow{\simeq} N(S^1 \otimes A)$$

in the category of simplicial R-modules to define bonding maps for induced spectrum objects $\Gamma(B)$ and $N(A)$, respectively. The first map, in particular, induces a natural sectionwise weak equivalence

$$S^1 \otimes \Gamma(C) \to \Gamma(C[-1])$$

for all chain complexes C. It is an exercise to show that a map $X \to Y$ of S^1-complexes is a stable equivalence if and only if the induced map $NX \to NY$ of spectrum objects in chain complexes is a stable equivalence in the sense of Proposition 8.16.

The stable category $\mathrm{Ho}\,(\mathbf{Spt}(s\mathbf{Pre}_R(\mathcal{C})))$ is therefore equivalent to the derived category of presheaves or sheaves of R-modules, as described in Sect. 8.2. See [61] for more detail.

We have the following analogue of Theorem 10.46 for T-complexes:

Theorem 10.89 *Suppose that T is compact up to f-equivalence and is cycle trivial. Then the composite*

$$A \to \mathbf{Hom}(T, A \otimes T) \xrightarrow{\eta_*} \mathbf{Hom}(T, Q_T(A \otimes T))$$

is a stable f-equivalence for all T-complexes A.

The T-complex $\mathbf{Hom}(T, E)$ is the "real" T-loop object for a T-complex E. The underlying T-spectrum of this object is the real T-loop object for the T-spectrum $u(E)$.

We shall use the *layer filtration* of a T-complex A: the nth layer $L_n A$ is the T-complex which is specified by the complexes

$$A^0, A^1, \ldots, A^n, T \otimes A^n, T^{\otimes 2} \otimes A^n, \ldots.$$

There is a natural stable equivalence

$$(H(R) \otimes A^n)[-n] \xrightarrow{\simeq} L_n(A),$$

where $H(R) \otimes A^n$ is the T-suspension object associated to the simplicial R-module A^n.

Proof The proof of Theorem 10.89 proceeds by analogy with the proof of Theorem 10.46.

If A and B are strictly f-fibrant T-complexes then a map $A \to B$ is a stable f-equivalence if and only if the induced map

$$\mathbf{Hom}(T, A) \to \mathbf{Hom}(T, B)$$

is a stable f-equivalence. This statement is the analogue for T-complexes of Lemma 10.48, and has the same proof.

It therefore suffices to show that the composite

$$A \to \mathbf{Hom}(T, A \otimes T) \xrightarrow{j_*} \mathbf{Hom}(T, F(A \otimes T)) \qquad (10.22)$$

is a stable f-equivalence, where $j : A \otimes T \to F(A \otimes T)$ is a strictly f-fibrant model of the T-complex $A \otimes T$.

The constructions at both ends of the composite (10.22) are invariants of stable f-equivalences, shifts and filtered colimits. It therefore suffices to show that the composite

$$H(E) \to \mathbf{Hom}(T, H(E) \otimes T) \xrightarrow{j_*} \mathbf{Hom}(T, F(H(E) \otimes T))$$

is a stable f-equivalence for each simplicial R-module E. This claim is the T-complex analogue of Lemma 10.47, and it has a similar proof.

We have already observed that Lemma 10.48 has an analogue for T-complexes (provided that T is compact up to f-equivalence). As in Corollary 10.49, it follows that the evaluation map

$$ev : \mathbf{Hom}(T, A) \otimes T \to A$$

is a stable f-equivalence of T-complexes if A is strictly f-fibrant.

We then have the following analogue of Theorem 10.50. This result asserts that the T-loops and T-suspension functors are inverse to each other, at the level of the f-local stable category for T-complexes, in the presence of the usual niceness conditions on the parameter object T.

10.7 T-Complexes

Theorem 10.90 *Suppose that T is compact up to f-equivalence and is cycle trivial. Then the T-suspension and T-loops functors induce a Quillen equivalence*

$$\otimes T : \mathbf{Spt}_T(s\,\mathrm{Mod}_R) \leftrightarrows \mathbf{Spt}_T(s\,\mathrm{Mod}_R) : \mathbf{Hom}(T,)$$

for the f-local stable model structure on the category of T-complexes.

The results and constructions of Sect. 10.5 apply to the f-local stable category of $(S^1 \wedge T)$-complexes, provided that T is compact up to f-equivalence and that the f-local model structure on simplicial presheaves satisfies inductive colimit descent.

In particular, an $(S^1 \wedge T)$-complex X has a bigraded system of presheaves $\pi_{s,t} X$ of stable homotopy groups. These groups are R-modules, whose associated sheaves measure stable equivalence.

Some things are easier for complexes; the additivity property (Lemma 10.68) is a complete triviality in the R-module context, where finite coproducts coincide with finite products.

One can use standard homological algebra to derive an analogue of Lemma 10.58: if $p : X \to Y$ is a strict fibration of $(S^1 \wedge T)$-complexes with fibre F then the canonical map

$$X/F \to Y$$

is a stable equivalence of $(S^1 \wedge T)$-complexes. Similar techniques are used to show that any $(S^1 \wedge T)$-complex can be desuspended in the S^1-direction, in that there is a natural stable equivalence

$$A \simeq C(A) \otimes S^1$$

for a functorially constructed cofibrant complex $C(A)$, as in the proof of Lemma 10.60. The coincidence of fibre and cofibre sequences in the stable f-local category of $(S^1 \wedge T)$-complexes, culminating in the analogue of Lemma 10.62, is essentially a formal consequence of these two observations.

Right properness for the f-local stable model structure on $(S^1 \wedge T)$-complexes also follows, as in Theorem 10.64.

Suppose that K is a pointed simplicial presheaf. There is a natural map

$$\gamma : u(H(R)) \wedge K \to u(H(R) \otimes K)$$

of $(S^1 \wedge T)$-spectra which consists of the maps

$$\gamma : (S^1 \otimes T)^{\otimes n} \wedge K \to (S^1 \otimes T)^{\otimes n} \otimes K.$$

Lemma 10.91 *The map*

$$\gamma : u(H(R)) \wedge K \to u(H(R) \otimes K)$$

is a stable equivalence of $(S^1 \wedge T)$-spectra for any pointed simplicial presheaf K.

Proof The object $H(R)$ is the diagonal of an (S^1, T)-bispectrum object $H(R)^{p,q}$ with

$$H(R)^{p,q} = (S^1)^{\otimes p} \otimes (T)^{\otimes q},$$

with obvious S^1-bonding map, and with T-bonding map given by the composite

$$T \otimes (S^1)^{\otimes p} \otimes T^{\otimes q} \xrightarrow{\tau \otimes 1} (S^1)^{\otimes p} \otimes T \otimes T^{\otimes q} \cong (S^1)^{\otimes p} \otimes T^{\otimes (q+1)},$$

where the isomorphism

$$\tau : T \otimes (S^1)^{\otimes p} \to (S^1)^{\otimes p} \otimes T$$

permutes factors. More generally, the object $H(R) \otimes K)$ is the diagonal of the (S^1, T)-bispectrum object $H(R_\bullet(K))^{*,*}$ with

$$H(R_\bullet(K))^{p,q} = (S^1)^{\otimes p} \otimes T^{\otimes q} \otimes K.$$

The map γ is the diagonal of the (S^1, T)-bispectrum map

$$\gamma : u(H(R)^{*,*}) \wedge K \to u(H(R_\bullet(K))^{*,*}),$$

which is defined by the canonical maps

$$u((S^1)^{\otimes p} \otimes T^{\otimes q}) \wedge K \to u((S^1)^{\otimes p} \otimes T^{\otimes q} \otimes K).$$

For a fixed q, these maps define the map of S^1-spectra

$$u(H(R) \otimes T^{\otimes q}) \wedge K \to u(H(R) \otimes T^{\otimes q} \otimes K))$$

which is a (sectionwise) stable equivalence: it is an exercise to show that the canonical map

$$\gamma : u(H(R) \otimes B) \wedge L \to u(H(R) \otimes B \otimes L))$$

is a stable equivalence for any simplicial abelian group B and pointed simplicial set L.

The desired result follows from Lemma 10.56. The map γ is even a sectionwise stable equivalence.

Corollary 10.92 *Suppose that the cofibration $K \to L$ is an f-local equivalence of pointed simplicial presheaves. Then the induced map*

$$u(H(R) \otimes K) \to u(H(R) \otimes L)$$

is a stable f-equivalence of $(S^1 \wedge T)$-spectra.

Proof The map

$$E \wedge K \to E \wedge L$$

10.7 T-Complexes

is a strict f-equivalence for any $(S^1 \wedge T)$-spectrum E. This is a consequence of Corollary 7.24.

Lemma 10.93 *Suppose that T is compact up to f-equivalence and that the f-local model structure on simplicial presheaves satisfies inductive colimit descent.*

Suppose that E is an $(S^1 \wedge T)$-complex in simplicial R-modules, and let $E \to L(E)$ be a stable f-fibrant model for E. Then the induced map $u(E) \to u(L(E))$ is a stable f-equivalence of $(S^1 \wedge T)$-spectra.

Proof Suppose that the cofibration $C \to D$ of simplicial presheaves is an f-equivalence. Suppose given a pushout diagram

$$\begin{array}{ccc} R_\bullet(\Sigma^\infty C_+[-n]) & \xrightarrow{i} & R_\bullet(\Sigma^\infty D_+[-n]) \\ \downarrow & & \downarrow \\ E & \xrightarrow{i_*} & F \end{array}$$

in $(S^1 \wedge T)$-complexes. The cofibrations i and i_* have the same cokernel H, and there is an induced comparison of strict fibre sequences

$$\begin{array}{ccccc} u(R_\bullet(\Sigma^\infty C_+[-n])) & \xrightarrow{u(i)} & u(R_\bullet(\Sigma^\infty D_+[-n])) & \longrightarrow & u(H) \\ \downarrow & & \downarrow & & \downarrow 1 \\ u(E) & \xrightarrow{u(i_*)} & u(F) & \longrightarrow & u(H) \end{array}$$

of $(S^1 \wedge T)$-spectra. This comparison can be replaced up to stable equivalence by a comparison of level cofibre sequences, by Corollary 10.59.

The map $u(i)$ is stable equivalent to the map

$$u(H(R)) \wedge C_+[-n] \to u(H(R)) \wedge D_+[-n]$$

and is therefore a stable f-equivalence. The object $u(H)$ is therefore stable f-equivalent to a point, and so the map $u(i_*)$ is a stable f-equivalence—both claims follow from Lemma 10.60.

Suppose that $X \to Y$ is a cofibration of $(S^1 \wedge T)$-spectra such that the map $X^k \to Y^k$ is a weak equivalence for $k \geq N$, for some N. Then the induced cofibration $R_\bullet(X) \to R_\bullet(Y)$ consists of weak equivalences $R_\bullet X^k \to R_\bullet Y^k$ for $k \geq N$, and is therefore a stable equivalence. If we have a pushout diagram

$$\begin{array}{ccc} R_\bullet(X) & \xrightarrow{j} & R_\bullet(Y) \\ \downarrow & & \downarrow \\ E & \xrightarrow{j_*} & F \end{array}$$

in $(S^1 \wedge T)$-complexes, then the induced map $u(j_*)$ consists of weak equivalences $u(E^k) \to u(F^k)$ for $k \geq N$, and is therefore a stable equivalence—see the proof of Theorem 10.32.

Proposition 10.94 *Suppose that T is compact up to f-equivalence and that the f-local model structure on simplicial presheaves satisfies inductive colimit descent.*

Then a map $E \to F$ of $(S^1 \wedge T)$-complexes is a stable f-equivalence if and only if the induced map $u(E) \to u(F)$ is a stable f-equivalence of $(S^1 \wedge T)$-spectra.

Proof Suppose that $j : E \to L(E)$ is a stable f-fibrant model for E in in $(S^1 \wedge T)$-complexes. Then the induced map $u(j) : u(E) \to u(L(E))$ is a stable f-fibrant model for $u(E)$ in the category of $(S^1 \wedge T)$-spectra. In effect, $u(L(E))$ is stable f-fibrant by formal nonsense and the map $u(j)$ is a stable f-equivalence by Lemma 10.93.

Then $E \to F$ is a stable f-equivalence if and only if the map $L(E) \to L(F)$ is a levelwise weak equivalence, which is true if and only if $u(E) \to u(F)$ is a stable f-equivalence.

Example 10.95 Suppose that k is a perfect field, and let $f : * \to \mathbb{A}^1$ be the 0-section of the affine line over k, on the smooth Nisnevich site $(Sm|_k)_{Nis}$. This is the general setup for Example 8.49, and we continue that discussion here.

Recall that precomposition with the graph functor $\gamma : Sm|_k \to Cor_k$ determines a forgetful functor

$$\gamma_* : s\mathbf{PST}(k) \to s\mathbf{Mod}_{\mathbb{Z}}(Sm|_k)_{Nis}$$

on the category of simplicial presheaves with transfers, which takes values in simplicial abelian presheaves.

It is a consequence of Proposition 10.94 that a map $E \to F$ of $(S^1 \wedge \mathbb{G}_m)$-complexes is a motivic stable equivalence if and only if the underlying map $u(E) \to u(F)$ is a motivic stable equivalence of $(S^1 \wedge \mathbb{G}_m)$-spectra.

The basic example of an $(S^1 \wedge \mathbb{G}_m)$-complex is the object which represents motivic cohomology. We assume that the multiplicative group $\mathbb{G}_m$ is pointed by the identity e.

There is a natural pairing of presheaves with transfers

$$\mathbb{Z}_{tr}(\mathbb{G}_m, e) \otimes \mathbb{Z}_{tr}(X) \to \mathbb{Z}_{tr}(\mathbb{G}_m \wedge (X_+), *),$$

which is induced by the pairing (8.12). Suppose that

$$j : \mathbb{Z}_{tr}(\mathbb{G}_m \wedge (X_+), *) \to F(\mathbb{Z}_{tr}(\mathbb{G}_m \wedge (X_+), *))$$

is a motivic fibrant model in the category of presheaves with transfers. The Voevodsky cancellation theorem [104] asserts that the induced composite

$$\gamma_*(\mathbb{Z}_{tr}(X)) \to \Omega_{\mathbb{G}_m} \gamma_*(\mathbb{Z}_{tr}(\mathbb{G}_m \wedge (X_+), *)) \xrightarrow{j_*} \Omega_{\mathbb{G}_m} \gamma_* F(\mathbb{Z}_{tr}(\mathbb{G}_m \wedge (X_+), *))$$

of simplicial abelian presheaves is a motivic weak equivalence for each smooth k-scheme X.

10.7 T-Complexes   427

One can also show, more easily, that the natural composite

$$\gamma_*(\mathbb{Z}_{tr}(X)) \to \Omega_{S^1}\gamma_*(\mathbb{Z}_{tr}(X) \otimes S^1) \xrightarrow{j_*} \Omega_{S^1}\gamma_*(F(\mathbb{Z}_{tr}(X) \otimes S^1))$$

is a motivic weak equivalence for all smooth k-schemes X. For this, we use the natural isomorphism

$$C_*(K) \otimes S^1 \xrightarrow{\cong} C_*(K \otimes S^1)$$

that is defined for all presheaves with transfers K, where $C_*(K)$ is the singular complex of Example 8.49.

It follows that a strict motivic fibrant model for the bispectrum object

$$\gamma_*(\mathbb{Z}_{tr}(\mathbb{G}_n^{\wedge p}, *) \otimes (S^1)^{\otimes q})$$

is motivic stably fibrant. In particular, the associated diagonal $(S^1 \wedge \mathbb{G}_m)$-spectrum, which is composed of the objects

$$F\gamma_*(\mathbb{Z}(n)) \simeq F\gamma_*(\mathbb{Z}_{tr}(\mathbb{G}_n^{\wedge n}, *) \otimes (S^1)^{\otimes n})$$

is motivic stable fibrant. This object is called the *motivic Eilenberg–Mac Lane spectrum* for the ring $\mathbb{Z}$. There are various notations for this object; Voevodsky denotes it by $\mathbf{H}_{\mathbb{Z}}$.

This object $\mathbf{H}_{\mathbb{Z}}$ represents motivic cohomology in the motivic stable category. There are isomorphisms

$$H^p(X, \mathbb{Z}(q)) \cong [X_+, \gamma_*(\mathbb{Z}(q))[-p]]$$
$$\cong [X_+, \Omega_{S^1}^{n-p}\Omega_{\mathbb{G}_m}^n \gamma_*(\mathbb{Z}(q+n))] \text{ for } n \geq p$$
$$\cong [(S^1)^{\wedge(n-p)} \wedge \mathbb{G}_n^{\wedge n} \wedge X_+, \gamma_*\mathbb{Z}(q+n)]$$
$$\cong [(S^1)^{\wedge(-p-q+(q+n))} \wedge (S_t^1)^{-q+(q+n)}) \wedge X_+, \gamma_*\mathbb{Z}(q+n)]$$
$$\cong \pi_{-p-q,-q}\mathbf{H}_{\mathbb{Z}}(X)$$

for all smooth k-schemes X, where the square brackets denote morphisms in the motivic homotopy category.

We return to the general setup of Sect. 8.6.

Suppose that R is a commutative ring with identity, and let $s\mathbf{Mod}_R$ be the category of simplicial R-modules. Suppose that $\mathcal{A}$ is a small category which is enriched in R-modules, and suppose that there is a functor $\phi : \mathcal{C} \to \mathcal{A}$ which is the identity on objects, where $\mathcal{C}$ is a small site.

Recall that an $\mathcal{A}$-linear simplicial presheaf is a functor $X : \mathcal{A}^{op} \to s\mathbf{Mod}_R$ which respects the $\mathcal{A}$-linear structure. The category of $\mathcal{A}$-linear simplicial presheaves and $\mathcal{A}$-linear natural transformations is denoted by $s\mathbf{Mod}_R^{\mathcal{A}}$.

Lemma 8.44 says that the category $s\mathbf{Mod}_R^{\mathcal{A}}$ of $\mathcal{A}$-linear simplicial presheaves, together with sectionwise equivalence and sectionwise fibrations, satisfies the conditions for a proper closed simplicial model category. This is the projective model structure for the category of $\mathcal{A}$-linear simplicial presheaves. The cofibrations for this model structure are the projective cofibrations.

The class of projective cofibrations includes all map $\phi^* R(A) \to \phi^* R(B)$ which are induced by projective cofibrations $A \to B$ of simplicial presheaves, where ϕ^* is the left adjoint of the forgetful functor

$$\phi_* : s\mathbf{Mod}_R^{\mathcal{A}} \to s\mathbf{Pre}_R$$

which is induced by precomposition with the functor $\phi : \mathcal{C} \to \mathcal{A}$. More generally, there is a Quillen adjunction

$$\phi^* \cdot R : s\mathbf{Pre}(\mathcal{C}) \leftrightarrows s\mathbf{Mod}_R^{\mathcal{A}} : u \cdot \phi_*,$$

between the respective projective model structures, where R is the free R-module functor and the forgetful functor u is its right adjoint.

Suppose that X is an $\mathcal{A}$-linear simplicial presheaf and that T is a projective cofibrant pointed simplicial presheaf. Write

$$X \otimes T = X \otimes_R \phi^* R_\bullet(T),$$

where $R_\bullet(T) = R(T)/R(*)$ is the reduced free simplicial R-module associated to T. The functor $X \mapsto X \otimes T$ preserves weak equivalences for the projective model structure on $\mathcal{A}$-linear simplicial presheaves.

A T-spectrum X in the category $s\mathbf{Mod}_R^{\mathcal{A}}$ consists of $\mathcal{A}$-linear simplicial presheaves X^n, $n \geq 0$, together with bonding maps

$$T \otimes X^n \to X^{n+1}.$$

A morphism $X \to Y$ of T-spectra is the obvious thing. It consists of $\mathcal{A}$-linear maps $X^n \to Y^n$ which respect the bonding maps. Write $\mathbf{Spt}_T(s\mathbf{Mod}_R^{\mathcal{A}})$ for the resulting category of T-spectra in $\mathcal{A}$-linear simplicial presheaves.

The category $\mathbf{Spt}_T(s\mathbf{Mod}_R^{\mathcal{A}})$ of T-spectra has a *sectionwise strict model structure*, for which the weak equivalences and fibrations are defined levelwise in the projective model structure for simplicial $\mathcal{A}$-modules, by the usual method (Lemma 8.15, Proposition 10.15). The category $\mathbf{Spt}_T(\mathcal{C})$ also has a sectionwise strict model structure, with fibrations and weak equivalences defined levelwise and sectionwise. The composite functor $\phi^* \cdot R$ and its right adjoint $u \cdot \phi_*$ determine a Quillen adjunction

$$\phi^* \cdot R : \mathbf{Spt}_T(\mathcal{C}) \leftrightarrows \mathbf{Spt}_T(s\mathbf{Mod}_R^{\mathcal{A}}) : u \cdot \phi_*$$

between the respective sectionwise strict model structures.

Now suppose that $\mathcal{F}$ is a set of cofibrations $A \to B$ for the sectionwise strict model structure on T-spectra. Suppose further that

C1 The T-spectrum A is cofibrant for all maps $A \to B$ in $\mathcal{F}$.
C2 If the map $A \to B$ is in $\mathcal{F}$, then so are all maps

$$(A \wedge \Delta_+^n) \cup (B \wedge \partial \Delta_+^n) \to B \wedge \Delta_+^n.$$

10.7 T-Complexes

Say that an object Z of $\mathbf{Spt}_T(s\mathbf{Mod}_R^{\mathcal{A}})$ is $\mathcal{F}$-injective if the map $Z \to 0$ has the right lifting property with respect to the maps $\phi^*R(A) \to \phi^*R(B)$ associated to all members $A \to B$ of $\mathcal{F}$. Say that a map $E \to F$ of T-spectrum objects is an $\mathcal{F}$-equivalence if some projective cofibrant replacement $E_c \to F_c$ induces a weak equivalence

$$\mathbf{hom}(F_c, Z) \to \mathbf{hom}(E_c, Z)$$

for all $\mathcal{F}$-injective objects Z. The $\mathcal{F}$-fibrations are defined by the right lifting property with respect to maps which are projective cofibrations and $\mathcal{F}$-equivalences.

Then, by analogy with Theorem 8.48, we have the following:

Theorem 10.96 *Suppose that $\phi : \mathcal{C} \to \mathcal{A}$ is a functor, where $\mathcal{C}$ is a small Grothendieck site, $\mathcal{A}$ is a small R-linear category, and the functor ϕ is the identity on objects. Suppose that $\mathcal{F}$ is a set of cofibrations for the sectionwise strict structure on T-spectra which satisfies conditions* **C1** *and* **C2** *above.*

Then the category $\mathbf{Spt}_T(s\mathbf{Mod}_R^{\mathcal{A}})$ of T-spectrum objects in $\mathcal{A}$-linear simplicial presheaves, together with the classes of sectionwise strict cofibrations, $\mathcal{F}$-equivalences and $\mathcal{F}$-fibrations, satisfies the axioms for a left proper closed simplicial model category. This model structure is cofibrantly generated.

The model structure of Theorem 10.96 is the $\mathcal{F}$-*local model structure* on the category of T-spectrum objects in $\mathcal{A}$-linear simplicial presheaves.

Proof One constructs a functor $X \mapsto L_{\mathcal{F}}X$ for T-spectrum objects, by analogy with the construction in Sect. 8.6. The T-spectrum analogue of Lemma 8.46 is automatic.

One proves a bounded monomorphism property for the strict sectionwise model structure on T-spectrum objects by using the method of Lemma 10.18. The bounded monomorphism property for $\mathcal{F}$-equivalences is a formal consequence, as in Lemma 10.19. The desired result is a formal consequence of this bounded monomorphism property—see the proof of Theorem 8.48.

Example 10.97 Suppose that k is a perfect field, $\mathcal{C}$ is the smooth Nisnevich site $(Sm|_k)_{Nis}$ over k, and that

$$T = S^1 \wedge \mathbb{G}_m,$$

as in Examples 8.49 and 10.95. Recall that the graph functor $\gamma : Sm|_k \to Cor_k$ is the identity on objects.

Suppose that the set of projective cofibrations $\mathcal{F}$ contains the set $(S_T \wedge A)[-n] \to (S_T \wedge B)[-n]$ associated to the set $A \to B$ of generators for the motivic model structure on the category $s\mathbf{PST}(k)$ of simplicial presheaves with transfers over k, together with the set of projective cofibrant replacements of the maps

$$(S_T \wedge T)[-n-1] \to S_T[-n], \ n \geq 0,$$

where S_T is the sphere spectrum for T-spectra. Suppose that these two sets of maps generate $\mathcal{F}$ in the sense that $\mathcal{F}$ is the smallest set of cofibration containing these two sets, such that the closure property **C2** is satisfied.

The model structure for the category $\mathbf{Spt}_T(s\mathbf{PST}(k))$ which is given by Theorem 10.96 in this case is the motivic stable model structure for spectrum objects in simplicial presheaves with transfers. The corresponding homotopy category is Voevodsky's *big category* $\mathbf{DM}_k$ *of motives over* k. See [89].

An object Z of this category of T-spectra is $\mathcal{F}$-fibrant if and only if the underlying T-complex $\gamma_*(Z)$ is motivic stable fibrant. This means that the constituent simplicial presheaves with transfers Z^n are motivic fibrant, and the maps $Z^n \to \Omega_T Z^{n+1}$ are motivic (hence sectionwise) weak equivalences.

Suppose that X is a T-spectrum object in simplicial presheaves with transfers. The compactness of T and motivic descent together imply that the T-spectrum object $Q_T X$ is stable $\mathcal{F}$-fibrant and that the canonical map $X \to Q_T X$ is an $\mathcal{F}$-equivalence. It follows that a map $X \to Y$ of T-spectra in simplicial presheaves with transfers is an $\mathcal{F}$-equivalence if and only if the underlying map $\gamma_*(X) \to \gamma_*(Y)$ of T-complexes is a motivic stable equivalence.

Proposition 10.94 then implies that the map $X \to Y$ is an $\mathcal{F}$-equivalence if and only if the induced map $u\gamma_*(X) \to u\gamma_*(Y)$ is a motivic stable equivalence of T-spectra.

Chapter 11
Symmetric T-spectra

The étale topological variants of the algebraic K-theory spectrum that were developed in the 1980s suffered from two basic technical afflictions:

1) An explicit functorial model for the full algebraic K-theory spectrum did not quite exist.
2) The theory of products for the K-theory spectrum was difficult to construct. More generally, there was no good theory of products for the combinatorial machinery underlying the K-theory spectrum.

At the time, one constructed the K-theory spectrum by extracting a spectrum object in pointed simplicial sets from a pseudo-functorial symmetric monoidal category arising from Quillen's Q-construction. Thus, to construct the K-theory presheaf of spectra, particularly if one wanted information about the groups $K_0(S)$ for schemes S at the section level, one had to invoke a homotopy coherence machine, and such machines tend to complicate geometric section-level calculations. This was the approach taken in [56]—it worked, but it was awkward. The problem was removed finally [68], with the realization that Waldhausen's construction produces a K-theory presheaf of (symmetric) spectra on big enough sites, provided that one puts in a suitable category of big site vector bundles.

One cared about this, because the stable homotopy theory of presheaves of spectra was available early [51] (see also Sect. 10.1), and it was clear by the late 1980s that fundamental problems in algebraic K-theory, such as the Lichtenbaum–Quillen conjecture, could be reformulated in terms of maps of presheaves of spectra as descent questions.

The definition and analysis of multiplication by the Bott element, as in Thomason's work on étale descent for Bott periodic K-theory [99], requires an adequate theory of products for both algebraic K-theory and its topological variants. One early method for describing such products was based on a theory of presheaves of bispectra. This technique was adequate for the tasks at hand when it was first introduced, but it was again awkward.

The theory of bispectra, as it appeared in [56], was a formalization of Adams' handicrafted smash product construction from classical stable homotopy theory. Handicrafted smash products were long known to be inadequate for serious work with higher order products, and the search for an alternative, homotopically well-behaved smash product construction for spectra took many years to complete. On the combinatorial/algebraic side of the subject, a model for the full stable category with a symmetric monoidal smash product finally appeared in the breakthrough work of Hovey, Shipley and Smith on symmetric spectra—their paper [46] was published in 2000. A symmetric spectrum X is a spectrum object in pointed simplicial sets (as in the Bousfield–Friedlander paper [13]), which is also equipped with symmetric group actions $\Sigma_n \times X^n \to X^n$ on the level spaces, which actions are compatible with twists of smash factors in the simplicial spheres

$$S^p = (S^1)^{\wedge p}$$

under iterated bonding maps. The smash product $X \wedge_\Sigma Y$ of symmetric spectra X and Y is defined by a coequalizer that is modelled on the tensor product of graded modules—the underlying ring in this case is the sphere spectrum. The main results of [46] include the existence of a stable model structure on the category of symmetric spectra which is Quillen equivalent to the stable model structure of Bousfield and Friedlander, and that the smash product functor

$$(X, Y) \mapsto X \wedge_\Sigma Y$$

is monoidal for that structure.

The theory of symmetric spectra is a combinatorial model for the ordinary stable category, with a functorial smash product. As a rule, combinatorial homotopy theory constructions can be promoted in some form to the topos theoretic level, and it was obvious when the Hovey–Shipley–Smith paper was first circulated that such a transport of structure should be done for symmetric spectra.

A local homotopy theory version of the Hovey–Shipley–Smith results for presheaves of symmetric spectra appeared in [58]. The main features of the results of [46] survive: there is a stable model structure on the category of presheaves of symmetric spectra (over any small Grothendieck site) which is Quillen equivalent to the stable model structure for presheaves of spectra, and the Hovey–Shipley–Smith smash product defines a monoidal smash product for presheaves of symmetric spectra.

The local form of the theory of symmetric spectra is now a fundamental device. A smash product pairing $X \wedge_\Sigma Y \to Z$ of presheaves of symmetric spectra induces a smash product pairing for any of the Grothendieck topological variants (i.e. stable fibrant models) for X, Y and Z. In particular, the symmetric spectrum ring structure on the algebraic K-theory presheaf of spectra induces ring spectrum structures on all topologized versions, such as étale K-theory. Presheaves of symmetric spectra also appear in a foundational role in the theory of derived schemes and topological modular forms—see the Séminaire Bourbaki talk of Goerss [34] for an introduction to this subject.

One can similarly define symmetric spectrum objects in presheaves of simplicial modules [61] (see also Sect. 11.7), and a corresponding stable model structure for such objects whose associated homotopy theory is the full derived category of R-modules. This category of symmetric spectrum objects has a tensor product which is defined by analogy with the symmetric spectrum smash product, and which is derived monoidal for that structure. We thus have a homotopy theoretic tensor product for the derived category that extends the tensor products for simplicial abelian groups and chain complexes in a natural way.

These are, so far, rather "traditional" uses of symmetric spectra in a presheaf theoretic setting, in which the parameter object for the spectrum objects is the simplicial circle S^1. As we see in Chap. 10, there are also stable homotopy categories for T-spectra for alternate choices of parameter objects T, in f-local settings where f is a cofibration of simplicial presheaves which is formally inverted.

The first (and motivating) example is the motivic stable category of Voevodsky [102] that corresponds to the theory of presheaves of T-spectra on the smooth Nisnevich site of a scheme, in which the affine line is formally contracted to a point. In this case, the parameter object T is the projective line $\mathbb{P}^1$, or equivalently the smash product $S^1 \wedge \mathbb{G}_m$.

Symmetric T-spectra are defined by analogy with the Hovey–Shipley–Smith definition of symmetric spectra. A symmetric T-spectrum X is a T-spectrum which carries symmetric group actions $\Sigma_n \times X^n \to X^n$ on the pointed simplicial presheaves in the various levels, which actions are compatible with twists in the smash products $T^{\wedge p}$ under iterated bonding maps. The smash product $X \wedge_\Sigma Y$ of two symmetric T-spectra is defined as a suitable coequalizer, as is the smash product for symmetric spectra. There is a motivic stable model structure for the category of such things, which is Quillen equivalent to the motivic stable category of T-spectra, and such that the smash $(X, Y) \mapsto X \wedge_\Sigma Y$ is monoidal for that model structure.

These statements about symmetric T-spectra and the motivic stable category were proved in [57]. Calculations related to product structures in motivic stable homotopy theory now usually involve this theory. See, for example, [90], [21] and [84].

The technical issues which appear in the course of producing the theory of motivic symmetric spectra have already appeared in the description of the motivic stable category and its generalizations in Chap. 10. Such generalizations of the motivic stable constructions involve categories of T-spectra and symmetric T-spectra in an f-local setting on a potentially arbitrary site, such that the following conditions are satisfied:

A1 The parameter object T is compact for the f-local model structure in a suitable sense.
A2 The f-local model structure satisfies an inductive colimit descent condition.
A3 The shuffle permutation $c_{1,2}$ acts trivially on the 3-fold wedge $T^{\wedge 3}$ in the f-local model structure for simplicial presheaves.

These statements are colloquial versions of conditions having the same names, which appear and are discussed at some length in Chap. 10. The stable f-local model structure on the category of T-spectra is constructed by formally inverting a set of maps in a basic strict model structure for T-spectra. The conditions **A1**–**A3** on the parameter object T and the underlying f-local structure are necessary for the resulting stable f-local model structure to behave something like ordinary stable homotopy theory. The extra assumption that the parameter object is a topological suspension $S^1 \wedge T$ of a compact object T leads to a further tightening of that analogy.

These conditions hold in motivic model structures by geometric arguments involving Nisnevich descent for **A1** and **A2**, and the structure of the special linear group $Sl_3(\mathbb{Z})$ for **A3**. The parameter object $\mathbb{P}^1 \simeq S^1 \wedge \mathbb{G}_m$ is a topological suspension in that category. The familiar basic features of the motivic stable category are consequences of these phenomena.

These same conditions enable the creation of a well-behaved f-local stable model structure on the category of symmetric T-spectra on a site $\mathcal{C}$. This chapter presents the formal homotopy theory which is associated with this construction.

The f-local stable model structure on the category of symmetric T-spectra is constructed by localizing an underlying injective structure (Proposition 11.10) at the same set of maps that one uses to produce the f-local stable structure on T-spectra. See Theorem 11.13 and Example 11.15.

Provided that T is compact (assumption **A1**), the forgetful functor U from symmetric T-spectra to T-spectra reflects stable f-equivalences, in the way that one expects from symmetric spectra and motivic symmetric spectra. This is proved in Theorem 11.19.

If the parameter object is a suspension $S^1 \wedge T$, and if we also assume the compactness assumption **A1** for T and the descent assumption **A2** (so that, in particular, the circle S^1 is compact), then we have a coincidence of fibre and cofibre sequences which is displayed in Corollary 11.22. It follows that the f-local stable model structure on symmetric $(S^1 \wedge T)$-spectra is proper (Theorem 11.23).

Finally, if the compactness condition **A1** and the cycle triviality condition **A3** hold, then the forgetful functor U and its left adjoint V define a Quillen equivalence between the f-local stable model structures for T-spectra and symmetric T-spectra. This statement is Theorem 11.35.

Section 11.6 contains a discussion of the properties of the smash product construction within the f-local stable structure on symmetric T-spectra. Its monoidal aspects are displayed in Theorem 11.44 and in its corollaries.

The f-local stable model structure on symmetric T-spectra may not itself be monoidal as the term is normally defined. This stable model structure is constructed in a way which differs slightly from the stable model structures of [46] and [57], but it is monoidal in a derived sense.

The section finishes with a comparison of the naive smash product $X \wedge_n Y$ of T-spectra (which arises from the T-bispectrum object $X^p \wedge Y^q$) and the stable f-fibrant model $(VX \wedge_\Sigma VY)_s$ of the smash $VX \wedge_\Sigma VY$ of their associated symmetric spectra: Theorem 11.48 says that these objects coincide in the f-local stable category, in the presence of assumptions **A1** and **A3**, and if X or Y is cofibrant.

Section 11.7 presents an introduction to the f-local stable homotopy theory of symmetric T-complexes. A symmetric T-complex X it a T-spectrum object on presheaves of simplicial R-modules, in other words a T-complex in the language of Chap. 10, which is equipped with linear symmetric group actions $\Sigma_n \times X^n \to X^n$

Symmetric T-spectra

which respect the twists in the tensor products

$$T^{\otimes n} = R_\bullet(T^{\wedge n})$$

under iterated bonding maps. As in Chap. 10, the object $R_\bullet(X)$ is the free reduced simplicial R-module which is associated to a pointed simplicial presheaf X.

The motivic cohomology spectrum $\mathbf{H}_\mathbb{Z}$, which consists of the free presheaves with transfers

$$\mathbb{Z}_{tr}((\mathbb{G}_m \wedge S^1)^{\wedge n})$$

in the various levels (see Example 10.95), defines a symmetric $(S^1 \wedge \mathbb{G}_m)$-complex in presheaves of simplicial abelian groups on the smooth Nisnevich site of a field.

More generally, the Eilenberg–Mac Lane object $H(R)$, which is defined in levels by

$$H(R)^n = R_\bullet(T^{\wedge n}),$$

is a symmetric T-complex. This is the sphere spectrum object for the category of symmetric T-complexes over the ring object R.

Once again, the f-local stable model structure for such objects is produced in gross generality by a localization argument applied to a base injective model structure—this is Theorem 11.54, as interpreted in Example 11.57.

As in the nonabelian case, the forgetful functor to T-complexes reflects stable f-equivalences if T is compact (Theorem 11.60), and the categories of symmetric T-complexes and T-complexes have Quillen equivalent f-stable model structures if the parameter object T is compact and cycle trivial (Theorem 11.61). It is proved in Theorem 11.66 that the tensor product for symmetric T-complexes, which is defined by analogy with the smash product for symmetric T-spectra, is monoidal in a derived sense.

The results of this chapter and of Chap. 10 are quite general, and specialize to results about T-spectra and symmetric T-spectra in the unlocalized setting where one does not formally invert a cofibration, as well as to results about the f-local stable model structure for presheaves of symmetric spectra and presheaves of symmetric spectra. A presheaf of symmetric spectra is a symmetric S^1-spectrum in the language developed here.

In the unlocalized setting, the compactness of the parameter object T in the assumption **A1** can still be an issue, but not if T is a finite pointed simplicial set, such as the simplicial circle S^1. The injective model structure on simplicial presheaves automatically satisfies the descent condition **A2**. The cycle triviality condition **A3** is again an issue for arbitrary parameter objects T, but it is a well known consequence of a degree calculation for the simplicial circle S^1 and for any of the simplicial spheres $S^n = (S^1)^{\wedge n}$. We recover, in particular, the results of [58] about the local stable homotopy theory of presheaves of symmetric spectra.

The only assumption that one needs for presheaves of symmetric spectra in the f-local setting is the descent condition **A2** for the f-local model structure on simplicial presheaves. The compactness for the circle S^1 follows immediately. The descent condition can be serious business: it is proved for motivic model structures by using a Nisnevich descent argument.

11.1 Symmetric Spaces

We start with an arbitrary small Grothendieck site $\mathcal{C}$, and we choose a regular cardinal α such that $\alpha > |\operatorname{Mor}(\mathcal{C})|$.

A *symmetric space* X consists of pointed simplicial presheaves X^n, $n \geq 0$, on the site $\mathcal{C}$ with symmetric group actions

$$\Sigma_n \times X^n \to X^n.$$

A *morphism* $f : X \to Y$ of symmetric spaces consists of pointed simplicial presheaf morphisms $X^n \to Y^n$, $n \geq 0$, which respect the symmetric group actions in all levels. Following [46], the category of symmetric spaces is denoted by $s\mathbf{Pre}(\mathcal{C})_*^{\Sigma}$. See also [58].

The pointed simplicial presheaf T is a fixed choice of parameter object, which we assume to be α-bounded.

Example 11.1 The sequence of pointed simplicial presheaves

$$S^0,\ T,\ T \wedge T,\ T^{\wedge 3},\ \dots$$

forms a symmetric space, which is denoted by S_T. This is the symmetric space underlying the sphere T-spectrum.

If X is a symmetric space and K is a pointed simplicial presheaf, then the object $X \wedge K$ which consists of the pointed simplicial presheaves $X^n \wedge K$ is a symmetric space. In particular, the suspension object

$$\Sigma_T^\infty K = S_T \wedge K,$$

is the symmetric space which consists of the objects $T^{\wedge n} \wedge K$.

Example 11.2 Write Γ for the category of finite pointed sets and pointed functions between them. A Γ-*space* is a functor $A : \Gamma \to s\mathbf{Pre}(\mathcal{C})_*$ which is defined on the category of finite pointed sets and takes values in pointed simplicial presheaves. The list of spaces

$$dA(S^0),\ dA(S^1),\ dA(S^2),\ \dots$$

associated to a Γ-space A forms a symmetric space, where dX is the diagonal of a bisimplicial (or multisimplicial) set X, and Σ_n acts on $S^n = S^1 \wedge \cdots \wedge S^1$ by permuting smash factors.

Any pair of finite pointed sets K, L determines a canonical map

$$K \wedge A(L) \to A(K \wedge L),$$

11.1 Symmetric Spaces

and it follows that there are canonical maps

$$S^k \wedge A(S^n) \to A(S^{k+n}),$$

which induce pointed simplicial presheaf maps

$$S^k \wedge dA(S^n) \to dA(S^{k+n}). \tag{11.1}$$

In particular, the list of spaces $dA(S^n)$, $n \geq 0$, has the structure of an S^1-spectrum.

More is true: the map (11.1) is $(\Sigma_k \times \Sigma_n)$-equivariant. The list of spaces $dA(S^n)$, $n \geq 0$ therefore forms a symmetric S^1-spectrum, in the sense described below. One writes $A(S)$ for either the S^1-spectrum or the symmetric S^1-spectrum.

There is a Γ-space $\Phi(S, Y)$ that is associated to a spectrum Y, which is defined by

$$\Phi(S, Y)(\underline{n}) = \mathbf{hom}(S^{\times n}, Y).$$

The functor $\Phi(S, \ )$ is right adjoint to the functor $A \mapsto A(S)$, and for suitable stable model structures these two functors determine an equivalence of the stable homotopy category of connective S^1-spectra. See also [13].

The stable model structure for Γ-spaces is a special case of the stable model structure for presheaves of Γ-spaces, with an associated homotopy category which is equivalent to the stable category of connective presheaves of S^1-spectra [7].

Some basic examples of presheaves of S^1-spectra arise naturally from presheaves of Γ-spaces, in the way described here. These include the algebraic K-theory presheaf of spectra [68].

There is a functorial tensor product construction for symmetric spaces. Given symmetric spaces X and Y, their *tensor product* $X \otimes Y$ is specified in degree n by

$$(X \otimes Y)^n = \bigvee_{r+s=n} \Sigma_n \otimes_{\Sigma_r \times \Sigma_s} (X^r \wedge Y^s).$$

Here,

$$\Sigma_n \otimes_{\Sigma_r \times \Sigma_s} (X^r \wedge Y^s)$$

has the Σ_n-structure which is induced from the $(\Sigma_r \times \Sigma_s)$-structure on $X^r \wedge Y^s$ along the canonical inclusion

$$i : \Sigma_r \times \Sigma_s \subset \Sigma_{r+s} = \Sigma_n$$

This means that Σ_n-equivariant maps

$$\Sigma_n \otimes_{\Sigma_r \times \Sigma_s} (X^r \wedge Y^s) \to W$$

can be identified with $(\Sigma_r \times \Sigma_s)$-equivariant maps

$$X^r \wedge Y^s \to i_*W,$$

where i_*W denotes restriction of the Σ_n-structure of Z to a $(\Sigma_r \times \Sigma_s)$-structure along the inclusion i.

A map of symmetric spaces

$$X \otimes Y \to Z$$

therefore consists of $(\Sigma_r \times \Sigma_s)$-equivariant maps

$$X^r \wedge Y^s \to i_*Z^{r+s} = Z^{r+s}$$

for all $r, s \geq 0$. An example is the map

$$\otimes : S_T \otimes S_T \to S_T$$

which is defined by the canonical isomorphisms

$$T^r \wedge T^s \xrightarrow{\cong} T^{r+s}$$

Write $c_{r,s} \in \Sigma_{r+s}$ for the shuffle which is defined by

$$c_{r,s}(i) = \begin{cases} s+i & i \leq r \\ i-r & i > r \end{cases} \tag{11.2}$$

The *twist automorphism*

$$\tau : X \otimes Y \to Y \otimes X \tag{11.3}$$

is uniquely determined by the composites

$$X^r \wedge Y^s \xrightarrow{\tau} Y^s \wedge X^r \to (Y \otimes X)^{s+r} \xrightarrow{c_{s,r}} (Y \otimes X)^{r+s}.$$

We multiply by the shuffle $c_{s,r}$ to make this composite equivariant for the inclusion $\Sigma_r \times \Sigma_s \subset \Sigma_{r+s}$. One checks that the composite

$$X \otimes Y \xrightarrow{\tau} Y \otimes X \xrightarrow{\tau} X \otimes Y$$

is the identity.

The tensor product construction $(X, Y) \mapsto X \otimes Y$ is symmetric monoidal. The map

$$\otimes : S_T \otimes S_T \to S_T$$

gives the sphere object S_T the structure of an abelian monoid in the category of symmetric spaces.

A *symmetric T-spectrum* X is a symmetric space with the structure

$$m_X : S_T \otimes X \to X$$

11.1 Symmetric Spaces

of a module over S_T. Equivalently, the symmetric space X comes equipped with *bonding maps*

$$\sigma_{1,s} : T \wedge X^s \to X^{1+s}$$

which are defined by the module structure, such that all composite bonding maps

$$T^r \wedge X^s \to X^{r+s}$$

are equivariant for the inclusion $\Sigma^r \times \Sigma^s \subset \Sigma^{r+s}$. The symmetric T-spectra, with structure-preserving maps between them, form a category which we denote by $\mathbf{Spt}_T^\Sigma(\mathcal{C})$.

If X is a symmetric T-spectrum and A is a pointed simplicial presheaf then the levelwise smash product $X \wedge A$, which is defined by setting

$$(X \wedge A)^n = X^n \wedge A,$$

is a symmetric space that also has the structure of a symmetric T-spectrum.

The category of symmetric T-spectra has a symmetric monoidal smash product. Given symmetric T-spectra X, Y, the *smash product* $X \wedge_\Sigma Y$ is defined by the coequalizer

$$S_T \otimes X \otimes Y \rightrightarrows X \otimes Y \to X \wedge_\Sigma Y.$$

in symmetric spaces, where the arrows in the picture

$$S_T \otimes X \otimes Y \rightrightarrows X \otimes Y$$

are $m_X \otimes Y$ and the composite

$$S_T \otimes X \otimes Y \xrightarrow{\tau \otimes Y} X \otimes S_T \otimes Y \xrightarrow{X \otimes m_Y} X \otimes Y.$$

Remark 11.3 The tensor product functor $A \otimes B$ preserves colimits for symmetric spaces A and B, and it follows that the smash product $X \wedge_\Sigma Y$ for symmetric T-spectra preserves colimits in both X and Y.

If X is a symmetric space, then the tensor product $S_T \otimes X$ has the structure of a symmetric T-spectrum. The object $S_T \otimes X$ is the *free symmetric T-spectrum* on X, and there are natural bijections

$$\hom_{\mathbf{Spt}_T^\Sigma(\mathcal{C})} (S_T \otimes X, Y) \cong \hom_{s\mathbf{Pre}(\mathcal{C})_*^\Sigma} (X, Y).$$

If K is a pointed simplicial presheaf and $n \geq 0$ there is a symmetric space $G_n K$ with

$$(G_n K)^r = \begin{cases} * & r \neq n \\ \Sigma_n \otimes K = \bigvee_{\Sigma_n} K & r = n. \end{cases}$$

There is a natural bijection

$$\hom_{s\mathbf{Pre}(\mathcal{C})_*^\Sigma} (G_n K, W) \cong \hom_{s\mathbf{Pre}(\mathcal{C})_*} (K, W^n).$$

It follows that if we define a symmetric T-spectrum $F_n K$ by setting

$$F_n K = S_T \otimes G_n K,$$

then there is a natural bijection

$$\hom_{\mathbf{Spt}_T^\Sigma(\mathcal{C})}(F_n K, Z) \cong \hom_{s\mathbf{Pre}(\mathcal{C})_*}(K, Z^n).$$

This holds for all $n \geq 0$, pointed simplicial presheaves K and symmetric T-spectra Z.

There is a natural identification

$$F_0 K = S_T \wedge K = S_T \wedge K,$$

for pointed simplicial presheaves K.

There is a functor

$$U : \mathbf{Spt}_T^\Sigma(\mathcal{C}) \to \mathbf{Spt}_T(\mathcal{C})$$

which forgets the symmetric group actions. The functor U has a left adjoint

$$V : \mathbf{Spt}_T(\mathcal{C}) \to \mathbf{Spt}_T^\Sigma(\mathcal{C}),$$

which is constructed inductively, by using the layer filtration of (10.11).

In effect, for a shifted suspension T-spectrum $S_T \wedge K[-n]$, if the functor V is left adjoint to the forgetful functor U, then it must be the case that there are isomorphisms

$$\hom_{\mathbf{Spt}_T^\Sigma(\mathcal{C})}(V(S_T \wedge K[-n]), Z) \cong \hom_{\mathbf{Spt}_T(\mathcal{C})}(S_T \wedge K[-n], U(Z))$$

$$\cong \hom_{s\mathbf{Pre}(\mathcal{C})_*}(K, Z^n),$$

so there are natural isomorphisms

$$V(S_T \wedge K[-n]) \cong F_n K.$$

It follows that the symmetric T-spectra $V(L_n X)$ can be inductively specified on the layers $L_n X$ of a T-spectrum X by the pushouts

$$\begin{array}{ccc} F_{n+1}(T \wedge X^n) & \longrightarrow & V(L_n X) \\ \downarrow & & \downarrow \\ F_{n+1}(X^{n+1}) & \longrightarrow & V(L_{n+1} X) \end{array}$$

Write

$$VX = \varinjlim_n V(L_n X).$$

There is an identification

$$(F_n K)^n = \Sigma_n \otimes K,$$

and the map $e : K \to \Sigma_n \otimes K$ given by the inclusion of the summand corresponding to the identity $e \in \Sigma_n$ induces a natural map

$$\eta : S_T \wedge K[-n] \to U(F_n K). \tag{11.4}$$

The map η satisfies a universal property: given a map $f : K \to Z^n$ where Z is a symmetric T-spectrum, there is a unique map $f_* : F_n K \to Z$ of symmetric T-spectra such that the diagram

$$\begin{array}{ccc}
\Sigma^\infty K[-n] & \xrightarrow{\eta} & U(F_n K) \\
& \searrow{f_*} & \downarrow{Uf_*} \\
& & U(Z)
\end{array}$$

commutes.

The canonical maps $\eta : L_n X \to UV(L_n X)$ are inductively defined by the maps (11.4) and the diagrams (10.12), and the unit map

$$\eta : X \to UV(X)$$

is defined on the filtered colimit of the layers.

11.2 First Model Structures

A map $X \to Y$ of symmetric T-spectra is said to be a *level weak equivalence* if all component maps $X^n \to Y^n$ are local weak equivalences of (pointed) simplicial presheaves.

A *level cofibration* is a map $A \to B$ of symmetric T-spectra such that all maps $A^n \to B^n$ are cofibrations of pointed simplicial presheaves.

Say that a map $p : X \to Y$ of symmetric T-spectra is a *projective fibration* if all maps $p : X^n \to Y^n$ are injective fibrations of pointed simplicial presheaves. A map $i : A \to B$ is said to be a *projective cofibration* if it has the left lifting property with respect to all maps which are projective fibrations and level weak equivalences.

If the map $A \to B$ is a cofibration of pointed simplicial presheaves, then the induced map $F_n A \to F_n B$ is a projective cofibration of symmetric T-spectra for all n. It is an exercise to show that the map $F_n A \to F_n B$ is a level cofibration.

The induced map $F_n A \to F_n B$ is a level weak equivalence if the map $A \to B$ is a local weak equivalence. It follows that the functors F_n take trivial cofibrations to trivial projective cofibrations.

Lemma 11.4 *The category* $\mathbf{Spt}_T^\Sigma(\mathcal{C})$, *with the projective fibrations, projective cofibrations and level weak equivalences, satisfies the conditions for a proper closed simplicial model category. This model structure is cofibrantly generated. Every projective cofibration is a level cofibration.*

Proof A map $p : X \to Y$ of symmetric T-spectra is a projective fibration if and only if it has the right lifting property with respect to all maps $F_n A \to F_n B$ which are induced by α-bounded trivial cofibrations $A \to B$ of pointed simplicial presheaves. The map p is a projective trivial fibration if and only if it has the right lifting property with respect to all maps $F_n C \to F_n D$ which are induced by α-bounded cofibrations $C \to D$ of pointed simplicial presheaves. It follows that any map $f : X \to Y$ of symmetric T-spectra has factorizations

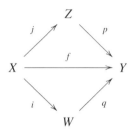

such that p is a projective fibration and j is a projective trivial cofibration which has the left lifting property with respect to all projective fibrations, and q is a projective trivial fibration and i is a projective cofibration. We have therefore verified the factorization axiom **CM5**.

It follows that if $i : A \to B$ is a projective trivial cofibration, then i has left lifting property with respect to all projective fibrations, by the usual argument. This proves the lifting axiom **CM4**.

The remaining closed model axioms are easily verified.

The function complex construction $\mathbf{hom}(X, Y)$ is the obvious one, with n-simplices given by the maps

$$X \wedge \Delta_+^n \to Y.$$

The axiom **SM7** follows from the corresponding axiom for the the injective model structure on pointed simplicial presheaves—see also Proposition 10.15 and Lemma 10.16.

Every projective cofibration is in the saturation of the maps $F_n A \to F_n B$ which are induced by cofibrations $A \to B$ of pointed simplicial presheaves. The maps $F_n A \to F_n B$ are level cofibrations, and so every projective cofibration is a level cofibration.

Properness is an easy consequence of this last observation, together with properness of the injective model structure for pointed simplicial presheaves.

The maps $F_n A \to F_n B$ which are induced by α-bounded cofibrations (respectively α-bounded trivial cofibrations) $A \to B$ generate the projective cofibrations (respectively trivial projective cofibrations) of $\mathbf{Spt}_T^\Sigma(\mathcal{C})$.

11.2 First Model Structures

The model structure of Lemma 11.4 is the *projective model structure* for the category of T-spectra.

Corollary 11.5 *The functor V takes cofibrations (respectively strict trivial cofibrations) of T-spectra to projective cofibrations (respectively trivial projective cofibrations) of symmetric T-spectra.*

The various statements of Corollary 11.5 are consequences of the proof of Lemma 11.4. Compare with the proof Lemma 3 of [58].

There is also a model structure, called the *injective structure*, on the category $\mathbf{Spt}_T^\Sigma(\mathcal{C})$ of symmetric T-spectra, for which the cofibrations are the level cofibrations and the weak equivalences are the level weak equivalences. This result is proved in Proposition 11.10 below, after some preliminary results. We begin by establishing the bounded monomorphism property for level cofibrations.

Say that a symmetric T-spectrum X is α-bounded if all component simplicial presheaves X^n are α-bounded.

Lemma 11.6 *Suppose that α is a regular cardinal such that $\alpha > |\mathrm{Mor}(\mathcal{C})|$. Suppose given level cofibrations*

such that the vertical map i is a level equivalence and the object A is α-bounded. Then there is an α-bounded subobject $B \subset Y$ with $A \subset B$ such that the induced map $B \cap X \to B$ is a level equivalence.

The proof of this result is essentially the same as that of the corresponding result for simplicial presheaves, which is Lemma 5.2.

Given a symmetric T-spectrum X, one can use Kan's Ex^∞-construction (see Sect. 4.1) to construct a symmetric T-spectrum $\mathrm{Ex}^\infty X$ with $(\mathrm{Ex}^\infty X)^n = \mathrm{Ex}^\infty(X^n)$, together with a natural level weak equivalence $X \to \mathrm{Ex}^\infty X$. The maps $X^n \to \mathrm{Ex}^\infty(X^n)$ are sectionwise equivalences.

Proof There is a natural commutative diagram in symmetric T-spectra

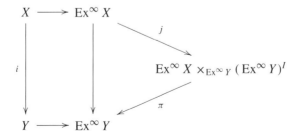

such that the map j is a sectionwise cofibration and weak equivalence and π is a sectionwise fibration, in all levels. Form the pullback of this factorization of $\mathrm{Ex}^\infty X \to \mathrm{Ex}^\infty Y$ along the map $Y \to \mathrm{Ex}^\infty Y$ to find a factorization

$$\begin{array}{ccc} X & \xrightarrow{j_Y} & Z_Y \\ {\scriptstyle i}\downarrow & \swarrow{\scriptstyle \pi_Y} & \\ Y & & \end{array}$$

such that j_Y is a sectionwise cofibration and a weak equivalence in all levels and π_Y is sectionwise fibration in all levels. This construction respects all filtered colimits of maps i, so that

$$Z_Y = \varinjlim_{B \subset Y} Z_B,$$

where $\pi_B : Z_B \to B$ is the corresponding replacement for the map $B \cap X \to B$. The object Z_B is α-bounded if B is α-bounded.

The maps $Z_Y^n \to Y^n$ are local equivalences and local fibrations, and therefore have the local right lifting property with respect to all inclusions $\partial \Delta^m \subset \Delta^m$. Then it follows that every lifting problem

$$\begin{array}{ccc} \partial \Delta^m & \longrightarrow & Z_A^n(U) \\ \downarrow & & \downarrow \\ \Delta^m & \longrightarrow & A^m(U) \end{array}$$

has a local solution over some α-bounded $B' \subset Y$ with $A \subset B'$. There are at most α such lifting problems, so there is an α-bounded subobject $B_1 \subset Y$ with $A \subset B_1$ such that every lifting problem as above has a local solution over B_1.

Repeat inductively to produce a countable sequence

$$A = B_0 \subset B_1 \subset B_2 \subset \ldots$$

such that every lifting problem

$$\begin{array}{ccc} \partial \Delta^m & \longrightarrow & Z_{B_i}^n(U) \\ \downarrow & & \downarrow \\ \Delta^m & \longrightarrow & B_i^m(U) \end{array}$$

has a local solution over B_{i+1}. Let $B = \cup_i B_i$. Each map $\pi_B : Z_B^n \to B^n$ is therefore a local trivial Kan fibration, so that the map $B^n \cap X^n \to B^n$ is a local weak equivalence.

11.2 First Model Structures

Say that a map $p : X \to Y$ of symmetric T-spectra is an *injective fibration* if it has the right lifting property with respect to all level trivial cofibrations.

Lemma 11.7

1) *The map $p : X \to Y$ is an injective fibration if and only if it has the right lifting property with respect to all α-bounded level trivial cofibrations.*
2) *A map $q : Z \to W$ has the right lifting property with respect to all level cofibrations if and only if it has the right lifting property with respect to all α-bounded level cofibrations. The map q must also be a projective fibration and a level equivalence.*

Proof Statement 1) is a formal consequence of Lemma 11.6—see the proof of Lemma 5.4. The first part of statement 2) has a similar proof.

For these arguments, one needs to know that every section $\sigma \in X^n(U)$ of a symmetric T-spectrum X is a member of an α-bounded subobject $A \subset X$. This follows, in part, from the assumption that T is α-bounded.

For the second claim of statement 2), the map q has the right lifting property with respect to all maps $F_n A \to F_n B$ induced by cofibrations $A \to B$ of pointed simplicial presheaves, so that it is a trivial injective fibration of simplicial presheaves in all levels.

It follows from Lemma 11.7 that, if a map $q : Z \to W$ has the right lifting property with respect to all α-bounded level cofibrations, then it is an injective fibration and a level weak equivalence. The following result is the key to the proof of the converse statement, which appears in Corollary 11.9 below.

Lemma 11.8 *Suppose that the map $p : X \to Y$ has the right lifting property with respect to all α-bounded level trivial cofibrations and that p is a level equivalence. Then p has the right lifting property with respect to all level cofibrations.*

The outline of the argument for this result has appeared in multiple contexts, most explicitly in the proof of Lemma 7.3. We show that the map p has the right lifting property with respect to all α-bounded level cofibrations, so that we can invoke statement 2) of Lemma 11.7.

Proof A small object argument based on Lemma 11.7 shows that the map p has a factorization

where π has the right lifting property with respect to all cofibrations, and i is a level cofibration. The map π is a level equivalence, again by Lemma 11.7, so that the map

i is a level equivalence. Given a diagram

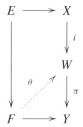

with $E \to F$ an α-bounded cofibration, the dotted lifting θ exists, and the image $\theta(F) \subset W$ is α-bounded. By Lemma 11.6 there is an α-bounded object $D \subset W$ with $\theta(F) \subset D$ such that the induced map $D \cap X \to D$ is a level equivalence. It follows that any diagram

with $E \to F$ an α-bounded cofibration has a factorization

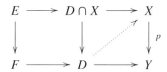

where the dotted lifting exists since $D \cap X \to D$ is an α-bounded trivial level cofibration.

Corollary 11.9 *A map $q : Z \to W$ of symmetric T-spectra is an injective fibration and a level weak equivalence if and only if it has the right lifting property with respect to all α-bounded level cofibrations.*

One forms the function complex $\mathbf{hom}(X, Y)$ for symmetric T-spectra X, Y by requiring that
$$\mathbf{hom}(X, Y)_n = \hom(X \wedge \Delta^n_+, Y).$$
The argument for the following result is an application of Lemma 11.7 and Corollary 11.9, and is left as an exercise.

Proposition 11.10 *The level weak equivalences, level cofibrations and injective fibrations give the category $\mathbf{Spt}^\Sigma_T(\mathcal{C})$ of symmetric T-spectra the structure of a proper closed simplicial model category. This model structure is cofibrantly generated, by the α-bounded level cofibrations and the α-bounded level trivial cofibrations.*

The model structure of Proposition 11.10 is the *injective model structure* for the category of symmetric T-spectra.

11.3 Localized Model Structures

This section gives the method for formally inverting a set of cofibrations within the injective model structure of Proposition 11.10. This method is based on the definitions and results of Sect. 7.2.

Suppose that $\mathcal{F}$ is a set of level cofibrations of symmetric T-spectra which includes the set J of α-bounded level trivial cofibrations. Suppose that all induced maps

$$(E \wedge D) \cup (F \wedge C) \to F \wedge D$$

are in $\mathcal{F}$ for each morphism $E \to F$ in $\mathcal{F}$ and each α-bounded cofibration $C \to D$ of pointed simplicial presheaves.

Suppose that β is a regular cardinal such that $\beta > |\operatorname{Mor}(\mathcal{C})|$. Suppose also that $\beta > |B|$ for all morphisms $i : A \to B$ appearing in the set $\mathcal{F}$ and that $\beta > |\mathcal{F}|$. Recall the assumption that $|T| < \alpha$.

Choose a cardinal λ such that $\lambda > 2^\beta$.

Every morphism $g : X \to Y$ of symmetric T-spectra has a functorial factorization

$$X \xrightarrow{i} E_\lambda(g)$$
$$g \searrow \quad \downarrow p$$
$$Y$$

where the cofibration i is in the saturation of the set $\mathcal{F}$, and the map p has the right lifting property with respect to all members of $\mathcal{F}$. As in Sect. 7.2, one constructs this factorization with a small object argument which terminates after λ steps.

Write

$$L_\mathcal{F}(X) = E_\lambda(X \to *)$$

for the result of this construction when applied to the canonical map $X \to *$. Then we have the following:

Lemma 11.11

1) *Suppose that $t \mapsto X_t$ is a diagram of level cofibrations indexed by any cardinal $\gamma > 2^\beta$. Then the natural map*

$$\varinjlim_{t<\gamma} L_\mathcal{F}(X_t) \to L_\mathcal{F}(\varinjlim_{t<\gamma} X_t)$$

is an isomorphism.

2) *Suppose that ζ is a cardinal with $\zeta > \beta$, and let $\mathcal{F}_\zeta(X)$ denote the filtered system of subobjects of X having cardinality less than ζ. Then the map*

$$\varinjlim_{Y \in \mathcal{F}_\zeta(X)} L_\mathcal{F}(Y) \to L_\mathcal{F}(X)$$

is an isomorphism.
3) *The functor $X \mapsto L_{\mathcal{F}}(X)$ preserves level cofibrations.*
4) *Suppose that U, V are subobjects of X. Then the natural map*

$$L_{\mathcal{F}}(U \cap V) \to L_{\mathcal{F}}(U) \cap L_{\mathcal{F}}(V)$$

is an isomorphism.
5) *If $|X| \leq 2^\mu$ where $\mu \geq \lambda$ then $|L_{\mathcal{F}}(X)| \leq 2^\mu$.*

Lemma 11.11 is a consequence of Lemma 7.16, which is the corresponding result for simplicial presheaves.

A map is said to be $\mathcal{F}$-*injective* if it has the right lifting property with respect to all members of $\mathcal{F}$, and an object X is $\mathcal{F}$-*injective* if the map $X \to *$ is $\mathcal{F}$-injective. All $\mathcal{F}$-injective objects are injective fibrant, ie. fibrant for the model structure of Proposition 11.10. The object $L_{\mathcal{F}}X$ is $\mathcal{F}$-injective.

Say that a morphism $f : X \to Y$ of $\mathbf{Spt}_T^\Sigma(\mathcal{C})$ is an $L_{\mathcal{F}}$-*equivalence* if it induces a weak equivalence

$$f^* : \mathbf{hom}(Y, Z) \to \mathbf{hom}(X, Z)$$

of simplicial sets for all $\mathcal{F}$-injective objects Z. Equivalently, $f : X \to Y$ is an $L_{\mathcal{F}}$-equivalence if and only if the induced map $f_* : L_{\mathcal{F}}(X) \to L_{\mathcal{F}}(Y)$ is a level equivalence of symmetric T-spectra—see the preamble to Theorem 7.10.

Every level equivalence is an $L_{\mathcal{F}}$-equivalence.

Following the pattern of Sect. 7.1, we say that a map $X \to Y$ of symmetric T-spectra is an $L_{\mathcal{F}}$-*fibration* if it has the right lifting property with respect to all maps which are level cofibrations and $L_{\mathcal{F}}$-equivalences.

Let κ be the successor cardinal for 2^μ, where μ is cardinal of statement 5) of Lemma 11.11. Then κ is a regular cardinal, and Lemma 11.11 implies that if a symmetric T-spectrum X is κ-bounded then $L_{\mathcal{F}}(X)$ is κ-bounded. The following result establishes the bounded monomorphism property for $L_{\mathcal{F}}$-equivalences:

Lemma 11.12 *Suppose given a level cofibration $i : X \to Y$ which is an $L_{\mathcal{F}}$-equivalence, and suppose that $A \subset Y$ is a κ-bounded subobject. Then there is a κ-bounded subobject $B \subset Y$ with $A \subset B$, and such that the level cofibration $B \cap X \to B$ is an $L_{\mathcal{F}}$-equivalence.*

Lemma 11.12 is a consequence of Lemma 11.6 (which is the bounded monomorphism property for the injective model structure of Proposition 11.10), in the same way that Lemma 7.17 follows from Lemma 5.2.

The following result is a consequence of Theorem 7.10. The left properness statement follows from Lemma 7.8.

Theorem 11.13 *The category $\mathbf{Spt}_T^\Sigma(\mathcal{C})$ of symmetric T-spectra, with the classes of level cofibrations, $L_{\mathcal{F}}$-equivalences and $L_{\mathcal{F}}$-fibrations, satisfies the axioms for a left proper closed simplicial model category.*

11.3 Localized Model Structures

The model structure of Theorem 11.13 is said to be a *stable model structure* if all maps of symmetric T-spectra which are obtained by applying the functor V to the stabilization maps

$$(S_T \wedge T)[-1-n] \to S_T[-n], \tag{11.5}$$

are $L_\mathcal{F}$-equivalences. Recall that S_T is the sphere spectrum for the T-spectrum category.

We can force the model structure of Theorem 11.13 to be a stable model structure by adding cofibrant replacements of the maps

$$V((S_T \wedge T)[-1-n]) \to V(S_T[-n]) \tag{11.6}$$

to the defining set $\mathcal{F}$.

Every symmetric T-spectrum is cofibrant in the model structure of Theorem 11.13. A cofibrant replacement of a map $\alpha : X \to Y$ of symmetric T-spectra is therefore found by taking the cofibration i in a factorization

of the map α, in which i is a level cofibration and the map θ is level weak equivalence.

The T-spectra $(S_T \wedge T)[-1-n]$ and $S_T[-n]$ are cofibrant. It follows from Corollary 11.5 that the functor V takes level weak equivalences between cofibrant T-spectra to level weak equivalences of symmetric T-spectra. Thus, given a cofibrant replacement

$$(S_T \wedge T)[-1-n] \xrightarrow{j} W \xrightarrow{\pi} S_T[-n]$$

for the stabilization map (11.5) in the strict model structure for T-spectra (with j a cofibration and π a strict equivalence), then the induced level cofibration $V(j)$ is a cofibrant replacement for the map (11.6) in the category of symmetric T-spectra, in the model structure of Proposition 11.10.

When we say that the set $\mathcal{F}$ is *generated by a set S* of level cofibrations, this means that $\mathcal{F}$ is the smallest set of level cofibrations which contains S and satisfies the *closure property* that the map

$$(E \wedge D) \cup (F \wedge C) \to F \wedge D$$

is in $\mathcal{F}$ for each morphism $E \to F$ in $\mathcal{F}$ and each α-bounded cofibration $C \to D$ of pointed simplicial presheaves.

This closure property is analogous to the property **C3** for spectra that appears in Sect. 10.3. The analogue of property **C1** is automatic in the present context (specifically, in the model structure of Proposition 11.10), and the analogue of property **C2** is one of the assumptions that we make for the set $\mathcal{F}$.

Example 11.14 The stable model structure for symmetric T-spectra arises, via Theorem 11.13, by formally inverting the set $\mathcal{F}$ of level cofibrations, where $\mathcal{F}$ is generated by the set J of α-bounded level trivial cofibrations, together with a set of cofibrant replacements of the maps (11.6).

Say that an $L_{\mathcal{F}}$-equivalence for this theory is a *stable equivalence*, and that an $L_{\mathcal{F}}$-fibration is a *stable fibration*. The cofibrations for this theory are the level cofibrations.

Example 11.15 Suppose that $f : A \to B$ is a cofibration of simplicial presheaves. Following Example 10.23, let J_f be the set of all maps

$$F_n(C_+) \to F_n(D_+)$$

which are induced by a fixed set of generators $C \to D$ for the trivial cofibrations of the f-local model structure. Let S_f be the set of cofibrations which is generated by the set J_f, together with cofibrant replacements of the maps (11.6). The corresponding model structure which is given by Theorem 11.13 is the *f-local stable model structure* for symmetric T-spectra.

An $L_{\mathcal{F}}$-equivalence for this theory is a *stable f-equivalence*, and an $L_{\mathcal{F}}$-fibration is a *stable f-fibration*. We shall say that the trivial cofibrations and trivial fibrations for this theory are *stable f-trivial*.

If a symmetric T-spectrum Z is stable f-fibrant, then it follows from Corollary 7.12 that the T-spectrum underlying Z must be stable f-fibrant. Corollary 10.26 implies that all level objects Z^n are f-fibrant simplicial presheaves and all maps $\sigma_* : Z^n \to \Omega_T Z^{n+1}$ are local (even sectionwise) weak equivalences of pointed simplicial presheaves.

Example 11.16 Start with the setup of Example 11.15, namely that $f : A \to B$ is a cofibration of simplicial presheaves, and that J_f is the set of all maps $F_n(C_+) \to F_n(D_+)$ which are induced by a fixed set of generators $C \to D$ for the trivial cofibrations of the f-local model structure.

Let S'_f be the set of cofibrations which is generated by the set J_f alone.

The corresponding model structure which is given by Theorem 11.13 is the *f-injective structure* for symmetric T-spectra. The fibrant objects for this structure, or the f-injective symmetric T-spectra are those objects Z which are level f-fibrant in the sense that all constituent simplicial presheaves Z^n are f-fibrant. The trivial fibrations for this theory are trivial fibrations for the injective structure, and are therefore level weak equivalences.

The weak equivalences for the f-injective model structure are the level f-equivalences, i.e. maps $X \to Y$ such that all maps $X^n \to Y^n$ are f-equivalences of simplicial presheaves. This follows from the fact that the set S'_f consists of cofibrations which are level f-equivalences. Any map $g : X \to Y$ of symmetric T-spectra then sits in a commutative diagram

$$\begin{array}{ccc} X & \xrightarrow{j_X} & L_{\mathcal{F}}X \\ g \downarrow & & \downarrow g_* \\ Y & \xrightarrow{j_Y} & L_{\mathcal{F}}Y \end{array}$$

in which the horizontal maps are f-injective (or f-fibrant) models, and are therefore level f-equivalences. All objects $L_{\mathcal{F}}X^n$ and $L_{\mathcal{F}}Y^n$ are f-fibrant simplicial presheaves. Thus, g is a levelwise f-equivalence if and only if the induced map $L_{\mathcal{F}}X \to L_{\mathcal{F}}Y$ of f-injective models is a level equivalence.

11.4 Stable Homotopy Theory of Symmetric Spectra

Suppose that $f : A \to B$ is a cofibration of pointed simplicial presheaves, and that S_f is the set of level cofibrations of Example 11.15. We shall work, throughout this section, within the resulting f-local stable model structure on the category $\mathbf{Spt}_T^\Sigma(\mathcal{C})$ of symmetric T-spectra.

It is an exercise to show that a symmetric T-spectrum Z is stable f-fibrant if and only if Z is injective (i.e. fibrant for the model structure of Proposition 11.10) and the underlying T-spectrum UZ is stable f-fibrant.

Lemma 11.17

1) The functor

$$V : \mathbf{Spt}_T(\mathcal{C}) \to \mathbf{Spt}_T^\Sigma(\mathcal{C})$$

takes cofibrations to level cofibrations, and preserves stable f-trivial cofibrations.
2) The functor V takes strict f-equivalences of cofibrant T-spectra to level f-equivalences of symmetric T-spectra.

Proof We know from Lemma 11.4 and Corollary 11.5 that the functor V takes cofibrations to projective cofibrations, and hence to level cofibrations.

There is a natural isomorphism of function complexes

$$\mathbf{hom}(V(A), Z) \cong \mathbf{hom}(A, U(Z)),$$

and the T-spectrum $U(Z)$ is stable f-fibrant if Z is stable f-fibrant. Thus, if the map $i : A \to B$ is a cofibration and a stable f-equivalence of T-spectra, then the simplicial set map

$$V(i)^* : \mathbf{hom}(V(B), Z) \to \mathbf{hom}(V(A), Z)$$

is a weak equivalence for all stable f-fibrant objects Z, and it follows that the map $V(i)$ is a stable f-equivalence as well as a level cofibration.

For statement 2), the functor V takes the generators $(S_T \wedge A)[-n] \to (S_T \wedge B)[-n]$ of the class of trivial cofibrations for the strict f-local model structure (Remark 10.24) to maps $F_n(A) \to F_n(B)$ of symmetric T-spectra which are level f-equivalences and level cofibrations, and it follows that V takes all trivial cofibrations for the strict f-local model structure to maps which are level cofibrations and level f-equivalences.

The image $V(j)$ of a cofibrant replacement $j : X \to U$ of a strict f-equivalence $g : X \to Y$ between cofibrant T-spectra is therefore a level f-equivalence. The functor V takes level equivalences between cofibrant T-spectra to level equivalences of symmetric T-spectra by Corollary 11.5, so the map $V(g)$ is a level f-equivalence.

Corollary 11.18 *The adjoint pair of functors*

$$V : \mathbf{Spt}_T(\mathcal{C}) \leftrightarrows \mathbf{Spt}_T^\Sigma(\mathcal{C}) : U$$

defines a Quillen adjunction between the respective f-local stable model structures.

It follows from Lemma 11.17 that every map $g : X \to Y$ of symmetric T-spectra has a natural factorization

where i is a stable f-trivial cofibration and $U(p_s)$ is a stable f-fibration. Applying this construction to the map $X \to *$ determines a natural stable f-trivial cofibration $i : X \to X_s$ such that $U(X_s)$ is stable f-fibrant.

Consider the composite

$$X \xrightarrow{i} X_s \xrightarrow{j} I(X_s) \tag{11.7}$$

where $j : X_s \to I(X_s)$ is the natural f-injective model (Example 11.16). The map $j : X_s \to I(X_s)$ is a level equivalence, so that the T-spectrum $U(I(X_s))$ is stable f-fibrant by Corollary 10.26. The composite ji is a stable f-equivalence, and therefore determines a natural stable f-fibrant model for symmetric T-spectra X.

More generally, if Z is a symmetric T-spectrum such that $U(Z)$ is stable f-fibrant, then the f-injective model $I(Z)$ is a stable f-fibrant model for Z in symmetric T-spectra.

It follows that a map $X \to Y$ of symmetric T-spectra is a stable f-equivalence if and only if the induced map $I(X_s) \to I(Y_s)$ is a level f-equivalence, or even a level sectionwise equivalence.

11.4 Stable Homotopy Theory of Symmetric Spectra

We now describe a construction for symmetric T-spectra which has no analogue for T-spectra, namely a natural map $\tilde{\sigma} : X \to \Omega_T X[1]$ where $\Omega_T X$ is a real (not fake) loop space and $\Omega_T X[1]$ is a shifted object that is to be defined.

In general, suppose that K is a pointed simplicial presheaf and X is a symmetric T spectrum. Then $\Omega_K X = \mathbf{Hom}(K, X)$ is the symmetric T-spectrum defined by internal function complexes, with

$$\Omega_K X^n = \mathbf{Hom}(K, X^n).$$

The bonding map

$$T^p \wedge \mathbf{Hom}(K, X^n) \xrightarrow{\sigma} \mathbf{Hom}(K, X^{p+n})$$

is defined by its adjoint: it is the unique map such that the diagram

$$\begin{array}{ccc}
T^p \wedge \mathbf{Hom}(K, X^n) \wedge K & \xrightarrow{\sigma \wedge K} & \mathbf{Hom}(K, X^{p+n}) \wedge K \\
{\scriptstyle T^p \wedge ev} \downarrow & & \downarrow {\scriptstyle ev} \\
T^p \wedge X^n & \xrightarrow{\sigma} & X^{p+n}
\end{array}$$

commutes. The symmetric group Σ_n has the evident induced action on the simplicial presheaf $\mathbf{Hom}(K, X^n)$.

There is a bijection

$$\hom(X \wedge K, Y) \cong \hom(X, \Omega_K Y)$$

which is natural in symmetric T-spectra X and Y and pointed simplicial presheaves K.

Suppose that X is a symmetric T-spectrum and that $n > 0$. The *shifted symmetric T-spectrum* $X[n]$ has $X[n]^k = X^{n+k}$, and $\alpha \in \Sigma_k$ acts on $X[n]^k$ as the element $1 \oplus \alpha \in \Sigma_{n+k}$. The bonding map $\sigma : T^p \wedge X[n]^k \to X[n]^{p+k}$ is defined to be the composite

$$T^p \wedge X^{n+k} \xrightarrow{\sigma} X^{p+n+k} \xrightarrow{c(p,n) \oplus 1} X^{n+p+k}$$

where $c(p, n)$ is the shuffle permutation of Σ_{p+n} which moves the first p letters past the last n letters, in order.

Every element $\gamma \in \Sigma_n$ induces an isomorphism $\gamma \oplus 1 : X^{n+k} \to X^{n+k}$, and all diagrams

$$\begin{array}{ccccc}
T^p \wedge X^{n+k} & \xrightarrow{\sigma} & X^{p+n+k} & \xrightarrow{c(p,n) \oplus 1} & X^{n+p+k} \\
{\scriptstyle T^p \wedge (\gamma \oplus 1)} \downarrow & & \downarrow {\scriptstyle 1 \oplus \gamma \oplus 1} & & \downarrow {\scriptstyle \gamma \oplus 1 \oplus 1} \\
T^p \wedge X^{n+k} & \xrightarrow{\sigma} & X^{p+n+k} & \xrightarrow{c(p,n) \oplus 1} & X^{n+p+k}
\end{array}$$

commute. It follows that each $\gamma \in \Sigma_n$ induces a natural isomorphism
$$\gamma : X[n] \to X[n]$$
of shifted symmetric T-spectra.

The map $\tilde{\sigma} : X^n \to \Omega_T X[1]^n = \Omega_T X^{1+n}$ in level n is the adjoint σ_* of the bonding map
$$\sigma : T \wedge X^n \to X^{1+n},$$
in the usual sense that the diagram

$$\begin{CD} T \wedge X^n @>{T \wedge \sigma_*}>> T \wedge \Omega_T X^{1+n} @>{\tau}>{\cong}> \Omega_T X^{1+n} \wedge T \\ @. @VV{\sigma}V @VV{ev}V \\ @. @. X^{1+n} \end{CD}$$

commutes. One shows that the diagram

$$\begin{CD} T^p \wedge X^n @>{T^p \wedge \sigma_*}>> T^p \wedge \Omega^T X^{1+n} \\ @VV{\sigma}V @VV{\sigma}V \\ X^{p+n} @>{\sigma_*}>> \Omega_T X^{1+p+n} \end{CD}$$

commutes by checking adjoints, and the morphisms σ_* assemble to define a natural map
$$\tilde{\sigma} : X \to \Omega_T X[1]$$
of symmetric T-spectra.

Suppose that X is a symmetric T-spectrum. Define a system $k \mapsto (Q_T^\Sigma X)(k)$, $k \geq 0$ of symmetric T-spectra by specifying that
$$(Q_T^\Sigma X)(k) = \Omega_T^k I(X)[k], \ k \geq 0.$$
In particular, $(Q_T^\Sigma X)(0) = I(X)$, where $j : X \to I(X)$ is the natural choice of f-injective model for X. There is a natural map $(Q_T^\Sigma X)(k) \to (Q_T^\Sigma X)(k+1)$ of symmetric T-spectra, which is the map
$$\Omega_T^k \tilde{\sigma}[k] : \Omega_T^k I(X)[k] \to \Omega_T^k \Omega_T I(X)[1][k]$$
that is induced by $\tilde{\sigma}$. Set
$$Q_T^\Sigma X = I(\varinjlim_k (Q_T^\Sigma X)(k)),$$

11.4 Stable Homotopy Theory of Symmetric Spectra

and write $\eta : X \to Q_T^\Sigma X$ for the natural composite

$$X \xrightarrow{j} I(X) = (Q_T^\Sigma X)(0) \to \varinjlim_k (Q_T^\Sigma X)(k) \xrightarrow{j} I(\varinjlim_k (Q_T^\Sigma X(k))).$$

Theorem 11.19 *Suppose that the object T is compact up to f-equivalence. Suppose that $g : X \to Y$ is a map of symmetric T-spectra such that the induced map $g_* : U(X) \to U(Y)$ is a stable f-equivalence of underlying T-spectra. Then g is a stable f-equivalence.*

Proof The compactness assumption guarantees that the natural map $\eta : Y \to Q_T Y$ is a stable f-fibrant model for each T-spectrum Y, by Theorem 10.32.

In level n, the symmetric T-spectrum $\varinjlim_k (Q_T^\Sigma X)(k)$ is given by the filtered colimit of the system

$$I(X)^n \xrightarrow{\sigma_*} \Omega_T I(X)^{n+1} \xrightarrow{\Omega_T \sigma_*} \Omega_T^2 I(X)^{n+2} \xrightarrow{\Omega_T^2 \sigma_*} \cdots$$

The object $U(I(X))$ is strict f-fibrant and the map $U(X) \to U(I(X))$ is a strict f-equivalence, so that the comparison $F(U(X)) \to U(I(X))$ is a strict equivalence and induces a commutative diagram

$$\begin{array}{ccccccc}
F(U(X))^n & \longrightarrow & \Omega_T F(U(X))^{n+1} & \longrightarrow & \Omega_T^2 F(U(X))^{n+2} & \longrightarrow & \cdots \\
\downarrow & & \downarrow & & \downarrow & & \\
U(I(X))^n & \longrightarrow & U(\Omega_T I(X))^{n+1} & \longrightarrow & U(\Omega_T^2 I(X))^{n+2} & \longrightarrow & \cdots
\end{array}$$

in which all the vertical maps are weak equivalences of pointed simplicial presheaves. It follows that there are induced natural f-equivalences

$$\begin{array}{ccc}
\varinjlim_k \Omega_T^k F(U(X))^{n+k} & \xrightarrow[\simeq]{j} & Q_T U(X)^n \\
\simeq \downarrow & & \\
U(\varinjlim_k \Omega_T^k I(X)^{n+k}) & \xrightarrow[Uj]{\simeq} & U(Q_T^\Sigma X)^n
\end{array}$$

Thus if a map of symmetric T-spectra $g : X \to Y$ induces a stable f-equivalence $U(g)$, meaning a level f-equivalence $Q_T U(X) \to Q_T U(Y)$ of T-spectra, then the map of symmetric T-spectra $g_* : Q_T^\Sigma X \to Q_T^\Sigma Y$ is a level equivalence.

If a symmetric T-spectrum Z is stable f-fibrant then all objects $(Q_T^\Sigma Z)(k)$ are stable f-fibrant and all maps $Z \to (Q_T^\Sigma Z)(k)$ are level f-equivalences, and it follows from the compactness of T that the natural map $\eta : Z \to Q_T^\Sigma Z$ is a level f-equivalence.

Finally, take a stable f-fibrant model $X \to L_{\mathcal{F}}X$ for a symmetric T-spectrum X and consider the diagram

$$\begin{array}{ccc} X & \xrightarrow{\simeq} & L_{\mathcal{F}}X \\ \eta \downarrow & & \simeq \downarrow \eta \\ Q_T^{\Sigma}X & \longrightarrow & Q_T^{\Sigma}L_{\mathcal{F}}X \end{array}$$

The indicated maps are stable f-equivalences, so that each symmetric T-spectrum X is a natural retract of the associated object $Q_T^{\Sigma}X$ in the f-local stable homotopy category.

Thus, if the map $g : X \to Y$ induces a stable f-equivalence $g_* : U(X) \to U(Y)$ of T-spectra, then the induced map $Q_T^{\Sigma}X \to Q_T^{\Sigma}Y$ is a level and hence stable f-equivalence, so that g is a stable f-equivalence.

Results about stable categories of T-spectra create results for stable categories of symmetric T-spectra, via Theorem 11.19. In particular, the full calculus of fibre and cofibre sequences for $(S^1 \wedge T)$-spectra from Sect. 10.5 has an analogue in symmetric $(S^1 \wedge T)$-spectra, provided that we assume that T is compact up to f-equivalence and that the f-local model structure on simplicial presheaves satisfies inductive colimit descent.

Recall that the inductive colimit descent condition implies that the simplicial circle S^1 is compact up to f-equivalence (Lemma 10.35), so that the smash $S^1 \wedge T$ is compact up to f-equivalence (Lemma 10.28).

Corollary 11.20 *Suppose that T is compact up to f-equivalence, and that the f-local model structure on simplicial presheaves satisfies inductive colimit descent.*

Suppose that there is a level f-fibre sequence

$$F \xrightarrow{i} X \xrightarrow{p} Y$$

of symmetric $(S^1 \wedge T)$-spectra. Then the canonical map $X/F \to Y$ is a stable f-equivalence.

Proof The map $U(X/F) \to U(Y)$ is a stable f-equivalence of $(S^1 \wedge T)$-spectra by Corollary 10.59. Now use Theorem 11.19.

Lemma 11.21 *Suppose that T is compact up to f-equivalence, and that the f-local model structure on the category of simplicial presheaves satisfies inductive colimit descent.*

Suppose given a comparison of level cofibre sequences

$$\begin{array}{ccccc} A_1 & \longrightarrow & B_1 & \longrightarrow & B_1/A_1 \\ g_1 \downarrow & & g_2 \downarrow & & g_3 \downarrow \\ A_2 & \longrightarrow & B_2 & \longrightarrow & B_2/A_2 \end{array}$$

11.4 Stable Homotopy Theory of Symmetric Spectra

of symmetric $(S^1 \wedge T)$-spectra. If any two of g_1, g_2 and g_3 are stable f-equivalences, then so is the third.

Proof There is a natural isomorphism

$$\Omega_{S^1 \wedge T} Z[1] \cong \Omega_{S^1} \Omega_T Z[1]$$

and the canonical map $Z \to \Omega_{S^1 \wedge T} Z[1]$ is a level equivalence if Z is stable f-fibrant. It follows that the induced diagram

$$\begin{array}{ccc}
\mathbf{hom}(B_2/A_2, Z) & \longrightarrow \mathbf{hom}(B_2, Z) & \longrightarrow \mathbf{hom}(A_2, Z) \\
\downarrow & \downarrow & \downarrow \\
\mathbf{hom}(B_1/A_1, Z) & \longrightarrow \mathbf{hom}(B_1, Z) & \longrightarrow \mathbf{hom}(A_1, Z)
\end{array}$$

is a comparison of fibre sequences of infinite loop spaces for each stable f-fibrant object Z. Thus, if any two of the vertical maps is a weak equivalence then so is the third.

Corollary 11.22 *Suppose that T is compact up to f-equivalence and that the f-local model structure on simplicial presheaves satisfies inductive colimit descent.*

Suppose that the map $i : A \to B$ is a cofibration of symmetric $(S^1 \wedge T)$-spectra, and take a factorization

such that j is a cofibration and an f-equivalence and p is an f-injective fibration. Let F be the fibre of p. Then the induced map $A \to F$ is a stable f-equivalence.

Proof It follows from Lemma 10.62 that the map $U(A) \to U(F)$ is a stable f-equivalence of $(S^1 \wedge T)$-spectra. Use Theorem 11.19.

The following properness result for symmetric $(S^1 \wedge T)$-spectra is now evident:

Theorem 11.23 *Suppose that T is compact up to f-equivalence, and that the f-local model structure on simplicial presheaves satisfies inductive colimit descent.*
Then the f-local stable model structure for symmetric $(S^1 \wedge T)$-spectra is proper.

Proof The proof is the same as that of Theorem 10.64. Use Corollary 11.20 and Lemma 11.21 in place of Corollary 10.59 and Lemma 10.60, respectively.

Generally, a symmetric T-spectrum is stable f-fibrant if and only if is f-injective, and all simplicial presheaf maps $\sigma_* : Z^n \to \Omega_T Z^{n+1}$ are local weak equivalences.

This observation can be promoted to a recognition principle for stable f-fibrations of $(S^1 \wedge T)$-spectra, in the presence of the usual assumptions on the object T and the the f-local structure on simplicial presheaves. The formal statement appears in Lemma 11.25 below. The proof uses the calculus of fibre and cofibre sequences for $(S^1 \wedge T)$-symmetric spectra of Corollary 11.20 through Corollary 11.22. The proof of Lemma 11.25 also requires the following technical result:

Lemma 11.24 *Suppose that T is compact up to f-equivalence, and that the f-local model structure for simplicial presheaves satisfies inductive colimit descent.*

Suppose that a map $p : X \to Y$ is a stable f-equivalence of symmetric $(S^1 \wedge T)$-spectra and that the underlying map $U(p) : U(X) \to U(Y)$ is a stable f-fibration of $(S^1 \wedge T)$-spectra. Then p is a level weak equivalence.

Proof The map p is a level f-fibration. Let F be the fibre of p and consider the fibre sequence
$$F \xrightarrow{i} X \xrightarrow{p} Y.$$
The canonical map $X/F \to Y$ is a stable f-equivalence of symmetric $(S^1 \wedge T)$-spectra by Corollary 11.20, so that the map $X \to X/F$ is a stable f-equivalence. Lemma 11.21 and the existence of the comparison of cofibre sequences

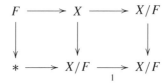

together show that the map $F \to *$ is a stable f-equivalence of symmetric $(S^1 \wedge T)$-spectra. It follows that F is levelwise contractible, since the f-injective model $I(F)$ is stable f-fibrant (see Example 11.16 and the construction in (11.7)).

The induced map $U(X) \to U(X/F)$ is a stable f-equivalence of $(S^1 \wedge T)$-spectra by Lemma 10.60, so that the map $U(p) : U(X) \to U(Y)$ is a stable f-equivalence of $(S^1 \wedge T)$-spectra as well as a stable f-fibration. It follows that the map $U(p) : U(X) \to U(Y)$ is a level equivalence.

Lemma 11.25 *Suppose that T is compact up to f-equivalence, and that the f-local model structure on simplicial presheaves satisfies inductive colimit descent.*

Suppose that the map $p : X \to Y$ is an injective fibration of symmetric $(S^1 \wedge T)$-spectra. Then p is a stable f-fibration if and only if the induced map $U(p) : U(X) \to U(Y)$ is a stable f-fibration of $(S^1 \wedge T)$-spectra.

Proof If the map $p : X \to Y$ is a stable f-fibration, then the underlying map $U(p) : U(X) \to U(Y)$ is a stable f-fibration on account of the Quillen adjunction of Corollary 11.18.

11.4 Stable Homotopy Theory of Symmetric Spectra

Suppose that the map $i : A \to B$ is a cofibration and a stable f-equivalence. Then i has a factorization

such that q is an injective fibration and $U(q)$ is a stable f-fibration, and j is a cofibration which is a stable f-equivalence and has the left lifting property with respect to all maps p such that p is an injective fibration and $U(p)$ is a stable f-fibration.

In effect, there are two factorizations

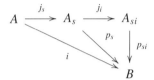

In the first of these, p_s is a map such that $U(p_s)$ is a stable f-fibration and j_s is a stable f-trivial cofibration which has the left lifting property with respect to all maps p such that $U(p)$ is a stable f-fibration. The map j_i is a level trivial cofibration and p_{si} is an injective fibration. It follows that $U(p_{si})$ is a strict fibration which is strict equivalent to a stable f-fibration, so that $U(p_{si})$ is a stable f-fibration by Theorem 10.65. Set $q = p_{si}$ and $j = j_i j_s$.

The map q is also a stable f-equivalence, so it is a level weak equivalence by Lemma 11.24. Thus, q is a trivial injective fibration, and therefore has the right lifting property with respect to all cofibrations. But then the map i is a retract of j and therefore has the left lifting property with respect to all injective fibrations p such that $U(p)$ is a stable fibration.

It follows that, if p is an injective fibration such that $U(p)$ is a stable f-fibration, then p has the right lifting property with respect to all maps which are cofibrations and stable f-equivalences.

In the original description of the stable model structure for symmetric spectra of [46], a stable fibration is defined to be a map of symmetric spectra $p : X \to Y$ such that the underlying map $U(p) : U(X) \to U(Y)$ of spectra is a stable fibration. The definition of stable equivalence for symmetric spectra of [46] is a special case of the definition of stable f-equivalence for symmetric $(S^1 \wedge T)$-spectra which appears here. See also [58]. There is an analogue of the model structure for symmetric spectra of [46] in the f-local context for $(S^1 \wedge T)$-spectra, called the HSS-structure, which we now describe. Once it is in place, the HSS-structure is easily seen to be Quillen equivalent to the stable f-local model structure for symmetric $(S^1 \wedge T)$-spectra—this is discussed below.

Say that the map $p : X \to Y$ of symmetric $(S^1 \wedge T)$-spectra is an *HSS-fibration* if the underlying map $U(p) : U(X) \to U(Y)$ is a stable f-fibration of $(S^1 \wedge T)$-spectra. Say that the map $i : A \to B$ is an *HSS-cofibration* if it has the left lifting property with respect to all maps which are stable f-equivalences and HSS-fibrations.

Lemma 11.24 implies that every map $p : X \to Y$ which is both an HSS-fibration and a stable f-equivalence must be a level f-equivalence, and hence a trivial strict fibration. It follows that the class of HSS-cofibrations includes all maps $F_n A \to F_n B$ which are induced by all cofibrations $A \to B$ of pointed simplicial presheaves.

Theorem 11.26 *Suppose that T is compact up to equivalence and that the f-local model structure on simplicial presheaf satisfies inductive colimit descent.*

Then the category $\mathbf{Spt}^{\Sigma}_{S^1 \wedge T}(\mathcal{C})$ of symmetric $(S^1 \wedge T)$-spectra, with the classes of stable f-equivalences, HSS-fibrations and HSS-cofibrations, satisfies the axioms for a proper closed simplicial model category. All HSS-cofibrations are level cofibrations.

Proof The f-local stable model structure on $(S^1 \wedge T)$-spectra is cofibrantly generated. It follows from Lemma 11.17 that every map $g : X \to Y$ of symmetric $(S^1 \wedge T)$-spectra has factorizations

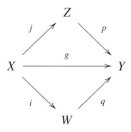

where p is an HHS-fibration and j is an HSS-cofibration which is a stable f-equivalence and has the left lifting property with respect to all HHS-fibrations, and i is an HSS-cofibration and q is an HSS-fibration such that $U(q)$ is a trivial stable f-fibration. The map q is a stable f-equivalence, by Theorem 11.19. We have verified the factorization axiom **CM5**.

The maps i and j are both level cofibrations, by Lemma 11.17. Every HSS-cofibration is a retract of a map of the form i, and is therefore a level cofibration. Every HSS-cofibration which is a stable f-equivalence is a retract of a map of the form j, and therefore has the left lifting property with respect to all HSS fibrations. We have thus proved the lifting axiom **CM4**.

The function complex construction $\mathbf{hom}(X, Y)$ is the one that we know: an n-simplex of the simplicial set $\mathbf{hom}(X, Y)$ is a map $X \wedge \Delta^n_+ \to Y$ of symmetric $(S^1 \wedge T)$-spectra. The simplicial model property **SM7** is inherited from $(S^1 \wedge T)$-spectra. In effect, if $p : X \to Y$ is an HSS-fibration and $i : K \to L$ is a cofibration of simplicial sets, then the map

$$\mathbf{hom}(L, X) \to \mathbf{hom}(K, X) \times_{\mathbf{hom}(K,Y)} \mathbf{hom}(L, Y)$$

11.5 Equivalence of Stable Categories

is an HSS-fibration, which is a stable f-equivalence if either p is a stable f-equivalence or i is a trivial cofibration of simplicial sets, because the same is true for the maps $U(p)$ and i.

Every HSS-fibration is level f-fibration. Right properness for the present model structure on symmetric $(S^1 \wedge T)$-spectra is is a consequence of Corollary 11.20 and Lemma 11.21, as in the proof of the corresponding result for the stable f-local model structure in Theorem 11.23. Left properness for the HSS-structure follows from the fact that every HSS-cofibration is a level cofibration, along with the left properness for the f-local stable model structure on symmetric $(S^1 \wedge T)$-spectra.

The model structure of Theorem 11.26 is the (f-local) *HSS-structure* for the category of symmetric $(S^1 \wedge T)$-spectra.

It is a consequence of Lemma 11.17 that every stable f-fibration is an HSS-fibration. The weak equivalences for the stable f-local model structure and the HSS-structure coincide, by definition. It follows that the identity functor on the category of symmetric $(S^1 \wedge T)$-spectra defines a Quillen equivalence between the two model structures. This is subject, of course, to having the compactness and inductive descent conditions in place.

Example 11.27 Suppose that $(Sm|_S)_{Nis}$ is the smooth Nisnevich site of a (decent) scheme S, and let $f : * \to \mathbb{A}^1$ be a rational point of the affine line over S.

Theorem 11.26 specializes to Theorem 4.15 of [57], which establishes the motivic stable model structure for symmetric $(S^1 \wedge \mathbb{G}_m)$-spectra on the smooth Nisnevich site of a Noetherian scheme S of finite dimension. The model structure for symmetric $(S^1 \wedge \mathbb{G}_m)$-spectra on smooth S-schemes of [57] is, in present terms, the motivic (or f-local) HSS-structure for that category.

The HHS-structure of Theorem 11.26 does not coincide with the stable model structure of Theorem 11.13 and Example 11.15. In particular, the motivic variant of the stable model structure for symmetric T-spectra which is given by Theorem 11.13 is different from that of [57]. In the structure of [57], an object X is stable f-fibrant if and only the underlying T-spectrum UX is stable f-fibrant. The stable f-fibrant objects for the structure of Theorem 11.13 must also be fibrant for the injective model structure of Proposition 11.10. The composite ji of (11.7) is a variant of the stable fibrant model construction of [57, p. 509], which makes implicit use of motivic descent.

The motivic stable model structure for the category of $(S^1 \wedge \mathbb{G}_m)$-spectra on smooth S-schemes which arises from Theorem 11.13 and Example 11.15 is Quillen equivalent to the HSS-structure of [57]. One naturally refers to both of these model structures as "the" motivic stable model structure for the category of symmetric $(S^1 \wedge \mathbb{G}_m)$-spectra on the smooth Nisnevich site.

11.5 Equivalence of Stable Categories

As usual, the map $f : A \to B$ is a cofibration of pointed simplicial presheaves which is formally inverted.

We shall require the assumption that T is compact up to f-equivalence throughout this section. Later on, starting with Proposition 11.29, we shall also need the assumption that the parameter object T is cycle trivial, in the sense that the automorphism

$$c_{1,2} : T^{\wedge 3} \to T^{\wedge 3},$$

which shuffles smash factors according to the rule

$$x_1 \wedge x_2 \wedge x_3 \mapsto x_3 \wedge x_1 \wedge x_2,$$

induces the identity in the f-local homotopy category on pointed simplicial presheaves.

Recall that the compactness and cycle triviality of T are the conditions for Theorem 10.46, which asserts that the T-suspension functor $X \mapsto X \wedge T$ and and the T-loops functor $Y \mapsto \mathbf{Hom}(T, Y)$ are inverse to each other in the f-local stable category.

We begin with a discussion of T-bispectrum objects. These objects are used in the proof of Proposition 11.29, which is the key technical step in proving the main result of this section, which is Theorem 11.36. This theorem asserts, subject to the compactness and cycle triviality of T, that the forgetful functor U and its left adjoint V define a Quillen equivalence between the f-local stable structures for T-spectra and symmetric T-spectra.

A *T-bispectrum* X is a T-spectrum object in T-spectra, or a (T, T)-bispectrum. In other words, X consists of T-spectra X^s, $s \geq 0$, together with maps of T-spectra $\sigma_v : X^s \wedge T \to X^{s+1}$.

Each T-spectrum X^s consists of pointed simplicial presheaves $X^{r,s}$, with bonding maps $\sigma_h : T \wedge X^{r,s} \to X^{r+1,s}$, and the diagrams

$$\begin{CD} T \wedge X^{r,s} \wedge T @>{\sigma_h \wedge T}>> X^{r+1,s} \wedge T \\ @V{T \wedge \sigma_v}VV @VV{\sigma_v}V \\ T \wedge X^{r,s+1} @>>{\sigma_h}> X^{r+1,s+1} \end{CD} \qquad (11.8)$$

commute. Following [56], we could write σ_{v*} for the composite

$$T \wedge X^{r,s} \xrightarrow{\tau} X^{r,s} \wedge T \xrightarrow{\sigma_v} X^{r,s+1},$$

11.5 Equivalence of Stable Categories

and then the commutativity of (11.8) is equivalent to the commutativity of the diagram

$$
\begin{array}{ccc}
T \wedge T \wedge X^{r,s} & \xrightarrow{T \wedge \sigma_h} & T \wedge X^{r+1,s} \\
{\scriptstyle \tau \wedge X^{r,s}} \downarrow & & \downarrow {\scriptstyle \sigma_{v*}} \\
T \wedge T \wedge X^{r,s} & & \\
{\scriptstyle T \wedge \sigma_{v*}} \downarrow & & \\
T \wedge X^{r,s+1} & \xrightarrow{\sigma_h} & X^{r+1,s+1}
\end{array}
\tag{11.9}
$$

Thus, a T-bispectrum X can be described as a collection of pointed simplicial presheaves $X^{r,s}$, together with maps σ_h and σ_{v*} such that all diagrams (11.9) commute. This is consistent with the theory of bispectra which is presented in [56].

Every T-bispectrum X has an associated *diagonal* $(T \wedge T)$-*spectrum* $d(X)$, with $d(X)^r = X^{r,r}$ and with bonding maps given by the composites

$$ T \wedge T \wedge X^{r,r} \xrightarrow{T \wedge \sigma_h} T \wedge X^{r+1,r} \xrightarrow{\sigma_{v*}} X^{r+1,r+1} $$

A map $g : X \to Y$ of T-bispectra consists of pointed simplicial presheaf maps $g : X^{r,s} \to Y^{r,s}$ which respect all structure. Every such map g induces a map $g_* : d(X) \to d(Y)$ of diagonal $(T \wedge T)$-spectra. This construction defines a functor from T-bispectra to $(T \wedge T)$-spectra.

As in Sect. 10.5, we say that a map g of T-bispectra is a *stable f-equivalence* of T-bispectra if the diagonal map $g_* : d(X) \to d(Y)$ is a stable f-equivalence of $(T \wedge T)$-spectra.

A *strict f-equivalence* $X \to Y$ of T-bispectra consists of f-equivalences $X^{r,s} \to Y^{r,s}$ in all bidegrees. Every strict f-equivalence of T-bispectra is a stable f-equivalence.

Since we assume that T is compact up to f-equivalence, it follows from Lemma 10.56 that a map $X \to Y$ of T-bispectra is a stable f-equivalence if either all maps $X^{r,*} \to Y^{r,*}$ or all maps $X^{*,s} \to Y^{*,s}$ are stable f-equivalences of T-spectra.

Say that a T-bispectrum Z is *stable f-fibrant* if all simplicial presheaves $X^{r,s}$ are f-fibrant and all adjoint bonding maps $X^{r,s} \to \Omega_T X^{r+1,s}$ and $X^{r,s} \to \Omega_T X^{r,s+1}$ are f-equivalences.

Lemma 11.28 *Suppose that $g : Z \to W$ is a map of stable f-fibrant T-bispectra. Then g is a stable f-equivalence if and only if all maps $g : Z^{r,s} \to W^{r,s}$ are f-equivalences.*

Proof The diagonal $(T \wedge T)$-spectra $d(Z)$ and $d(W)$ are stable f-fibrant, and so the induced map $g_* : d(Z) \to d(W)$ is a stable f-equivalence if and only if all maps $Z^{r,r} \to W^{r,r}$ are f-equivalences. Every $Z^{r,s}$ is f-equivalent to an interated T-loop space $\Omega_T^k Z^{n,n}$ for some k and n since Z is a stable f-fibrant bispectrum, and this identification is natural in Z for all such Z. Thus, if g_* is a stable f-equivalence of stable f-fibrant $(T \wedge T)$-spectra, then all maps $Z^{r,s} \to W^{r,s}$ are f-equivalences.

We need another formal observation before proving the main results of this section. Suppose that

$$X \to LX = L_{\mathcal{F}}X$$

is the stable f-fibrant model construction for T-spectra X which is produced in the proof Theorem 10.20, where the the set $\mathcal{F}$ is generated by S_f. The closure property **C3** for $\mathcal{F}$ guarantees that there is a map

$$\gamma : L(X) \wedge K \to L(X \wedge K)$$

which is natural in T-spectra X and pointed simplicial presheaves K. There is also a natural commutative diagram

$$\begin{array}{ccc} X \wedge K & & \\ {\scriptstyle j \wedge K} \downarrow & \searrow^{j} & \\ L(X) \wedge K & \xrightarrow{\gamma} & L(X \wedge K) \end{array}$$

in the category of T-spectra.

Suppose that X is a T-bispectrum, or a T-spectrum object in T-spectra. Then the composite maps

$$L(X^r) \wedge T \xrightarrow{\gamma} L(X^r \wedge T) \xrightarrow{L\sigma} L(X^{r+1})$$

define a T-bispectrum $L(X)$ with $L(X)^r = L(X^r)$, together with a map of T-bispectra $j : X \to L(X)$ such that each map $X^r \to L(X^r)$ is a stable f-equivalence.

Proposition 11.29 *Suppose that T is compact up to f-equivalence and is cycle trivial. Suppose that K is a pointed simplicial presheaf, and that $j : S_T \wedge K \to Z$ is a stable f-fibrant model in the symmetric T-spectrum category. Then the underlying map $U(S_T \wedge K) \to U(Z)$ is a stable f-equivalence of T-spectra.*

Proof It is enough to find one example of a stable f-fibrant model $S_T \wedge K \to Z$ such that the conclusion holds. The proof is achieved by stabilizing the object K in two directions.

Explicitly, suppose that $X(K)$ consists of the objects

$$X(K)^{r,s} = T^{\wedge r} \wedge K \wedge T^{\wedge s}.$$

Holding s fixed and letting r vary gives the symmetric spectrum

$$X(K)^{*,s} = S_T \wedge K \wedge T^{\wedge s},$$

11.5 Equivalence of Stable Categories

while holding r fixed and letting s vary gives T-spectra

$$X(K)^{r,*} = T^{\wedge r} \wedge K \wedge T^{\wedge *}.$$

The object $X(K)$ can be interpreted as either a T-spectrum object in symmetric T-spectra, or a symmetric T-spectrum object in T-spectra. Forgetting the symmetric spectrum structures gives a T-bispectrum $UX(K)$.

Applying the stable f-fibrant model construction $X \to L(X)$ for T-spectra in each r defines T-spectra $L(X(K)^{r,*})$, together with maps

$$T^{\wedge k} \wedge L(X(K)^{r,*}) \to L(X(K)^{k+r,*})$$

which give the object $L(X(K)^{*,*})$ the structure of a symmetric T-spectrum object in T-spectra. Write

$$L(X(K)) = L(X(K)^{*,*}).$$

The stable f-equivalences $X(K)^{r,*} \to L(X(K)^{r,*})$ assemble to give a map

$$X(K) \to L(X(K))$$

of symmetric T-spectrum objects in T-spectra. It follows from Lemma 10.56 that the underlying map

$$UX(K) \to UL(X(K))$$

is a stable f-equivalence of T-bispectra.

The object $L(X(K))^{*,0}$ is a symmetric T-spectrum by construction. It follows from Theorem 10.46 that the T-bispectrum $UL(X(K))$ is stable f-fibrant.

By stabilizing in the other index, we construct a stable f-fibrant model for $UX(K)$ in the T-bispectrum category, by taking stable f-fibrant models

$$UX(K)^{*,s} \to L(UX(K)^{*,s})$$

in T-spectra for each s. One uses Theorem 10.46 again to conclude that the resulting T-bispectrum $L(UX(K))$ is stable f-fibrant.

Form the diagram of T-bispectra

$$\begin{array}{ccc} UX(K) & \longrightarrow & L(UX(K)) \\ \downarrow & & \downarrow \\ U(LX(K)) & \longrightarrow & L(U(LX(K))) \end{array}$$

All maps in the diagram are stable f-equivalences of T-bispectra, and so there are level f-equivalences

$$U(L(X(K))^{*,0} \xrightarrow{\simeq} L(U(LX(K)))^{*,0} \xleftarrow{\simeq} L(UX(K))^{*,0}$$

by Lemma 11.28. The map $UX(K)^{*,0} \to L(UX(K))^{*,0}$ is a stable f-equivalence by construction. It follows that the map
$$S_T \wedge K = X(K)^{*,0} \to L(X(K))^{*,0}$$
is a map of symmetric T-spectra which induces a stable f-equivalence of underlying T-spectra.

The underlying T-spectrum $U(L(X(K)))^{*,0}$ is also stable f-fibrant. The f-injective model $IL(X(K))^{*,0}$ therefore determines a stable f-fibrant model
$$S_T \wedge K = X(K)^{*,0} \to L(X(K))^{*,0} \xrightarrow{\simeq} IL(X(K))^{*,0}$$
in symmetric T-spectra, as in (11.7), such that the underlying map of T-spectra is a stable f-equivalence.

If Z is a stable f-fibrant symmetric T-spectrum and $n \geq 0$, then shifted object $Z[n]$ is stable f-fibrant, and the canonical map
$$Z \to \Omega_T Z[1]$$
is a level f-equivalence.

Lemma 11.30 *A map $g : X \to Y$ of symmetric T-spectra is a stable f-equivalence if and only if the induced map $g \wedge T : X \wedge T \to Y \wedge T$ is a stable f-equivalence.*

Proof There is a natural isomorphism of function complexes
$$\mathbf{hom}(X \wedge T, Z) \cong \mathbf{hom}(X, \Omega_T Z),$$
and the T-loops object $\Omega_T Z$ is stable f-fibrant if Z is stable f-fibrant. Thus, if the map $g : X \to Y$ is a stable f-equivalence and Z is stable f-fibrant, then the induced simplicial set map
$$\mathbf{hom}(Y, \Omega_T Z) \to \mathbf{hom}(X, \Omega_T Z)$$
is a weak equivalence, and so the map
$$\mathbf{hom}(Y \wedge T, Z) \to \mathbf{hom}(X \wedge T, Z)$$
is a weak equivalence. This is true for all stable f-fibrant objects Z, so that the map $g \wedge T : X \wedge T \to Y \wedge T$ is a stable f-equivalence if g is a stable f-equivalence.

Suppose that $g \wedge T$ is a stable f-equivalence, and form the diagram of simplicial set maps

$$\begin{array}{ccc}
\mathbf{hom}(Y, Z) & \xrightarrow{g^*} & \mathbf{hom}(X, Z) \\
\simeq \downarrow & & \downarrow \simeq \\
\mathbf{hom}(Y, \Omega_T Z[1]) & \longrightarrow & \mathbf{hom}(X, \Omega_T Z[1]) \\
\cong \downarrow & & \downarrow \cong \\
\mathbf{hom}(Y \wedge T, Z[1]) & \xrightarrow[(g \wedge T)^*]{\simeq} & \mathbf{hom}(X \wedge T, Z[1])
\end{array}$$

11.5 Equivalence of Stable Categories

It follows that the map g^* is a weak equivalence. This is true for all stable f-fibrant objects Z, so that g is a stable f-equivalence.

We now have the following analogue and consequence of Theorem 10.46 for symmetric T-spectra:

Lemma 11.31 *Suppose that T is compact up to f-equivalence and is cycle trivial, and let X be a symmetric T-spectrum. Suppose that the map $j : X \wedge T \to (X \wedge T)_s$ is a stable f-fibrant model for $X \wedge T$. Then the composite map*

$$X \xrightarrow{\eta} \Omega_T(X \wedge T) \xrightarrow{j} \Omega_T(X \wedge T)_s$$

is a stable f-equivalence.

Proof Write η_* for the displayed composite in the statement of the Lemma. It suffices, by Lemma 11.30, to show that the map $\eta_* \wedge T$ is a stable f-equivalence. There is a commutative diagram

$$\begin{array}{ccc} X \wedge T & \xrightarrow{\eta_* \wedge T} & \Omega_T(X \wedge T)_s \wedge T \\ & \searrow{j} & \downarrow{ev} \\ & & (X \wedge T)_s \end{array}$$

The evaluation map ev induces a stable f-equivalence of the underlying T-spectra by Corollary 10.49, and is therefore a stable f-equivalence of symmetric T-spectra by Theorem 11.19. It follows that the map $\eta_* \wedge T$ is a stable f-equivalence of symmetric T-spectra.

Corollary 11.32 *Suppose that T is compact up to f-equivalence and is cycle trivial. Suppose that Z is a stable f-fibrant symmetric T-spectrum. Then any stable fibrant model $j : Z \wedge T \to (Z \wedge T)_s$ induces a stable f-equivalence $U(Z \wedge T) \to U((Z \wedge T)_s)$ of underlying T-spectra.*

Proof The composite

$$\eta_* : Z \xrightarrow{\eta} \Omega_T(Z \wedge T) \xrightarrow{j} \Omega_T(Z \wedge T)_s$$

is a stable f-equivalence of stable f-fibrant symmetric T-spectra by Lemma 11.31, and is therefore a level f-equivalence. In the diagram

$$\begin{array}{ccc} Z \wedge T & \xrightarrow{\eta_* \wedge T} & \Omega_T(Z \wedge T)_s \wedge T \\ & \searrow{j} & \downarrow{ev} \\ & & (Z \wedge T)_s \end{array}$$

the map $\eta_* \wedge T$ is a level f-equivalence, and the map ev induces a stable f-equivalence of underlying T-spectra by Corollary 10.49.

Lemma 11.33 *Suppose that T is compact up to f-equivalence and is cycle trivial. Then the map $\eta_* : X \to U((VX)_s)$ is a stable f-equivalence of T-spectra if and only if the map $\eta_* : X \wedge T \to U(V(X \wedge T)_s)$ is a stable f-equivalence.*

Proof There is a commutative diagram

$$\begin{array}{ccc}
X \wedge T & \xrightarrow{\eta \wedge T} U(V(X)) \wedge T \xrightarrow{U(j) \wedge T} & U((VX)_s) \wedge T \\
& \downarrow \cong & \downarrow \cong \\
& U(V(X) \wedge T) \xrightarrow{U(j \wedge T)} & U((VX)_s \wedge T) \\
\eta \searrow & \downarrow \cong & \downarrow U\tilde{j} \\
& U(V(X \wedge T)) \xrightarrow{U(j)} & U(V(X \wedge T)_s)
\end{array}$$

where the map $\tilde{j}$ is chosen such that the diagram

$$\begin{array}{ccc}
V(X) \wedge T & \xrightarrow{j \wedge T} & V(X)_s \wedge T \\
\cong \downarrow & & \downarrow \tilde{j} \\
V(X \wedge T) & \xrightarrow{j} & (V(X \wedge T))_s
\end{array}$$

commutes. It suffices to show that the map $U\tilde{j}$ is a stable f-equivalence of T-spectra.

The map $\tilde{j}$ is a stable f-fibrant model for the symmetric T-spectrum $V(X)_s \wedge T$ and the object $V(X)_s$ is stable f-fibrant, so the desired result follows from Corollary 11.32.

Corollary 11.34 *Suppose that T is compact up to f-equivalence and is cycle trivial.*

1) Suppose that K is a pointed simplicial presheaf. Then the map

$$\eta_* : (S_T \wedge K)[n] \to U(V((S_T \wedge K)[n])_s)$$

is a stable f-equivalence for all $n \in \mathbb{Z}$.

2) Suppose that X is a cofibrant T-spectrum. Then the map

$$\eta_* : L_n X \to U((V(L_n X))_s)$$

is a stable f-equivalence for all $n \geq 0$.

11.5 Equivalence of Stable Categories

Proof If $n \geq 0$, then there is an isomorphism

$$(S_T \wedge K)[n] \cong S_T \wedge T^n \wedge K,$$

If $m > 0$ there is a canonical stable equivalence of cofibrant T-spectra

$$(S_T \wedge K)[-m] \wedge T^{\wedge m} \to S_T \wedge K.$$

Statement 1) therefore reduces to Proposition 11.29 in all cases.

In general, the question of whether or not the map

$$\eta_* : X \xrightarrow{\eta} U(V(X)) \xrightarrow{U(j)} U(V(X)_s)$$

can be a stable f-equivalence is insensitive to stable f-equivalences in cofibrant T-spectra X, since the functor V preserves stable f-equivalences between such objects by Corollary 11.18. Statement 2) follows, since the layer $L_n X$ is cofibrant and there is a stable f-equivalence

$$(S_T \wedge X^n)[-n] \to L_n X.$$

If $s \mapsto Z_s$ is an inductive system of stable f-fibrant symmetric T-spectra, then the f-injective model $I(\varinjlim_s Z_s)$ is stable f-fibrant. In effect, there is a commutative diagram

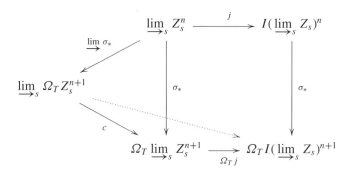

corresponding to the injective model $j : \varinjlim_s Z_s \to I(\varinjlim_s Z_s)$, in which the dotted arrow is an f-equivalence (by the compactness of T), as are the maps $\varinjlim \sigma_*$ and j. It follows that the map

$$\sigma_* : I(\varinjlim_s Z_s)^n \to \Omega_T I(\varinjlim_s Z_s)^{n+1}$$

is an f-equivalence.

Theorem 11.35 *Suppose that T is compact up to f-equivalence and is cycle trivial. Suppose that X is a cofibrant T-spectrum. Then the map*

$$\eta_* : X \xrightarrow{\eta} UV(X) \xrightarrow{Uj} U(V(X)_s)$$

is a stable f-equivalence.

Proof We can choose stable f-fibrant models $V(L_n X) \to V(L_n X)_s$ such that the diagram

$$\begin{array}{ccccccc} V(L_0 X) & \longrightarrow & V(L_1 X) & \longrightarrow & V(L_2 X) & \longrightarrow & \cdots \\ \downarrow & & \downarrow & & \downarrow & & \\ V(L_0 X)_s & \longrightarrow & V(L_1 X)_s & \longrightarrow & V(L_2 X)_s & \longrightarrow & \cdots \end{array}$$

commutes. The composites

$$L_n X \xrightarrow{\eta} U(V(L_n X)) \to U(V(L_n X)_s)$$

are stable f-equivalences of T-spectra by Corollary 11.34. It follows that the induced map

$$X \cong \varinjlim_n L_n X \to \varinjlim_n U(V(L_n X)) \to \varinjlim_n U(V(L_n X)_s)$$

is also a stable f-equivalence. The f-injective model

$$j : \varinjlim_n V(L_n X)_s \to I(\varinjlim_n V(L_n X)_s)$$

is a level f-equivalence, and is a stable f-fibrant model for the colimit $\varinjlim_n V(L_n X)_s$ in symmetric T-spectra.

The composite

$$X \to U(V(X)) \cong U(\varinjlim_n V(L_n X)) \to U(\varinjlim_n V(L_n X)_s) \to U(I(\varinjlim_n V(L_n X)_s))$$

is therefore a stable f-equivalence, while the composite

$$V(X) \cong \varinjlim_n V(L_n X) \to \varinjlim_n V(L_n X)_s \to I(\varinjlim_n V(L_n X)_s)$$

is a stable f-fibrant model for $V(X)$ in symmetric T-spectra.

We now have the main result of this section:

Theorem 11.36 *Suppose that the parameter object T is compact up to f-equivalence and is cycle trivial. Then the adjoint functors*

$$V : \mathbf{Spt}_T(\mathcal{C}) \leftrightarrows \mathbf{Spt}_T^\Sigma(\mathcal{C}) : U$$

11.5 Equivalence of Stable Categories

define a Quillen equivalence between the respective f-stable model structures.

Proof It is shown in Corollary 11.18 that these functors define a Quillen adjunction. Theorem 11.35 says that the composite

$$X \xrightarrow{\eta} U(V(X)) \xrightarrow{Uj} U(Z)$$

is a stable f-equivalence of T spectra for any cofibrant T-spectrum X and any choice of stable f-fibrant model $j : V(X) \to Z$ in symmetric T-spectra.

Suppose that Z is a stable f-fibrant symmetric T-spectrum, and let $\pi : X \to U(Z)$ be a cofibrant model for the underlying T-spectrum $U(Z)$. It remains to show that the composite

$$V(X) \xrightarrow{V(\pi)} V(U(Z)) \xrightarrow{\epsilon} Z$$

is a stable f-equivalence of symmetric T-spectra.

Suppose that $j : Y \to Y_s$ is a (natural) stable f-fibrant model in symmetric T-spectra. Then there is a commutative diagram

$$\begin{array}{ccccc}
V(X)_s & \xrightarrow{V(\pi)_s} & V(U(Z))_s & \xrightarrow{\epsilon_*} & Z \\
{\scriptstyle j}\uparrow & & {\scriptstyle j}\uparrow & \nearrow {\scriptstyle \epsilon} & \\
V(X) & \xrightarrow{V(\pi)} & V(U(Z)) & &
\end{array}$$

and it suffices to show that the top composite is a stable f-equivalence. Applying the functor U gives a commutative diagram

$$\begin{array}{ccccc}
U(V(X)_s) & \xrightarrow{U(V(\pi)_s)} & U((V(U(Z)))_s) & \xrightarrow{U(\epsilon_*)} & U(Z) \\
{\scriptstyle U(j)}\uparrow & & {\scriptstyle U(j)}\uparrow & \nearrow {\scriptstyle U(\epsilon)} & \\
U(V(X)) & \xrightarrow{U(V(\pi))} & U(V(U(Z))) & & \\
{\scriptstyle \eta}\uparrow & & {\scriptstyle \eta}\uparrow & \nearrow {\scriptstyle 1} & \\
X & \xrightarrow{\pi} & U(Z) & &
\end{array}$$

The left vertical composite is a stable f-equivalence by Theorem 11.35, and it follows that the composite

$$U(V(X)_s) \xrightarrow{U(V(\pi_s))} U((V(U(Z)))_s) \xrightarrow{U\epsilon_*} U(Z)$$

is a stable f-equivalence of T-spectra. Theorem 11.19 implies that the composite

$$V(X)_s \xrightarrow{V(\pi)_s} (V(U(Z)))_s \xrightarrow{\epsilon_*} Z$$

is a stable f-equivalence of symmetric T-spectra.

Example 11.37 The object
$$\mathbb{P}^1 \simeq S^1 \wedge \mathbb{G}_m$$
is compact and cycle trivial for the motivic model structure on the category of simplicial presheaves on the smooth Nisnevich site $(Sm|_S)_{Nis}$ of a scheme. See Example 10.45.

In this case, Theorem 11.36 implies that the adjoint functors
$$V : \mathbf{Spt}_{S^1 \wedge \mathbb{G}_m}(Sm|_S)_{Nis} \leftrightarrows \mathbf{Spt}^{\Sigma}_{S^1 \wedge \mathbb{G}_m}(Sm|_S)_{Nis} : U$$
define a Quillen equivalence between the motivic stable categories for $(S^1 \wedge \mathbb{G}_m)$-spectra and symmetric $(S^1 \wedge \mathbb{G}_m)$-spectra.

This statement for motivic stable categories originally appeared in [57], and was the first result having the form of Theorem 11.36.

11.6 The Smash Product

Recall from Sect. 11.1 that the tensor product $S_T \otimes X$ is the free symmetric T-spectrum associated to symmetric space X.

If Y a symmetric T-spectrum and X is a symmetric space, then the symmetric space $Y \otimes X$ inherits the structure of an S_T-module from that of Y, via the morphism
$$S_T \otimes Y \otimes X \xrightarrow{\sigma \otimes X} Y \otimes X,$$
where the map $\sigma : S_T \otimes Y \to Y$ defines the symmetric T-spectrum structure of Y.

Lemma 11.38 *Suppose that Y is a symmetric spectrum and that X is a symmetric space. Then there is a canonical isomorphism of symmetric spectra*
$$Y \wedge_\Sigma (S_T \otimes X) \cong Y \otimes X.$$
In particular, there are natural isomorphisms
$$S_T \wedge_\Sigma Y \cong Y \wedge_\Sigma S_T \cong Y$$
for symmetric spectra Y.

Proof The composite
$$Y \otimes S_T \otimes X \xrightarrow{\tau \otimes X} S_T \otimes Y \otimes X \xrightarrow{m} Y \otimes X$$
induces a natural map $Y \wedge_\Sigma (S_T \otimes X) \to Y \otimes X$. The unit of S_T induces a map of symmetric spaces $X \to S_T \otimes X$ which then induces a map of symmetric spectra $Y \otimes X \to Y \wedge_\Sigma (S_T \otimes X)$. These two maps are inverse to each other.

11.6 The Smash Product

Suppose that $n \geq 0$, and recall the definition of the shift $X[n]$ of a symmetric T-spectrum X from Sect. 11.4. There is a similarly defined shift $Y[n]$ for a symmetric space Y, where

$$Y[n]^k = Y^{n+k},$$

and $\alpha \in \Sigma_k$ acts as the permutation $1 \oplus \alpha \in \Sigma_{n+k}$. The forgetful functor from symmetric T-spectra to symmetric spaces preserves shifts.

Lemma 11.39 *Suppose that $n \geq 0$. There is a natural bijection*

$$\hom(G_n(S^0) \otimes X, Y) \cong \hom(X, Y[n])$$

for morphisms of symmetric T-spectra.

Proof A morphism of symmetric T-spectra

$$h : G_n S^0 \otimes X \to Y$$

can be identified with a sequence of pointed maps $h : X^k \to Y^{n+k}$ which are equivariant for the inclusions $\Sigma_k \to \Sigma_{n+k}$ defined by $\alpha \mapsto 1 \oplus \alpha$, so that the diagrams

$$\begin{array}{ccc} T^p \wedge X^k & \xrightarrow{T^p \wedge h} & T^p \wedge Y^{n+k} \\ \sigma \downarrow & & \downarrow \sigma \\ X^{p+k} \xrightarrow{h} Y^{n+p+k} & \xrightarrow{c(n,p)\oplus 1} & Y^{p+n+k} \end{array} \quad (11.10)$$

commute.

A morphism $h : X \to Y[n]$ can be identified with a sequence of pointed maps $h : X^k \to Y^{n+k}$ which are equivariant for the same inclusions $\Sigma_k \to \Sigma_{n+k}$, so that the diagrams

$$\begin{array}{ccc} T^p \wedge X^k & \xrightarrow{T^p \wedge h} & T^p \wedge Y^{n+k} \\ & & \downarrow \sigma \\ \sigma \downarrow & & Y^{p+n+k} \\ & & \downarrow c(p,n)\oplus 1 \\ X^{p+k} & \xrightarrow{h} & Y^{n+p+k} \end{array} \quad (11.11)$$

commute.

The conditions which are represented by diagrams (11.10) and (11.11) coincide, since $c(n, p) = c(p, n)^{-1}$ in the symmetric group Σ_{n+p+k}.

The proof of the following result is a manipulation of coequalizers of symmetric spaces.

Lemma 11.40 *There is a natural isomorphism*

$$F_n(A) \wedge_\Sigma F_m(B) \cong F_{n+m}(A \wedge B).$$

Proof It follows from Lemma 11.38 that there are isomorphisms

$$(G_m(S^0) \otimes S_T) \wedge_\Sigma X \cong G_m(S^0) \otimes (S_T \wedge_\Sigma X) \cong G_m(S^0) \otimes X,$$

which are natural in symmetric T-spectra X.

There are isomorphisms

$$F_n A \wedge_\Sigma F_m B \cong ((G_n(S^0) \otimes S_T) \wedge A)) \wedge_\Sigma ((G_m(S^0) \otimes S_T) \wedge B))$$
$$\cong ((G_n(S^0) \otimes G_m(S^0)) \otimes S_T) \wedge (A \wedge B)$$

There are isomorphisms of maps of symmetric spectra

$$\hom(G_n(S^0) \otimes G_m(S^0) \otimes S_T, Y) \cong \hom(G_m(S^0) \otimes S_T, Y[n])$$
$$\cong \hom(S_T, Y[m+n])$$
$$\cong \hom(G_{n+m}(S^0) \otimes S_T, Y)$$

by Lemma 11.39. It follows that there is an isomorphism of symmetric T-spectra

$$G_n(S^0) \otimes G_m(S^0) \otimes S_T \cong G_{m+n}(S^0) \otimes S_T.$$

Suppose, as before, that $f : A \to B$ is a fixed cofibration of pointed simplicial presheaves.

Corollary 11.41 *The functor $X \mapsto X[n]$ preserves f-injective fibrations and trivial f-injective fibrations.*

Proof The functor $Y \mapsto G_n(S^0) \otimes Y$ preserves level cofibrations and level f-equivalences.

Lemma 11.42 *There is an isomorphism of symmetric T-spectra*

$$V(E[-n]) \cong F_n(S^0) \wedge_\Sigma V(E),$$

11.6 The Smash Product

which is natural in T-spectra E.

Proof There are natural isomorphisms

$$\hom(V(E[-n]), X) \cong \hom(E, U(X)[n])$$

and

$$\hom(F_n(S^0) \wedge V(E), X) \cong \hom(E, U(X[n]))$$

for T-spectra E and symmetric T-spectra X. It remains to show that there is an isomorphism of T-spectra

$$U(X)[n] \xrightarrow{\cong} U(X[n])$$

which is natural in symmetric T-spectra X.

The T-spectrum $U(X)[n]$ consists of the spaces X^{n+k} and has bonding maps $\sigma : T \wedge X^{n+k} \to X^{n+k+1}$ which come from the symmetric T-spectrum structure for X. The T-spectrum $U(X[n])$ consists of the same spaces, but has bonding maps given by the composites

$$T \wedge X^{n+k} \xrightarrow{\sigma} X^{n+k+1} \xrightarrow{\theta_{k+1}} X^{n+k+1}$$

where θ_{k+1} is induced by an element of the same name in the symmetric group Σ_{n+k+1}. The element θ_{k+1} is a shuffle permutation. Write θ_n for the identity element of Σ_n. Then the desired isomorphism $U(X)[n] \to U(X[n])$ is defined in level k by the composite isomorphism

$$X^{n+k} \xrightarrow{1 \oplus \theta_n} X^{n+k} \xrightarrow{1 \oplus \theta_{n+1}} \cdots \xrightarrow{\theta_{n+k}} X^{n+k},$$

which is induced by the indicated members of the symmetric group Σ_{n+k}.

Corollary 11.43 *The stabilization map*

$$V((S_T \wedge T)[-1-m]) \to V(S_T[-m])$$

is isomorphic to the map

$$1 \wedge_\Sigma s : F_m(S^0) \wedge_\Sigma F_1(T) \to F_m(S^0) \wedge_\Sigma S_T$$

where the map $s : F_1(T) \to S_T$ is induced by the identity on T in level 1.

Recall from Sect. 11.2 that a map $p : X \to Y$ of symmetric T-spectra is a projective fibration if all maps $p : X^n \to Y^n$ are injective fibrations of pointed simplicial presheaves, or equivalently if the induced map $U(p) : U(X) \to U(Y)$ is a strict fibration of T-spectra. A projective cofibration is a map which has the left lifting property with respect to all maps which are projective fibrations and level weak equivalences.

The projective model structure of Lemma 11.4 is cofibrantly generated, and the morphisms $F_n(A) \to F_n(B)$, $n \geq 0$, which are induced by α-bounded cofibrations $A \to B$ of pointed simplicial presheaves generate the projective cofibrations.

If the map $j : E \to F$ is a cofibration of T-spectra, then the induced map $V(j) : V(E) \to V(F)$ is a projective cofibration of symmetric T-spectra.

Theorem 11.44 *Suppose that the map $i : A \to B$ is a projective cofibration and that $j : C \to D$ is a level cofibration of symmetric T-spectra. Then the map*

$$(j,i)_* : (D \wedge_\Sigma A) \cup_{(C \wedge_\Sigma A)} (C \wedge_\Sigma B) \to D \wedge_\Sigma B \qquad (11.12)$$

of symmetric T-spectra is a level cofibration. If j is a projective cofibration, then the map $(j,i)_*$ is a projective cofibration. If j is a stable f-equivalence (respectively level f-equivalence), then $(j,i)_*$ is a stable f-equivalence (respectively level f-equivalence).

Proof For a fixed level cofibration j, the class of projective cofibrations i such that the statements of the theorem hold is closed under composition, pushout and retract (i.e. saturation). These statements are exercises—use Remark 11.3.

It therefore suffices to assume that the projective cofibration i is a map $F_n K \to F_n L$ which is induced by a cofibration $i : K \to L$ of pointed simplicial presheaves, with $n \geq 0$.

The functor

$$X \mapsto G_n(S^0) \otimes X$$

preserves level cofibrations and level trivial cofibrations of symmetric T-spectra X.

There are isomorphisms of symmetric T-spectra

$$F_n(K) = V((S_T \wedge K)[-n]) \cong (S_T \otimes G_n(S^0)) \wedge K,$$

which are natural in pointed simplicial presheaves K. The functor $K \mapsto F_n(K)$ therefore takes local weak equivalences in pointed simplicial presheaves K to level weak equivalences. This functor also takes f-equivalences to level f-equivalences, by the left properness of the f-local model structure on simplicial presheaves. The functor $K \mapsto F_n(K)$ takes cofibrations to level cofibrations.

Lemma 11.40 implies that there are isomorphisms

$$C \wedge_\Sigma F_n K \cong (C \otimes G_n(S^0)) \wedge K$$

which are natural in symmetric T-spectra C and pointed simplicial presheaves K. It follows that the map

$$(j,i)_* : (D \wedge_\Sigma F_n K) \cup (C \wedge_\Sigma F_n L) \to D \wedge_\Sigma F_n L$$

is a level cofibration, and that this map is a level f-equivalence if j is a level f-equivalence (use Corollary 7.24).

11.6 The Smash Product

If the map $j : C \to D$ is a projective cofibration, then it can be approximated by cofibrations of the form $j : F_m A \to F_m B$, in which case the map $(j, i)_*$ is the result of applying the functor F_{m+n} to the cofibration of pointed simplicial presheaves

$$(B \wedge K) \cup_{(A \wedge K)} (A \wedge L) \to B \wedge L,$$

by Lemma 11.40, and is therefore a projective cofibration. It follows that the map (11.12) is a projective cofibration if $j : C \to D$ is a projective cofibration.

To show that the map $(j, i)_*$ is a stable f-equivalence if j is a stable f-equivalence, we first show that the map $(j, i)_*$ is a stable f-equivalence if j is in the saturation of the set S_f.

Suppose that the level cofibration $j : C \to D$ induces a stable f-trivial cofibration

$$(D \wedge_\Sigma F_n A) \cup (C \wedge_\Sigma F_n B) \to D \wedge_\Sigma F_n B,$$

for all cofibrations $A \to B$ of pointed simplicial presheaves and all $n \geq 0$. If $E \to F$ is an α-bounded cofibration of pointed simplicial presheaves then the map

$$((D \wedge_\Sigma F_n B) \wedge E) \cup (((D \wedge_\Sigma F_n A) \cup (C \wedge_\Sigma F_n B)) \wedge F) \to (D \wedge_\Sigma F_n B) \wedge F$$

coincides with the map

$$(D \wedge_\Sigma F_n((B \wedge E) \cup (A \wedge F))) \cup (C \wedge_\Sigma F_n(B \wedge F)) \to D \wedge_\Sigma F_n(B \wedge F)$$

up to isomorphism, as well as with the map

$$((D \wedge F) \wedge_\Sigma F_n A) \cup (((D \wedge E) \cup (C \wedge F)) \wedge_\Sigma F_n B) \to (D \wedge F) \wedge_\Sigma F_n(B).$$

It follows that if $j : C \to D$ induces a stable f-trivial cofibration

$$(D \wedge_\Sigma F_n A) \cup (C \wedge_\Sigma F_n B) \to D \wedge_\Sigma F_n B,$$

for each cofibration $A \to B$ of pointed simplicial presheaves and all $n \geq 0$, then the map

$$(D \wedge E) \cup (C \wedge F) \to D \wedge F$$

has the same property, for each cofibration $E \to F$ of pointed simplicial presheaves.

Each stabilization map

$$F_1(T) \wedge_\Sigma F_n(S^0) \to S_T \wedge_\Sigma F_n(S^0) \tag{11.13}$$

(from Corollary 11.43) is a stable f-equivalence by assumption. If K is a pointed simplicial presheaf, then the induced map

$$(F_1(T) \wedge_\Sigma F_n(S^0)) \wedge K \to (S_T \wedge_\Sigma F_n(S^0)) \wedge K,$$

or equivalently the map
$$F_1(T) \wedge_\Sigma F_n(K) \to S_T \wedge_\Sigma F_n(K),$$
is also a stable f-equivalence.

Suppose that the map $F_1(T) \to A$ be a cofibrant replacement of the map $F_1(K) \to S_T$. Then all of the level cofibrations
$$F_1(T) \wedge_\Sigma F_n(K) \to A \wedge_\Sigma F_n(K)$$
are stable f-equivalences, since the map $A \to S_T$ is a level equivalence by construction and the object $F_n(K)$ is projective cofibrant. The map
$$F_1 T \wedge_\Sigma F_n(S^0) \to A \wedge_\Sigma F_n(S^0)$$
is also a cofibrant replacement of the stabilization map (11.13).

Suppose that $K \to L$ is a cofibration of pointed simplicial presheaves. Then the diagram
$$\begin{array}{ccc} F_1 T \wedge_\Sigma F_n(S^0) \wedge_\Sigma F_m(K) & \longrightarrow & F_1 T \wedge_\Sigma F_n(S^0) \wedge_\Sigma F_m(L) \\ \downarrow & & \downarrow \\ A \wedge_\Sigma F_n(S^0) \wedge_\Sigma F_m(K) & \longrightarrow & A \wedge_\Sigma F_n(S^0) \wedge_\Sigma F_m(L) \end{array}$$
is isomorphic to a diagram of level cofibrations
$$\begin{array}{ccc} F_1(T) \wedge_\Sigma F_{n+m}(K) & \longrightarrow & F_1(T) \wedge_\Sigma F_{n+m}(L) \\ \downarrow & & \downarrow \\ A \wedge_\Sigma F_{n+m}(K) & \longrightarrow & A \wedge_\Sigma F_{n+m}(L) \end{array}$$
in which the vertical maps are stable f-equivalences. It follows from left properness for the f-local stable model structure that the map
$$(A \wedge_\Sigma F_n(S^0) \wedge_\Sigma F_m(K)) \cup (F_1(T) \wedge_\Sigma F_n(S^0) \wedge_\Sigma F_m(L)) \to A \wedge_\Sigma F_n(S^0) \wedge_\Sigma F_m(L)$$
is a stable f-equivalence.

We have shown that the map $(j, i)_*$ is a stable f-equivalence if the level cofibration j is in the saturation of the set S_f.

Suppose now that the level cofibration j is a stable f-equivalence. It follows that all stabilization cofibrations $C \to LC$ (which are in the saturation of S_f) induce stable f-equivalences
$$C \wedge F_n K \to LC \wedge F_n K$$

11.6 The Smash Product

for all pointed simplicial presheaves K. These maps induce stable f-equivalences

$$
\begin{array}{ccccc}
D \wedge F_n K & \longleftarrow & C \wedge F_n K & \longrightarrow & C \wedge F_n L \\
\simeq \downarrow & & \simeq \downarrow & & \simeq \downarrow \\
LD \wedge F_n K & \longleftarrow & LC \wedge F_n K & \longrightarrow & LC \wedge F_n L
\end{array}
$$

in which the vertical maps are stable f-equivalences and the maps induced by the level cofibration $F_n K \to F_n L$ are level cofibrations. It follows from left properness for the f-local stable model structure on symmetric T-spectra that the map of pushouts

$$(D \wedge F_n K) \cup (C \wedge F_n L) \to (LD \wedge F_n K) \cup (LC \wedge F_n L)$$

is a stable f-equivalence. There is a commutative diagram

$$
\begin{array}{ccc}
(D \wedge F_n K) \cup (C \wedge F_n L) & \longrightarrow & D \wedge F_n L \\
\simeq \downarrow & & \downarrow \simeq \\
(LD \wedge F_n K) \cup (LC \wedge F_n L) & \longrightarrow & LD \wedge F_n L
\end{array}
$$

The bottom horizontal map is a level f-equivalence since $C \to D$ is a stable f-equivalence, and the desired result follows.

The map $(j, i)_*$ associated to the maps $j : C \to D$ and $i : A \to B$ of Theorem 11.44 is often called the *pushout smash product* of j and i [46].

Corollary 11.45 *Suppose that the map $g : X \to Y$ is a stable f-equivalence and that A is projective cofibrant. Then the induced map $g \wedge_\Sigma A : X \wedge_\Sigma A \to Y \wedge_\Sigma A$ is a stable f-equivalence.*

Proof Any stable f-trivial cofibration $j : C \to D$ induces a stable f-equivalence $C \wedge_\Sigma A \to D \wedge_\Sigma A$ since A is projective cofibrant, by Theorem 11.44. The result then follows from a standard factorization argument: the map g has a factorization $g = q \cdot j$ where j is a level cofibration and the map q is a trivial injective fibration. The map q therefore has a section by a level cofibration which is a level equivalence.

Corollary 11.46 *Suppose that the maps $i : A \to B$ and $j : C \to D$ are projective cofibrations. Then the induced map*

$$(i, j)_* : (B \wedge_\Sigma C) \cup_{(A \wedge_\Sigma C)} (A \wedge_\Sigma D) \to B \wedge_\Sigma D$$

is a projective cofibration which is stable f-trivial if either i or j is a stable f-equivalence.

Remove the adjective "projective" in the statement of Corollary 11.46, and you have the description of what it means for a model structure to be monoidal, subject to having a symmetric monoidal smash product and provided that the unit is cofibrant [44].

For example, the pointed simplicial set (or presheaf) category with the obvious smash product is monoidal.

The sphere spectrum S_T is projective cofibrant, and is therefore cofibrant in all of the model structures for symmetric T-spectra that we have discussed here.

We then have the following:

Corollary 11.47 *Suppose that T is compact up to f-equivalence, and that the f-local model structure on simplicial presheaves satisfies inductive colimit descent. Then the HSS-stucture on the category of symmetric $(S^1 \wedge T)$-spectra is monoidal.*

Proof The cofibrations for the HSS-structure are the projective cofibrations.

The HSS-structure on symmetric T-spectra is monoidal when it exists, which we have demonstrated (Theorem 11.26) only in the case where T is a suspension of an object which is compact up to f-equivalence.

On the other hand, both the stable f-local structure for symmetric T-spectra and Corollary 11.46 obtain in extreme generality, and there is a universal description of a derived smash product $L(X \wedge_\Sigma Y)$ for symmetric T-spectra X and Y: set

$$L(X \wedge_\Sigma Y) = X_c \wedge_\Sigma Y_c$$

where $\pi_X : X_c \to X$ and $\pi_Y : Y_c \to Y$ are projective cofibrant models for X and Y, respectively.

Suppose that X and Y are T-spectra, and let $X \wedge_n Y$ be the following particular choice of *naive smash product*:

$$(X \wedge_n Y)^{2n} = X^n \wedge Y^n, \quad (X \wedge_n Y)^{2n+1} = X^{n+1} \wedge Y^n,$$

and the bonding maps are specified by:

$$\begin{cases} T \wedge X^n \wedge Y^n \xrightarrow{\sigma \wedge 1} X^{n+1} \wedge Y^n \\ T \wedge X^{n+1} \wedge Y^n \xrightarrow{\tau \wedge 1} X^{n+1} \wedge T \wedge Y^n \xrightarrow{1 \wedge \sigma} X^{n+1} \wedge Y^{n+1} \end{cases}$$

We now formalize a construction from the proof of Lemma 11.42. Suppose that Z is a symmetric T-spectrum, and choose permutations $\theta_n \in \Sigma_n$ for $n \geq 1$. Then there is a spectrum $\theta_*(UZ)$ with level spaces $\theta_*(UZ)^n = Z^n$, and having bonding maps σ_θ given by the composites

$$T \wedge Z^n \xrightarrow{\sigma} Z^{n+1} \xrightarrow{\theta_{n+1}} Z^{n+1}$$

There is a natural isomorphism of T-spectra

$$\nu_\theta : UZ \to \theta_* UZ$$

11.6 The Smash Product

which is defined by finding inductively elements $v_{\theta,n} \in \Sigma_n$. We require that $v_{\theta,n}$ is the identity in levels 0 and 1. Then there is a commutative diagram

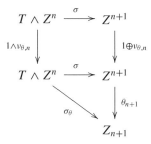

and so

$$v_{\theta,n+1} = \theta_{n+1}(1 \oplus v_{\theta,n}) = \theta_{n+1}(1 \oplus \theta_n)(1 \oplus \theta_{n-1}) \cdots (1 \oplus \theta_2).$$

Suppose now that X and Y are symmetric T-spectra. There are commutative diagrams

$$\begin{array}{ccc} T \wedge X^n \wedge Y^n & \xrightarrow{\sigma \wedge 1} & X^{n+1} \wedge Y^n \\ {\scriptstyle T \wedge c} \downarrow & & \downarrow {\scriptstyle c} \\ T \wedge (X \wedge_\Sigma Y)^{2n} & \xrightarrow{\sigma} & (X \wedge_\Sigma Y)^{2n+1} \end{array}$$

and

$$\begin{array}{ccccc} T \wedge X^{n+1} \wedge Y^n & \xrightarrow{\tau \wedge 1} & X^{n+1} \wedge T \wedge Y^n & \xrightarrow{1 \wedge \sigma} & X^{n+1} \wedge Y^n \\ {\scriptstyle 1 \wedge c} \downarrow & & & & \downarrow {\scriptstyle c} \\ T \wedge (X \wedge_\Sigma Y)^{2n+1} & \xrightarrow{\sigma} & (X \wedge_\Sigma Y)^{2n+2} & \xrightarrow{c_{1,n+1} \oplus 1} & (X \wedge_\Sigma Y)^{2n+2} \end{array}$$

Here, $c : X^p \wedge Y^q \to (X \wedge_\Sigma Y)^{p+q}$ is the canonical map, and $c_{1,n+1} \in \Sigma_{n+2}$ is the shuffle map which moves the number 1 past the numbers $2, \ldots, n+2$ in order.

It follows that the canonical maps $c : X^p \wedge Y^q \to (X \wedge_\Sigma Y)^{p+q}$ determine a natural map of T-spectra

$$c : UX \wedge_n UY \to \theta_* U(X \wedge_\Sigma Y),$$

where

$$\theta_i = \begin{cases} 1 & \text{if } i = 2n+1, \\ c_{1,n+1} \oplus 1 & \text{if } i = 2n+2. \end{cases}$$

Theorem 11.48 *Suppose that T is compact up to f-equivalence and is cycle trivial. Then the natural composite*

$$X \wedge_n Y \to UVX \wedge_n UVY \xrightarrow{c} \theta_*U(VX \wedge_\Sigma VY) \xrightarrow{\theta_*Uj} \theta_*U((VX \wedge_\Sigma VY)_s) \tag{11.14}$$

is a stable f-equivalence if either of the T-spectra X or Y is cofibrant.

Proof Suppose that X is cofibrant.

The functors on both ends of the composition (11.14) respect stable f-equivalences of cofibrant T-spectra Y, and also preserve filtered colimits in Y. Thus, by taking a layer filtration for Y, it is enough to verify that the map (11.14) is a stable f-equivalence when $Y = (S_T \wedge L)[-m]$ for some pointed simplicial presheaf L.

The object Y is now cofibrant, so the same argument says that it suffices to assume that $X = (S_T \wedge K)[-n]$ for some pointed simplicial presheaf K.

It is a consequence of Theorem 10.46 that a map $A \to B$ of cofibrant T-spectra is a stable f-equivalence if and only if the induced map $A \wedge T \to B \wedge T$ is a stable f-equivalence. Thus, by smashing with the object $T^{\wedge(n+m)}$ the map (11.14) is a stable f-equivalence for $X = (S_T \wedge K)[-n]$ and $Y = (S_T \wedge L)[-m]$ if and only if it is a stable f-equivalence when $X = S_T \wedge K$ and $Y = S_T \wedge L$.

The map (11.14) in this last case is stable equivalent to the smash of the composite

$$S_T \wedge_n S_T \to UV(S_T) \wedge_n UV(S_T) \xrightarrow{c} \theta_*U(V(S_T) \wedge_\Sigma V(S_T)) \\ \xrightarrow{\theta_*Uj} \theta_*U((V(S_T) \wedge_\Sigma V(S_T))_s) \tag{11.15}$$

with the pointed simplicial presheaf $K \wedge L$, and it suffices to show that the composite (11.15) is a stable f-equivalence.

The object $V(S_T) = S_T$ is the sphere object for symmetric T-spectra, and the monoid structure of S_T defines an isomorphism

$$m : S_T \wedge S_T \xrightarrow{\cong} S_T.$$

The composite

$$S_T \wedge S_T \xrightarrow[\cong]{m} S_T \xrightarrow{j} (S_T)_s$$

is a stable fibrant model for the symmetric T-spectrum $S_T \wedge S_T$. There is an identification of T-spectra

$$\theta_*U(S_T) = S_T \wedge_n S_T,$$

and the composite

$$\theta_*U(S_T) = S_T \wedge_n S_T \to \theta_*U(S_T \wedge S_T) \xrightarrow[\cong]{\theta_*Um} \theta_*U(S_T)$$

is the identity. The map

$$S_T = U(S_T) \xrightarrow{Uj} U((S_T)_s)$$

is a stable f-equivalence of T-spectra by Proposition 11.29.

11.7 Symmetric T-complexes

Suppose, as in Sect. 10.7, that R is a presheaf of commutative rings with identity, and let T be a pointed simplicial presheaf on the site $\mathcal{C}$.

A simplicial R-module is a simplicial object in the category of presheaves of R-modules. The category of such things is denoted by $s\mathbf{Pre}_R$, following the practice established in Sect. 8.1.

Again as in Sect. 10.7, if A is a simplicial R-module, and K is a pointed simplicial presheaf, we write

$$K \otimes A := R_\bullet(K) \otimes A$$

and

$$A \otimes K := A \otimes R_\bullet(K),$$

where $R_\bullet(K) = R(K)/R(*)$ is the reduced free simplicial R-module which is associated to K. There is a canonical isomorphism

$$K \otimes L \otimes A \cong (K \wedge L) \otimes A \tag{11.16}$$

for simplicial R-modules A and pointed simplicial presheaves K and L, which is natural in all variables.

A *symmetric T-complex* A consists of simplicial R-modules A^s, $s \geq 0$, together with *bonding morphisms*

$$\sigma : T^{\otimes r} \otimes A^s \to A^{r+s},$$

such that the object A^s has a Σ_s-action for $s \geq 0$, and the bonding morphisms σ is equivariant for the block diagonal inclusion $\Sigma_r \times \Sigma_s \subset \Sigma_{r+s}$. A morphism $g : A \to B$ of symmetric T-complexes consists of simplicial R-module maps $g : A^s \to B^s$, $s \geq 0$, which preserve structure. The resulting category $\mathbf{Spt}_T^\Sigma(s\mathbf{Pre}_R)$ is the category of symmetric T-complexes in simplicial R-modules. It is the category of *symmetric T-spectrum objects* in simplicial R-modules, in a suitable sense.

The forgetful functor

$$u : s\mathbf{Pre}_R \to s\mathbf{Pre}_*(\mathcal{C}),$$

taking values in pointed simplicial presheaves, and its left adjoint

$$R_\bullet : s\mathbf{Pre}(\mathcal{C}) \to s\mathbf{Pre}_R$$

together determine an adjoint pair of functors

$$R_\bullet : \mathbf{Spt}_T^\Sigma(\mathcal{C}) \leftrightarrows \mathbf{Spt}_T^\Sigma(s\mathbf{Pre}_R) : u$$

In particular, the forgetful functor $A \mapsto u(A)$ is defined in levels by $u(A)^s = u(A^s)$, and the bonding maps for the symmetric T-spectrum $u(A)$ are the composites

$$T^{\wedge r} \wedge u(A^s) \xrightarrow{\gamma} u((T^{\wedge r}) \otimes A^s) \cong u(T^{\otimes r} \otimes A^s) \xrightarrow{u(\sigma)} u(A^{r+s}).$$

Recall from (10.21) that the natural pointed map

$$\gamma : K \wedge u(A) \to u(K \otimes A)$$

is defined, for pointed simplicial presheaves K and simplicial R-modules A, by the assignment $x \wedge a \mapsto x \otimes a$ in sections. The functor $R_\bullet$ is defined levelwise by $R_\bullet(X)^s = R_\bullet(X^s)$ for a symmetric T-spectrum X, and instances of the natural isomorphism (11.16) are used to define the bonding maps.

Example 11.49 The symmetric T-complex, which is defined by

$$H(R) := R_\bullet(S_T)$$

is the *sphere T-spectrum object* for the category of symmetric T-complexes.

More generally, suppose that A is a simplicial R-module. Then the *Eilenberg–Mac Lane object* $H(A)$ is the symmetric T-complex

$$H(A) := S_T \otimes A = R_\bullet(S_T) \otimes A.$$

The object $H(A)$ is the suspension spectrum object associated to A in the category of symmetric T-complexes, in that there is a natural isomorphism

$$\hom(H(A), B) \cong \hom(A, B^0)$$

for simplicial R-modules and symmetric T-complexes B.

The T-spectrum $u(H(A))$ is the Eilenberg–Mac Lane T-complex for A, as defined in Example 10.79.

The category of symmetric T-complexes inherits a model structure from the injective model structure on symmetric T-spectra (Proposition 11.10), by transfer of structure. We have used similar arguments for simplicial modules (Theorem 8.6) and T-complexes (Proposition 10.80).

Explicitly, say that a map $g : A \to B$ is a *level equivalence* if the underlying map $u(g) : u(A) \to u(B)$ of symmetric T-spectra is a level equivalence. Say that g is an *injective fibration* if the underlying map $u(g)$ is an injective fibration of symmetric T-spectra. An *injective cofibration* of T-complexes is a map which has the left lifting property with respect to all trivial injective fibrations.

Lemma 11.50 *The category* $\mathbf{Spt}_T^\Sigma(s\mathbf{Pre}_R)$ *of symmetric T-complexes, together with the classes of level weak equivalences, injective fibrations and injective cofibrations, satisfies the conditions for a proper closed simplicial model category. This model structure is cofibrantly generated. Every injective cofibration is a monomorphism.*

Proof Suppose that α is a regular cardinal which is sufficiently large that $\alpha > |\operatorname{Mor}(\mathcal{C})|$, $\alpha > |R|$ for the presheaf of rings R, and $\alpha > |T|$. Recall

11.7 Symmetric T-complexes

that the injective model structure on the category of symmetric T-spectra is cofibrantly generated, by the α-bounded level cofibrations and the α-bounded level trivial cofibrations.

The functor
$$R_\bullet : \mathbf{Spt}_T^\Sigma(\mathcal{C}) \to \mathbf{Spt}_T^\Sigma(s\mathbf{Pre}_R)$$
preserves level weak equivalences, by Lemma 8.2.

The functor $R_\bullet$ takes level cofibrations $A \to B$ to injective cofibrations, by an adjointness argument. Each map $R_\bullet A \to R_\bullet B$ also consists of cofibrations $R_\bullet(A^s) \to R_\bullet(B^s)$ in the model structure for simplicial R-modules of Theorem 8.6.

Thus, if we suppose that $i : A \to B$ is a trivial injective cofibration of symmetric T-spectra and the diagram

$$\begin{array}{ccc} R_\bullet A & \longrightarrow & X \\ R_\bullet(i) \downarrow & & \downarrow i_* \\ R_\bullet B & \longrightarrow & Y \end{array}$$

is a pushout of symmetric T-complexes, then the map i_* is an injective cofibration and a level equivalence of symmetric T-complexes.

It follows that every map $g : A \to B$ of symmetric T-complexes has factorizations

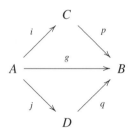

such that p is an injective fibration, i is a trivial injective cofibration which has the left lifting property with respect to all injective fibrations, q is an injective fibration and a level equivalence, and j is an injective cofibration. This proves the factorization axiom **CM5**, and the lifting axiom **CM4** follows by a standard argument. The remaining closed model axioms are easily verified.

The n-simplices of the function complex $\mathbf{hom}(A, B)$ are the maps
$$A \otimes (\Delta_+^n) \to B,$$
and the simplicial model axiom **SM7** is a consequence of the corresponding statement for the injective model structure on symmetric T-spectra.

This model structure for symmetric T-complexes is cofibrantly generated by the α-bounded injective cofibrations and the α-bounded trivial injective cofibrations.

Every injective cofibration $A \to B$ of symmetric T-complexes consists of cofibrations $A^s \to B^s$ of simplicial R-modules, and is therefore a monomorphism. In effect, all maps $R_\bullet C \to R_\bullet D$ which are induced by level cofibrations $C \to D$ of symmetric T-spectra have this property, and the class of cofibrations having this property is closed under pushout, composition and retraction—it therefore includes all injective cofibrations.

The claim about properness follows from properness for the model structure on simplicial R-modules of Theorem 8.6.

The model structure of Lemma 11.50 is the *injective model structure* for symmetric T-complexes.

We have a bounded monomorphism property:

Lemma 11.51 *Suppose that α is a regular cardinal which is sufficiently large in the sense described above. Suppose given a diagram of monomorphisms*

of symmetric T-complexes such that the map i is a level equivalence and A is α-bounded. Then there is a factorization of j by monomorphisms $A \subset B \subset Y$ such that B is α-bounded and the map $B \cap X \to B$ is a strict equivalence.

Proof The proof is essentially the same as that of Lemma 8.11, which is the corresponding statement for simplicial R-modules.

The quotient symmetric T-complex Y/X is a filtered colimit of the quotients $B/(B \cap X)$, where B is α-bounded. Every homology class $v \in H_k(A/(A \cap X))(U)$ maps to 0 in the sheaf $\tilde{H}_k(Y/X)$ and therefore maps to 0 (locally) in some group $H_k(A'/(A' \cap Y))(V)$ along some covering sieve $V \to U$ for U, where $A \subset A' \subset Y$ and A' is α-bounded.

This is true for all homology classes $v \in H_k(A/(A \cap X))(U)$, for all k and U. This is a α-bounded list of elements, so there is a α-bounded subobject A_1 of Y with $A \subset A_1$ such that the induced map

$$\tilde{H}_*(A/(A \cap X)) \to \tilde{H}_*(A_1/(A_1 \cap X))$$

is the 0-map.

Proceed inductively to find an increasing family

$$A \subset A_1 \subset A_2 \subset \dots$$

of α-bounded subobjects of Y such that the induced maps

$$\tilde{H}_*(A_k/(A_k \cap X)) \to \tilde{H}_*(A_{k+1}/(A_{k+1} \cap X))$$

11.7 Symmetric T-complexes

are 0. Set $B = \varinjlim A_i$. Then B is α-bounded and $\tilde{H}_*(B/(B \cap X)) = 0$, so that the map $B \cap X \to B$ is a level equivalence of symmetric T-complexes.

Suppose (as in Sect. 10.3) that $\mathcal{F}$ is a set of cofibrations of symmetric T-spectra which includes the set J of α-bounded level trivial cofibrations. Suppose also that the induced maps

$$(E \wedge D) \cup (F \wedge C) \to F \wedge D \tag{11.17}$$

are in $\mathcal{F}$ for each morphism $E \to F$ in $\mathcal{F}$ and each α-bounded cofibration $C \to D$ of pointed simplicial presheaves, as in Sect. 10.3.

Write $R_\bullet(\mathcal{F})$ for the set of maps of symmetric T-complexes $R_\bullet A \to R_\bullet B$ which are induced by members $A \to B$ of the set $\mathcal{F}$.

Choose a regular cardinal β which is sufficiently large as above, and such that $\beta > |\mathcal{F}|$ and $\beta > |B|$ for any cofibration $A \to B$ of $\mathcal{F}$. Choose a cardinal λ such that $\lambda > 2^\beta$.

Every map $g : X \to Y$ of symmetric T-complexes has a functorial factorization

such that the map i is a cofibration which is in the saturation of $R_\bullet(\mathcal{F})$ and the map p has the right lifting property with respect to all members of $R_\bullet(\mathcal{F})$. The construction of this factorization uses a small object argument which terminates after λ steps.

Set

$$L_\mathcal{F}(X) = E_\lambda(X \to 0)$$

for all symmetric T-complexes X.

Say that a map $X \to Y$ of symmetric T-complexes is an $L_\mathcal{F}$-equivalence if the induced map $L_\mathcal{F}X \to L_\mathcal{F}Y$ is a level equivalence of symmetric T-complexes.

Then we have the following:

Lemma 11.52

1) Suppose that the assignment $t \mapsto X_t$ defines a diagram of monomorphisms indexed by $t < \gamma$ where γ is a cardinal such that $\gamma > 2^\beta$. Then the natural map

$$\varinjlim_{t<\gamma} L_\mathcal{F}(X_t) \to L_\mathcal{F}(\varinjlim_{t<\gamma} X_t)$$

is an isomorphism.

2) Suppose that ζ is a cardinal with $\zeta > \beta$, and let $B_\zeta(X)$ denote the filtered system of subobjects of X having cardinality less than ζ. Then the natural map

$$\varinjlim_{Y \in B_\zeta(X)} L_\mathcal{F}(Y) \to L_\mathcal{F}(X)$$

is an isomorphism.
3) The functor $X \mapsto L_{\mathcal{F}}(X)$ preserves monomorphisms.
4) Suppose that U, V are subobjects of a T-complex X. Then the natural map

$$L_{\mathcal{F}}(U \cap V) \to L_{\mathcal{F}}(U) \cap L_{\mathcal{F}}(V)$$

is an isomorphism.
5) If $|X| \leq 2^\mu$ where $\mu \geq \lambda$ then $|L_{\mathcal{F}}(X)| \leq 2^\mu$.

Proof This result is a consequence of Lemma 8.38, just like Lemma 10.81.

Let κ be the successor cardinal for 2^μ, where μ is cardinal of statement 5) of Lemma 11.52. Then κ is a regular cardinal, and Lemma 11.52 implies that if a symmetric T-spectrum X is κ-bounded then $L_{\mathcal{F}}(X)$ is κ-bounded. We now have a bounded monomorphism property for $L_{\mathcal{F}}$-equivalences:

Lemma 11.53 *Suppose given a monomorphism $i : X \to Y$ of symmetric T-complexes which is an $L_{\mathcal{F}}$-equivalence, and suppose that $A \subset Y$ is a κ-bounded subobject. Then there is a κ-bounded subobject $B \subset Y$ with $A \subset B$, such that the map $B \cap X \to B$ is an $L_{\mathcal{F}}$-equivalence.*

Proof This result is a consequence of Lemma 11.51 and Lemma 11.52, via a proof which is formally the same as that of Lemma 7.17. See the proof of Lemma 8.38 also.

A map $p : X \to Y$ of symmetric T-complexes is said to be $\mathcal{F}$-*injective* if it has the right lifting property with respect to all members of the set $R_\bullet(\mathcal{F})$. An adjointness argument says that $p : X \to Y$ is $\mathcal{F}$-injective if and only if the underlying map $p_* : u(X) \to u(Y)$ is an $\mathcal{F}$-injective map of symmetric T-spectra. A symmetric T-complex Z is $\mathcal{F}$-injective if the map $Z \to 0$ is $\mathcal{F}$-injective, or equivalently if the underlying symmetric T-spectrum $u(Z)$ is $\mathcal{F}$-injective.

Following Sect. 7.2, a map of symmetric T-complexes $X \to Y$ is an $\mathcal{F}$-*equivalence* if and only if the map

$$\mathbf{hom}(Y_c, Z) \to \mathbf{hom}(X_c, Z)$$

is a weak equivalence of simplicial sets for every $\mathcal{F}$-injective object Z, and where $X_c \to X$ is a natural injective cofibrant model construction. It is a consequence of Lemma 7.6 that a map $X \to Y$ is an $\mathcal{F}$-equivalence if and only if it is an $L_{\mathcal{F}}$-equivalence.

An $\mathcal{F}$-*fibration* is a map which has the right lifting property with respect to all cofibrations of symmetric T-complexes which are $\mathcal{F}$-equivalences. The $\mathcal{F}$-fibrations coincide with the $L_{\mathcal{F}}$-fibrations of Theorem 7.10, for the category of symmetric T-complexes.

We have shown, in Lemma 11.50, that every injective cofibration is a monomorphism. We can therefore use the methods of Sect. 7.1 to prove the following:

11.7 Symmetric T-complexes

Theorem 11.54 *Suppose that $\mathcal{F}$ is a set of cofibrations of symmetric T-spectra which satisfies the closure condition of (11.17). Then the category*

$$\mathbf{Spt}_T^\Sigma(s\mathbf{Pre}_R(\mathcal{C}))$$

of symmetric T-complexes, together with the classes of injective cofibrations, $\mathcal{F}$-equivalences and $\mathcal{F}$-fibrations, satisfies the axioms for a cofibrantly generated closed simplicial model category. This model structure is left proper.

Example 11.55 The *stable model structure* for symmetric T-complexes arises, via Theorem 11.54, by formally inverting the set $\mathcal{F}$ which is generated by the set J of α-bounded level trivial cofibrations together with a set of injective cofibrant replacements of the (stabilization) maps of symmetric T-spectra

$$V(S_T \wedge T[-1-n]) \to V(S_T[-n]). \tag{11.18}$$

Generation is with respect to the closure property (11.17), by analogy with the corresponding property for sets of cofibrations of T-spectra in Sect. 10.3.

An $L_{\mathcal{F}}$-equivalence for this theory is called a *stable equivalence*, and an $L_{\mathcal{F}}$-fibration is a *stable fibration*.

Example 11.56 Suppose that $f : A \to B$ is a cofibration of simplicial presheaves. Following Example 11.15, let J_f be the set of all maps

$$F_n(C_+) \to F_n(D_+)$$

which are induced by a fixed set of generators $C \to D$ for the trivial cofibrations of the f-local model structure on simplicial presheaves. Let S_f be the set of cofibrations of symmetric T-spectra which is generated by the set J_f, together with injective cofibrant replacements of all maps (11.18). The model structure for the set S_f which is given by Theorem 11.54 is the *f-local stable model structure* for symmetric T-complexes.

An $L_{\mathcal{F}}$-equivalence for this theory is called a *stable f-equivalence*, and an $L_{\mathcal{F}}$-fibration is a *stable f-fibration*.

It follows from Corollary 7.12 that the stable f-fibrant symmetric T-complexes Z are those objects whose underlying symmetric T-spectra are stable f-fibrant. The underlying T-spectrum of a stable f-fibrant symmetric T-spectrum is also stable f-fibrant. It follows (see Example 11.15 and Remark 8.40) that all simplicial R-modules Z^n are f-fibrant, and all simplicial R-module maps $\sigma_* : Z^n \to \Omega_T Z^{n+1}$ are local (even sectionwise) weak equivalences for such objects Z.

Example 11.57 Following Example 11.16, suppose that S'_f is the set of cofibrations of symmetric T-spectra which is generated by the set J_f alone. The model structure for symmetric T-complex which corresponds to the set S'_f is the *f-injective model structure*. The fibrant objects for this structure and those symmetric T-complexes Z which are level f-fibrant in the sense that all simplicial R-modules are fibrant for the f-local model structure of Theorem 8.39. The trivial fibrations for this structure are level weak equivalences.

The free module functor

$$R : s\mathbf{Pre}(\mathcal{C}) \to s\mathbf{Pre}_R$$

preserves f-equivalences (see Remark 8.40). All members of the set S'_f are level f-equivalences, so that all members of the set $R_\bullet(S'_f)$ are level f-equivalences and level cofibrations of symmetric T-complexes. It follows (see the argument for Lemma 11.50) that any f-injective model $j : A \to LA$ is a level f-equivalence. One concludes that the weak equivalences for the f-injective model structure on the category of symmetric T-complexes are the level f-equivalences.

Following Sect. 11.1, say that a *symmetric complex X* consists of simplicial R-modules X^n, $n \geq 0$, together with R-linear symmetric group actions

$$\Sigma_n \times X^n \to X^n.$$

A morphism $g : X \to Y$ of symmetric complexes consists of morphisms of simplicial R-modules $X^n \to Y^n$, $n \geq 0$, which respect the symmetric group actions.

The tensor product $X \otimes Y$ of symmetric complexes X and Y is the symmetric complex which is defined in level n by

$$(X \otimes Y)^n = \bigoplus_{r+s=n} \Sigma_n \otimes_{\Sigma_r \times \Sigma_s} X^r \otimes Y^s.$$

There is a natural twist isomorphism

$$\tau : X \otimes Y \xrightarrow{\cong} Y \otimes X$$

which is defined in bidegree (r, s) by the composite

$$X^r \otimes Y^s \xrightarrow{\tau} Y^s \otimes X^r \xrightarrow{in} (Y \otimes X)^{s+r} \xrightarrow{c_{r,s}} (Y \otimes X)^{r+s},$$

where $c_{r,s} \in \Sigma_{r+s}$ is shuffle defined in (11.2).

From this point of view, a symmetric T-complex is a symmetric complex X, together with an action

$$H(R) \otimes X \to X$$

in the category of symmetric complexes.

Every simplicial R-module A determines a symmetric complex $G_n(A)$ with

$$G_n(A)^k = \begin{cases} \Sigma_n \otimes A & \text{if } k = n, \text{ and} \\ 0 & \text{otherwise.} \end{cases}$$

There is a natural isomorphism

$$\hom(G_n(A), X) \cong \hom(A, X^n)$$

that relates morphisms $G_n(A) \to X$ of symmetric complexes to morphisms $A \to X^n$ of simplicial R-modules.

11.7 Symmetric T-complexes

The symmetric T-complex

$$F_n(A) := H(R) \otimes G_n(A)$$

is the free symmetric T-complex associated to the symmetric complex $G_n(A)$, and it follows that there is a natural isomorphism

$$\hom(F_n(A), Y) \cong \hom(A, Y^n)$$

relating morphisms of symmetric T-complexes to morphisms of simplicial R-modules.

There is a forgetful functor

$$U_R : \mathbf{Spt}_T^\Sigma(s\mathbf{Pre}_R) \to \mathbf{Spt}_T(s\mathbf{Pre}_R)$$

from symmetric T-complexes to T-complexes. This functor has a left adjoint

$$V_R : \mathbf{Spt}_T(s\mathbf{Pre}_R) \to \mathbf{Spt}_T^\Sigma(s\mathbf{Pre}_R)$$

which is defined, following the definition of the functor V of Sect. 11.1, by the requirement that

$$V_R(H(A)[-n]) = F_n(A).$$

See also Example 10.79.

We shall now focus on the f-local stable model structure for symmetric T-complexes of Example 11.56.

Lemma 11.58 *The adjoint pair of functors*

$$V_R : \mathbf{Spt}_T(s\mathbf{Pre}_R) \leftrightarrows \mathbf{Spt}_T^\Sigma(s\mathbf{Pre}_R) : U_R$$

defines a Quillen adjunction between the respective stable f-local model structures.

Proof If the symmetric T-complex Z is stable f-fibrant then underlying T-complex $U_R(Z)$ is stable f-fibrant.

If $p : X \to Y$ is an injective fibration of symmetric T-complexes, then the underlying map of symmetric T-spectra is an injective fibration, which implies that the pointed simplicial presheaf maps underlying all level maps $X^n \to Y^n$ are injective fibrations. This means that all maps $X^n \to Y^n$ are injective fibrations of simplicial R-modules, and so the map $U_R(p)$ is a strict fibration of T-complexes.

If p is a trivial stable f-fibration of symmetric T-complexes, then all maps $X^n \to Y^n$ are local weak equivalences, and so the map $U_R(p)$ is a trivial stable f-fibration of T-complexes. It follows that the functor V_R takes cofibrations of T-complexes to injective cofibrations of symmetric T-complexes.

Suppose that $i : A \to B$ is a cofibration of T-complexes and that Z is a stable f-fibrant symmetric T-complex. Then the induced map

$$i^* : \mathbf{hom}(V_R(B), Z) \to \mathbf{hom}(V_R(A), Z)$$

of function complexes is isomorphic to the map

$$i^* : \mathbf{hom}(B, U_R(Z)) \to \mathbf{hom}(A, U_R(Z)),$$

and so one instance of i^* is a weak equivalence if and only if the other one is. The object $U_R(Z)$ is stable f-fibrant, so both maps i^* are weak equivalences if i is a stable f-trivial cofibration. This is true for all stable f-fibrant symmetric T-complexes Z, so that the map $i_* : V_R(A) \to V_R(B)$ is a stable f-trivial cofibration if the map i is a stable f-trivial cofibration.

Suppose that $n > 0$. The *shift* $A[n]$ of a symmetric T-complex A is defined as for symmetric T-spectra. We have

$$A[n]^k = A^{n+k}.$$

and bonding maps $T^p \otimes A[n]^k \to A[n]^{p+k}$ defined by the composite

$$T^p \otimes A^{n+k} \xrightarrow{\sigma} A^{p+n+k} \xrightarrow{c(p,n)\oplus 1} A^{n+p+k}.$$

The symmetric T-complex $\Omega_T A$ is also defined by analogy with the corresponding construction for symmetric T-spectra:

$$\Omega_T A^n = \mathbf{Hom}(T, A^n),$$

with bonding maps

$$T^p \otimes \mathbf{Hom}(T, A^n) \to \mathbf{Hom}(T, A^{p+n})$$

defined by adjunction. The adjoint bonding maps $A^n \to \Omega_T A^{1+n}$ define a natural map of symmetric T-complexes

$$\tilde{\sigma} : A \to \Omega_T A[1],$$

by analogy with the corresponding map for symmetric T-spectra in (11.8).

The object $Q_T^\Sigma A$ is defined as a symmetric R-complex, by the assignment

$$Q_T^\Sigma A = I(\varinjlim_n \Omega_T^n (IA)[n]),$$

where $A \to IA$ is a natural choice of f-injective fibrant model as in Example 11.57. This construction is functorial in symmetric T-complexes A, and there is a natural map

$$\eta : A \to Q_T^\Sigma A.$$

11.7 Symmetric T-complexes

The map η is a level equivalence if A is stable f-fibrant.

Remark 11.59 It is not clear that the natural f-injective fibrant model $A \to IA$ for symmetric T-complexes A restricts to an f-injective fibrant model of the underlying symmetric T-spectrum $u(A)$. The object $u(IA)$ is certainly f-injective fibrant in symmetric T-spectra, but the map $u(A) \to u(IA)$ may not be a level f-equivalence of symmetric T-spectra in general. See Remark 8.40.

On the other hand, there are natural isomorphisms of symmetric T-spectra

$$u(\Omega_T A) \cong \Omega_T(u(A))$$

and

$$u(A[n]) \cong u(A)[n]$$

for all symmetric T-complexes A, and the canonical map $\tilde\sigma : u(A) \to \Omega_T u(A)[1]$ is naturally isomorphic to the image of the map of symmetric T-complexes $\tilde\sigma : A \to \Omega_T A[1]$ under the forgetful functor

$$u : \mathbf{Spt}_T^\Sigma(s\mathbf{Pre}_R) \to \mathbf{Spt}_T^\Sigma(\mathcal{C}).$$

We have the following analogue of Theorem 11.19 for symmetric T-complexes:

Theorem 11.60 *Suppose that T is compact up to f-equivalence. Suppose that $g : A \to B$ is map of symmetric T-complexes such that the map $U_R(g) : U_R(A) \to U_R(B)$ is a stable f-equivalence of T-complexes. Then g is a stable f-equivalence.*

Proof This result has the same proof as does Theorem 11.19. We use Theorem 10.87 in place of Theorem 10.32.

We have the Quillen adjunction

$$V_R : \mathbf{Spt}_T(s\mathbf{Pre}_R) \leftrightarrows \mathbf{Spt}_T^\Sigma(s\mathbf{Pre}_R) : U_R$$

of stable f-local model structures from Lemma 11.58, and we know from Theorem 11.60 that the forgetful functor U_R reflects stable f-equivalences if the parameter object T is compact up to f-equivalence.

We now proceed to demonstrate that this Quillen adjunction is a Quillen equivalence under the additional assumption that T is cycle trivial. More formally, we have the following result:

Theorem 11.61 *Suppose that the parameter object T is compact up to f-equivalence and is cycle trivial. Then the adjoint functors*

$$V_R : \mathbf{Spt}_T(s\mathbf{Pre}_R) \leftrightarrows \mathbf{Spt}_T^\Sigma(s\mathbf{Pre}_R) : U_R$$

define a Quillen equivalence between the respective f-stable model structures.

The proof is by analogy with the proof of Theorem 11.36, and will only be sketched. It follows from Proposition 11.64 below in the same way that Theorem 11.36 follows from Proposition 11.29.

There is a category and a theory of T-bicomplexes: a *T-bicomplex* A is a T-complex object in T-complexes. Alternatively A consists of T-complexes A^s together with bonding maps $\sigma_v : A^s \otimes T \to A^{s+1}$ of T-complexes. The structure therefore consists of simplicial R-modules $A^{r,s}$ together with maps $\sigma_h : T \otimes A^{r,s} \to A^{r+1,s}$ and $\sigma_{v*} : T \otimes A^{r,s} \to A^{r,s+1}$ such that the diagrams

$$\begin{array}{ccc}
T \otimes T \otimes A^{r,s} & \xrightarrow{T \otimes \sigma_h} & T \otimes A^{r+1,s} \\
{\scriptstyle \tau \otimes A^{r,s}} \downarrow & & \downarrow {\scriptstyle \sigma_{v*}} \\
T \otimes T \otimes A^{r,s} & & \\
{\scriptstyle \tau \otimes \sigma_{v*}} \downarrow & & \\
T \otimes A^{r,s+1} & \xrightarrow{\sigma_h} & A^{r+1,s+1}
\end{array}$$

commute, where $\tau : T \otimes T \to T \otimes T$ is the automorphism which flips tensor factors. A map $A \to B$ of T-bicomplexes is the obvious thing.

The diagonal $d(A)$ is the $(T \wedge T)$-complex with $d(A)^n = A^{n,n}$ and with bonding maps defined by the composites

$$(T \wedge T) \otimes A^{n,n} \cong T \otimes T \otimes A^{n,n} \xrightarrow{T \otimes \sigma_h} T \otimes A^{n+1,n} \xrightarrow{\sigma_{v*}} A^{n+1,n+1}.$$

A map $g : A \to B$ of T-bicomplexes is *strict f-equivalence* if all maps $A^{r,s} \to B^{r,s}$ are f-equivalences of simplicial R-modules. The map g is said to be a *stable f-equivalence* if the map $d(A) \to d(B)$ is a stable equivalence of $(T \wedge T)$-complexes. Every strict f-equivalence of T-bicomplexes is a stable f-equivalence.

We also have the following analogue of Lemma 10.56, with the same proof:

Lemma 11.62 *Suppose that T is compact up to f-equivalence. Suppose that $g : A \to B$ is a map of T-bicomplexes such that all constituent maps of T-complexes $A^s \to B^s$ are stable f-equivalences of T-complexes. Then the map g is a stable f-equivalence of T-bicomplexes.*

A T-bicomplex Z is said to be *stable f-fibrant* if all simplicial R-modules $Z^{r,s}$ are f-fibrant and the adjoint bonding maps $Z^{r,s} \to \Omega_T Z^{r+1,s}$ and $Z^{r,s} \to \Omega_T Z^{r,s+1}$ are f-equivalences. Then Lemma 11.28 has the following consequence:

Corollary 11.63 *Suppose that $g : Z \to W$ is a map of stable f-fibrant T-bicomplexes. Then g is a stable f-equivalence if and only if all simplicial R-module maps $g : Z^{r,s} \to W^{r,s}$ are f-equivalences.*

The following result is the R-linear analogue of Proposition 11.29:

Proposition 11.64 *Suppose that T is compact up to f-equivalence and is cycle trivial. Suppose that A is a simplicial R-module, and that $j : H(A) \to Z$ is a*

11.7 Symmetric T-complexes

stable f-fibrant model in the symmetric T-complex category. Then the underlying map $U_R(H(A)) \to U_R(Z)$ is a stable f-equivalence of T-complexes.

Proof The proof is by analogy with the proof of Proposition 11.29.

Let $X(A)$ be the T-bicomplex with

$$X(A)^{r,s} = T^{\otimes r} \otimes A \otimes T^{\otimes s},$$

and observe that $X(A)$ can be intepreted as a T-spectrum object in symmetric T-complexes. Specifically, $X(A)^{*,s}$ is the symmetric T-complex $H(A) \otimes T^{\otimes s}$.

The stable f-fibrant models

$$X(A)^{r,*} \to L(X(A)^{r,*})$$

give a map $j : X(A) \to L(X(A))$ of symmetric T-spectrum objects in T-complexes which induces a stable f-equivalence of T-bicomplexes by Lemma 11.62. The stable f-fibrant models $X(A)^{*,s} \to L(X(A)^{*,s})$ of T-complexes also define a stable f-equivalence $U_R(X(A))L(U_R X(A))$ of T-bicomplexes.

The T-bicomplexes $U_R(L(A))$ and $L(U_R(X(A)))$ are stable f-fibrant by Theorem 10.89, and there are stable f-equivalences of T-complexes

$$U_R(L(X(A))^{*,0} \xrightarrow{\simeq} L(U_R(L(X(A))))^{*,0} \xleftarrow{\simeq} L(U_R(X(A)))^{*,0}$$

by Corollary 11.63. The map $U_R(H(A)) \to L(U_R(X(A)))^{*,0}$ is a stable f-equivalence of T-complexes by construction, so the map

$$U_R(H(A)) \to U_R(L(X(A)^{*,0}))$$

is a stable f-equivalence of T-complexes. The map $H(A) \to L(X(A))^{*,0}$ is therefore a map of symmetric T-complexes whose underlying map of T-complexes is a stable f-equivalence. The T-complex underlying $L(X(A))^{*,0}$ is stable f-fibrant, so an f-injective model $L(X(A))^{*,0} \to I(L(X(A))^{*,0})$ defines a stable f-fibrant object $I(L(X(A))^{*,0})$ in symmetric T-complexes. The composite

$$H(A) \to L(X(A))^{*,0} \to I(L(X(A))^{*,0})$$

has an underlying map of T-complexes which is a stable f-equivalence. This composite is therefore a stable f-equivalence of symmetric R-complexes by Theorem 11.60.

One proves the analogue of Theorem 11.35, which asserts that the map

$$X \to U_R V_R(X) \xrightarrow{U_R(j)} U_R(V_r(X)_s)$$

is a stable f-equivalence of T-complexes for all cofibrant objects X. This result is a formal consequence of Proposition 11.64 which uses the compactness of T and the fact that the functor V_R preserves stable f-equivalences between cofibrant T-complexes.

The final argument for Theorem 11.61 is now formal, as is the corresponding argument for Theorem 11.36.

The tensor product for symmetric T-complexes is defined by analogy with the smash product construction for symmetric T-spectra. Specifically, if A and B are symmetric T-complexes, then the tensor product $A \otimes_\Sigma B$ is defined in the category of symmetric complexes by the coequalizer

$$H(R) \otimes A \otimes B \rightrightarrows A \otimes B \to A \otimes_\Sigma B.$$

The two arrows of the picture are $m_A \otimes B$ and the composite

$$H(R) \otimes A \otimes B \xrightarrow{\tau \otimes B} A \otimes H(R) \otimes B \xrightarrow{A \otimes m_B} A \otimes B$$

where the maps $m_A : H(R) \otimes A \to A$ and $m_N : H(R) \otimes B \to B$ define the symmetric T-complex structures of A and B, respectively.

There is a naive tensor product $A \otimes_n B$ of T-complexes A and B, which is defined by analogy with the naive smash product of T-spectra. Explicitly,

$$(A \otimes B)^{2n} = A^n \otimes B^n, \; (A \otimes_n B)^{2n+1} = A^{n+1} \otimes B^n,$$

and the bonding maps are specified by:

$$\begin{cases} T \otimes A^n \otimes B^n \xrightarrow{\sigma \otimes 1} A^{n+1} \otimes B^n, \\ T \otimes A^{n+1} \otimes B^n \xrightarrow{\tau \otimes 1} A^{n+1} \otimes T \otimes B^n \xrightarrow{1 \otimes \sigma} A^{n+1} \otimes B^{n+1}. \end{cases}$$

As in the nonabelian case, there is a natural map

$$c : U_R(X) \otimes_n U_R(Y) \to \theta_* U(X \otimes_\Sigma Y)$$

for symmetric T-complexes X and Y. The map c is defined by the canonical maps $c : X^p \otimes Y^q \to (X \otimes_\Sigma Y)^{p+q}$, and $\theta_* U_R(X \otimes_\Sigma Y)$ is the T-complex with bonding maps

$$T \otimes (X \otimes_\Sigma Y)^n \to (X \otimes_\Sigma Y)^{1+n} \xrightarrow{\theta_{1+n}} (X \otimes_\Sigma Y)^{1+n},$$

where θ_i is the permutation with

$$\theta_i = \begin{cases} 1 & \text{if } i = 2n+1, \\ c_{1,n+1} & \text{if } i = 2n+2. \end{cases}$$

As in the nonabelian case, there is a natural isomorphism of T-complexes

$$\theta_* U_R(X \otimes_\Sigma Y) \cong U_R(X \otimes_\Sigma Y).$$

Then we have the following analogue of Theorem 11.48:

11.7 Symmetric T-complexes

Theorem 11.65 *Suppose that T is compact up to f-equivalence and is cycle trivial. Then the natural composite*

$$A \otimes_n B \to U_R V_R(A) \otimes_n U_R V_R(B) \xrightarrow{c} \theta_* U_R(V_R(A) \otimes_\Sigma V_R(B))$$

$$\xrightarrow{U_R(j)} \theta_* U_R((V_R(A) \otimes_\Sigma V_R(B))_s)$$

is a stable f-equivalence of T-complexes if either A or B is a cofibrant T-complex.

In the statement of Theorem 11.65, the map

$$j : V_R(A) \otimes_\Sigma V_R(B) \to (V_R(A) \otimes_\Sigma V_R(B))_s$$

is a stable f-fibrant model in symmetric T-complexes.

The proof of Theorem 11.65 is analogous to that of Theorem 11.48. For this proof, we need to know that an analogue of Corollary 11.45 holds, namely that the functor $X \mapsto X \otimes_\Sigma Y$ preserves stable f-equivalences of symmetric T-complexes if Y is projective cofibrant in a suitable sense.

In fact, one proves the following symmetric T-complex analogue of Theorem 11.44:

Theorem 11.66 *Suppose that $i : A \to B$ is a projective cofibration and that $j : C \to D$ is a level cofibration of symmetric T-complexes. Then the map*

$$(j, i)_* : (D \wedge_\Sigma A) \cup_{(C \wedge_\Sigma A)} (C \wedge_\Sigma B) \to D \wedge_\Sigma B$$

of symmetric T-complexes is a level cofibration. If j is a projective cofibration then the map $(j, i)_*$ is a projective cofibration. If j is a stable f-equivalence (respectively level f-equivalence), then $(j, i)_*$ is a stable f-equivalence (respectively level f-equivalence).

A projective cofibration is a map $A \to B$ of symmetric T-complexes which has the left lifting property with respect to all maps $p : X \to Y$ of symmetric T-complexes such that all maps $X^n \to Y^n$ are trivial fibrations of simplicial R-modules. If A is a cofibrant T-complex (in the sense of Proposition 10.80), then $V_R(A)$ is a projective cofibrant symmetric T-complex.

Theorem 11.66 also enables a definition of a derived tensor product $L(A \otimes_\Sigma B)$ for symmetric T-complexes A and B in the f-local stable model structure. Set

$$L(A \otimes_\Sigma B) = A_c \otimes_\Sigma B_c,$$

where $A_c \to A$ and $B_c \to B$ are projective cofibrant models for A and B, respectively. This construction specializes to a derived tensor product for symmetric S^1-complexes, and hence gives a derived tensor product for presheaves of unbounded chain complexes, as in [61].

References

1. *Séminaire de géométrie algébrique du Bois Marie 1960/61 (SGA 1), dirigé par Alexander Grothendieck. Augmenté de deux exposés de M. Raynaud. Revêtements étales et groupe fondamental. Exposés I à XIII* . Lecture Notes in Mathematics. 224. Berlin-Heidelberg-New York: Springer-Verlag, 1971.
2. *Théorie des topos et cohomologie étale des schémas. Tome 1: Théorie des topos*. Lecture Notes in Mathematics, Vol. 269. Springer-Verlag, Berlin, 1972. Séminaire de Géométrie Algébrique du Bois-Marie 1963–1964 (SGA 4), Dirigé par M. Artin, A. Grothendieck, et J. L. Verdier. Avec la collaboration de N. Bourbaki, P. Deligne et B. Saint-Donat.
3. M. Artin and B. Mazur. On the van Kampen theorem. *Topology*, 5:179–189, 1966.
4. M. Artin and B. Mazur. *Etale homotopy*. Lecture Notes in Mathematics, No. 100. Springer-Verlag, Berlin, 1969.
5. Michael Barr. Toposes without points. *J. Pure Appl. Algebra*, 5:265–280, 1974.
6. Clark Barwick. On left and right model categories and left and right Bousfield localizations. *Homology, Homotopy Appl.*, 12(2):245–320, 2010.
7. Tibor Beke. Sheafifiable homotopy model categories. *Math. Proc. Cambridge Philos. Soc.*, 129(3):447–475, 2000.
8. Håkon Schad Bergsaker. Homotopy theory of presheaves of Γ-spaces. *Homology, Homotopy Appl.*, 11(1):35–60, 2009.
9. Georg Biedermann. On the homotopy theory of n-types. *Homology, Homotopy Appl.*, 10(1):305–325, 2008.
10. Benjamin A. Blander. Local projective model structures on simplicial presheaves. *K-Theory*, 24(3):283–301, 2001.
11. Francis Borceux. *Handbook of categorical algebra. 1*, volume 50 of *Encyclopedia of Mathematics and its Applications*. Cambridge University Press, Cambridge, 1994. Basic category theory.
12. A. K. Bousfield. The localization of spaces with respect to homology. *Topology*, 14:133–150, 1975.
13. A. K. Bousfield. On the telescopic homotopy theory of spaces. *Trans. Amer. Math. Soc.*, 353(6):2391–2426 (electronic), 2001.
14. A. K. Bousfield and E. M. Friedlander. Homotopy theory of Γ-spaces, spectra, and bisimplicial sets. In *Geometric applications of homotopy theory (Proc. Conf., Evanston, Ill., 1977), II*, volume 658 of *Lecture Notes in Math.*, pages 80–130. Springer, Berlin, 1978.
15. A. K. Bousfield and D. M. Kan. *Homotopy limits, completions and localizations*. Springer-Verlag, Berlin, 1972. Lecture Notes in Mathematics, Vol. 304.
16. Lawrence Breen. Extensions du groupe additif. *Inst. Hautes Études Sci. Publ. Math.*, 48:39–125, 1978.
17. Kenneth S. Brown. Abstract homotopy theory and generalized sheaf cohomology. *Trans. Amer. Math. Soc.*, 186:419–458, 1974.

18. Kenneth S. Brown and Stephen M. Gersten. Algebraic K-theory as generalized sheaf cohomology. In *Algebraic K-theory, I: Higher K-theories (Proc. Conf., Battelle Memorial Inst., Seattle, Wash., 1972)*, pages 266–292. Lecture Notes in Math., Vol. 341. Springer, Berlin, 1973.
19. Antonio M. Cegarra and Josué Remedios. The behaviour of the $\overline{W}$-construction on the homotopy theory of bisimplicial sets. *Manuscripta Math.*, 124(4):427–457, 2007.
20. Denis-Charles Cisinski. Les préfaisceaux comme modèles des types d'homotopie. *Astérisque*, 308:xxiv+390, 2006.
21. D. Dugger. Classification spaces of maps in model categories. Preprint, http://www.uoregon.edu/ ddugger/, 2006.
22. Daniel Dugger and Daniel C. Isaksen. The motivic Adams spectral sequence. *Geom. Topol.*, 14(2):967–1014, 2010.
23. W. G. Dwyer and D. M. Kan. Function complexes in homotopical algebra. *Topology*, 19(4):427–440, 1980.
24. W. G. Dwyer and D. M. Kan. Homotopy theory and simplicial groupoids. *Nederl. Akad. Wetensch. Indag. Math.*, 46(4):379–385, 1984.
25. William G. Dwyer and Eric M. Friedlander. Algebraic and etale K-theory. *Trans. Amer. Math. Soc.*, 292(1):247–280, 1985.
26. Franc Forstnerič and Finnur Lárusson.Survey of Oka theory. *New York J. Math.*, 17A:11–38, 2011.
27. Eric M. Friedlander. *Étale homotopy of simplicial schemes*, volume 104 of *Annals of Mathematics Studies*. Princeton University Press, Princeton, N.J., 1982.
28. Eric M. Friedlander. Étale K-theory. II. Connections with algebraic K-theory. *Ann. Sci. École Norm. Sup. (4)*, 15(2):231–256, 1982.
29. Rudolf Fritsch and Renzo A. Piccinini. *Cellular structures in topology*, volume 19 of *Cambridge Studies in Advanced Mathematics*. Cambridge University Press, Cambridge, 1990.
30. Ofer Gabber.K-theory of Henselian local rings and Henselian pairs. In *Algebraic K-theory, commutative algebra, and algebraic geometry (Santa Margherita Ligure, 1989)*, volume 126 of *Contemp. Math.*, pages 59–70. Amer. Math. Soc., Providence, RI, 1992.
31. P. Gabriel and M. Zisman. *Calculus of fractions and homotopy theory*. Ergebnisse der Mathematik und ihrer Grenzgebiete, B and 35. Springer-Verlag New York, Inc., New York, 1967.
32. Jean Giraud. *Cohomologie non abélienne*. Springer-Verlag, Berlin, 1971. Die Grundlehren der mathematischen Wissenschaften, Band 179.
33. P. G. Goerss and J. F. Jardine. *Simplicial Homotopy Theory*, volume 174 of *Progress in Mathematics*. Birkhäuser Verlag, Basel, 1999.
34. Paul G. Goerss. Homotopy fixed points for Galois groups. In *The Čech centennial (Boston, MA, 1993)*, volume 181 of *Contemp. Math.*, pages 187–224. Amer. Math. Soc., Providence, RI, 1995.
35. Paul G. Goerss. Topological modular forms [after Hopkins, Miller and Lurie]. *Astérisque*, (332):Exp. No. 1005, viii, 221–255, 2010. Séminaire Bourbaki. Volume 2008/2009. Exposés 997–1011.
36. A. Grothendieck.Classes de Chern et représentations linéaires des groupes discrets. In *Dix Exposés sur la Cohomologie des Schémas*, pages 215–305. North-Holland, Amsterdam, 1968.
37. Alexander Grothendieck. Sur quelques points d'algèbre homologique. *Tôhoku Math. J. (2)*, 9:119–221, 1957.
38. Alex Heller. Homotopy theories. *Mem. Amer. Math. Soc.*, 71(383):vi+78, 1988.
39. Michael A. Hill.The equivariant slice filtration: a primer. *Homology Homotopy Appl.*, 14(2):143–166, 2012.
40. Michael A. Hill, Michael J. Hopkins, and Douglas C. Ravenel. On the non-existence of elements of Kervaire invariant one. Preprint, http://arxiv.org/abs/0908.3724, 2010.

References

41. Michael A. Hill, Michael J. Hopkins, and Douglas C. Ravenel.The Arf-Kervaire problem in algebraic topology: sketch of the proof. In *Current developments in mathematics, 2010*, pages 1–43. Int. Press, Somerville, MA, 2011.
42. Philip S. Hirschhorn. *Model categories and their localizations*, volume 99 of *Mathematical Surveys and Monographs*. American Mathematical Society, Providence, RI, 2003.
43. André Hirschowitz and Carlos Simpson. Descente pour les n-champs. 2001, math/9807049.
44. Sharon Hollander.A homotopy theory for stacks. *Israel J. Math.*, 163:93–124, 2008.
45. Mark Hovey. *Model categories*, volume 63 of *Mathematical Surveys and Monographs*. American Mathematical Society, Providence, RI, 1999.
46. Mark Hovey.Spectra and symmetric spectra in general model categories. *J. Pure Appl. Algebra*, 165(1):63–127, 2001.
47. Mark Hovey, Brooke Shipley, and Jeff Smith. Symmetric spectra. *J. Amer. Math. Soc.*, 13(1):149–208, 2000.
48. Luc Illusie.*Complexe cotangent et déformations. I*. Lecture Notes in Mathematics, Vol. 239. Springer-Verlag, Berlin, 1971.
49. Luc Illusie.*Complexe cotangent et déformations. II*. Lecture Notes in Mathematics, Vol. 283. Springer-Verlag, Berlin, 1972.
50. J. F. Jardine. Simplicial objects in a Grothendieck topos. In *Applications of algebraic K-theory to algebraic geometry and number theory, Part I, II (Boulder, Colo., 1983)*, pages 193–239. Amer. Math. Soc., Providence, RI, 1986.
51. J. F. Jardine. Simplicial presheaves. *J. Pure Appl. Algebra*, 47(1):35–87, 1987.
52. J. F. Jardine. Stable homotopy theory of simplicial presheaves. *Canad. J. Math.*, 39(3):733–747, 1987.
53. J. F. Jardine. The Leray spectral sequence. *J. Pure Appl. Algebra*, 61(2):189–196, 1989.
54. J. F. Jardine. Universal Hasse-Witt classes. In *Algebraic K-theory and algebraic number theory (Honolulu, HI, 1987)*, pages 83–100. Amer. Math. Soc., Providence, RI, 1989.
55. J. F. Jardine. Higher spinor classes. *Mem. Amer. Math. Soc.*, 110(528):vi+88, 1994.
56. J. F. Jardine. Boolean localization, in practice. *Doc. Math.*, 1:No. 13, 245–275 (electronic), 1996.
57. J. F. Jardine. Generalized Étale Cohomology Theories, volume 146 of *Progress in Mathematics*. Birkhäuser Verlag, Basel, 1997.
58. J. F. Jardine. Motivic symmetric spectra. *Doc. Math.*, 5:445–553 (electronic), 2000.
59. J. F. Jardine. Presheaves of symmetric spectra. *J. Pure Appl. Algebra*, 150(2):137–154, 2000.
60. J. F. Jardine. Stacks and the homotopy theory of simplicial sheaves. *Homology Homotopy Appl.*, 3(2):361–384 (electronic), 2001. Equivariant stable homotopy theory and related areas (Stanford, CA, 2000).
61. J. F. Jardine. Finite group torsors for the qfh topology. *Math. Z.*, 244(4):859–871, 2003.
62. J. F. Jardine. Presheaves of chain complexes. *K-Theory*, 30(4):365–420, 2003. Special issue in honor of Hyman Bass on his seventieth birthday. Part IV.
63. J. F. Jardine. Simplicial approximation. *Theory Appl. Categ.*, 12:No. 2, 34–72 (electronic), 2004.
64. J. F. Jardine. Categorical homotopy theory. *Homology, Homotopy Appl.*, 8(1):71–144 (electronic), 2006.
65. J. F. Jardine. Intermediate model structures for simplicial presheaves. *Canad. Math. Bull.*, 49(3):407–413, 2006.
66. J. F. Jardine. Cocycle categories. In *Algebraic topology*, volume 4 of *Abel Symp.*, pages 185–218. Springer, Berlin, 2009.
67. J. F. Jardine. Homotopy classification of gerbes. *Publ. Mat.*, 54(1):83–111, 2010.
68. J. F. Jardine. The Verdier hypercovering theorem. *Canad. Math. Bull.*, 55(2):319–328, 2012.
69. J.F. Jardine. The K-theory presheaf of spectra. In *New topological contexts for Galois theory and algebraic geometry (BIRS 2008)*, volume 16 of *Geom. Topol. Monogr.*, pages 151–178. Geom. Topol. Publ., Coventry, 2009.
70. J.F. Jardine. Galois descent criteria. Preprint, http://www.math.uwo.ca/ jardine, 2013.

71. A. Joyal. Letter to A. Grothendieck, 1984.
72. André Joyal and Myles Tierney.Strong stacks and classifying spaces. In *Category theory (Como, 1990)*, volume 1488 of *Lecture Notes in Math.*, pages 213–236. Springer, Berlin, 1991.
73. Masaki Kashiwara and Pierre Schapira. *Categories and sheaves*, volume 332 of *Grundlehren der Mathematischen Wissenschaften [Fundamental Principles of Mathematical Sciences]*. Springer-Verlag, Berlin, 2006.
74. Zhiming Luo. *Presheaves of simplicial groupoids*. ProQuest LLC, Ann Arbor, MI, 2004. Thesis (Ph.D.)–The University of Western Ontario (Canada).
75. J. Lurie.A survey of elliptic cohomology. In *Algebraic topology*, volume 4 of *Abel Symp.*, pages 219–277. Springer, Berlin, 2009.
76. Jacob Lurie. *Higher topos theory*, volume 170 of *Annals of Mathematics Studies*. Princeton University Press, Princeton, NJ, 2009.
77. Saunders Mac Lane. *Categories for the Working Mathematician*, volume 5 of *Graduate Texts in Mathematics*. Springer-Verlag, New York, second edition, 1998.
78. Saunders Mac Lane and Ieke Moerdijk. *Sheaves in geometry and logic*. Universitext. Springer-Verlag, New York, 1994. A first introduction to topos theory, Corrected reprint of the 1992 edition.
79. Carlo Mazza, Vladimir Voevodsky, and Charles Weibel. *Lecture notes on motivic cohomology*, volume 2 of *Clay Mathematics Monographs*. American Mathematical Society, Providence, RI, 2006.
80. James S. Milne. *Étale Cohomology*, volume 33 of *Princeton Mathematical Series*. Princeton University Press, Princeton, N.J., 1980.
81. Ieke Moerdijk.Bisimplicial sets and the group-completion theorem. In *Algebraic K-theory: connections with geometry and topology (Lake Louise, AB, 1987)*, volume 279 of *NATO Adv. Sci. Inst. Ser. C Math. Phys. Sci.*, pages 225–240. Kluwer Acad. Publ., Dordrecht, 1989.
82. Fabien Morel.On the Friedlander-Milnor conjecture for groups of small rank. In *Current developments in mathematics, 2010*, pages 45–93. Int. Press, Somerville, MA, 2011.
83. Fabien Morel and Vladimir Voevodsky. $\mathbf{A}^1$-homotopy theory of schemes. *Inst. Hautes Études Sci. Publ. Math.*, 90:45–143 (2001), 1999.
84. Ye. A. Nisnevich. The completely decomposed topology on schemes and associated descent spectral sequences in algebraic K-theory. In *Algebraic K-theory: connections with geometry and topology (Lake Louise, AB, 1987)*, volume 279 of *NATO Adv. Sci. Inst. Ser. C Math. Phys. Sci.*, pages 241–342. Kluwer Acad. Publ., Dordrecht, 1989.
85. Kyle M. Ormsby. Motivic invariants of p-adic fields. *J. K-Theory*, 7(3):597–618, 2011.
86. Pablo Pelaez. Multiplicative properties of the slice filtration.*Astérisque*, (335):xvi+289, 2011.
87. Daniel Quillen. Rational homotopy theory. *Ann. of Math. (2)*, 90:205–295, 1969.
88. Daniel Quillen. Higher algebraic K-theory. I. In *Algebraic K-theory, I: Higher K-theories (Proc. Conf., Battelle Memorial Inst., Seattle, Wash., 1972)*, pages 85–147. Lecture Notes in Math., Vol. 341. Springer, Berlin, 1973.
89. Daniel G. Quillen. The geometric realization of a Kan fibration is a Serre fibration. *Proc. Amer. Math. Soc.*, 19:1499–1500, 1968.
90. Oliver Röndigs and Paul Arne Østvær.Modules over motivic cohomology. *Adv. Math.*, 219(2):689–727, 2008.
91. Oliver Röndigs and Paul Arne Østvær.Modules over motivic cohomology. *Adv. Math.*, 219(2):689–727, 2008.
92. Horst Schubert. *Categories*. Springer-Verlag, New York, 1972. Translated from the German by Eva Gray.
93. Jean-Pierre Serre.*Cohomologie Galoisienne*. Lecture Notes in Mathematics, Vol. 5. Springer-Verlag, Berlin-New York, 1973. Cours au Collège de France, Paris, 1962–1963, Avec des textes inédits de J. Tate et de Jean-Louis Verdier, Quatrième édition.
94. Carlos Simpson. *Homotopy theory of higher categories*, volume 19 of *New Mathematical Monographs*. Cambridge University Press, Cambridge, 2012.

95. Danny Stevenson.Décalage and Kan's simplicial loop group functor. *Theory Appl. Categ.*, 26:768–787, 2012.
96. A. Suslin.On the K-theory of algebraically closed fields. *Invent. Math.*, 73(2):241–245, 1983.
97. Andrei Suslin and Vladimir Voevodsky.Bloch-Kato conjecture and motivic cohomology with finite coefficients. In *The arithmetic and geometry of algebraic cycles (Banff, AB, 1998)*, volume 548 of *NATO Sci. Ser. C Math. Phys. Sci.*, pages 117–189. Kluwer Acad. Publ., Dordrecht, 2000.
98. Andrei Suslin and Vladimir Voevodsky. Relative cycles and Chow sheaves. In *Cycles, transfers, and motivic homology theories*, volume 143 of *Ann. of Math. Stud.*, pages 10–86. Princeton Univ. Press, Princeton, NJ, 2000.
99. Andrei A. Suslin. On the K-theory of local fields. In *Proceedings of the Luminy conference on algebraic K-theory (Luminy, 1983)*, volume 34, pages 301–318, 1984.
100. R. W. Thomason. Algebraic K-theory and étale cohomology. *Ann. Sci. École Norm. Sup. (4)*, 18(3):437–552, 1985.
101. R. W. Thomason and Thomas Trobaugh. Higher algebraic K-theory of schemes and of derived categories. In *The Grothendieck Festschrift, Vol. III*, volume 88 of *Progr. Math.*, pages 247–435. Birkhäuser Boston, Boston, MA, 1990.
102. D. H. Van Osdol. Simplicial homotopy in an exact category. *Amer. J. Math.*, 99(6):1193–1204, 1977.
103. Vladimir Voevodsky. $\mathbf{A}^1$-homotopy theory. In *Proceedings of the International Congress of Mathematicians, Vol. I (Berlin, 1998)*, number Extra Vol. I, pages 579–604 (electronic), 1998.
104. Vladimir Voevodsky.Open problems in the motivic stable homotopy theory. I. In *Motives, polylogarithms and Hodge theory, Part I (Irvine, CA, 1998)*, volume 3 of *Int. Press Lect. Ser.*, pages 3–34. Int. Press, Somerville, MA, 2002.
105. Vladimir Voevodsky. Cancellation theorem. *Doc. Math.*, (Extra volume: Andrei A. Suslin sixtieth birthday):671–685, 2010.
106. Vladimir Voevodsky.Homotopy theory of simplicial sheaves in completely decomposable topologies. *J. Pure Appl. Algebra*, 214(8):1384–1398, 2010.
107. Vladimir Voevodsky.Unstable motivic homotopy categories in Nisnevich and cdh-topologies. *J. Pure Appl. Algebra*, 214(8):1399–1406, 2010.
108. Vladimir Voevodsky. On motivic cohomology with $\mathbf{Z}/l$-coefficients. *Ann. of Math. (2)*, 174(1):401–438, 2011.

Index

(S, T)-bispectrum, 392
2-groupoid, 313
H-functor, 260
I, cofibrations, 27, 99
J, trivial cofibrations, 27, 99
K-theory, étale, 357
$K(D, n)$, chain complex D, 212
K_+, disjoint base point, 347
L-equivalence, 162
L-fibrant, 162
L-fibration, 162
L_U, adjoint of U-sections, 25, 81
P-simplex, 285
S^1-spectrum, 360
T-bicomplex, 494
T-bispectrum, 393
T-complex, 413
T-loops, 361
T-spectrum, 360
T-spectrum, underlying, 415
T-suspension, 363
T-suspension, fake, 363
$T^{\wedge n}$, smash power, 363
Ex^∞ construction, 62, 65
Γ-space, 436
α-bounded, 94
$\mathbb{A}^1$-homotopy theory, 180
$\mathbb{A}^1$-local, 245
$\mathbf{Ext}(H, \mathcal{F})$, 324
$\mathbf{Iso}(\mathcal{F})$, 321
$\mathcal{F}$-equivalence, 166, 175, 228, 242, 418
$\mathcal{F}$-fibration, 171, 179, 232, 242, 418
$\mathcal{F}$-injective, 175
$\overline{W}$, Eilenberg-Mac Lane functor, 290
$\overline{W}$-fibration, 296, 304
$\pi(Z, X)$, simplicial homotopy classes, 151
c_r, s, shuffle permutatation, 438

cd-topology, 32
f-equivalence, 179, 234
f-equivalence, stable, 419, 450
f-equivalences, stable, 369
f-fibration, 179, 234
f-fibration, stable, 369, 419, 450
n-equivalence, 131, 310, 406
n-fibration, 134, 310
n-stack, 316
n-type, 131, 310

A
Axiom of Choice, 49

B
base points, local, 68
BG-property, 121
bonding map, 203, 345, 360, 413, 439, 483
bonding map, adjoint, 353, 361, 420
Boolean algebra, 47
Boolean localization, 30, 49, 77, 196
bounded monomorphism property, 23, 95, 128, 162, 201, 231, 366
Brauer group, 257

C
cardinal, regular, 95
category, cocycle, 143
category, derived, 206
category, orbit, 356
chain complex, Moore, 194
chain complex, normalized, 195
classifying space, nerve, 18
classifying topos, 42, 356
cochain complex, 212
cocycle, 143
cocycle conjugate, 295
cocycle, bounded, 146

cocycle, canonical, 262
coequivalence, 409
cofibrant replacement, 166
cofibration, 2-groupoids, 313
cofibration, T-complexes, 414
cofibration, T-spectra, 361
cofibration, chain complexes, 200
cofibration, injective, 484
cofibration, level, 363, 441
cofibration, projective, 25, 126, 441
cofibration, simplical modules, 197
cofibration, simplicial presheaves, 84
cofibration, simplicial sheaves, 99
cofibration, spectra, 205, 346
cofibration, trivial, 96
cohomology, mixed, 218
cohomology, sheaf, 216
cohomology, simplicial presheaf, 217
colocal, 409
compact, up to f-equivalence, 372
complex, shifted, 202
constant sheaf, 37
correspondence, Dold-Kan, 195, 200, 421
covering family, 31
cup product, 225

D

décalage functor, 286
degree, stable homotopy groups, 403
descent, 92, 102, 116, 214
descent, Brown-Gersten, 118
descent, cohomological, 104
descent, effective, 277
descent, Galois, 105
descent, motivic, 181
descent, Nisnevich, 116, 125, 357
descent, spectra, 357
diagonal, of a bispectrum, 393, 463
direct image, 43

E

Eilenberg-Mac Lane spectrum, 414
elementary distinguished square, 120
epimorphism, local, 38
equivalence, Morita, 273
equivalence, stable, 205, 419, 463
essentially surjective, 323

F

fake T-loops, 374
family, R-compatible, 34
family, groups, 320, 321
fat point, 49
fibrant model, injective, 101

fibration, 2-groupoids, 313
fibration, $\overline{W}$, 296
fibration, injective, 96, 99, 197, 200, 268, 445, 484
fibration, level, 361
fibration, local, 71
fibration, local trivial, 72, 82
fibration, motivic, 234
fibration, projective, 441
fibration, sectionwise, 25, 71, 101
fibration, stable, 205, 347, 369, 419
fibration, strict, 205, 346, 361, 414
finite correspondences, 242
free simplicial module, reduced, 195, 412
free symmetric spectrum, 439
function complex, 19, 99, 199, 346
function complex, internal, 106, 367
functor, internal, 260
fundamental 2-groupoid, 315

G

gerbe, 318, 331
global section, 34
graph functor, 242
Grothendieck construction, 326
group, profinite, 32, 357
groupoid, Čech, 73
groupoid, enriched in simplicial sets, 280
groupoid, of torsors, 252, 261

H

hammock localization, 145
Hasse-Witt class, 256
homology sheaf, 194, 217
homology sheaf, reduced, 224
homotopy category, pointed, 224
homotopy fixed points, 105
homotopy group sheaf, 68
HSS-fibration, 460
Hurewicz homomorphism, 211
hypercohomology, 221, 222
hypercover, 72

I

Illusie conjecture, 196
image, groupoid morphism, 322
inductive colimit descent, 377
interval, in a poset, 286
inverse image, 43
isomorphism, local, 39

K

kernel, groupoid morphism, 323

Index 507

L
layer filtration, 384, 422
level, spectrum, 345, 360
loop groupoid, 290

M
model structure, 2-groupoids, 314
model structure, $\mathcal{F}$-local, 179, 232, 242, 367, 418
model structure, $\overline{W}$, 296, 304
model structure, f-injective, 489
model structure, f-local, 179, 234, 369, 419, 450, 489
model structure, n-equivalence, 137, 313, 407
model structure, Dwyer-Kan, 281
model structure, enriched, 106
model structure, HSS, 461
model structure, injective, 98–100, 199, 200, 270, 443, 446, 486
model structure, intermediate, 128
model structure, motivic, 180, 234, 244
model structure, projective, 25, 126, 237, 238, 443
model structure, projective local, 128, 183
model structure, projective motivic, 128, 184, 188
model structure, stable, 206, 347, 369, 419, 449, 450, 489
model structure, strict, 205, 347, 362, 370, 393, 415, 428
monomorphism, local, 38
morphism, geometric, 42, 107
morphism, site, 108
motive $\mathbb{Z}(n)$, 246
motives, effective, 244
motivic cohomology, 246, 427, 435

O
object, Eilenberg-Mac Lane, 212

P
parameter object, 360
parameter object, cycle trivial, 384
path component groupoid, 297
path component sheaf, 68
point, geometric, 44
point, topos, 44
points, enough, 46
poset join, 285
Postnikov section, simplicial groupoid, 309
Postnikov section, simplicial presheaf, 131
Postnikov section, simplicial set, 131
Postnikov section, spectrum, 405
presheaf of spectra, 345

presheaf with transfers, 242
presheaf, additive, 356

Q
quasi-isomorphism, 197
quotient stack, 272

R
resolution 2-groupoid, 322
resolution, Čech, 73
resolution, Godement, 102
restriction, 66
right lifting property, local, 69

S
sheaf, homotopy invariant, 245
sheaf, of homotopy groups, 68
sheaf, of path components, 68
sheaf, of stable homotopy groups, 347
simplicial presheaf, 33
simplicial circle, 337
simplicial presheaf, linear, 236
simplicial presheaf, pointed, 155, 223
simplicial sheaf, 35
singular complex, 245
site morphism, 45
site, big, 32, 111
site, fibred, 112, 223
site, Grothendieck, 31
slice category, 33
slice filtration, 406, 411
smash product, naive, 480
smash product, pointed simplicial presheaves, 224
smash product, pushout, 479
smash product, symmetric spectra, 439
spectral sequence, descent, 358
spectrum, chain complexes, 203
sphere spectrum, 345, 363, 414
stabilization maps, 369
stable equivalence, 347
stable homotopy groups, bigraded, 401
stack, 271
suspension spectrum, 362
symmetric T-complex, 483
symmetric T-complex, shifted, 492
symmetric T-spectrum, 438
symmetric T-spectrum, bounded, 443
symmetric T-spectrum, shifted, 453
symmetric complex, 490
symmetric space, 436

T

tensor product, 198
tensor product, symmetric spaces, 437
topology, chaotic, 32
torsor, 251, 261
torsor, trivial, 252, 265, 266
total simplicial set, 287
truncation, good, 202
twist automorphism, 438

U

universal coefficients, 218

V

Verdier hypercovering theorem, 152

W

weak equivalence, level, 361, 441, 484
weak equivalence, local, 64, 99, 197, 200, 268, 304, 313
weak equivalence, motivic, 234
weak equivalence, sectionwise, 25, 64, 72
weak equivalence, strict, 205, 346, 361, 393, 414, 463
weight, stable homotopy groups, 403

Printed by Printforce, the Netherlands